INTRODUCTION TO
PROBABILITY AND STATISTICS
FOR ENGINEERS AND SCIENTISTS

Fifth Edition

INTRODUCTION TO
PROBABILITY AND STATISTICS
FOR ENGINEERS AND SCIENTISTS

■ Fifth Edition ■

Sheldon M. Ross
University of Southern California, Los Angeles, USA

ELSEVIER

AMSTERDAM · BOSTON · HEIDELBERG · LONDON
NEW YORK · OXFORD · PARIS · SAN DIEGO
SAN FRANCISCO · SINGAPORE · SYDNEY · TOKYO
Academic Press is an imprint of Elsevier

Academic Press is an imprint of Elsevier

32 Jamestown Road, London NW1 7BY, UK
525 B Street, Suite 1800, San Diego, CA 92101-4495, USA
225 Wyman Street, Waltham, MA 02451, USA
The Boulevard, Langford Lane, Kidlington, Oxford OX5 1GB, UK

Fifth Edition 2014
Copyright © 2014, 2009, 2004, 1999 Elsevier Inc. All rights reserved.

Notices

Knowledge and best practice in this field are constantly changing. As new research and experience broaden our understanding, changes in research methods or professional practices, may become necessary.

Practitioners and researchers must always rely on their own experience and knowledge in evaluating and using any information or methods described here in. In using such information or methods they should be mindful of their own safety and the safety of others, including parties for whom they have a professional responsibility.

To the fullest extent of the law, neither the Publisher nor the authors, contributors, or editors, assume any liability for any injury and/or damage to persons or property as a matter of products liability, negligence or otherwise, or from any use or operation of any methods, products, instructions, or ideas contained in the material herein.

ISBN: 978-0-12-394811-3

Library of Congress Cataloging-in-Publication Data
Ross, Sheldon M.
   Introduction to probability and statistics for engineers and scientists / Sheldon M. Ross, Department of Industrial Engineering and Operations Research, University of California, Berkeley. Fifth edition.
      pages cm.
   Includes index.
   ISBN 978-0-12-394811-3
1. Probabilities. 2. Mathematical statistics. I. Title.
   TA340.R67 2014
   519.5–dc23

2014011941

British Library Cataloguing in Publication Data
A catalogue record for this book is available from the British Library

For information on all Academic Press publications
visit our web site at store.elsevier.com

Printed and bound in the United States of America

For
Elise

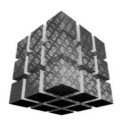

Contents

---

* Denotes optional material.

Preface

The fifth edition of this book continues to demonstrate how to apply probability theory to gain insight into real, everyday statistical problems and situations. As in the previous editions, carefully developed coverage of probability motivates probabilistic models of real phenomena and the statistical procedures that follow. This approach ultimately results in an intuitive understanding of statistical procedures and strategies most often used by practicing engineers and scientists.

This book has been written for an introductory course in statistics or in probability and statistics for students in engineering, computer science, mathematics, statistics, and the natural sciences. As such it assumes knowledge of elementary calculus.

ORGANIZATION AND COVERAGE

Chapter 1 presents a brief introduction to statistics, presenting its two branches of descriptive and inferential statistics, and a short history of the subject and some of the people whose early work provided a foundation for work done today.

The subject matter of descriptive statistics is then considered in **Chapter 2**. Graphs and tables that describe a data set are presented in this chapter, as are quantities that are used to summarize certain of the key properties of the data set.

To be able to draw conclusions from data, it is necessary to have an understanding of the data's origination. For instance, it is often assumed that the data constitute a "random sample" from some population. To understand exactly what this means and what its consequences are for relating properties of the sample data to properties of the entire population, it is necessary to have some understanding of probability, and that is the subject of **Chapter 3**. This chapter introduces the idea of a probability experiment, explains the concept of the probability of an event, and presents the axioms of probability.

Our study of probability is continued in **Chapter 4**, which deals with the important concepts of random variables and expectation, and in **Chapter 5**, which considers some special types of random variables that often occur in applications. Such random variables as the binomial, Poisson, hypergeometric, normal, uniform, gamma, chi-square, t, and F are presented.

In **Chapter 6**, we study the probability distribution of such sampling statistics as the sample mean and the sample variance. We show how to use a remarkable theoretical result of probability, known as the central limit theorem, to approximate the probability distribution of the sample mean. In addition, we present the joint probability distribution of the sample mean and the sample variance in the important special case in which the underlying data come from a normally distributed population.

Chapter 7 shows how to use data to estimate parameters of interest. For instance, a scientist might be interested in determining the proportion of Midwestern lakes that are afflicted by acid rain. Two types of estimators are studied. The first of these estimates the quantity of interest with a single number (for instance, it might estimate that 47 percent of Midwestern lakes suffer from acid rain), whereas the second provides an estimate in the form of an interval of values (for instance, it might estimate that between 45 and 49 percent of lakes suffer from acid rain). These latter estimators also tell us the "level of confidence" we can have in their validity. Thus, for instance, whereas we can be pretty certain that the exact percentage of afflicted lakes is not 47, it might very well be that we can be, say, 95 percent confident that the actual percentage is between 45 and 49.

Chapter 8 introduces the important topic of statistical hypothesis testing, which is concerned with using data to test the plausibility of a specified hypothesis. For instance, such a test might reject the hypothesis that fewer than 44 percent of Midwestern lakes are afflicted by acid rain. The concept of the p-value, which measures the degree of plausibility of the hypothesis after the data have been observed, is introduced. A variety of hypothesis tests concerning the parameters of both one and two normal populations are considered. Hypothesis tests concerning Bernoulli and Poisson parameters are also presented.

Chapter 9 deals with the important topic of regression. Both simple linear regression — including such subtopics as regression to the mean, residual analysis, and weighted least squares — and multiple linear regression are considered.

Chapter 10 introduces the analysis of variance. Both one-way and two-way (with and without the possibility of interaction) problems are considered.

Chapter 11 is concerned with goodness of fit tests, which can be used to test whether a proposed model is consistent with data. In it we present the classical chi-square goodness of fit test and apply it to test for independence in contingency tables. The final section of this chapter introduces the Kolmogorov–Smirnov procedure for testing whether data come from a specified continuous probability distribution.

Chapter 12 deals with nonparametric hypothesis tests, which can be used when one is unable to suppose that the underlying distribution has some specified parametric form (such as normal).

Chapter 13 considers the subject matter of quality control, a key statistical technique in manufacturing and production processes. A variety of control charts, including not only the Shewhart control charts but also more sophisticated ones based on moving averages and cumulative sums, are considered.

Chapter 14 deals with problems related to life testing. In this chapter, the exponential, rather than the normal, distribution plays the key role.

In **Chapter 15**, we consider the statistical inference techniques of bootstrap statistical methods and permutation tests. We first show how probabilities can be obtained by simulation and then how to utilize simulation in these statistical inference approaches.

The fifth edition contains a multitude of small changes designed to even further increase the clarity of the text's presentations and arguments. There are also many new examples and problems. In addition, this edition includes new subsections on

- The Pareto Distribution (subsection 5.6.2)
- Prediction Intervals (subsection 7.3.2)
- Dummy Variables for Categorical Data (subsection 9.10.2)
- Testing the Equality of Multiple Probability Distributions (subsection 12.4.2)

SUPPLEMENTAL MATERIALS

Solutions manual and software useful for solving text examples and problems are available at: textbooks.elsevier.com/web/Manuals.aspx?isbn=9780123948113.

ACKNOWLEDGMENTS

We thank the following people for their helpful comments on material of the fifth edition:

- Gideon Weiss, Uniferisty of Haifa
- N. Balakrishnan, McMaster University
- Mark Brown, Columbia University
- Rohitha Goonatilake, Texas A and M University
- Steve From, University of Nebraska at Omaha
- Subhash Kochar, Portland State University

as well as all those reviewers who asked to remain anonymous.

INTRODUCTION TO STATISTICS

1.1   INTRODUCTION

It has become accepted in today's world that in order to learn about something, you must first collect data. *Statistics* is the art of learning from data. It is concerned with the collection of data, its subsequent description, and its analysis, which often leads to the drawing of conclusions.

1.2   DATA COLLECTION AND DESCRIPTIVE STATISTICS

Sometimes a statistical analysis begins with a given set of data: For instance, the government regularly collects and publicizes data concerning yearly precipitation totals, earthquake occurrences, the unemployment rate, the gross domestic product, and the rate of inflation. Statistics can be used to describe, summarize, and analyze these data.

In other situations, data are not yet available; in such cases statistical theory can be used to design an appropriate experiment to generate data. The experiment chosen should depend on the use that one wants to make of the data. For instance, suppose that an instructor is interested in determining which of two different methods for teaching computer programming to beginners is most effective. To study this question, the instructor might divide the students into two groups, and use a different teaching method for each group. At the end of the class the students can be tested and the scores of the members of the different groups compared. If the data, consisting of the test scores of members of each group, are significantly higher in one of the groups, then it might seem reasonable to suppose that the teaching method used for that group is superior.

It is important to note, however, that in order to be able to draw a valid conclusion from the data, it is essential that the students were divided into groups in such a manner that neither group was more likely to have the students with greater natural aptitude for programming. For instance, the instructor should not have let the male class members be one group and the females the other. For if so, then even if the women scored significantly higher than the men, it would not be clear whether this was due to the method used to teach them, or to the fact that women may be inherently better than men at learning programming skills. The accepted way of avoiding this pitfall is to divide the class members into the two groups "at random." This term means that the division is done in such a manner that all possible choices of the members of a group are equally likely.

At the end of the experiment, the data should be described. For instance, the scores of the two groups should be presented. In addition, summary measures such as the average score of members of each of the groups should be presented. This part of statistics, concerned with the description and summarization of data, is called *descriptive statistics*.

1.3   INFERENTIAL STATISTICS AND PROBABILITY MODELS

After the preceding experiment is completed and the data are described and summarized, we hope to be able to draw a conclusion about which teaching method is superior. This part of statistics, concerned with the drawing of conclusions, is called *inferential statistics*.

To be able to draw a conclusion from the data, we must take into account the possibility of chance. For instance, suppose that the average score of members of the first group is quite a bit higher than that of the second. Can we conclude that this increase is due to the teaching method used? Or is it possible that the teaching method was not responsible for the increased scores but rather that the higher scores of the first group were just a chance occurrence? For instance, the fact that a coin comes up heads 7 times in 10 flips does not necessarily mean that the coin is more likely to come up heads than tails in future flips. Indeed, it could be a perfectly ordinary coin that, by chance, just happened to land heads 7 times out of the total of 10 flips. (On the other hand, if the coin had landed heads 47 times out of 50 flips, then we would be quite certain that it was not an ordinary coin.)

To be able to draw logical conclusions from data, we usually make some assumptions about the chances (or *probabilities*) of obtaining the different data values. The totality of these assumptions is referred to as a *probability model* for the data.

Sometimes the nature of the data suggests the form of the probability model that is assumed. For instance, suppose that an engineer wants to find out what proportion of computer chips, produced by a new method, will be defective. The engineer might select a group of these chips, with the resulting data being the number of defective chips in this group. Provided that the chips selected were "randomly" chosen, it is reasonable to suppose that each one of them is defective with probability p, where p is the unknown proportion of all the chips produced by the new method that will be defective. The resulting data can then be used to make inferences about p.

In other situations, the appropriate probability model for a given data set will not be readily apparent. However, careful description and presentation of the data sometimes enable us to infer a reasonable model, which we can then try to verify with the use of additional data.

Because the basis of statistical inference is the formulation of a probability model to describe the data, an understanding of statistical inference requires some knowledge of the theory of probability. In other words, statistical inference starts with the assumption that important aspects of the phenomenon under study can be described in terms of probabilities; it then draws conclusions by using data to make inferences about these probabilities.

1.4 POPULATIONS AND SAMPLES

In statistics, we are interested in obtaining information about a total collection of elements, which we will refer to as the *population*. The population is often too large for us to examine each of its members. For instance, we might have all the residents of a given state, or all the television sets produced in the last year by a particular manufacturer, or all the households in a given community. In such cases, we try to learn about the population by choosing and then examining a subgroup of its elements. This subgroup of a population is called a *sample*.

If the sample is to be informative about the total population, it must be, in some sense, representative of that population. For instance, suppose that we are interested in learning about the age distribution of people residing in a given city, and we obtain the ages of the first 100 people to enter the town library. If the average age of these 100 people is 46.2 years, are we justified in concluding that this is approximately the average age of the entire population? Probably not, for we could certainly argue that the sample chosen in this case is probably not representative of the total population because usually more young students and senior citizens use the library than do working-age citizens.

In certain situations, such as the library illustration, we are presented with a sample and must then decide whether this sample is reasonably representative of the entire population. In practice, a given sample generally cannot be assumed to be representative of a population unless that sample has been chosen in a random manner. This is because any specific nonrandom rule for selecting a sample often results in one that is inherently biased toward some data values as opposed to others.

Thus, although it may seem paradoxical, we are most likely to obtain a representative sample by choosing its members in a totally random fashion without any prior considerations of the elements that will be chosen. In other words, we need not attempt to deliberately choose the sample so that it contains, for instance, the same gender percentage and the same percentage of people in each profession as found in the general population. Rather, we should just leave it up to "chance" to obtain roughly the correct percentages. Once a random sample is chosen, we can use statistical inference to draw conclusions about the entire population by studying the elements of the sample.

1.5 A BRIEF HISTORY OF STATISTICS

A systematic collection of data on the population and the economy was begun in the Italian city-states of Venice and Florence during the Renaissance. The term *statistics*, derived from the word *state*, was used to refer to a collection of facts of interest to the state. The idea of collecting data spread from Italy to the other countries of Western Europe. Indeed, by the first half of the 16th century it was common for European governments to require parishes to register births, marriages, and deaths. Because of poor public health conditions this last statistic was of particular interest.

The high mortality rate in Europe before the 19th century was due mainly to epidemic diseases, wars, and famines. Among epidemics, the worst were the plagues. Starting with the Black Plague in 1348, plagues recurred frequently for nearly 400 years. In 1562, as a way to alert the King's court to consider moving to the countryside, the City of London began to publish weekly bills of mortality. Initially these mortality bills listed the places of death and whether a death had resulted from plague. Beginning in 1625 the bills were expanded to include all causes of death.

In 1662 the English tradesman John Graunt published a book entitled *Natural and Political Observations Made upon the Bills of Mortality*. Table 1.1, which notes the total number of deaths in England and the number due to the plague for five different plague years, is taken from this book.

TABLE 1.1   *Total Deaths in England*

Year	Burials	Plague Deaths
1592	25,886	11,503
1593	17,844	10,662
1603	37,294	30,561
1625	51,758	35,417
1636	23,359	10,400

Source: John Graunt, Observations Made upon the Bills of Mortality.
3rd ed. London: John Martyn and James Allestry (1st ed. 1662).

Graunt used London bills of mortality to estimate the city's population. For instance, to estimate the population of London in 1660, Graunt surveyed households in certain London parishes (or neighborhoods) and discovered that, on average, there were approximately 3 deaths for every 88 people. Dividing by 3 shows that, on average, there was roughly 1 death for every 88/3 people. Because the London bills cited 13,200 deaths in London for that year, Graunt estimated the London population to be about

$$13,200 \times 88/3 = 387,200$$

Graunt used this estimate to project a figure for all England. In his book he noted that these figures would be of interest to the rulers of the country, as indicators of both the number of men who could be drafted into an army and the number who could be taxed.

Graunt also used the London bills of mortality — and some intelligent guesswork as to what diseases killed whom and at what age — to infer ages at death. (Recall that the bills of mortality listed only causes and places at death, not the ages of those dying.) Graunt then used this information to compute tables giving the proportion of the population that

TABLE 1.2 *John Graunt's Mortality Table*

Age at Death	Number of Deaths per 100 Births
0–6	36
6–16	24
16–26	15
26–36	9
36–46	6
46–56	4
56–66	3
66–76	2
76 and greater	1

Note: The categories go up to but do not include the right-hand value. For instance, 0–6 means all ages from 0 up through 5.

dies at various ages. Table 1.2 is one of Graunt's mortality tables. It states, for instance, that of 100 births, 36 people will die before reaching age 6, 24 will die between the age of 6 and 15, and so on.

Graunt's estimates of the ages at which people were dying were of great interest to those in the business of selling annuities. Annuities are the opposite of life insurance in that one pays in a lump sum as an investment and then receives regular payments for as long as one lives.

Graunt's work on mortality tables inspired further work by Edmund Halley in 1693. Halley, the discoverer of the comet bearing his name (and also the man who was most responsible, by both his encouragement and his financial support, for the publication of Isaac Newton's famous *Principia Mathematica*), used tables of mortality to compute the odds that a person of any age would live to any other particular age. Halley was influential in convincing the insurers of the time that an annual life insurance premium should depend on the age of the person being insured.

Following Graunt and Halley, the collection of data steadily increased throughout the remainder of the 17th and on into the 18th century. For instance, the city of Paris began collecting bills of mortality in 1667, and by 1730 it had become common practice through-out Europe to record ages at death.

The term *statistics*, which was used until the 18th century as a shorthand for the descriptive science of states, became in the 19th century increasingly identified with numbers. By the 1830s the term was almost universally regarded in Britain and France as being synonymous with the "numerical science" of society. This change in meaning was caused by the large availability of census records and other tabulations that began to be systematically collected and published by the governments of Western Europe and the United States beginning around 1800.

Throughout the 19th century, although probability theory had been developed by such mathematicians as Jacob Bernoulli, Karl Friedrich Gauss, and Pierre-Simon Laplace, its use in studying statistical findings was almost nonexistent, because most social statisticians at the time were content to let the data speak for themselves. In particular, statisticians of that time were not interested in drawing inferences about individuals, but rather were concerned with the society as a whole. Thus, they were not concerned with sampling but rather tried to obtain censuses of the entire population. As a result, probabilistic inference from samples to a population was almost unknown in 19th century social statistics.

It was not until the late 1800s that statistics became concerned with inferring conclusions from numerical data. The movement began with Francis Galton's work on analyzing hereditary genius through the uses of what we would now call regression and correlation analysis (see Chapter 9), and obtained much of its impetus from the work of Karl Pearson. Pearson, who developed the chi-square goodness of fit tests (see Chapter 11), was the first director of the Galton Laboratory, endowed by Francis Galton in 1904. There Pearson originated a research program aimed at developing new methods of using statistics in inference. His laboratory invited advanced students from science and industry to learn statistical methods that could then be applied in their fields. One of his earliest visiting researchers was W. S. Gosset, a chemist by training, who showed his devotion to Pearson by publishing his own works under the name "Student." (A famous story has it that Gosset was afraid to publish under his own name for fear that his employers, the Guinness brewery, would be unhappy to discover that one of its chemists was doing research in statistics.) Gosset is famous for his development of the t-test (see Chapter 8).

Two of the most important areas of applied statistics in the early 20th century were population biology and agriculture. This was due to the interest of Pearson and others at his laboratory and also to the remarkable accomplishments of the English scientist Ronald A. Fisher. The theory of inference developed by these pioneers, including among others

TABLE 1.3   *The Changing Definition of Statistics*

Statistics has then for its object that of presenting a faithful representation of a state at a determined epoch. (Quetelet, 1849)

Statistics are the only tools by which an opening can be cut through the formidable thicket of difficulties that bars the path of those who pursue the Science of man. (Galton, 1889)

Statistics may be regarded (i) as the study of populations, (ii) as the study of variation, and (iii) as the study of methods of the reduction of data. (Fisher, 1925)

Statistics is a scientific discipline concerned with collection, analysis, and interpretation of data obtained from observation or experiment. The subject has a coherent structure based on the theory of Probability and includes many different procedures which contribute to research and development throughout the whole of Science and Technology. (E. Pearson, 1936)

Statistics is the name for that science and art which deals with uncertain inferences — which uses numbers to find out something about nature and experience. (Weaver, 1952)

Statistics has become known in the 20th century as the mathematical tool for analyzing experimental and observational data. (Porter, 1986)

Statistics is the art of learning from data. (this book, 2014)

Karl Pearson's son Egon and the Polish born mathematical statistician Jerzy Neyman, was general enough to deal with a wide range of quantitative and practical problems. As a result, after the early years of the 20th century a rapidly increasing number of people in science, business, and government began to regard statistics as a tool that was able to provide quantitative solutions to scientific and practical problems (see Table 1.3).

Nowadays the ideas of statistics are everywhere. Descriptive statistics are featured in every newspaper and magazine. Statistical inference has become indispensable to public health and medical research, to engineering and scientific studies, to marketing and quality control, to education, to accounting, to economics, to meteorological forecasting, to polling and surveys, to sports, to insurance, to gambling, and to all research that makes any claim to being scientific. Statistics has indeed become ingrained in our intellectual heritage.

Problems

1. An election will be held next week and, by polling a sample of the voting population, we are trying to predict whether the Republican or Democratic candidate will prevail. Which of the following methods of selection is likely to yield a representative sample?

   (a) Poll all people of voting age attending a college basketball game.
   (b) Poll all people of voting age leaving a fancy midtown restaurant.
   (c) Obtain a copy of the voter registration list, randomly choose 100 names, and question them.
   (d) Use the results of a television call-in poll, in which the station asked its listeners to call in and name their choice.
   (e) Choose names from the telephone directory and call these people.

2. The approach used in Problem 1(e) led to a disastrous prediction in the 1936 presidential election, in which Franklin Roosevelt defeated Alfred Landon by a landslide. A Landon victory had been predicted by the *Literary Digest*. The magazine based its prediction on the preferences of a sample of voters chosen from lists of automobile and telephone owners.

   (a) Why do you think the *Literary Digest*'s prediction was so far off?
   (b) Has anything changed between 1936 and now that would make you believe that the approach used by the *Literary Digest* would work better today?

3. A researcher is trying to discover the average age at death for people in the United States today. To obtain data, the obituary columns of the *New York Times* are read for 30 days, and the ages at death of people in the United States are noted. Do you think this approach will lead to a representative sample?

4. To determine the proportion of people in your town who are smokers, it has been decided to poll people at one of the following local spots:

   (a) the pool hall;
   (b) the bowling alley;
   (c) the shopping mall;
   (d) the library.

   Which of these potential polling places would most likely result in a reasonable approximation to the desired proportion? Why?

5. A university plans on conducting a survey of its recent graduates to determine information on their yearly salaries. It randomly selected 200 recent graduates and sent them questionnaires dealing with their present jobs. Of these 200, however, only 86 were returned. Suppose that the average of the yearly salaries reported was $75,000.

   (a) Would the university be correct in thinking that $75,000 was a good approximation to the average salary level of all of its graduates? Explain the reasoning behind your answer.
   (b) If your answer to part (a) is no, can you think of any set of conditions relating to the group that returned questionnaires for which it would be a good approximation?

6. An article reported that a survey of clothing worn by pedestrians killed at night in traffic accidents revealed that about 80 percent of the victims were wearing dark-colored clothing and 20 percent were wearing light-colored clothing. The conclusion drawn in the article was that it is safer to wear light-colored clothing at night.

   (a) Is this conclusion justified? Explain.
   (b) If your answer to part (a) is no, what other information would be needed before a final conclusion could be drawn?

7. Critique Graunt's method for estimating the population of London. What implicit assumption is he making?

8. The London bills of mortality listed 12,246 deaths in 1658. Supposing that a survey of London parishes showed that roughly 2 percent of the population died that year, use Graunt's method to estimate London's population in 1658.

9. Suppose you were a seller of annuities in 1662 when Graunt's book was published. Explain how you would make use of his data on the ages at which people were dying.

10. Based on Graunt's mortality table:

    (a) What proportion of people survived to age 6?
    (b) What proportion survived to age 46?
    (c) What proportion died between the ages of 6 and 36?

Chapter 2

DESCRIPTIVE STATISTICS

2.1 INTRODUCTION

In this chapter we introduce the subject matter of descriptive statistics, and in doing so learn ways to describe and summarize a set of data. Section 2.2 deals with ways of describing a data set. Subsections 2.2.1 and 2.2.2 indicate how data that take on only a relatively few distinct values can be described by using frequency tables or graphs, whereas Subsection 2.2.3 deals with data whose set of values is grouped into different intervals. Section 2.3 discusses ways of summarizing data sets by use of statistics, which are numerical quantities whose values are determined by the data. Subsection 2.3.1 considers three statistics that are used to indicate the "center" of the data set: the sample mean, the sample median, and the sample mode. Subsection 2.3.2 introduces the sample variance and its square root, called the sample standard deviation. These statistics are used to indicate the spread of the values in the data set. Subsection 2.3.3 deals with sample percentiles, which are statistics that tell us, for instance, which data value is greater than 95 percent of all the data. In Section 2.4 we present Chebyshev's inequality for sample data. This famous inequality gives an upper bound to the proportion of the data that can differ from the sample mean by more than k times the sample standard deviation. Whereas Chebyshev's inequality holds for all data sets, we can in certain situations, which are discussed in Section 2.5, obtain more precise estimates of the proportion of the data that is within k sample standard deviations of the sample mean. In Section 2.5 we note that when a graph of the data follows a bell-shaped form the data set is said to be approximately normal, and more precise estimates are given by the so-called empirical rule. Section 2.6 is concerned with situations in which the data consist of paired values. A graphical technique, called the scatter diagram, for presenting such data is introduced, as is the sample correlation coefficient, a statistic that indicates the degree to which a large value of the first member of the pair tends to go along with a large value of the second.

2.2 DESCRIBING DATA SETS

The numerical findings of a study should be presented clearly, concisely, and in such a manner that an observer can quickly obtain a feel for the essential characteristics of the data. Over the years it has been found that tables and graphs are particularly useful ways of presenting data, often revealing important features such as the range, the degree of concentration, and the symmetry of the data. In this section we present some common graphical and tabular ways for presenting data.

2.2.1 FREQUENCY TABLES AND GRAPHS

A data set having a relatively small number of distinct values can be conveniently presented in a *frequency table*. For instance, Table 2.1 is a frequency table for a data set consisting of the starting yearly salaries (to the nearest thousand dollars) of 42 recently graduated students with B.S. degrees in electrical engineering. Table 2.1 tells us, among other things, that the lowest starting salary of $57,000 was received by four of the graduates, whereas the highest salary of $70,000 was received by a single student. The most common starting salary was $62,000, and was received by 10 of the students.

TABLE 2.1   *Starting Yearly Salaries*

Starting Salary	Frequency
57	4
58	1
59	3
60	5
61	8
62	10
63	0
64	5
66	2
67	3
70	1

Data from a frequency table can be graphically represented by a *line graph* that plots the distinct data values on the horizontal axis and indicates their frequencies by the heights of vertical lines. A line graph of the data presented in Table 2.1 is shown in Figure 2.1.

When the lines in a line graph are given added thickness, the graph is called a *bar graph*. Figure 2.2 presents a bar graph.

Another type of graph used to represent a frequency table is the *frequency polygon*, which plots the frequencies of the different data values on the vertical axis, and then connects the plotted points with straight lines. Figure 2.3 presents a frequency polygon for the data of Table 2.1.

2.2.2 RELATIVE FREQUENCY TABLES AND GRAPHS

Consider a data set consisting of n values. If f is the frequency of a particular value, then the ratio f/n is called its *relative frequency*. That is, the relative frequency of a data value is

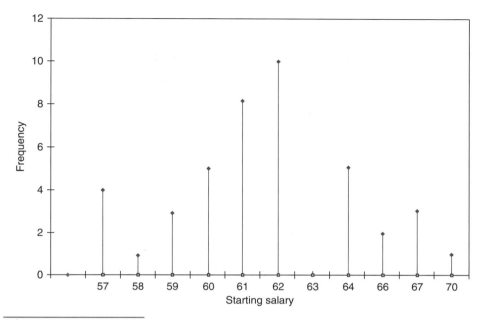

FIGURE 2.1   *Starting salary data.*

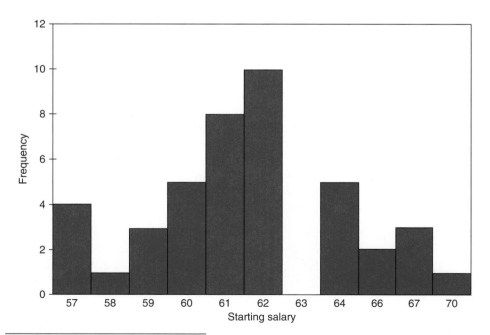

FIGURE 2.2   *Bar graph for starting salary data.*

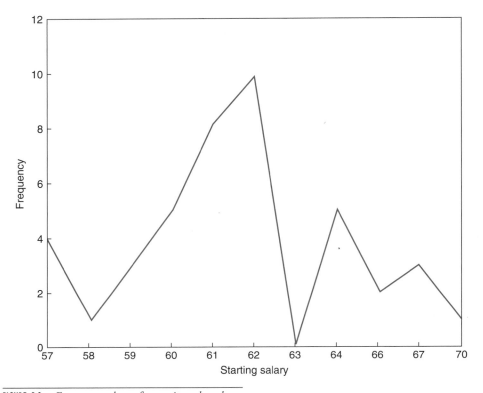

FIGURE 2.3   *Frequency polygon for starting salary data.*

the proportion of the data that have that value. The relative frequencies can be represented graphically by a relative frequency line or bar graph or by a relative frequency polygon. Indeed, these relative frequency graphs will look like the corresponding graphs of the absolute frequencies except that the labels on the vertical axis are now the old labels (that gave the frequencies) divided by the total number of data points.

EXAMPLE 2.2a   Table 2.2 is a relative frequency table for the data of Table 2.1. The relative frequencies are obtained by dividing the corresponding frequencies of Table 2.1 by 42, the size of the data set.   ■

A *pie chart* is often used to indicate relative frequencies when the data are not numerical in nature. A circle is constructed and then sliced into different sectors; one for each distinct type of data value. The relative frequency of a data value is indicated by the area of its sector, this area being equal to the total area of the circle multiplied by the relative frequency of the data value.

EXAMPLE 2.2b   The following data relate to the different types of cancers affecting the 200 most recent patients to enroll at a clinic specializing in cancer. These data are represented in the pie chart presented in Figure 2.4.   ■

TABLE 2.2

Starting Salary	Frequency
47	4/42 = .0952
48	1/42 = .0238
49	3/42
50	5/42
51	8/42
52	10/42
53	0
54	5/42
56	2/42
57	3/42
60	1/42

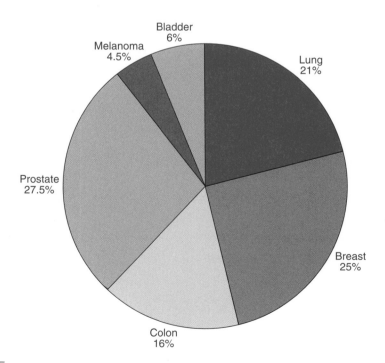

FIGURE 2.4

Type of Cancer	Number of New Cases	Relative Frequency
Lung	42	.21
Breast	50	.25
Colon	32	.16
Prostate	55	.275
Melanoma	9	.045
Bladder	12	.06

2.2.3 GROUPED DATA, HISTOGRAMS, OGIVES, AND STEM AND LEAF PLOTS

As seen in Subsection 2.2.2, using a line or a bar graph to plot the frequencies of data values is often an effective way of portraying a data set. However, for some data sets the number of distinct values is too large to utilize this approach. Instead, in such cases, it is useful to divide the values into groupings, or *class intervals*, and then plot the number of data values falling in each class interval. The number of class intervals chosen should be a trade-off between (1) choosing too few classes at a cost of losing too much information about the actual data values in a class and (2) choosing too many classes, which will result in the frequencies of each class being too small for a pattern to be discernible. Although 5 to 10

TABLE 2.3   *Life in Hours of 200 Incandescent Lamps*

				Item Lifetimes					
1,067	919	1,196	785	1,126	936	918	1,156	920	948
855	1,092	1,162	1,170	929	950	905	972	1,035	1,045
1,157	1,195	1,195	1,340	1,122	938	970	1,237	956	1,102
1,022	978	832	1,009	1,157	1,151	1,009	765	958	902
923	1,333	811	1,217	1,085	896	958	1,311	1,037	702
521	933	928	1,153	946	858	1,071	1,069	830	1,063
930	807	954	1,063	1,002	909	1,077	1,021	1,062	1,157
999	932	1,035	944	1,049	940	1,122	1,115	833	1,320
901	1,324	818	1,250	1,203	1,078	890	1,303	1,011	1,102
996	780	900	1,106	704	621	854	1,178	1,138	951
1,187	1,067	1,118	1,037	958	760	1,101	949	992	966
824	653	980	935	878	934	910	1,058	730	980
844	814	1,103	1,000	788	1,143	935	1,069	1,170	1,067
1,037	1,151	863	990	1,035	1,112	931	970	932	904
1,026	1,147	883	867	990	1,258	1,192	922	1,150	1,091
1,039	1,083	1,040	1,289	699	1,083	880	1,029	658	912
1,023	984	856	924	801	1,122	1,292	1,116	880	1,173
1,134	932	938	1,078	1,180	1,106	1,184	954	824	529
998	996	1,133	765	775	1,105	1,081	1,171	705	1,425
610	916	1,001	895	709	860	1,110	1,149	972	1,002

class intervals are typical, the appropriate number is a subjective choice, and of course, you can try different numbers of class intervals to see which of the resulting charts appears to be most revealing about the data. It is common, although not essential, to choose class intervals of equal length.

The endpoints of a class interval are called the *class boundaries*. We will adopt the *left-end inclusion convention*, which stipulates that a class interval contains its left-end but not its right-end boundary point. Thus, for instance, the class interval 20–30 contains all values that are both greater than *or equal to* 20 and less than 30.

Table 2.3 presents the lifetimes of 200 incandescent lamps. A class frequency table for the data of Table 2.3 is presented in Table 2.4. The class intervals are of length 100, with the first one starting at 500.

TABLE 2.4 *A Class Frequency Table*

Class Interval	Frequency (Number of Data Values in the Interval)
500–600	2
600–700	5
700–800	12
800–900	25
900–1000	58
1000–1100	41
1100–1200	43
1200–1300	7
1300–1400	6
1400–1500	1

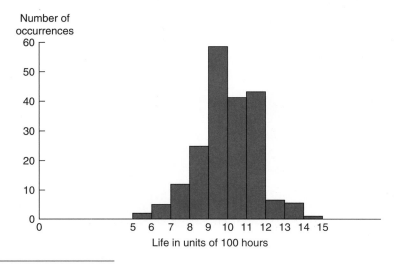

FIGURE 2.5 *A frequency histogram.*

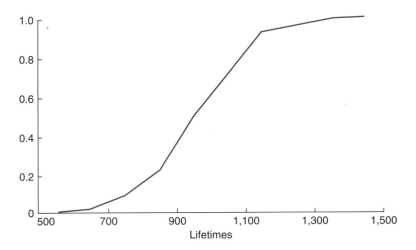

FIGURE 2.6   *A cumulative frequency plot.*

   A bar graph plot of class data, with the bars placed adjacent to each other, is called
a *histogram*. The vertical axis of a histogram can represent either the class frequency or the
relative class frequency; in the former case the graph is called a *frequency histogram* and
in the latter a *relative frequency histogram*. Figure 2.5 presents a frequency histogram of
the data in Table 2.4.
   We are sometimes interested in plotting a cumulative frequency (or cumulative relative
frequency) graph. A point on the horizontal axis of such a graph represents a possible
data value; its corresponding vertical plot gives the number (or proportion) of the data
whose values are less than or equal to it. A cumulative relative frequency plot of the data
of Table 2.3 is given in Figure 2.6. We can conclude from this figure that 100 percent
of the data values are less than 1,500, approximately 40 percent are less than or equal to
900, approximately 80 percent are less than or equal to 1,100, and so on. A cumulative
frequency plot is called an *ogive*.
   An efficient way of organizing a small- to moderate-sized data set is to utilize a *stem
and leaf plot*. Such a plot is obtained by first dividing each data value into two parts —
its stem and its leaf. For instance, if the data are all two-digit numbers, then we could let
the stem part of a data value be its tens digit and let the leaf be its ones digit. Thus, for
instance, the value 62 is expressed as

Stem	Leaf
6	2

and the two data values 62 and 67 can be represented as

Stem	Leaf
6	2, 7

EXAMPLE 2.2c Table 2.5 gives the monthly and yearly average daily minimum temperatures in 35 U.S. cities.

The annual average daily minimum temperatures from Table 2.5 are represented in the following stem and leaf plot.

```
7 | 0.0
6 | 9.0
5 | 1.0, 1.3, 2.0, 5.5, 7.1, 7.4, 7.6, 8.5, 9.3
4 | 0.0, 1.0, 2.4, 3.6, 3.7, 4.8, 5.0, 5.2, 6.0, 6.7, 8.1, 9.0, 9.2
3 | 3.1, 4.1, 5.3, 5.8, 6.2, 9.0, 9.5, 9.5
2 | 9.0, 9.8   ■
```

2.3 SUMMARIZING DATA SETS

Modern-day experiments often deal with huge sets of data. For instance, in an attempt to learn about the health consequences of certain common practices, in 1951 the medical statisticians R. Doll and A. B. Hill sent questionnaires to all doctors in the United Kingdom and received approximately 40,000 replies. Their questions dealt with age, eating habits, and smoking habits. The respondents were then tracked for the ensuing 10 years and the causes of death for those who died were monitored. To obtain a feel for such a large amount of data, it is useful to be able to summarize it by some suitably chosen measures. In this section we present some summarizing *statistics*, where a statistic is a numerical quantity whose value is determined by the data.

2.3.1 SAMPLE MEAN, SAMPLE MEDIAN, AND SAMPLE MODE

In this section we introduce some statistics that are used for describing the center of a set of data values. To begin, suppose that we have a data set consisting of the n numerical values $x_1, x_2, \ldots, x_n$. The sample mean is the arithmetic average of these values.

Definition

The *sample mean*, designated by $\bar{x}$, is defined by

$$\bar{x} = \sum_{i=1}^{n} x_i/n$$

The computation of the sample mean can often be simplified by noting that if for constants a and b

$$y_i = ax_i + b, \qquad i = 1, \ldots, n$$

TABLE 2.5 *Normal Daily Minimum Temperature — Selected Cities*

[In Fahrenheit degrees. Airport data except as noted. Based on standard 30-year period, 1961 through 1990]

State	Station	Jan.	Feb.	Mar.	Apr.	May	June	July	Aug.	Sept.	Oct.	Nov.	Dec.	Annual avg.
AL	Mobile	40.0	42.7	50.1	57.1	64.4	70.7	73.2	72.9	68.7	57.3	49.1	43.1	57.4
AK	Juneau	19.0	22.7	26.7	32.1	38.9	45.0	48.1	47.3	42.9	37.2	27.2	22.6	34.1
AZ	Phoenix	41.2	44.7	48.8	55.3	63.9	72.9	81.0	79.2	72.8	60.8	48.9	41.8	59.3
AR	Little Rock	29.1	33.2	42.2	50.7	59.0	67.4	71.5	69.8	63.5	50.9	41.5	33.1	51.0
CA	Los Angeles	47.8	49.3	50.5	52.8	56.3	59.5	62.8	64.2	63.2	59.2	52.8	47.9	55.5
	Sacramento	37.7	41.4	43.2	45.5	50.3	55.3	58.1	58.0	55.7	50.4	43.4	37.8	48.1
	San Diego	48.9	50.7	52.8	55.6	59.1	61.9	65.7	67.3	65.6	60.9	53.9	48.8	57.6
	San Francisco	41.8	45.0	45.8	47.2	49.7	52.6	53.9	55.0	55.2	51.8	47.1	42.7	49.0
CO	Denver	16.1	20.2	25.8	34.5	43.6	52.4	58.6	56.9	47.6	36.4	25.4	17.4	36.2
CT	Hartford	15.8	18.6	28.1	37.5	47.6	56.9	62.2	60.4	51.8	40.7	32.8	21.3	39.5
DE	Wilmington	22.4	24.8	33.1	41.8	52.2	61.6	67.1	65.9	58.2	45.7	37.0	27.6	44.8
DC	Washington	26.8	29.1	37.7	46.4	56.6	66.5	71.4	70.0	62.5	50.3	41.1	31.7	49.2
FL	Jacksonville	40.5	43.3	49.2	54.9	62.1	69.1	71.9	71.8	69.0	59.3	50.2	43.4	57.1
	Miami	59.2	60.4	64.2	67.8	72.1	75.1	76.2	76.7	75.9	72.1	66.7	61.5	69.0
GA	Atlanta	31.5	34.5	42.5	50.2	58.7	66.2	69.5	69.0	63.5	51.9	42.8	35.0	51.3
HI	Honolulu	65.6	65.4	67.2	68.7	70.3	72.2	73.5	74.2	73.5	72.3	70.3	67.0	70.0
ID	Boise	21.6	27.5	31.9	36.7	43.9	52.1	57.7	56.8	48.2	39.0	31.1	22.5	39.1
IL	Chicago	12.9	17.2	28.5	38.6	47.7	57.5	62.6	61.6	53.9	42.2	31.6	19.1	39.5
	Peoria	13.2	17.7	29.8	40.8	50.9	60.7	65.4	63.1	55.2	43.1	32.5	19.3	41.0
IN	Indianapolis	17.2	20.9	31.9	41.5	51.7	61.0	65.2	62.8	55.6	43.5	34.1	23.2	42.4
IA	Des Moines	10.7	15.6	27.6	40.0	51.5	61.2	66.5	63.6	54.5	42.7	29.9	16.1	40.0
KS	Wichita	19.2	23.7	33.6	44.5	54.3	64.6	69.9	67.9	59.2	46.6	33.9	23.0	45.0
KY	Louisville	23.2	26.5	36.2	45.4	54.7	62.9	67.3	65.8	58.7	45.8	37.3	28.6	46.0
LA	New Orleans	41.8	44.4	51.6	58.4	65.2	70.8	73.1	72.8	69.5	58.7	51.0	44.8	58.5
ME	Portland	11.4	13.5	24.5	34.1	43.4	52.1	58.3	57.1	48.9	38.3	30.4	17.8	35.8
MD	Baltimore	23.4	25.9	34.1	42.5	52.6	61.8	66.8	65.7	58.4	45.9	37.1	28.2	45.2
MA	Boston	21.6	23.0	31.3	40.2	49.8	59.1	65.1	64.0	56.8	46.9	38.3	26.7	43.6
MI	Detroit	15.6	17.6	27.0	36.8	47.1	56.3	61.3	59.6	52.5	40.9	32.2	21.4	39.0
	Sault Ste. Marie	4.6	4.8	15.3	28.4	38.4	45.5	51.3	51.3	44.3	36.2	25.9	11.8	29.8
MN	Duluth	−2.2	2.8	15.7	28.9	39.6	48.5	55.1	53.3	44.5	35.1	21.5	4.9	29.0
	Minneapolis-St. Paul ..	2.8	9.2	22.7	36.2	47.6	57.6	63.1	60.3	50.3	38.8	25.2	10.2	35.3
MS	Jackson	32.7	35.7	44.1	51.9	60.0	67.1	70.5	69.7	63.7	50.3	42.3	36.1	52.0
MO	Kansas City	16.7	21.8	32.6	43.8	53.9	63.1	68.2	65.7	56.9	45.7	33.6	21.9	43.7
	St. Louis	20.8	25.1	35.5	46.4	56.0	65.7	70.4	67.9	60.5	48.3	37.7	26.0	46.7
MT	Great Falls	11.6	17.2	22.8	31.9	40.9	48.6	53.2	52.2	43.5	35.8	24.3	14.6	33.1

Source: U.S. National Oceanic and Atmospheric Administration, Climatography of the United States, No. 81.

then the sample mean of the data set $y_1, \ldots, y_n$ is

$$\bar{y} = \sum_{i=1}^{n}(ax_i + b)/n = \sum_{i=1}^{n} ax_i/n + \sum_{i=1}^{n} b/n = a\bar{x} + b$$

EXAMPLE 2.3a The winning scores in the U.S. Masters golf tournament in the years from 2004 to 2013 were as follows:

$$280, 278, 272, 276, 281, 279, 276, 281, 289, 280$$

Find the sample mean of these scores.

SOLUTION Rather than directly adding these values, it is easier to first subtract 280 from each one to obtain the new values $y_i = x_i - 280$:

$$0, -2, -8, -4, 1, -1, -4, 1, 9, 0$$

Because the arithmetic average of the transformed data set is

$$\bar{y} = -8/10$$

it follows that
$$\bar{x} = \bar{y} + 280 = 279.2 \quad \blacksquare$$

Sometimes we want to determine the sample mean of a data set that is presented in a frequency table listing the k distinct values $v_1, \ldots, v_k$ having corresponding frequencies $f_1, \ldots, f_k$. Since such a data set consists of $n = \sum_{i=1}^{k} f_i$ observations, with the value v_i appearing f_i times, for each $i = 1, \ldots, k$, it follows that the sample mean of these n data values is

$$\bar{x} = \sum_{i=1}^{k} v_i f_i / n$$

By writing the preceding as

$$\bar{x} = \frac{f_1}{n} v_1 + \frac{f_2}{n} v_2 + \cdots + \frac{f_k}{n} v_k$$

we see that the sample mean is a *weighted average* of the distinct values, where the weight given to the value v_i is equal to the proportion of the n data values that are equal to $v_i, i = 1, \ldots, k$.

EXAMPLE 2.3b The following is a frequency table giving the ages of members of a symphony orchestra for young adults.

Age	Frequency
15	2
16	5
17	11
18	9
19	14
20	13

Find the sample mean of the ages of the 54 members of the symphony.

SOLUTION

$$\bar{x} = (15 \cdot 2 + 16 \cdot 5 + 17 \cdot 11 + 18 \cdot 9 + 19 \cdot 14 + 20 \cdot 13)/54 \approx 18.24 \quad \blacksquare$$

Another statistic used to indicate the center of a data set is the *sample median*; loosely speaking, it is the middle value when the data set is arranged in increasing order.

Definition

Order the values of a data set of size n from smallest to largest. If n is odd, the *sample median* is the value in position $(n+1)/2$; if n is even, it is the average of the values in positions $n/2$ and $n/2 + 1$.

Thus the sample median of a set of three values is the second smallest; of a set of four values, it is the average of the second and third smallest.

EXAMPLE 2.3c Find the sample median for the data described in Example 2.3b.

SOLUTION Since there are 54 data values, it follows that when the data are put in increasing order, the sample median is the average of the values in positions 27 and 28. Thus, the sample median is 18.5. ■

The sample mean and sample median are both useful statistics for describing the central tendency of a data set. The sample mean makes use of all the data values and is affected by extreme values that are much larger or smaller than the others; the sample median makes use of only one or two of the middle values and is thus not affected by extreme values. Which of them is more useful depends on what one is trying to learn from the data. For instance, if a city government has a flat rate income tax and is trying to estimate its total revenue from the tax, then the sample mean of its residents' income would be a more useful statistic. On the other hand, if the city was thinking about constructing middle-income housing, and wanted to determine the proportion of its population able to afford it, then the sample median would probably be more useful.

EXAMPLE 2.3d In a study reported in Hoel, D. G., "A representation of mortality data by competing risks," *Biometrics*, **28**, pp. 475–488, 1972, a group of 5-week-old mice were each given a radiation dose of 300 rad. The mice were then divided into two groups; the first group was kept in a germ-free environment, and the second in conventional laboratory conditions. The numbers of days until death were then observed. The data for those whose death was due to thymic lymphoma are given in the following stem and leaf plots (whose stems are in units of hundreds of days); the first plot is for mice living in the germ-free conditions and the second for mice living under ordinary laboratory conditions.

Germ-Free Mice

```
1 | 58, 92, 93, 94, 95
2 | 02, 12, 15, 29, 30, 37, 40, 44, 47, 59
3 | 01, 01, 21, 37
4 | 15, 34, 44, 85, 96
5 | 29, 37
6 | 24
7 | 07
8 | 00
```

Conventional Mice

```
1 | 59, 89, 91, 98
2 | 35, 45, 50, 56, 61, 65, 66, 80
3 | 43, 56, 83
4 | 03, 14, 28, 32
```

Determine the sample means and the sample medians for the two sets of mice.

SOLUTION It is clear from the stem and leaf plots that the sample mean for the set of mice put in the germ-free setting is larger than the sample mean for the set of mice in the usual laboratory setting; indeed, a calculation gives that the former sample mean is 344.07, whereas the latter one is 292.32. On the other hand, since there are 29 data values for the germ-free mice, the sample median is the 15th largest data value, namely, 259; similarly, the sample median for the other set of mice is the 10th largest data value, namely, 265. Thus, whereas the sample mean is quite a bit larger for the first data set, the sample medians are approximately equal. The reason for this is that whereas the sample mean for the first set is greatly affected by the five data values greater than 500, these values have a much smaller effect on the sample median. Indeed, the sample median would remain unchanged if these values were replaced by any other five values greater than or equal to 259. It appears from the stem and leaf plots that the germ-free conditions probably improved the life span of the five longest living rats, but it is unclear what, if any, effect it had on the life spans of the other rats. ∎

Another statistic that has been used to indicate the central tendency of a data set is the *sample mode*, defined to be the value that occurs with the greatest frequency. If no single value occurs most frequently, then all the values that occur at the highest frequency are called *modal values*.

EXAMPLE 2.3e   The following frequency table gives the values obtained in 40 rolls of a die.

Value	Frequency
1	9
2	8
3	5
4	5
5	6
6	7

Find **(a)** the sample mean, **(b)** the sample median, and **(c)** the sample mode.

SOLUTION   **(a)** The sample mean is

$$\bar{x} = (9 + 16 + 15 + 20 + 30 + 42)/40 = 3.05$$

(b) The sample median is the average of the 20th and 21st smallest values, and is thus equal to 3. **(c)** The sample mode is 1, the value that occurred most frequently.   ■

2.3.2 SAMPLE VARIANCE AND SAMPLE STANDARD DEVIATION

Whereas we have presented statistics that describe the central tendencies of a data set, we are also interested in ones that describe the spread or variability of the data values. A statistic that could be used for this purpose would be one that measures the average value of the squares of the distances between the data values and the sample mean. This is accomplished by the sample variance, which for technical reasons divides the sum of the squares of the differences by $n - 1$ rather than n, where n is the size of the data set.

Definition

The *sample variance*, call it s^2, of the data set $x_1, \ldots, x_n$ is defined by

$$s^2 = \sum_{i=1}^{n} (x_i - \bar{x})^2/(n - 1)$$

EXAMPLE 2.3f   Find the sample variances of the data sets **A** and **B** given below.

$$\mathbf{A}: 3, 4, 6, 7, 10 \qquad \mathbf{B}: -20, 5, 15, 24$$

SOLUTION As the sample mean for data set **A** is $\bar{x} = (3 + 4 + 6 + 7 + 10)/5 = 6$, it follows that its sample variance is

$$s^2 = [(-3)^2 + (-2)^2 + 0^2 + 1^2 + 4^2]/4 = 7.5$$

The sample mean for data set **B** is also 6; its sample variance is

$$s^2 = [(-26)^2 + (-1)^2 + 9^2 + (18)^2]/3 \approx 360.67$$

Thus, although both data sets have the same sample mean, there is a much greater variability in the values of the **B** set than in the **A** set. ■

The following algebraic identity is often useful for computing the sample variance:

An Algebraic Identity

$$\sum_{i=1}^{n}(x_i - \bar{x})^2 = \sum_{i=1}^{n} x_i^2 - n\bar{x}^2$$

The identity is proven as follows:

$$\sum_{i=1}^{n}(x_i - \bar{x})^2 = \sum_{i=1}^{n}\left(x_i^2 - 2x_i\bar{x} + \bar{x}^2\right)$$

$$= \sum_{i=1}^{n}x_i^2 - 2\bar{x}\sum_{i=1}^{n}x_i + \sum_{i=1}^{n}\bar{x}^2$$

$$= \sum_{i=1}^{n}x_i^2 - 2n\bar{x}^2 + n\bar{x}^2$$

$$= \sum_{i=1}^{n}x_i^2 - n\bar{x}^2$$

The computation of the sample variance can also be eased by noting that if

$$y_i = a + bx_i, \qquad i = 1, \ldots, n$$

then $\bar{y} = a + b\bar{x}$, and so

$$\sum_{i=1}^{n}(y_i - \bar{y})^2 = b^2\sum_{i=1}^{n}(x_i - \bar{x})^2$$

That is, if s_y^2 and s_x^2 are the respective sample variances, then

$$s_y^2 = b^2 s_x^2$$

In other words, adding a constant to each data value does not change the sample variance; whereas multiplying each data value by a constant results in a new sample variance that is equal to the old one multiplied by the square of the constant. ■

EXAMPLE 2.3g The following data give the worldwide number of fatal airline accidents of commercially scheduled air transports in the years from 1997 to 2005.

Year	1997	1998	1999	2000	2001	2002	2003	2004	2005
Accidents	25	20	21	18	13	13	7	9	18

Source: National Safety Council.

Find the sample variance of the number of accidents in these years.

SOLUTION Let us start by subtracting 18 from each value, to obtain the new data set:

$$7, 2, 3, 0, -5, -5, -11, -9, 0$$

Calling the transformed data $y_1, \ldots, y_9$, we have

$$\bar{y} = \sum_{i=1}^{9} y_i/9 = -2, \qquad \sum_{i=1}^{n} y_i^2 = 49 + 4 + 9 + 25 + 25 + 121 + 81 = 314$$

Hence, since the sample variance of the transformed data is equal to that of the original data, upon using the algebraic identity we obtain

$$s^2 = \frac{314 - 9(4)}{8} = 34.75 \quad ■$$

Program 2.3 on the text disk can be used to obtain the sample variance for large data sets.

The positive square root of the sample variance is called the *sample standard deviation*.

Definition

The quantity s, defined by

$$s = \sqrt{\sum_{i=1}^{n} (x_i - \bar{x})^2/(n-1)}$$

is called the *sample standard deviation*.

The sample standard deviation is measured in the same units as the data.

2.3.3 SAMPLE PERCENTILES AND BOX PLOTS

Loosely speaking, the sample $100p$ percentile of a data set is that value such that $100p$ percent of the data values are less than or equal to it, $0 \leq p \leq 1$. More formally, we have the following definition.

Definition

The *sample 100p percentile* is that data value such that at least $100p$ percent of the data are less than or equal to it and at least $100(1-p)$ percent are greater than or equal to it. If two data values satisfy this condition, then the sample $100p$ percentile is the arithmetic average of these two values.

To determine the sample $100p$ percentile of a data set of size n, we need to determine the data values such that

1. At least np of the values are less than or equal to it.
2. At least $n(1-p)$ of the values are greater than or equal to it.

To accomplish this, first arrange the data in increasing order. Then, note that if np is not an integer, then the only data value that satisfies the preceding conditions is the one whose position when the data are ordered from smallest to largest is the smallest integer exceeding np. For instance, if $n = 22$, $p = .8$, then we require a data value such that at least 17.6 of the values are less than or equal to it, and at least 4.4 of them are greater than or equal to it. Clearly, only the 18th smallest value satisfies both conditions and this is the sample 80 percentile. On the other hand, if np is an integer, then it is easy to check that both the values in positions np and $np+1$ satisfy the preceding conditions, and so the sample $100p$ percentile is the average of these values. For instance, if we wanted the 90 percentile of a data set of size 20, then both the (18)th and (19)th smallest values would be such that at least 90 percent of the data values are less than or equal to them, and at least 10 percent of the data values are greater than or equal to them. Thus, the 90 percentile is the average of these two values.

EXAMPLE 2.3h Table 2.6 lists the populations of the 25 most populous U.S. cities for the year 1994. For this data set, find **(a)** the sample 10 percentile and **(b)** the sample 80 percentile.

SOLUTION **(a)** Because the sample size is 25 and $25(.10) = 2.5$, the sample 10 percentile is the third smallest value, equal to $590,763$.

(b) Because $25(.80) = 20$, the sample 80 percentile is the average of the twentieth and the twenty-first smallest values. Hence, the sample 80 percentile is

$$\frac{1,512,986 + 1,448,394}{2} = 1,480,690 \quad \blacksquare$$

The sample 50 percentile is, of course, just the sample median. Along with the sample 25 and 75 percentiles, it makes up the sample quartiles.

Definition

The sample 25 percentile is called the *first quartile*; the sample 50 percentile is called the sample median or the *second quartile*; the sample 75 percentile is called the *third quartile*.

TABLE 2.6   *Population of 25 Largest U.S. Cities, July 2006*

Rank	City	Population
1	New York, NY	8,250,567
2	Los Angeles, CA	3,849,378
3	Chicago, IL	2,833,321
4	Houston, TX	2,144,491
5	Phoenix, AR	1,512,986
6	Philadelphia, PA	1,448,394
7	San Antonio, TX	1,296,682
8	San Diego, CA	1,256,951
9	Dallas, TX	1,232,940
10	San Jose, CA	929,936
11	Detroit, MI	918,849
12	Jacksonville, FL	794,555
13	Indianapolis, IN	785,597
14	San Francisco, CA	744,041
15	Columbus, OH	733,203
16	Austin, TX	709,893
17	Memphis, TN	670,902
18	Fort Worth, TX	653,320
19	Baltimore, MD	640,961
20	Charlotte, NC	630,478
21	El Paso, TX	609,415
22	Milwaukee, WI	602,782
23	Boston, MA	590,763
24	Seattle, WA	582,454
25	Washington, DC	581,530

The quartiles break up a data set into four parts, with roughly 25 percent of the data being less than the first quartile, 25 percent being between the first and second quartile, 25 percent being between the second and third quartile, and 25 percent being greater than the third quartile.

EXAMPLE 2.3i   Noise is measured in decibels, denoted as dB. One decibel is about the level of the weakest sound that can be heard in a quiet surrounding by someone with good hearing; a whisper measures about 30 dB; a human voice in normal conversation is about 70 dB; a loud radio is about 100 dB. Ear discomfort usually occurs at a noise level of about 120 dB.

The following data give noise levels measured at 36 different times directly outside of Grand Central Station in Manhattan.

82, 89, 94, 110, 74, 122, 112, 95, 100, 78, 65, 60, 90, 83, 87, 75, 114, 85

69, 94, 124, 115, 107, 88, 97, 74, 72, 68, 83, 91, 90, 102, 77, 125, 108, 65

Determine the quartiles.

57   70

60   61.5   64

FIGURE 2.7   *A box plot.*

SOLUTION   A stem and leaf plot of the data is as follows:

6	0, 5, 5, 8, 9
7	2, 4, 4, 5, 7, 8
8	2, 3, 3, 5, 7, 8, 9
9	0, 0, 1, 4, 4, 5, 7
10	0, 2, 7, 8
11	0, 2, 4, 5
12	2, 4, 5

Because $36/4 = 9$, the first quartile is 74.5, the average of the 9th and 10th smallest data values; the second quartile is 89.5, the average of the 18th and 19th smallest values; the third quartile is 104.5, the average of the 27th and 28th smallest values.   ■

A *box plot* is often used to plot some of the summarizing statistics of a data set. A straight line segment stretching from the smallest to the largest data value is drawn on a horizontal axis; imposed on the line is a "box," which starts at the first and continues to the third quartile, with the value of the second quartile indicated by a vertical line. For instance, the 42 data values presented in Table 2.1 go from a low value of 57 to a high value of 70. The value of the first quartile (equal to the value of the 11th smallest on the list) is 60; the value of the second quartile (equal to the average of the 21st and 22nd smallest values) is 61.5; and the value of the third quartile (equal to the value of the 32nd smallest on the list) is 64. The box plot for this data set is shown in Figure 2.7.

The length of the line segment on the box plot, equal to the largest minus the smallest data value, is called the *range* of the data. Also, the length of the box itself, equal to the third quartile minus the first quartile, is called the *interquartile range*.

2.4 CHEBYSHEV'S INEQUALITY

Let $\bar{x}$ and s be the sample mean and sample standard deviation of a data set. Assuming that $s > 0$, Chebyshev's inequality states that for any value of $k \geq 1$, greater than $100(1 - 1/k^2)$ percent of the data lie within the interval from $\bar{x} - ks$ to $\bar{x} + ks$. Thus, by letting $k = 3/2$, we obtain from Chebyshev's inequality that greater than $100(5/9) = 55.56$ percent of the data from any data set lies within a distance $1.5s$ of the sample mean $\bar{x}$; letting $k = 2$ shows that greater than 75 percent of the data lies within $2s$ of the sample mean; and letting $k = 3$ shows that greater than $800/9 \approx 88.9$ percent of the data lies within 3 sample standard deviations of $\bar{x}$.

When the size of the data set is specified, Chebyshev's inequality can be sharpened, as indicated in the following formal statement and proof.

Chebyshev's Inequality

Let $\bar{x}$ and s be the sample mean and sample standard deviation of the data set consisting of the data $x_1, \ldots, x_n$, where $s > 0$. Let

$$S_k = \{i, 1 \leq i \leq n : |x_i - \bar{x}| < ks\}$$

and let $|S_k|$ be the number of elements in the set S_k. Then, for any $k \geq 1$,

$$\frac{|S_k|}{n} \geq 1 - \frac{n-1}{nk^2} > 1 - \frac{1}{k_2}$$

Proof

$$
\begin{aligned}
(n-1)s^2 &= \sum_{i=1}^{n}(x_i - \bar{x})^2 \\
&= \sum_{i \in S_k}(x_i - \bar{x})^2 + \sum_{i \notin S_k}(x_i - \bar{x})^2 \\
&\geq \sum_{i \notin S_k}(x_i - \bar{x})^2 \\
&\geq \sum_{i \notin S_k} k^2 s^2 \\
&= k^2 s^2 (n - |S_k|)
\end{aligned}
$$

where the first inequality follows because all terms being summed are nonnegative, and the second follows since $(x_1 - \bar{x})^2 \geq k^2 s^2$ when $i \notin S_k$. Dividing both sides of the preceding inequality by $nk^2 s^2$ yields that

$$\frac{n-1}{nk^2} \geq \frac{n - |S_k|}{n} = 1 - \frac{|S_k|}{n}$$

and the result is proven. ∎

Because Chebyshev's inequality holds universally, it might be expected for given data that the actual percentage of the data values that lie within the interval from $\bar{x} - ks$ to $\bar{x} + ks$ might be quite a bit larger than the bound given by the inequality.

EXAMPLE 2.4a Table 2.7 lists the 10 top-selling passenger cars in the United States in the month of June, 2013.

TABLE 2.7 *Top Selling Vehicles*

June 2013 Sales (in thousands of vehicles)	
Ford F Series	68.0
Chevrolet Silverado.	43.3
Toyoto Camry	35.9
Chevrolet Cruze	32.9
Honda Accord	31.7
Honda Civic.	29.7
Dodge Ram	29.6
Ford Escape	28.7
Nissan Altima	26.9
Honda CR – V	26.6

A simple calculation yields that the sample mean and sample standard deviation of these data are

$$\bar{x} = 35.33 \qquad s = 11.86$$

Thus Chebyshev's Inequality states that at least $100(5/9) = 55.55$ percent of the data lies in the interval

$$\left(\bar{x} - \frac{3}{2}s, \ \ \bar{x} + \frac{3}{2}s \right) = (17.54, \ \ 53.12)$$

whereas, in actuality, 90 percent of the data falls within these limits. ∎

Suppose now that we are interested in the fraction of data values that exceed the sample mean by at least k sample standard deviations, where k is positive. That is, suppose that $\bar{x}$ and s are the sample mean and the sample standard deviation of the data set $x_1, x_2, \ldots, x_n$. Then, with

$$N(k) = \text{number of } i : x_i - \bar{x} \geq ks$$

what can we say about $N(k)/n$? Clearly,

$$\frac{N(k)}{n} \leq \frac{\text{number of } i : |x_i - \bar{x}| \geq ks}{n}$$

$$\leq \frac{1}{k^2} \quad \text{by Chebyshev's inequality}$$

However, we can make a stronger statement, as is shown in the following one-sided version of Chebyshev's inequality.

The One-Sided Chebyshev Inequality

Let $\bar{x}$ and s be the sample mean and sample standard deviation of the data set consisting of the data $x_1, \ldots, x_n$. Suppose $s > 0$, and let $N(k) = \text{number of } i : x_i - \bar{x} \geq ks$. Then, for any $k > 0$,

$$\frac{N(k)}{n} \le \frac{1}{1+k^2}$$

Proof

Let $y_i = x_i - \bar{x}$, $i = 1, \ldots, n$. For any $b > 0$, we have that

$$\sum_{i=1}^{n} (y_i + b)^2 \ge \sum_{i:y_i \ge ks} (y_i + b)^2$$

$$\ge \sum_{i:y_i \ge ks} (ks + b)^2$$

$$= N(k)(ks + b)^2 \qquad (2.4.1)$$

where the first inequality follows because $(y_i + b)^2 \ge 0$, the second because both ks and b are positive, and the final equality because $N(k)$ is equal to the number of i such that $y_i \ge ks$. However,

$$\sum_{i=1}^{n} (y_i + b)^2 = \sum_{i=1}^{n} (y_i^2 + 2by_i + b^2)$$

$$= \sum_{i=1}^{n} y_i^2 + 2b \sum_{i=1}^{n} y_i + nb^2$$

$$= \sum_{i=1}^{n} y_i^2 + nb^2$$

$$= (n-1)s^2 + nb^2$$

where the next to last equation used that $\sum_{i=1}^{n} y_i = \sum_{i=1}^{n} (x_i - \bar{x}) = \sum_{i=1}^{n} x_i - n\bar{x} = 0$. Therefore, we obtain from Equation (2.4.1) that

$$N(k) \le \frac{(n-1)s^2 + nb^2}{(ks + b)^2} < \frac{ns^2 + nb^2}{(ks + b)^2}$$

implying that

$$\frac{N(k)}{n} \le \frac{s^2 + b^2}{(ks + b)^2}$$

Because the preceding is valid for all $b > 0$, we can set $b = s/k$ (which is the value of b that minimizes the right-hand side of the preceding) to obtain that

$$\frac{N(k)}{n} \le \frac{s^2 + s^2/k^2}{(ks + s/k)^2}$$

Multiplying the numerator and the denominator of the right side of the preceding by k^2/s^2 gives

$$\frac{N(k)}{n} \le \frac{k^2 + 1}{(k^2 + 1)^2} = \frac{1}{k^2 + 1}$$

and the result is proven. Thus, for instance, where the usual Chebyshev inequality shows that at most 25 percent of data values are at least 2 standard deviations greater than the sample mean, the one-sided Chebyshev inequality lowers the bound to "at most 20 percent." ■

2.5 NORMAL DATA SETS

Many of the large data sets observed in practice have histograms that are similar in shape. These histograms often reach their peaks at the sample median and then decrease on both sides of this point in a bell-shaped symmetric fashion. Such data sets are said to be *normal* and their histograms are called *normal histograms*. Figure 2.8 is the histogram of a normal data set.

If the histogram of a data set is close to being a normal histogram, then we say that the data set is *approximately normal*. For instance, we would say that the histogram given in Figure 2.9 is from an approximately normal data set, whereas the ones presented in Figures 2.10 and 2.11 are not (because each is too nonsymmetric). Any data set that is not approximately symmetric about its sample median is said to be *skewed*. It is "skewed to the right" if it has a long tail to the right and "skewed to the left" if it has a long tail to the left. Thus the data set presented in Figure 2.10 is skewed to the left and the one of Figure 2.11 is skewed to the right.

It follows from the symmetry of the normal histogram that a data set that is approximately normal will have its sample mean and sample median approximately equal.

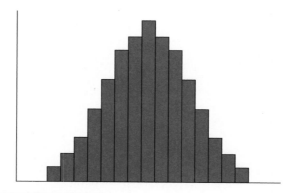

FIGURE 2.8 *Histogram of a normal data set.*

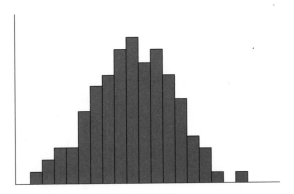

FIGURE 2.9   *Histogram of an approximately normal data set.*

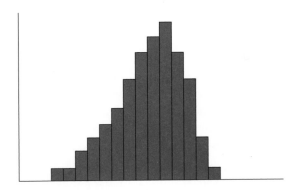

FIGURE 2.10   *Histogram of a data set skewed to the left.*

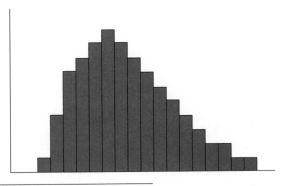

FIGURE 2.11   *Histogram of a data set skewed to the right.*

Suppose that $\bar{x}$ and s are the sample mean and sample standard deviation of an approximately normal data set. The following rule, known as the *empirical rule*, specifies the approximate proportions of the data observations that are within s, $2s$, and $3s$ of the sample mean $\bar{x}$.

The Empirical Rule

If a data set is approximately normal with sample mean $\bar{x}$ and sample standard deviation s, then the following statements are true.

1. Approximately 68 percent of the observations lie within

$$\bar{x} \pm s$$

2. Approximately 95 percent of the observations lie within

$$\bar{x} \pm 2s$$

3. Approximately 99.7 percent of the observations lie within

$$\bar{x} \pm 3s$$

EXAMPLE 2.5a The following stem and leaf plot gives the scores on a statistics exam taken by industrial engineering students.

```
9 | 0, 1, 4
8 | 3, 5, 5, 7, 8
7 | 2, 4, 4, 5, 7, 7, 8
6 | 0, 2, 3, 4, 6, 6
5 | 2, 5, 5, 6, 8
4 | 3, 6
```

By standing the stem and leaf plot on its side we can see that the corresponding histogram is approximately normal. Use it to assess the empirical rule.

SOLUTION A calculation gives that

$$\bar{x} \approx 70.571, \quad s \approx 14.354$$

Thus the empirical rule states that approximately 68 percent of the data are between 56.2 and 84.9; the actual percentage is $1,500/28 \approx 53.6$. Similarly, the empirical rule gives that approximately 95 percent of the data are between 41.86 and 99.28, whereas the actual percentage is 100. ∎

A data set that is obtained by sampling from a population that is itself made up of subpopulations of different types is usually not normal. Rather, the histogram from such a data set often appears to resemble a combining, or superposition, of normal histograms

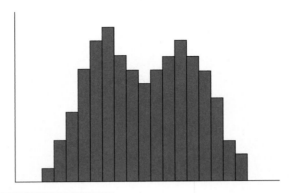

FIGURE 2.12   *Histogram of a bimodal data set.*

and thus will often have more than one local peak or hump. Because the histogram will be higher at these local peaks than at their neighboring values, these peaks are similar to modes. A data set whose histogram has two local peaks is said to be *bimodal*. The data set represented in Figure 2.12 is bimodal.

2.6 PAIRED DATA SETS AND THE SAMPLE CORRELATION COEFFICIENT

We are often concerned with data sets that consist of pairs of values that have some relationship to each other. If each element in such a data set has an x value and a y value, then we represent the ith data point by the pair (x_i, y_i). For instance, in an attempt to determine the relationship between the daily midday temperature (measured in degrees Celsius) and the number of defective parts produced during that day, a company recorded the data presented in Table 2.8. For this data set, x_i represents the temperature in degrees Celsius and y_i the number of defective parts produced on day i.

A useful way of portraying a data set of paired values is to plot the data on a two-dimensional graph, with the x-axis representing the x value of the data and the y-axis representing the y value. Such a plot is called a *scatter diagram*. Figure 2.13 presents a scatter diagram for the data of Table 2.8.

A question of interest concerning paired data sets is whether large x values tend to be paired with large y values, and small x values with small y values; if this is not the case, then we might question whether large values of one of the variables tend to be paired with small values of the other. A rough answer to these questions can often be provided by the scatter diagram. For instance, Figure 2.13 indicates that there appears to be some connection between high temperatures and large numbers of defective items. To obtain a quantitative measure of this relationship, we now develop a statistic that attempts to measure the degree to which larger x values go with larger y values and smaller x values with smaller y values.

TABLE 2.8   *Temperature and Defect Data*

Day	Temperature	Number of Defects
1	24.2	25
2	22.7	31
3	30.5	36
4	28.6	33
5	25.5	19
6	32.0	24
7	28.6	27
8	26.5	25
9	25.3	16
10	26.0	14
11	24.4	22
12	24.8	23
13	20.6	20
14	25.1	25
15	21.4	25
16	23.7	23
17	23.9	27
18	25.2	30
19	27.4	33
20	28.3	32
21	28.8	35
22	26.6	24

Suppose that the data set consists of the paired values (x_i, y_i), $i = 1, \ldots, n$. To obtain a statistic that can be used to measure the association between the individual values of a set of paired data, let $\bar{x}$ and $\bar{y}$ denote the sample means of the x values and the y values, respectively. For data pair i, consider $x_i - \bar{x}$ the deviation of its x value from the sample mean, and $y_i - \bar{y}$ the deviation of its y value from the sample mean. Now if x_i is a large x value, then it will be larger than the average value of all the x's, so the deviation $x_i - \bar{x}$ will be a positive value. Similarly, when x_i is a small x value, then the deviation $x_i - \bar{x}$ will be a negative value. Because the same statements are true about the y deviations, we can conclude the following:

> When large values of the x variable tend to be associated with large values of the y variable and small values of the x variable tend to be associated with small values of the y variable, then the signs, either positive or negative, of $x_i - \bar{x}$ and $y_i - \bar{y}$ will tend to be the same.

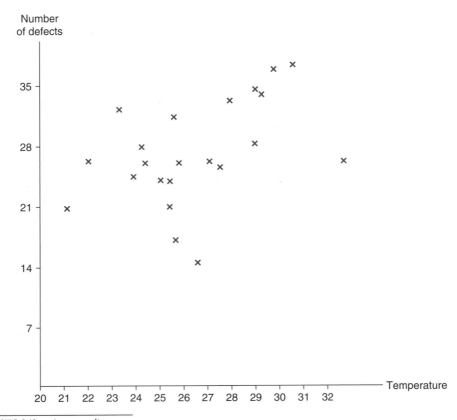

FIGURE 2.13   *A scatter diagram.*

Now, if $x_i - \bar{x}$ and $y_i - \bar{y}$ both have the same sign (either positive or negative), then their product $(x_i - \bar{x})(y_i - \bar{y})$ will be positive. Thus, it follows that when large x values tend to be associated with large y values and small x values are associated with small y values, then $\sum_{i=1}^{n}(x_i - \bar{x})(y_i - \bar{y})$ will tend to be a large positive number. [In fact, not only will all the products have a positive sign when large (small) x values are paired with large (small) y values, but it also follows from a mathematical result known as Hardy's lemma that the largest possible value of the sum of paired products will be obtained when the largest $x_i - \bar{x}$ is paired with the largest $y_i - \bar{y}$, the second largest $x_i - \bar{x}$ is paired with the second largest $y_i - \bar{y}$, and so on.] In addition, it similarly follows that when large values of x_i tend to be paired with small values of y_i then the signs of $x_i - \bar{x}$ and $y_i - \bar{y}$ will be opposite and so $\sum_{i=1}^{n}(x_i - \bar{x})(y_i - \bar{y})$ will be a large negative number.

To determine what it means for $\sum_{i=1}^{n}(x_i - \bar{x})(y_i - \bar{y})$ to be "large," we standardize this sum first by dividing by $n - 1$ and then by dividing by the product of the two sample standard deviations. The resulting statistic is called the *sample correlation coefficient.*

Definition

Consider the data pairs (x_i, y_i), $i = 1, \ldots, n$. and let s_x and s_y denote, respectively, the sample standard deviations of the x values and the y values. The *sample correlation coefficient,* call it r, of the data pairs (x_i, y_i), $i = 1, \ldots, n$ is defined by

$$r = \frac{\sum_{i=1}^{n}(x_i - \bar{x})(y_i - \bar{y})}{(n-1)s_x s_y}$$

$$= \frac{\sum_{i=1}^{n}(x_i - \bar{x})(y_i - \bar{y})}{\sqrt{\sum_{i=1}^{n}(x_i - \bar{x})^2 \sum_{i=1}^{n}(y_i - \bar{y})^2}}$$

When $r > 0$ we say that the sample data pairs are *positively correlated,* and when $r < 0$ we say that they are *negatively correlated.*

The following are properties of the sample correlation coefficient.

Properties of r

1. $$-1 \leq r \leq 1$$
2. If for constants a and b, with $b > 0$,

    $$y_i = a + bx_i, \qquad i = 1, \ldots, n$$

    then $r = 1$.
3. If for constants a and b, with $b < 0$,

    $$y_i = a + bx_i, \qquad i = 1, \ldots, n$$

    then $r = -1$.
4. If r is the sample correlation coefficient for the data pairs $x_i, y_i, i = 1, \ldots, n$ then it is also the sample correlation coefficient for the data pairs

    $$a + bx_i, \quad c + dy_i, \quad i = 1, \ldots, n$$

    provided that b and d are both positive or both negative.

Property 1 says that the sample correlation coefficient r is always between -1 and $+1$. Property 2 says that r will equal $+1$ when there is a straight line (also called a linear) relation between the paired data such that large y values are attached to large x values. Property 3 says that r will equal -1 when the relation is linear and large y values are attached to small x values. Property 4 states that the value of r is unchanged when a

38   Chapter 2: Descriptive Statistics

constant is added to each of the x variables (or to each of the y variables) or when each x variable (or each y variable) is multiplied by a positive constant. This property implies that r does not depend on the dimensions chosen to measure the data. For instance, the sample correlation coefficient between a person's height and weight does not depend on whether the height is measured in feet or in inches or whether the weight is measured in pounds or in kilograms. Also, if one of the values in the pair is temperature, then the sample correlation coefficient is the same whether it is measured in Fahrenheit or in Celsius.

The absolute value of the sample correlation coefficient r (that is, $|r|$, its value without regard to its sign) is a measure of the strength of the linear relationship between the x and the y values of a data pair. A value of $|r|$ equal to 1 means that there is a perfect linear relation — that is, a straight line can pass through all the data points $(x_i, y_i), i = 1, \ldots, n$. A value of $|r|$ of around .8 means that the linear relation is relatively strong; although there is no straight line that passes through all of the data points, there is one that is "close" to them all. A value for $|r|$ of around .3 means that the linear relation is relatively weak.

The sign of r gives the direction of the relation. It is positive when the linear relation is such that smaller y values tend to go with smaller x values and larger y values with larger x values (and so a straight line approximation points upward), and it is negative when larger y values tend to go with smaller x values and smaller y values with larger x values (and so a straight line approximation points downward). Figure 2.14 displays scatter diagrams for data sets with various values of r.

EXAMPLE 2.6a Find the sample correlation coefficient for the data presented in Table 2.8.

SOLUTION A computation gives the solution

$$r = .4189$$

thus indicating a relatively weak positive correlation between the daily temperature and the number of defective items produced that day. ■

EXAMPLE 2.6b The following data give the resting pulse rates (in beats per minute) and the years of schooling of 10 individuals. A scatter diagram of these data is presented in Figure 2.15. The sample correlation coefficient for these data is $r = -.7638$. This negative correlation indicates that for this data set a high pulse rate is strongly associated with a small number of years in school, and a low pulse rate with a large number of years in school. ■

Person	1	2	3	4	5	6	7	8	9	10
Years of School	12	16	13	18	19	12	18	19	12	14
Pulse Rate	73	67	74	63	73	84	60	62	76	71

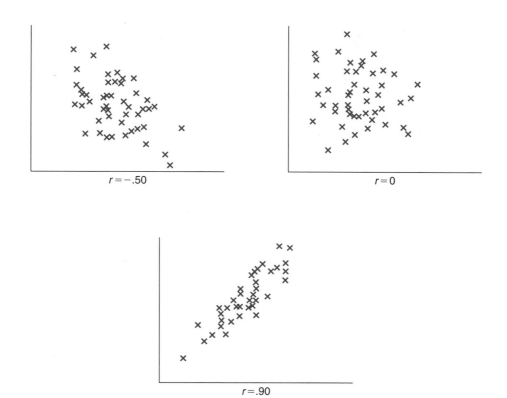

FIGURE 2.14   *Sample correlation coefficients.*

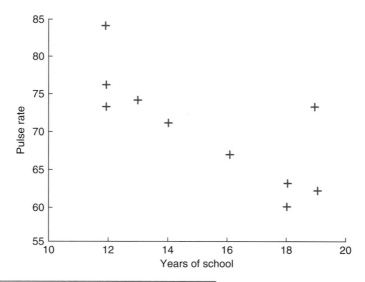

FIGURE 2.15   *Scatter diagram of years in school and pulse rate.*

---■---

Correlation Measures Association, Not Causation

The data set of Example 2.6b only considers 10 students and, as such, is not large enough for one to draw any firm conclusions about the relationship between years of school and pulse rate. Moreover, even if the data set were of larger size and with the same strong negative correlation between an individual's years of education and that individual's resting pulse rate, we would not be justified to conclude that additional years of school will directly reduce one's pulse rate. That is, whereas additional years of school tend to be associated with a lower resting pulse rate, this does not mean that it is a direct cause of it. Often, the explanation for such an association lies with an unexpressed factor that is related to both variables under consideration. In this instance, it may be that a person who has spent additional time in school is more aware of the latest findings in the area of health, and thus may be more aware of the importance of exercise and good nutrition; or it may be that it is not knowledge that is making the difference but rather it is that people who have had more education tend to end up in jobs that allow them more time for exercise and money for good nutrition. The strong negative correlation between years in school and resting pulse rate probably results from a combination of these as well as other underlying factors.

---■---

We will now prove the first three properties of the sample correlation coefficient r. That is, we will prove that $|r| \leq 1$ with equality when the data lie on a straight line. To begin, note that

$$\sum \left(\frac{x_i - \bar{x}}{s_x} - \frac{y_i - \bar{y}}{s_y} \right)^2 \geq 0 \qquad (2.6.1)$$

or

$$\sum \frac{(x_i - \bar{x})^2}{s_x^2} + \sum \frac{(y_i - \bar{y})^2}{s_y^2} - 2 \sum \frac{(x_i - \bar{x})(y_i - \bar{y})}{s_x s_y} \geq 0$$

or

$$n - 1 + n - 1 - 2(n-1)r \geq 0$$

showing that

$$r \leq 1$$

To see when $r = 1$, suppose first that the points (x_i, y_i), $i = 1, \ldots, n$ lie on the straight line

$$y_i = a + bx_i, \quad i = 1, \ldots, n$$

with positive slope b. If this is so, then

$$s_y^2 = b^2 s_x^2, \quad \bar{y} = a + b\bar{x},$$

showing that

$$b = \frac{s_y}{s_x}, \quad a = \bar{y} - \frac{s_y}{s_x}\bar{x}$$

Now, note also that $r = 1$ if and only if there is equality in Equation 2.6.1. That is, $r = 1$ if and only if for all i,

$$\frac{y_i - \bar{y}}{s_y} = \frac{x_i - \bar{x}}{s_x}$$

or, equivalently,

$$y_i = \bar{y} - \frac{s_y}{s_x}\bar{x} + \frac{s_y}{s_x}x_i$$

Consequently, $r = 1$ if and only if the data values (x_i, y_i) lie on a straight line having a positive slope.

To show that $r \geq -1$, with equality if and only if the data values (x_i, y_i) lie on a straight line having a negative slope, start with

$$\sum \left(\frac{x_i - \bar{x}}{s_x} + \frac{y_i - \bar{y}}{s_y} \right)^2 \geq 0$$

and use an argument analogous to the one just given.

Problems

1. The following is a sample of prices, rounded to the nearest cent, charged per gallon of standard unleaded gasoline in the San Francisco Bay area in June 1997.

   3.88, 3.90, 3.93, 3.90, 3.93, 3.96, 3.88, 3.94, 3.96, 3.88, 3.94, 3.99, 3.98

   Represent these data in

   (a) a frequency table;
   (b) a relative frequency line graph.

2. Explain how a pie chart can be constructed. If a data value had relative frequency r, at what angle would the lines defining its sector meet?

3. The following are the estimated oil reserves, in billions of barrels, for four regions in the Western Hemisphere.

United States   38.7
South America   22.6
Canada   8.8
Mexico   60.0

Represent these data in a pie chart.

4. The following table gives the average travel time to work for workers in each of the 50 states as well as the percentage of those workers who use public transportation.

   (a) Represent the data relating to the average travel times in a histogram.
   (b) Represent the data relating to the percentage of workers using public transportation in a stem and leaf plot.

Region, Division, and State	Means of Transportation to Work — Percent Using Public Transportation	Average Travel Time to Work[1] (minutes)
United States ..	5.3	22.4
Northeast	12.8	24.5
New England....	5.1	21.5
Maine......	0.9	19.0
New Hampshire .	0.7	21.9
Vermont.....	0.7	18.0
Massachusetts ..	8.3	22.7
Rhode Island...	2.5	19.2
Connecticut ...	3.9	21.1
Middle Atlantic...	15.7	25.7
New York	24.8	28.6
New Jersey....	8.8	25.3
Pennsylvania...	6.4	21.6
Midwest.......	3.5	20.7
East North Central ...	4.3	21.7
Ohio	2.5	20.7
Indiana	1.3	20.4
Illinois	10.1	25.1
Michigan	1.6	21.2
Wisconsin....	2.5	18.3
West North Central .	1.9	18.4
Minnesota....	3.6	19.1
Iowa	1.2	16.2
Missouri.....	2.0	21.6

(continued)

Region, Division, and State	Means of Transportation to Work	Average Travel Time to Work[1] (minutes)
	Percent Using Public Transportation	
North Dakota	0.6	13.0
South Dakota	0.3	13.8
Nebraska..........	1.2	15.8
Kansas............	0.6	17.2
South.................	**2.6**	**22.0**
South Atlantic	3.4	22.5
Delaware	2.4	20.0
Maryland	8.1	27.0
Virginia...........	4.0	24.0
West Virginia......	1.1	21.0
North Carolina	1.0	19.8
South Carolina	1.1	20.5
Georgia...........	2.8	22.7
Florida	2.0	21.8
East South Central ...	1.2	21.1
Kentucky	1.6	20.7
Tennessee	1.3	21.5
Alabama	0.8	21.2
Mississippi	0.8	20.6
West South Central ...	2.0	21.6
Arkansas	0.5	19.0
Louisiana	3.0	22.3
Oklahoma	0.6	19.3
Texas	2.2	22.2
West.................	**4.1**	**22.7**
Mountain	2.1	19.7
Montana..........	0.6	14.8
Idaho.............	1.9	17.3
Wyoming	1.4	15.4
Colorado	2.9	20.7
New Mexico	1.0	19.1
Arizona	2.1	21.6
Utah	2.3	18.9
Nevada	2.7	19.8
Pacific	4.8	23.8
Washington	4.5	22.0
Oregon	3.4	19.6
California.........	4.9	24.6
Alaska	2.4	16.7
Hawaii	7.4	23.8

[1] *Excludes persons who worked at home.*
Source: U.S. Bureau of the Census. Census of Population and Housing, 1990.

5. Choose a book or article and count the number of words in each of the first 100 sentences. Present the data in a stem and leaf plot. Now choose another book or article, by a different author, and do the same. Do the two stem and leaf plots look similar? Do you think this could be a viable method for telling whether different articles were written by different authors?

6. The following table gives the number of commercial airline accidents and the total number of resulting fatalities in the United States in the years from 1985 to 2006.

   (a) Represent the number of yearly airline accidents in a frequency table.
   (b) Give a frequency polygon graph of the number of yearly airline accidents.
   (c) Give a cumulative relative frequency plot of the number of yearly airline accidents.
   (d) Find the sample mean of the number of yearly airline accidents.
   (e) Find the sample median of the number of yearly airline accidents.
   (f) Find the sample mode of the number of yearly airline accidents.
   (g) Find the sample standard deviation of the number of yearly airline accidents.

U.S. Airline Safety, Scheduled Commercial Carriers, 1985–2006

Year	Departures (millions)	Accidents	Fatalities	Year	Departures (millions)	Accidents	Fatalities
1985	6.1	4	197	1996	7.9	3	342
1986	6.4	2	5	1997	9.9	3	3
1987	6.6	4	231	1998	10.5	1	1
1988	6.7	3	285	1999	10.9	2	12
1989	6.6	11	278	2000	11.1	2	89
1990	7.8	6	39	2001	10.6	6	531
1991	7.5	4	62	2002	10.3	0	0
1992	7.5	4	33	2003	10.2	2	22
1993	7.7	1	1	2004	10.8	1	13
1994	7.8	4	239	2005	10.9	3	22
1995	8.1	2	166	2006	11.2	2	50

Source: National Transportation Safety Board.

7. (Use the table from Problem 6.)

   (a) Represent the number of yearly airline fatalities in a histogram.
   (b) Represent the number of yearly airline fatalities in a stem and leaf plot.
   (c) Find the sample mean of the number of yearly airline fatalities.
   (d) Find the sample median of the number of yearly airline fatalities.
   (e) Find the sample standard deviation of the number of yearly airline fatalities.

8. The sample mean of the weights of the adult women of town A is larger than the sample mean of the weights of the adult women of town B. Moreover, the sample mean of the weights of the adult men of town A is larger than the sample mean

of the weights of the adult men of town B. Can we conclude that the sample mean of the weights of the adults of town A is larger than the sample mean of the weights of the adults of town B? Explain your answer.

9. Using the table given in Problem 4, find the sample mean and sample median of the average travel time for those states in the

   (a) northeast;
   (b) midwest;
   (c) south;
   (d) west.

10. A total of 100 people work at company A, whereas a total of 110 work at company B. Suppose the total employee payroll is larger at company A than at company B.

   (a) What does this imply about the median of the salaries at company A with regard to the median of the salaries at company B?
   (b) What does this imply about the average of the salaries at company A with regard to the average of the salaries at company B?

11. The sample mean of the initial 99 values of a data set consisting of 198 values is equal to 120, whereas the sample mean of the final 99 values is equal to 100. What can you conclude about the sample mean of the entire data set

   (a) Repeat when "sample mean" is replaced by "sample median."
   (b) Repeat when "sample mean" is replaced by "sample mode."

12. The following table gives the number of pedestrians, classified according to age group and sex, killed in fatal road accidents in England in 1922.

   (a) Approximate the sample means of the ages of the males.
   (b) Approximate the sample means of the ages of the females.
   (c) Approximate the quartiles of the males killed.
   (d) Approximate the quartiles of the females killed.

Age	Number of Males	Number of Females
0–5	120	67
5–10	184	120
10–15	44	22
15–20	24	15
20–30	23	25
30–40	50	22
40–50	60	40
50–60	102	76
60–70	167	104
70–80	150	90
80–100	49	27

13. The following are the percentages of ash content in 12 samples of coal found in close proximity:

$$9.2, 14.1, 9.8, 12.4, 16.0, 12.6, 22.7, 18.9, 21.0, 14.5, 20.4, 16.9$$

Find the

(a) sample mean, and
(b) sample standard deviation of these percentages.

14. The sample mean and sample variance of five data values are, respectively, $\bar{x} = 104$ and $s^2 = 16$. If three of the data values are 102, 100, 105, what are the other two data values?

15. Suppose you are given the average pay of all working people in each of the 50 states of the United States.

(a) Do you think that the sample mean of the averages for the 50 states will equal the value given for the entire United States?
(b) If the answer to part (a) is no, explain what other information aside from just the 50 averages would be needed to determine the sample mean salary for the entire country. Also, explain how you would use the additional information to compute this quantity.

16. The following data represent the lifetimes (in hours) of a sample of 40 transistors:

$$112, 121, 126, 108, 141, 104, 136, 134$$
$$121, 118, 143, 116, 108, 122, 127, 140$$
$$113, 117, 126, 130, 134, 120, 131, 133$$
$$118, 125, 151, 147, 137, 140, 132, 119$$
$$110, 124, 132, 152, 135, 130, 136, 128$$

(a) Determine the sample mean, median, and mode.
(b) Give a cumulative relative frequency plot of these data.

17. An experiment measuring the percent shrinkage on drying of 50 clay specimens produced the following data:

18.2	21.2	23.1	18.5	15.6
20.8	19.4	15.4	21.2	13.4
16.4	18.7	18.2	19.6	14.3
16.6	24.0	17.6	17.8	20.2
17.4	23.6	17.5	20.3	16.6
19.3	18.5	19.3	21.2	13.9
20.5	19.0	17.6	22.3	18.4
21.2	20.4	21.4	20.3	20.1
19.6	20.6	14.8	19.7	20.5
18.0	20.8	15.8	23.1	17.0

(a) Draw a stem and leaf plot of these data.

(b) Compute the sample mean, median, and mode.

(c) Compute the sample variance.

(d) Group the data into class intervals of size 1 percent starting with the value 13.0, and draw the resulting histogram.

(e) For the grouped data acting as if each of the data points in an interval was actually located at the midpoint of that interval, compute the sample mean and sample variance and compare this with the results obtained in parts (b) and (c). Why do they differ?

18. A computationally efficient way to compute the sample mean and sample variance of the data set $x_1, x_2, \ldots, x_n$ is as follows. Let

$$\bar{x}_j = \frac{\sum\limits_{i=1}^{j} x_i}{j}, \quad j = 1, \ldots, n$$

be the sample mean of the first j data values, and let

$$s_j^2 = \frac{\sum\limits_{i=1}^{j} (x_i - \bar{x}_j)^2}{j - 1}, \quad j = 2, \ldots, n$$

be the sample variance of the first $j, j \geq 2$, values. Then, with $s_1^2 = 0$, it can be shown that

$$\bar{x}_{j+1} = \bar{x}_j + \frac{x_{j+1} - \bar{x}_j}{j + 1}$$

and

$$s_{j+1}^2 = \left(1 - \frac{1}{j}\right) s_j^2 + (j + 1)(\bar{x}_{j+1} - \bar{x}_j)^2$$

(a) Use the preceding formulas to compute the sample mean and sample variance of the data values 3, 4, 7, 2, 9, 6.

(b) Verify your results in part (a) by computing as usual.

(c) Verify the formula given above for $\bar{x}_{j+1}$ in terms of $\bar{x}_j$.

19. Use the data of Table 2.5 to find the

(a) 90 percentile of the average temperature for January;

(b) 75 percentile of the average temperature for July.

20. Find the quartiles of the following ages at death as given in obituaries of the New York Times in the 2 weeks preceding 1 August 2013.

$$92, 90, 92, 74, 69, 80, 94, 98, 65, 96, 84, 69, 86, 91, 88$$

$$74, 97, 85, 88, 68, 77, 94, 88, 65, 76, 75, 60$$

$$69, 97, 92, 85, 70, 80, 93, 91, 68, 82, 78, 89$$

21. The universities having the largest number of months in which they ranked in the top 10 for the number of google searches over the past 114 months (as of June 2013) are as follows.

University	Number of Months in Top 10
Harvard University	114
University of Texas, Austin	114
University of Michigan	114
Stanford University	113
University of California Los Angeles (UCLA)	111
University of California Berkeley	97
Penn State University	94
Massachusetts Institute of Technology (MIT)	66
University of Southern California (USC)	63
Ohio State University	52
Yale University	48
University of Washington	33

   (a) Find the sample mean of the data.
   (b) Find the sample variance of the data.
   (c) Find the sample quartiles of the data.

22. Use the part of the table given in Problem 4 that gives the percentage of workers in each state that use public transportation to get to work to draw a box plot of these 50 percentages.

23. Represent the data of Problem 20 in a box plot.

24. The average particulate concentration, in micrograms per cubic meter, was measured in a petrochemical complex at 36 randomly chosen times, with the following concentrations resulting:

$$5, 18, 15, 7, 23, 220, 130, 85, 103, 25, 80, 7, 24, 6, 13, 65, 37, 25,$$

$$24, 65, 82, 95, 77, 15, 70, 110, 44, 28, 33, 81, 29, 14, 45, 92, 17, 53$$

(a) Represent the data in a histogram.

(b) Is the histogram approximately normal?

25. A chemical engineer desiring to study the evaporation rate of water from brine evaporation beds obtained data on the number of inches of evaporation in each of 55 July days spread over 4 years. The data are given in the following stem and leaf plot, which shows that the smallest data value was .02 inches, and the largest .56 inches.

```
.0 | 2, 6
.1 | 1, 4
.2 | 1, 1, 1, 3, 3, 4, 5, 5, 5, 6, 9
.3 | 0, 0, 2, 2, 2, 3, 3, 3, 3, 4, 4, 5, 5, 5, 6, 6, 7, 8, 9
.4 | 0, 1, 2, 2, 2, 3, 4, 4, 4, 5, 5, 5, 7, 8, 8, 8, 9, 9
.5 | 2, 5, 6
```

Find the

(a) sample mean;

(b) sample median;

(c) sample standard deviation of these data.

(d) Do the data appear to be approximately normal?

(e) What percentage of data values are within 1 standard deviation of the mean?

26. The following are the grade point averages of 30 students recently admitted to the graduate program in the Department of Industrial Engineering and Operations Research at the University of California at Berkeley.

3.46, 3.72, 3.95, 3.55, 3.62, 3.80, 3.86, 3.71, 3.56, 3.49, 3.96, 3.90, 3.70, 3.61,

3.72, 3.65, 3.48, 3.87, 3.82, 3.91, 3.69, 3.67, 3.72, 3.66, 3.79, 3.75, 3.93, 3.74,

3.50, 3.83

(a) Represent the preceding data in a stem and leaf plot.

(b) Calculate the sample mean $\bar{x}$.

(c) Calculate the sample standard deviation s.

(d) Determine the proportion of the data values that lies within $\bar{x} \pm 1.5s$ and compare with the lower bound given by Chebyshev's inequality.

(e) Determine the proportion of the data values that lies within $\bar{x} \pm 2s$ and compare with the lower bound given by Chebyshev's inequality.

27. Do the data in Problem 26 appear to be approximately normal? For parts (c) and (d) of this problem, compare the approximate proportions given by the empirical rule with the actual proportions.

28. Would you expect that a histogram of the weights of all the members of a health club would be approximately normal?

29. Use the data of Problem 16.

  (a) Compute the sample mean and sample median.
  (b) Are the data approximately normal?
  (c) Compute the sample standard deviation s.
  (d) What percentage of the data fall within $\bar{x} \pm 1.5s$?
  (e) Compare your answer in part (d) to that given by the empirical rule.
  (f) Compare your answer in part (d) to the bound given by Chebyshev's inequality.

30. The following are the heights and starting salaries of 12 law school classmates whose law school examination scores were roughly the same.

Height	Salary
64	91
65	94
66	88
67	103
69	77
70	96
72	105
72	88
74	122
74	102
75	90
76	114

  (a) Represent these data in a scatter diagram.
  (b) Find the sample correlation coefficient.

31. A random sample of individuals were rated as to their standing posture. In addition, the numbers of days of back pain each had experienced during the past year were also recorded. Surprisingly to the researcher these data indicated a positive correlation between good posture and number of days of back pain. Does this indicate that good posture causes back pain?

32. If for each of the fifty states we plot the paired data consisting of the average income of residents of the state and the number of foreign-born immigrants who reside in the state, then the data pairs will have a positive correlation. Can we conclude that immigrants tend to have higher incomes than native-born Americans? If not, how else could this phenomenon be explained?

33. A random group of 12 high school juniors were asked to estimate the average number of hours they study each week. The following give these hours along with the student's grade point average.

Hours	GPA
6	2.8
14	3.2
3	3.1
22	3.6
9	3.0
11	3.3
12	3.4
5	2.7
18	3.1
24	3.8
15	3.0
17	3.9

Find the sample correlation coefficient between hours reported and GPA.

34. Verify property **3** of the sample correlation coefficient.

35. Verify property **4** of the sample correlation coefficient.

36. In a study of children in grades 2 through 4, a researcher gave each student a reading test. When looking at the resulting data the researcher noted a positive correlation between a student's reading test score and height. The researcher concluded that taller children read better because they can more easily see the blackboard. What do you think?

37. A recent study yielded a positive correlation between breast-fed babies and scores on a vocabulary test taken at age 6. Discuss the potential difficulties in interpreting the results of this study.

Chapter 3

ELEMENTS OF PROBABILITY

3.1   INTRODUCTION

The concept of the probability of a particular event of an experiment is subject to various meanings or interpretations. For instance, if a geologist is quoted as saying that "there is a 60 percent chance of oil in a certain region," we all probably have some intuitive idea as to what is being said. Indeed, most of us would probably interpret this statement in one of two possible ways: either by imagining that

1. the geologist feels that, over the long run, in 60 percent of the regions whose outward environmental conditions are very similar to the conditions that prevail in the region under consideration, there will be oil; or
2. the geologist believes that it is more likely that the region will contain oil than it is that it will not; and in fact .6 is a measure of the geologist's belief in the hypothesis that the region will contain oil.

The two foregoing interpretations of the probability of an event are referred to as being the frequency interpretation and the subjective (or personal) interpretation of probability. In the *frequency interpretation*, the probability of a given outcome of an experiment is considered as being a "property" of that outcome. It is imagined that this property can be operationally determined by continual repetition of the experiment — the probability of the outcome will then be observable as being the proportion of the experiments that result in the outcome. This is the interpretation of probability that is most prevalent among scientists.

In the subjective interpretation, the probability of an outcome is not thought of as being a property of the outcome but rather is considered a statement about the beliefs of the person who is quoting the probability, concerning the chance that the outcome will occur. Thus, in this interpretation, probability becomes a subjective or personal concept and has no meaning outside of expressing one's degree of belief. This interpretation of probability is often favored by philosophers and certain economic decision makers.

Regardless of which interpretation one gives to probability, however, there is a consensus that the mathematics of probability are the same in either case. For instance, if you think that the probability that it will rain tomorrow is .3 and you feel that the probability that it will be cloudy but without any rain is .2, then you should feel that the probability that it will either be cloudy or rainy is .5 independently of your individual interpretation of the concept of probability. In this chapter, we present the accepted rules, or axioms, used in probability theory. As a preliminary to this, however, we need to study the concept of the sample space and the events of an experiment.

3.2   SAMPLE SPACE AND EVENTS

Consider an experiment whose outcome is not predictable with certainty in advance. Although the outcome of the experiment will not be known in advance, let us suppose that the set of all possible outcomes is known. This set of all possible outcomes of an experiment is known as the *sample space* of the experiment and is denoted by S. Some examples are the following.

1. If the outcome of an experiment consists in the determination of the sex of a newborn child, then

$$S = \{g, b\}$$

   where the outcome g means that the child is a girl and b that it is a boy.
2. If the experiment consists of the running of a race among the seven horses having post positions 1, 2, 3, 4, 5, 6, 7, then

$$S = \{\text{all orderings of } (1, 2, 3, 4, 5, 6, 7)\}$$

   The outcome $(2, 3, 1, 6, 5, 4, 7)$ means, for instance, that the number 2 horse is first, then the number 3 horse, then the number 1 horse, and so on.
3. Suppose we are interested in determining the amount of dosage that must be given to a patient until that patient reacts positively. One possible sample space for this experiment is to let S consist of all the positive numbers. That is, let

$$S = (0, \infty)$$

   where the outcome would be x if the patient reacts to a dosage of value x but not to any smaller dosage.

Any subset E of the sample space is known as an *event*. That is, an event is a set consisting of possible outcomes of the experiment. If the outcome of the experiment is contained in E, then we say that E has occurred. Some examples of events are the following.

In Example 1 if $E = \{g\}$, then E is the event that the child is a girl. Similarly, if $F = \{b\}$, then F is the event that the child is a boy.

In Example 2 if

$$E = \{\text{all outcomes in } S \text{ starting with a 3}\}$$

then E is the event that the number 3 horse wins the race.

For any two events E and F of a sample space S, we define the new event $E \cup F$, called the *union* of the events E and F, to consist of all outcomes that are either in E or in F or in both E and F. That is, the event $E \cup F$ will occur if *either* E or F occurs. For instance, in Example 1 if $E = \{g\}$ and $F = \{b\}$, then $E \cup F = \{g, b\}$. That is, $E \cup F$ would be the whole sample space S. In Example 2 if $E = \{\text{all outcomes starting with 6}\}$ is the event that the number 6 horse wins and $F = \{\text{all outcomes having 6 in the second position}\}$ is the event that the number 6 horse comes in second, then $E \cup F$ is the event that the number 6 horse comes in either first or second.

Similarly, for any two events E and F, we may also define the new event EF, sometimes written as $E \cap F$, called the *intersection* of E and F, to consist of all outcomes that are in both E and F. That is, the event EF will occur only if both E and F occur. For instance, in Example 3 if $E = (0, 5)$ is the event that the required dosage is less than 5 and $F = (2, 10)$ is the event that it is between 2 and 10, then $EF = (2, 5)$ is the event that the required dosage is between 2 and 5. In Example 2 if $E = \{\text{all outcomes ending in 5}\}$ is the event that horse number 5 comes in last and $F = \{\text{all outcomes starting with 5}\}$ is the event that horse number 5 comes in first, then the event EF does not contain any outcomes and hence cannot occur. To give such an event a name, we shall refer to it as the null event and denote it by $\emptyset$. Thus $\emptyset$ refers to the event consisting of no outcomes. If $EF = \emptyset$, implying that E and F cannot both occur, then E and F are said to be *mutually exclusive*.

For any event E, we define the event E^c, referred to as the *complement* of E, to consist of all outcomes in the sample space S that are not in E. That is, E^c will occur if and only if E does not occur. In Example 1 if $E = \{b\}$ is the event that the child is a boy, then $E^c = \{g\}$ is the event that it is a girl. Also note that since the experiment must result in some outcome, it follows that $S^c = \emptyset$.

For any two events E and F, if all of the outcomes in E are also in F, then we say that E is contained in F and write $E \subset F$ (or equivalently, $F \supset E$). Thus if $E \subset F$, then the occurrence of E necessarily implies the occurrence of F. If $E \subset F$ and $F \subset E$, then we say that E and F are equal (or identical) and we write $E = F$.

We can also define unions and intersections of more than two events. In particular, the union of the events $E_1, E_2, \ldots, E_n$, denoted either by $E_1 \cup E_2 \cup \cdots \cup E_n$ or by $\cup_1^n E_i$, is defined to be the event consisting of all outcomes that are in E_i for at least one $i = 1, 2, \ldots, n$. Similarly, the intersection of the events E_i, $i = 1, 2, \ldots, n$, denoted by $E_1 E_2 \cdots E_n$, is defined to be the event consisting of those outcomes that are in all of the events E_i, $i = 1, 2, \ldots, n$. In other words, the union of the E_i occurs when *at least* one of the events E_i occurs; the intersection occurs when *all* of the events E_i occur.

3.3   VENN DIAGRAMS AND THE ALGEBRA OF EVENTS

A graphical representation of events that is very useful for illustrating logical relations among them is the *Venn diagram*. The sample space S is represented as consisting of all the points in a large rectangle, and the events $E, F, G, \ldots$, are represented as consisting of all the points in given circles within the rectangle. Events of interest can then be *indicated* by shading appropriate regions of the diagram. For instance, in the three Venn diagrams shown in Figure 3.1, the shaded areas represent respectively the events $E \cup F$, EF, and E^c. The Venn diagram of Figure 3.2 indicates that $E \subset F$.

   The operations of forming unions, intersections, and complements of events obey certain rules not dissimilar to the rules of algebra. We list a few of these.

Commutative law $E \cup F = F \cup E$ $EF = FE$

Associative law $(E \cup F) \cup G = E \cup (F \cup G)$ $(EF)G = E(FG)$

Distributive law $(E \cup F)G = EG \cup FG$ $EF \cup G = (E \cup G)(F \cup G)$

These relations are verified by showing that any outcome that is contained in the event on the left side of the equality is also contained in the event on the right side and vice versa. One way of showing this is by means of Venn diagrams. For instance, the distributive law may be verified by the sequence of diagrams shown in Figure 3.3.

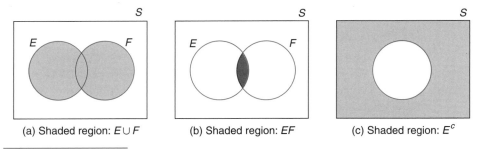

(a) Shaded region: $E \cup F$ (b) Shaded region: EF (c) Shaded region: E^c

FIGURE 3.1   *Venn diagrams.*

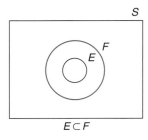

$E \subset F$

FIGURE 3.2   *Venn diagram.*

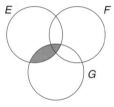   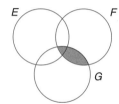   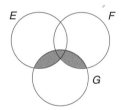

(a) Shaded region: *EG*   (b) Shaded region: *FG*   (c) Shaded region: $(E \cup F)G$
$(E \cup F)G = EG \cup FG$

FIGURE 3.3   *Proving the distributive law.*

The following useful relationship between the three basic operations of forming unions, intersections, and complements of events is known as *DeMorgan's laws*.

$$(E \cup F)^c = E^c F^c$$

$$(EF)^c = E^c \cup F^c$$

3.4 AXIOMS OF PROBABILITY

It appears to be an empirical fact that if an experiment is continually repeated under the exact same conditions, then for any event E, the proportion of time that the outcome is contained in E approaches some constant value as the number of repetitions increases. For instance, if a coin is continually flipped, then the proportion of flips resulting in heads will approach some value as the number of flips increases. It is this constant limiting frequency that we often have in mind when we speak of the probability of an event.

From a purely mathematical viewpoint, we shall suppose that for each event E of an experiment having a sample space S there is a number, denoted by $P(E)$, that is in accord with the following three axioms:

AXIOM 1
$$0 \leq P(E) \leq 1$$

AXIOM 2
$$P(S) = 1$$

AXIOM 3
For any sequence of mutually exclusive events $E_1, E_2, \ldots$ (that is, events for which $E_i E_j = \emptyset$ when $i \neq j$),

$$P\left(\bigcup_{i=1}^{n} E_i\right) = \sum_{i=1}^{n} P(E_i), \qquad n = 1, 2, \ldots, \infty$$

We call $P(E)$ the probability of the event E.

Thus, Axiom 1 states that the probability that the outcome of the experiment is contained in E is some number between 0 and 1. Axiom 2 states that, with probability 1, the outcome will be a member of the sample space S. Axiom 3 states that for any set of mutually exclusive events the probability that at least one of these events occurs is equal to the sum of their respective probabilities.

It should be noted that if we interpret $P(E)$ as the relative frequency of the event E when a large number of repetitions of the experiment are performed, then $P(E)$ would indeed satisfy the above axioms. For instance, the proportion (or frequency) of time that the outcome is in E is clearly between 0 and 1, and the proportion of time that it is in S is 1 (since all outcomes are in S). Also, if E and F have no outcomes in common, then the proportion of time that the outcome is in either E or F is the sum of their respective frequencies. As an illustration of this last statement, suppose the experiment consists of the rolling of a pair of dice and suppose that E is the event that the sum is 2, 3, or 12 and F is the event that the sum is 7 or 11. Then if outcome E occurs 11 percent of the time and outcome F 22 percent of the time, then 33 percent of the time the outcome will be either 2, 3, 12, 7, or 11.

These axioms will now be used to prove two simple propositions concerning probabilities. We first note that E and E^c are always mutually exclusive, and since $E \cup E^c = S$, we have by Axioms 2 and 3 that

$$1 = P(S) = P(E \cup E^c) = P(E) + P(E^c)$$

Or equivalently, we have the following:

PROPOSITION 3.4.1

$$P(E^c) = 1 - P(E)$$

In other words, Proposition 3.4.1 states that the probability that an event does not occur is 1 minus the probability that it does occur. For instance, if the probability of obtaining a head on the toss of a coin is $\frac{3}{8}$, the probability of obtaining a tail must be $\frac{5}{8}$.

Our second proposition gives the relationship between the probability of the union of two events in terms of the individual probabilities and the probability of the intersection.

PROPOSITION 3.4.2

$$P(E \cup F) = P(E) + P(F) - P(EF)$$

Proof

This proposition is most easily proven by the use of a Venn diagram as shown in Figure 3.4. As the regions I, II, and III are mutually exclusive, it follows that

$$P(E \cup F) = P(\text{I}) + P(\text{II}) + P(\text{III})$$
$$P(E) = P(\text{I}) + P(\text{II})$$
$$P(F) = P(\text{II}) + P(\text{III})$$

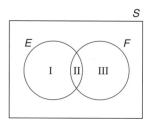

FIGURE 3.4   *Venn diagram.*

which shows that

$$P(E \cup F) = P(E) + P(F) - P(\text{II})$$

and the proof is complete since $\text{II} = EF$.   ■

EXAMPLE 3.4a   A total of 28 percent of males living in Nevada smoke cigarettes, 6 percent smoke cigars, and 3 percent smoke both cigars and cigarettes. What percentage of males smoke neither cigars nor cigarettes?

SOLUTION   Let E be the event that a randomly chosen male is a cigarette smoker and let F be the event that he is a cigar smoker. Then, the probability this person is either a cigarette or a cigar smoker is

$$P(E \cup F) = P(E) + P(F) - P(EF) = .28 + .06 - .03 = .31$$

Thus the probability that the person is not a smoker is $1 - .31 = .69$, implying that 69 percent of males smoke neither cigarettes nor cigars.   ■

The *odds* of an event A is defined by

$$\frac{P(A)}{P(A^c)} = \frac{P(A)}{1 - P(A)}$$

Thus the odds of an event A tells how much more likely it is that A occurs than that it does not occur. For instance, if $P(A) = 3/4$, then $P(A)/(1 - P(A)) = 3$, so the odds are 3. Consequently, it is 3 times as likely that A occurs as it is that it does not. (Common terminology is to say that the odds are 3 to 1 in favor of the event A.)

3.5   SAMPLE SPACES HAVING EQUALLY LIKELY OUTCOMES

For a large number of experiments, it is natural to assume that each point in the sample space is equally likely to occur. That is, for many experiments whose sample space S is a finite set, say $S = \{1, 2, \ldots, N\}$, it is often natural to assume that

$$P(\{1\}) = P(\{2\}) = \cdots = P(\{N\}) = p \quad (\text{say})$$

Now it follows from Axioms 2 and 3 that

$$1 = P(S) = P(\{1\}) + \cdots + P(\{N\}) = Np$$

which shows that

$$P(\{i\}) = p = 1/N$$

From this it follows from Axiom 3 that for any event E,

$$P(E) = \frac{\text{Number of points in } E}{N}$$

In words, if we assume that each outcome of an experiment is equally likely to occur, then the probability of any event E equals the proportion of points in the sample space that are contained in E.

Thus, to compute probabilities it is often necessary to be able to effectively count the number of different ways that a given event can occur. To do this, we will make use of the following rule.

BASIC PRINCIPLE OF COUNTING

Suppose that two experiments are to be performed. Then if experiment 1 can result in any one of m possible outcomes and if, for each outcome of experiment 1, there are n possible outcomes of experiment 2, then together there are mn possible outcomes of the two experiments.

Proof of the Basic Principle

The basic principle can be proven by enumerating all the possible outcomes of the two experiments as follows:

$$(1, 1), \ (1, 2), \ldots, (1, n)$$
$$(2, 1), \ (2, 2), \ldots, (2, n)$$
$$\vdots$$
$$(m, 1), \ (m, 2), \ldots, (m, n)$$

where we say that the outcome is (i, j) if experiment 1 results in its ith possible outcome and experiment 2 then results in the jth of its possible outcomes. Hence, the set of possible outcomes consists of m rows, each row containing n elements, which proves the result.   ■

EXAMPLE 3.5a   Two balls are "randomly drawn" from a bowl containing 6 white and 5 black balls. What is the probability that one of the drawn balls is white and the other black?

SOLUTION If we regard the order in which the balls are selected as being significant, then as the first drawn ball may be any of the 11 and the second any of the remaining 10, it follows that the sample space consists of $11 \cdot 10 = 110$ points. Furthermore, there are $6 \cdot 5 = 30$ ways in which the first ball selected is white and the second black, and similarly there are $5 \cdot 6 = 30$ ways in which the first ball is black and the second white. Hence, assuming that "randomly drawn" means that each of the 110 points in the sample space is equally likely to occur, then we see that the desired probability is

$$\frac{30 + 30}{110} = \frac{6}{11} \quad \blacksquare$$

When there are more than two experiments to be performed the basic principle can be generalized as follows:

Generalized Basic Principle of Counting

If r experiments that are to be performed are such that the first one may result in any of n_1 possible outcomes, and if for each of these n_1 possible outcomes there are n_2 possible outcomes of the second experiment, and if for each of the possible outcomes of the first two experiments there are n_3 possible outcomes of the third experiment, and if, ..., then there are a total of $n_1 \cdot n_2 \cdots n_r$ possible outcomes of the r experiments.

As an illustration of this, let us determine the number of different ways n distinct objects can be arranged in a linear order. For instance, how many different ordered arrangements of the letters a, b, c are possible? By direct enumeration we see that there are 6; namely, abc, acb, bac, bca, cab, cba. Each one of these ordered arrangements is known as a *permutation*. Thus, there are 6 possible permutations of a set of 3 objects. This result could also have been obtained from the basic principle, since the first object in the permutation can be any of the 3, the second object in the permutation can then be chosen from any of the remaining 2, and the third object in the permutation is then chosen from the remaining one. Thus, there are $3 \cdot 2 \cdot 1 = 6$ possible permutations.

Suppose now that we have n objects. Similar reasoning shows that there are

$$n(n-1)(n-2) \cdots 3 \cdot 2 \cdot 1$$

different permutations of the n objects. It is convenient to introduce the notation $n!$, which is read "n factorial," for the foregoing expression. That is,

$$n! = n(n-1)(n-2) \cdots 3 \cdot 2 \cdot 1$$

Thus, for instance, $1! = 1$, $2! = 2 \cdot 1 = 2$, $3! = 3 \cdot 2 \cdot 1 = 6$, $4! = 4 \cdot 3 \cdot 2 \cdot 1 = 24$, and so on. It is convenient to define $0! = 1$.

EXAMPLE 3.5b Mr. Jones has 10 books that he is going to put on his bookshelf. Of these, 4 are mathematics books, 3 are chemistry books, 2 are history books, and 1 is a language book. Jones wants to arrange his books so that all the books dealing with the same subject are together on the shelf. How many different arrangements are possible?

SOLUTION There are $4!\ 3!\ 2!\ 1!$ arrangements such that the mathematics books are first in line, then the chemistry books, then the history books, and then the language book. Similarly, for each possible ordering of the subjects, there are $4!\ 3!\ 2!\ 1!$ possible arrangements. Hence, as there are $4!$ possible orderings of the subjects, the desired answer is $4!\ 4!\ 3!\ 2!\ 1! = 6,912$. ∎

EXAMPLE 3.5c A class in probability theory consists of 6 men and 4 women. An exam is given and the students are ranked according to their performance. Assuming that no two students obtain the same score, **(a)** how many different rankings are possible? **(b)** If all rankings are considered equally likely, what is the probability that women receive the top 4 scores?

SOLUTION

  (a) Because each ranking corresponds to a particular ordered arrangement of the 10 people, we see the answer to this part is $10! = 3,628,800$.
  (b) Because there are $4!$ possible rankings of the women among themselves and $6!$ possible rankings of the men among themselves, it follows from the basic principle that there are $(6!)(4!) = (720)(24) = 17,280$ possible rankings in which the women receive the top 4 scores. Hence, the desired probability is

$$\frac{6!4!}{10!} = \frac{4 \cdot 3 \cdot 2 \cdot 1}{10 \cdot 9 \cdot 8 \cdot 7} = \frac{1}{210} \quad ∎$$

Suppose now that we are interested in determining the number of different groups of r objects that could be formed from a total of n objects. For instance, how many different groups of three could be selected from the five items A, B, C, D, E? To answer this, reason as follows. Since there are 5 ways to select the initial item, 4 ways to then select the next item, and 3 ways to then select the final item, there are thus $5 \cdot 4 \cdot 3$ ways of selecting the group of 3 when the order in which the items are selected is relevant. However, since every group of 3, say the group consisting of items A, B, and C, will be counted 6 times (that is, all of the permutations $ABC, ACB, BAC, BCA, CAB, CBA$ will be counted when the order of selection is relevant), it follows that the total number of different groups that can be formed is $(5 \cdot 4 \cdot 3)/(3 \cdot 2 \cdot 1) = 10$.

In general, as $n(n-1) \cdots (n-r+1)$ represents the number of different ways that a group of r items could be selected from n items when the order of selection is considered relevant (since the first one selected can be any one of the n, and the second selected any one of the remaining $n-1$, etc.), and since each group of r items will be counted $r!$ times in this count, it follows that the number of different groups of r items that could be formed from a set of n items is

$$\frac{n(n-1)\cdots(n-r+1)}{r!} = \frac{n!}{(n-r)!r!}$$

NOTATION AND TERMINOLOGY

We define $\binom{n}{r}$, for $r \le n$, by

$$\binom{n}{r} = \frac{n!}{(n-r)!r!}$$

and call $\binom{n}{r}$ the number of *combinations* of n objects taken r at a time.

Thus $\binom{n}{r}$ represents the number of different groups of size r that can be selected from a set of size n when the order of selection is not considered relevant. For example, there are

$$\binom{8}{2} = \frac{8 \cdot 7}{2 \cdot 1} = 28$$

different groups of size 2 that can be chosen from a set of 8 people, and

$$\binom{10}{2} = \frac{10 \cdot 9}{2 \cdot 1} = 45$$

different groups of size 2 that can be chosen from a set of 10 people. Also, since $0! = 1$, note that

$$\binom{n}{0} = \binom{n}{n} = 1$$

EXAMPLE 3.5d A committee of size 5 is to be selected from a group of 6 men and 9 women. If the selection is made randomly, what is the probability that the committee consists of 3 men and 2 women?

SOLUTION Let us assume that "randomly selected" means that each of the $\binom{15}{5}$ possible combinations is equally likely to be selected. Hence, since there are $\binom{6}{3}$ possible choices of 3 men and $\binom{9}{2}$ possible choices of 2 women, it follows from the basic principle of counting that the desired probability is given by

$$\frac{\binom{6}{3}\binom{9}{2}}{\binom{15}{5}} = \frac{240}{1001} \quad \blacksquare$$

EXAMPLE 3.5e From a set of n items a random sample of size k is to be selected. What is the probability a given item will be among the k selected?

SOLUTION Because there is $\binom{1}{1}$ way of choosing the given item and $\binom{n-1}{k-1}$ different choices of $k-1$ of the other $n-1$ items, it follows from the basic principle of counting that there are $\binom{1}{1}\binom{n-1}{k-1} = \binom{n-1}{k-1}$ different subsets of k of the n items that include the given item. As there are a total of $\binom{n}{k}$ different choices of k of the n items, it follows that the probability that a particular item is among the k selected is

$$\binom{n-1}{k-1} \Big/ \binom{n}{k} = \frac{(n-1)!}{(n-k)!(k-1)!} \Big/ \frac{n!}{(n-k)!k!} = \frac{k}{n} \quad \blacksquare$$

EXAMPLE 3.5f A basketball team consists of 6 black and 6 white players. The players are to be paired in groups of two for the purpose of determining roommates. If the pairings are done at random, what is the probability that none of the black players will have a white roommate?

SOLUTION Let us start by imagining that the 6 pairs are numbered — that is, there is a first pair, a second pair, and so on. Since there are $\binom{12}{2}$ different choices of a first pair; and for each choice of a first pair there are $\binom{10}{2}$ different choices of a second pair; and for each choice of the first 2 pairs there are $\binom{8}{2}$ choices for a third pair; and so on, it follows from the generalized basic principle of counting that there are

$$\binom{12}{2}\binom{10}{2}\binom{8}{2}\binom{6}{2}\binom{4}{2}\binom{2}{2} = \frac{12!}{(2!)^6}$$

ways of dividing the players into a *first* pair, a *second* pair, and so on. Hence there are $(12)!/2^6 6!$ ways of dividing the players into 6 (unordered) pairs of 2 each. Furthermore, since there are, by the same reasoning, $6!/2^3 3!$ ways of pairing the white players among themselves and $6!/2^3 3!$ ways of pairing the black players among themselves, it follows that there are $(6!/2^3 3!)^2$ pairings that do not result in any black–white roommate pairs. Hence, if the pairings are done at random (so that all outcomes are equally likely), then the desired probability is

$$\left(\frac{6!}{2^3 3!}\right)^2 \Big/ \frac{(12)!}{2^6 6!} = \frac{5}{231} = .0216$$

Hence, there are roughly only two chances in a hundred that a random pairing will not result in any of the white and black players rooming together. $\blacksquare$

EXAMPLE 3.5g If n people are present in a room, what is the probability that no two of them celebrate their birthday on the same day of the year? How large need n be so that this probability is less than $\frac{1}{2}$?

SOLUTION Because each person can celebrate his or her birthday on any one of 365 days, there are a total of $(365)^n$ possible outcomes. (We are ignoring the possibility of someone having been born on February 29.) Furthermore, there are $(365)(364)(363)\cdot(365-n+1)$ possible outcomes that result in no two of the people having the same birthday. This is so because the first person could have any one of 365 birthdays, the next person any of the remaining 364 days, the next any of the remaining 363, and so on. Hence, assuming that each outcome is equally likely, we see that the desired probability is

$$\frac{(365)(364)(363)\cdots(365-n+1)}{(365)^n}$$

It is a rather surprising fact that when $n \geq 23$, this probability is less than $\frac{1}{2}$. That is, if there are 23 or more people in a room, then the probability that at least two of them have the same birthday exceeds $\frac{1}{2}$. Many people are initially surprised by this result, since 23 seems so small in relation to 365, the number of days of the year. However, every pair of individuals has probability $\frac{365}{(365)^2} = \frac{1}{365}$ of having the same birthday, and in a group of 23 people there are $\binom{23}{2} = 253$ different pairs of individuals. Looked at this way, the result no longer seems so surprising. ∎

3.6 CONDITIONAL PROBABILITY

In this section, we introduce one of the most important concepts in all of probability theory — that of conditional probability. Its importance is twofold. In the first place, we are often interested in calculating probabilities when some partial information concerning the result of the experiment is available, or in recalculating them in light of additional information. In such situations, the desired probabilities are conditional ones. Second, as a kind of a bonus, it often turns out that the easiest way to compute the probability of an event is to first "condition" on the occurrence or nonoccurrence of a secondary event.

As an illustration of a conditional probability, suppose that one rolls a pair of dice. The sample space S of this experiment can be taken to be the following set of 36 outcomes

$$S = \{(i,j), \quad i = 1,2,3,4,5,6, \quad j = 1,2,3,4,5,6\}$$

where we say that the outcome is (i,j) if the first die lands on side i and the second on side j. Suppose now that each of the 36 possible outcomes is equally likely to occur and thus has probability $\frac{1}{36}$. (In such a situation we say that the dice are fair.) Suppose further that we observe that the first die lands on side 3. Then, given this information, what is the probability that the sum of the two dice equals 8? To calculate this probability, we reason as follows: Given that the initial die is a 3, there can be at most 6 possible outcomes of our experiment, namely, (3, 1), (3, 2), (3, 3), (3, 4), (3, 5), and (3, 6). In addition, because each of these outcomes originally had the same probability of occurring, they should still have equal probabilities. That is, given that the first die is a 3, then the (conditional) probability of each of the outcomes (3, 1), (3, 2), (3, 3), (3, 4), (3, 5), (3, 6) is $\frac{1}{6}$, whereas the (conditional) probability of the other 30 points in the sample space is 0. Hence, the desired probability will be $\frac{1}{6}$.

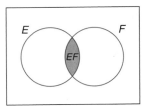

FIGURE 3.5　$P(E|F) = \frac{P(EF)}{P(F)}$.

If we let E and F denote, respectively, the event that the sum of the dice is 8 and the event that the first die is a 3, then the probability just obtained is called the conditional probability of E given that F has occurred, and is denoted by

$$P(E|F)$$

A general formula for $P(E|F)$ that is valid for all events E and F is derived in the same manner as just described. Namely, if the event F occurs, then in order for E to occur it is necessary that the actual occurrence be a point in both E and F; that is, it must be in EF. However, because we know that F has occurred, it follows that we can regard F as the new sample space and hence the probability that the event EF occurs will equal the probability of EF relative to the probability of F. That is,

$$P(E|F) = \frac{P(EF)}{P(F)} \tag{3.6.1}$$

Note that Equation 3.6.1 is well defined only when $P(F) > 0$ and hence $P(E|F)$ is defined only when $P(F) > 0$. (See Figure 3.5.)

The definition of conditional probability given by Equation 3.6.1 is consistent with the interpretation of probability as being a long-run relative frequency. To see this, suppose that a large number n of repetitions of the experiment are performed. Then, since $P(F)$ is the long-run proportion of experiments in which F occurs, it follows that F will occur approximately $nP(F)$ times. Similarly, in approximately $nP(EF)$ of these experiments, both E and F will occur. Hence, of the approximately $nP(F)$ experiments whose outcome is in F, approximately $nP(EF)$ of them will also have their outcome in E. That is, for those experiments whose outcome is in F, the proportion whose outcome is also in E is approximately

$$\frac{nP(EF)}{nP(F)} = \frac{P(EF)}{P(F)}$$

Since this approximation becomes exact as n becomes larger and larger, it follows that (3.6.1) gives the appropriate definition of the conditional probability of E given that F has occurred.

EXAMPLE 3.6a　A bin contains 5 defective (that immediately fail when put in use), 10 partially defective (that fail after a couple of hours of use), and 25 acceptable transistors.

A transistor is chosen at random from the bin and put into use. If it does not immediately fail, what is the probability it is acceptable?

SOLUTION   Since the transistor did not immediately fail, we know that it is not one of the 5 defectives and so the desired probability is:

$$P\{\text{acceptable}|\text{not defective}\}$$

$$= \frac{P\{\text{acceptable, not defective}\}}{P\{\text{not defective}\}}$$

$$= \frac{P\{\text{acceptable}\}}{P\{\text{not defective}\}}$$

where the last equality follows since the transistor will be both acceptable and not defective if it is acceptable. Hence, assuming that each of the 40 transistors is equally likely to be chosen, we obtain that

$$P\{\text{acceptable}|\text{not defective}\} = \frac{25/40}{35/40} = 5/7$$

It should be noted that we could also have derived this probability by working directly with the reduced sample space. That is, since we know that the chosen transistor is not defective, the problem reduces to computing the probability that a transistor, chosen at random from a bin containing 25 acceptable and 10 partially defective transistors, is acceptable. This is clearly equal to $\frac{25}{35}$.   ∎

EXAMPLE 3.6b   The organization that Jones works for is running a father–son dinner for those employees having at least one son. Each of these employees is invited to attend along with his youngest son. If Jones is known to have two children, what is the conditional probability that they are both boys given that he is invited to the dinner? Assume that the sample space S is given by $S = \{(b, b), (b, g), (g, b), (g, g)\}$ and all outcomes are equally likely [(b, g) means, for instance, that the younger child is a boy and the older child is a girl].

SOLUTION   The knowledge that Jones has been invited to the dinner is equivalent to knowing that he has at least one son. Hence, letting B denote the event that both children are boys, and A the event that at least one of them is a boy, we have that the desired probability $P(B|A)$ is given by

$$P(B|A) = \frac{P(BA)}{P(A)}$$

$$= \frac{P(\{(b, b)\})}{P(\{(b, b), (b, g), (g, b)\})}$$

$$= \frac{\frac{1}{4}}{\frac{3}{4}} = \frac{1}{3}$$

Many readers incorrectly reason that the conditional probability of two boys given at least one is $\frac{1}{2}$, as opposed to the correct $\frac{1}{3}$, since they reason that the Jones child not attending the dinner is equally likely to be a boy or a girl. Their mistake, however, is in assuming that these two possibilities are equally likely. Remember that initially there were four equally likely outcomes. Now the information that at least one child is a boy is equivalent to knowing that the outcome is not (g, g). Hence we are left with the three equally likely outcomes (b, b), (b, g), (g, b), thus showing that the Jones child not attending the dinner is twice as likely to be a girl as a boy. ■

By multiplying both sides of Equation 3.6.1 by $P(F)$ we obtain that

$$P(EF) = P(F)P(E|F) \qquad (3.6.2)$$

In words, Equation 3.6.2 states that the probability that both E and F occur is equal to the probability that F occurs multiplied by the conditional probability of E given that F occurred. Equation 3.6.2 is often quite useful in computing the probability of the intersection of events. This is illustrated by the following example.

EXAMPLE 3.6c Ms. Perez figures that there is a 30 percent chance that her company will set up a branch office in Phoenix. If it does, she is 60 percent certain that she will be made manager of this new operation. What is the probability that Perez will be a Phoenix branch office manager?

SOLUTION If we let B denote the event that the company sets up a branch office in Phoenix and M the event that Perez is made the Phoenix manager, then the desired probability is $P(BM)$, which is obtained as follows:

$$P(BM) = P(B)P(M|B)$$
$$= (.3)(.6)$$
$$= .18$$

Hence, there is an 18 percent chance that Perez will be the Phoenix manager. ■

3.7 BAYES' FORMULA

Let E and F be events. We may express E as

$$E = EF \cup EF^c$$

for, in order for a point to be in E, it must either be in both E and F or be in E but not in F. (See Figure 3.6.) As EF and EF^c are clearly mutually exclusive, we have by Axiom 3 that

$$P(E) = P(EF) + P(EF^c)$$
$$= P(E|F)P(F) + P(E|F^c)P(F^c)$$
$$= P(E|F)P(F) + P(E|F^c)[1 - P(F)] \qquad (3.7.1)$$

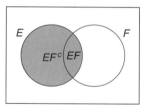

FIGURE 3.6 $E = EF \cup EF^c$.

Equation 3.7.1 states that the probability of the event E is a weighted average of the conditional probability of E given that F has occurred and the conditional probability of E given that F has not occurred, with each conditional probability being given as much weight as the event it is conditioned on has of occurring. It is an extremely useful formula, for its use often enables us to determine the probability of an event by first "conditioning" on whether or not some second event has occurred. That is, there are many instances where it is difficult to compute the probability of an event directly, but it is straightforward to compute it once we know whether or not some second event has occurred.

EXAMPLE 3.7a   An insurance company believes that people can be divided into two classes — those that are accident prone and those that are not. Their statistics show that an accident-prone person will have an accident at some time within a fixed 1-year period with probability .4, whereas this probability decreases to .2 for a non-accident-prone person. If we assume that 30 percent of the population is accident prone, what is the probability that a new policy holder will have an accident within a year of purchasing a policy?

SOLUTION   We obtain the desired probability by first conditioning on whether or not the policy holder is accident prone. Let A_1 denote the event that the policy holder will have an accident within a year of purchase; and let A denote the event that the policy holder is accident prone. Hence, the desired probability, $P(A_1)$, is given by

$$P(A_1) = P(A_1|A)P(A) + P(A_1|A^c)P(A^c)$$
$$= (.4)(.3) + (.2)(.7) = .26 \quad \blacksquare$$

In the next series of examples, we will indicate how to reevaluate an initial probability assessment in the light of additional (or new) information. That is, we will show how to incorporate new information with an initial probability assessment to obtain an updated probability.

EXAMPLE 3.7b   Twins can either be identical or fraternal. Identical, also called monozygotic, twins form when a single fertilized egg splits into two genetically identical parts. Consequently, identical twins always have the same set of genes. Fraternal, also called dizygotic, twins develop when two separate eggs are fertilized and implant in the uterus. The genetic connection of fraternal twins is no more or less the same as siblings born at separate times. A Los Angeles county scientist wishing to know the current fraction of twin pairs born in the county that are identical twins has assigned a county statistician to study this issue. The statistician initially requested each hospital in the county to record all twin births, indicating whether the resulting twins were identical or not. The hospitals, however, told her that to determine whether newborn twins were identical was not a simple task, as it involved the permission of the twins's parents to perform complicated and expensive DNA studies that the hospitals could not afford. After some deliberation, the statistician just asked the hospitals for data listing all twin births along with an indication as to whether the twins were of the same sex. When such data indicated that approximately 64 percent of twin births were same-sexed, the statistician declared that approximately 28 percent of all twins were identical. How did she come to this conclusion?

SOLUTION   The statistician reasoned that identical twins are always of the same sex, whereas fraternal twins, having the same relationship to each other as any pair of siblings, will have probability $\frac{1}{2}$ of being of the same sex. Letting I be the event that a pair of twins are identical, and SS be the event that a pair of twins are of the same sex, she computed the probability $P(SS)$ by conditioning on whether the twin pair was identical. This gave

$$P(SS) = P(SS|I)P(I) + P(SS|I^c)P(I^c)$$

or

$$P(SS) = 1 \times P(I) + \frac{1}{2} \times [1 - P(I)] = \frac{1}{2} + \frac{1}{2} P(I)$$

which, using that $P(SS) \approx .64$ yielded the result

$$P(I) \approx .28 \qquad \blacksquare$$

EXAMPLE 3.7c   Reconsider Example 3.7a and suppose that a new policy holder has an accident within a year of purchasing his policy. What is the probability that he is accident prone?

SOLUTION   Initially, at the moment when the policy holder purchased his policy, we assumed there was a 30 percent chance that he was accident prone. That is, $P(A) = .3$. However, based on the fact that he has had an accident within a year, we now reevaluate his probability of being accident prone as follows.

$$
\begin{aligned}
P(A|A_1) &= \frac{P(AA_1)}{P(A_1)} \\
&= \frac{P(A)P(A_1|A)}{P(A_1)} \\
&= \frac{(.3)(.4)}{.26} = \frac{6}{13} = .4615 \quad \blacksquare
\end{aligned}
$$

EXAMPLE 3.7d   In answering a question on a multiple-choice test, a student either knows the answer or she guesses. Let p be the probability that she knows the answer and $1 - p$

the probability that she guesses. Assume that a student who guesses at the answer will be correct with probability $1/m$, where m is the number of multiple-choice alternatives. What is the conditional probability that a student knew the answer to a question given that she answered it correctly?

SOLUTION Let C and K denote, respectively, the events that the student answers the question correctly and the event that she actually knows the answer. To compute

$$P(K|C) = \frac{P(KC)}{P(C)}$$

we first note that

$$P(KC) = P(K)P(C|K)$$
$$ = p \cdot 1$$
$$ = p$$

To compute the probability that the student answers correctly, we condition on whether or not she knows the answer. That is,

$$P(C) = P(C|K)P(K) + P(C|K^c)P(K^c)$$
$$ = p + (1/m)(1-p)$$

Hence, the desired probability is given by

$$P(K|C) = \frac{p}{p+(1/m)(1-p)} = \frac{mp}{1+(m-1)p}$$

Thus, for example, if $m = 5, p = \frac{1}{2}$, then the probability that a student knew the answer to a question she correctly answered is $\frac{5}{6}$. ∎

EXAMPLE 3.7e A laboratory blood test is 99 percent effective in detecting a certain disease when it is, in fact, present. However, the test also yields a "false positive" result for 1 percent of the healthy persons tested. (That is, if a healthy person is tested, then, with probability .01, the test result will imply he or she has the disease.) If .5 percent of the population actually has the disease, what is the probability a person has the disease given that his test result is positive?

SOLUTION Let D be the event that the tested person has the disease and E the event that his test result is positive. The desired probability $P(D|E)$ is obtained by

$$P(D|E) = \frac{P(DE)}{P(E)}$$

$$= \frac{P(E|D)P(D)}{P(E|D)P(D) + P(E|D^c)P(D^c)}$$

$$= \frac{(.99)(.005)}{(.99)(.005) + (.01)(.995)}$$

$$= .3322$$

Thus, only 33 percent of those persons whose test results are positive actually have the disease. Since many students are often surprised at this result (because they expected this figure to be much higher since the blood test seems to be a good one), it is probably worthwhile to present a second argument which, though less rigorous than the foregoing, is probably more revealing. We now do so.

Since .5 percent of the population actually has the disease, it follows that, on the average, 1 person out of every 200 tested will have it. The test will correctly confirm that this person has the disease with probability .99. Thus, on the average, out of every 200 persons tested, the test will correctly confirm that .99 person has the disease. On the other hand, out of the (on the average) 199 healthy people, the test will incorrectly state that (199) (.01) of these people have the disease. Hence, for every .99 diseased person that the test correctly states is ill, there are (on the average) 1.99 healthy persons that the test incorrectly states are ill. Hence, the proportion of time that the test result is correct when it states that a person is ill is

$$\frac{.99}{.99 + 1.99} = .3322 \quad \blacksquare$$

Equation 3.7.1 is also useful when one has to reassess one's (personal) probabilities in the light of additional information. For instance, consider the following examples.

EXAMPLE 3.7f   At a certain stage of a criminal investigation, the inspector in charge is 60 percent convinced of the guilt of a certain suspect. Suppose now that a *new* piece of evidence that shows that the criminal has a certain characteristic (such as left-handedness, baldness, brown hair, etc.) is uncovered. If 20 percent of the population possesses this characteristic, how certain of the guilt of the suspect should the inspector now be if it turns out that the suspect is among this group?

SOLUTION   Letting G denote the event that the suspect is guilty and C the event that he possesses the characteristic of the criminal, we have

$$P(G|C) = \frac{P(GC)}{P(C)}$$

Now

$$P(GC) = P(G)P(C|G)$$
$$= (.6)(1)$$
$$= .6$$

To compute the probability that the suspect has the characteristic, we condition on whether or not he is guilty. That is,

$$P(C) = P(C|G)P(G) + P(C|G^c)P(G^c)$$
$$= (1)(.6) + (.2)(.4)$$
$$= .68$$

where we have supposed that the probability of the suspect having the characteristic if he is, in fact, innocent is equal to .2, the proportion of the population possessing the characteristic. Hence

$$P(G|C) = \frac{60}{68} = .882$$

and so the inspector should now be 88 percent certain of the guilt of the suspect. ■

EXAMPLE 3.7f (continued)   Let us now suppose that the new evidence is subject to different possible interpretations, and in fact only shows that it is 90 percent likely that the criminal possesses this certain characteristic. In this case, how likely would it be that the suspect is guilty (assuming, as before, that he has this characteristic)?

SOLUTION   In this case, the situation is as before with the exception that the probability of the suspect having the characteristic given that he is guilty is now .9 (rather than 1). Hence,

$$P(G|C) = \frac{P(GC)}{P(C)}$$
$$= \frac{P(G)P(C|G)}{P(C|G)P(G) + P(C|G^c)P(G^c)}$$
$$= \frac{(.6)(.9)}{(.9)(.6) + (.2)(.4)}$$
$$= \frac{54}{62} = .871$$

which is slightly less than in the previous case (why?). ■

Equation 3.7.1 may be generalized in the following manner. Suppose that $F_1, F_2, \ldots, F_n$ are mutually exclusive events such that

$$\bigcup_{i=1}^{n} F_i = S$$

In other words, exactly one of the events $F_1, F_2, \ldots, F_n$ must occur. By writing

$$E = \bigcup_{i=1}^{n} EF_i$$

and using the fact that the events $EF_i, i = 1, \ldots, n$ are mutually exclusive, we obtain that

$$P(E) = \sum_{i=1}^{n} P(EF_i)$$

$$= \sum_{i=1}^{n} P(E|F_i)P(F_i) \qquad (3.7.2)$$

Thus, Equation 3.7.2 shows how, for given events $F_1, F_2, \ldots, F_n$ of which one and only one must occur, we can compute $P(E)$ by first "conditioning" on which one of the F_i occurs. That is, it states that $P(E)$ is equal to a weighted average of $P(E|F_i)$, each term being weighted by the probability of the event on which it is conditioned.

Suppose now that E has occurred and we are interested in determining which one of F_j also occurred. By Equation 3.7.2, we have that

$$P(F_j|E) = \frac{P(EF_j)}{P(E)}$$

$$= \frac{P(E|F_j)P(F_j)}{\sum_{i=1}^{n} P(E|F_i)P(F_i)} \qquad (3.7.3)$$

Equation 3.7.3 is known as *Bayes' formula*, after the English philosopher Thomas Bayes. If we think of the events F_j as being possible "hypotheses" about some subject matter, then Bayes' formula may be interpreted as showing us how opinions about these hypotheses held before the experiment [that is, the $P(F_j)$] should be modified by the evidence of the experiment.

EXAMPLE 3.7g A plane is missing and it is presumed that it was equally likely to have gone down in any of three possible regions. Let $1 - \alpha_i$ denote the probability the plane will be found upon a search of the ith region when the plane is, in fact, in that region, $i = 1, 2, 3$. (The constants α_i are called *overlook probabilities* because they represent the probability of overlooking the plane; they are generally attributable to the geographical and environmental conditions of the regions.) What is the conditional probability that the plane is in the ith region, given that a search of region 1 is unsuccessful, $i = 1, 2, 3$?

SOLUTION   Let R_i, $i = 1, 2, 3$, be the event that the plane is in region i; and let E be the event that a search of region 1 is unsuccessful. From Bayes' formula, we obtain

$$P(R_1|E) = \frac{P(ER_1)}{P(E)}$$

$$= \frac{P(E|R_1)P(R_1)}{\sum\limits_{i=1}^{3} P(E|R_i)P(R_i)}$$

$$= \frac{(\alpha_1)(1/3)}{(\alpha_1)(1/3) + (1)(1/3) + (1)(1/3)}$$

$$= \frac{\alpha_1}{\alpha_1 + 2}$$

For $j = 2, 3$,

$$P(R_j|E) = \frac{P(E|R_j)P(R_j)}{P(E)}$$

$$= \frac{(1)(1/3)}{(\alpha_1)1/3 + 1/3 + 1/3}$$

$$= \frac{1}{\alpha_1 + 2}, \qquad j = 2, 3$$

Thus, for instance, if $\alpha_1 = .4$, then the conditional probability that the plane is in region 1 given that a search of that region did not uncover it is $\frac{1}{6}$, whereas the conditional probabilities that it is in region 2 and that it is in region 3 are both equal to $\frac{1}{2.4} = \frac{5}{12}$. ∎

3.8 INDEPENDENT EVENTS

The previous examples in this chapter show that $P(E|F)$, the conditional probability of E given F, is not generally equal to $P(E)$, the unconditional probability of E. In other words, knowing that F has occurred generally changes the chances of E's occurrence. In the special cases where $P(E|F)$ does in fact equal $P(E)$, we say that E is independent of F. That is, E is independent of F if knowledge that F has occurred does not change the probability that E occurs.

Since $P(E|F) = P(EF)/P(F)$, we see that E is independent of F if

$$P(EF) = P(E)P(F) \tag{3.8.1}$$

Since this equation is symmetric in E and F, it shows that whenever E is independent of F so is F of E. We thus have the following.

Definition

Two events E and F are said to be *independent* if Equation 3.8.1 holds. Two events E and F that are not independent are said to be *dependent*.

EXAMPLE 3.8a   A card is selected at random from an ordinary deck of 52 playing cards. If A is the event that the selected card is an ace and H is the event that it is a heart, then A and H are independent, since $P(AH) = \frac{1}{52}$, while $P(A) = \frac{4}{52}$ and $P(H) = \frac{13}{52}$.   ■

EXAMPLE 3.8b   If we let E denote the event that the next president is a Republican and F the event that there will be a major earthquake within the next year, then most people would probably be willing to assume that E and F are independent. However, there would probably be some controversy over whether it is reasonable to assume that E is independent of G, where G is the event that there will be a recession within the next two years.   ■

We now show that if E is independent of F then E is also independent of F^c.

PROPOSITION 3.8.1   If E and F are independent, then so are E and F^c.

Proof

Assume that E and F are independent. Since $E = EF \cup EF^c$, and EF and EF^c are obviously mutually exclusive, we have that

$$P(E) = P(EF) + P(EF^c)$$
$$= P(E)P(F) + P(EF^c) \qquad \text{by the independence of } E \text{ and } F$$

or equivalently,

$$P(EF^c) = P(E)(1 - P(F))$$
$$= P(E)P(F^c)$$

and the result is proven.   ■

Thus if E is independent of F, then the probability of E's occurrence is unchanged by information as to whether or not F has occurred.

Suppose now that E is independent of F and is also independent of G. Is E then necessarily independent of FG? The answer, somewhat surprisingly, is no. Consider the following example.

EXAMPLE 3.8c   Two fair dice are thrown. Let E_7 denote the event that the sum of the dice is 7. Let F denote the event that the first die equals 4 and let T be the event that the second die equals 3. Now it can be shown (see Problem 36) that E_7 is independent of F and that E_7 is also independent of T; but clearly E_7 is not independent of FT [since $P(E_7|FT) = 1$]. ∎

It would appear to follow from the foregoing example that an appropriate definition of the independence of three events E, F, and G would have to go further than merely assuming that all of the $\binom{3}{2}$ pairs of events are independent. We are thus led to the following definition.

Definition

The three events E, F, and G are said to be independent if

$$P(EFG) = P(E)P(F)P(G)$$
$$P(EF) = P(E)P(F)$$
$$P(EG) = P(E)P(G)$$
$$P(FG) = P(F)P(G)$$

It should be noted that if the events E, F, G are independent, then E will be independent of any event formed from F and G. For instance, E is independent of $F \cup G$ since

$$
\begin{aligned}
P(E(F \cup G)) &= P(EF \cup EG) \\
&= P(EF) + P(EG) - P(EFG) \\
&= P(E)P(F) + P(E)P(G) - P(E)P(FG) \\
&= P(E)[P(F) + P(G) - P(FG)] \\
&= P(E)P(F \cup G)
\end{aligned}
$$

Of course we may also extend the definition of independence to more than three events. The events $E_1, E_2, \ldots, E_n$ are said to be independent if for every subset $E_{1'}, E_{2'}, \ldots, E_{r'}$, $r \leq n$, of these events

$$P(E_{1'}E_{2'} \cdots E_{r'}) = P(E_{1'})P(E_{2'}) \cdots P(E_{r'})$$

It is sometimes the case that the probability experiment under consideration consists of performing a sequence of subexperiments. For instance, if the experiment consists of continually tossing a coin, then we may think of each toss as being a subexperiment. In many cases it is reasonable to assume that the outcomes of any group of the subexperiments have no effect on the probabilities of the outcomes of the other subexperiments. If such is the case, then we say that the subexperiments are independent.

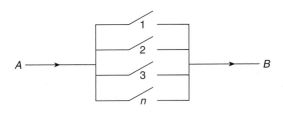

FIGURE 3.7   *Parallel system: functions if current flows from A to B.*

EXAMPLE 3.8d   A system composed of n separate components is said to be a parallel system if it functions when at least one of the components functions. (See Figure 3.7.) For such a system, if component i, independent of other components, functions with probability $p_i, i = 1, \ldots, n$, what is the probability the system functions?

SOLUTION   Let A_i denote the event that component i functions. Then

$$P\{\text{system functions}\} = 1 - P\{\text{system does not function}\}$$
$$= 1 - P\{\text{all components do not function}\}$$
$$= 1 - P\left(A_1^c A_2^c \cdots A_n^c\right)$$
$$= 1 - \prod_{i=1}^{n}(1 - p_i) \quad \text{by independence} \quad \blacksquare$$

EXAMPLE 3.8e   A set of k coupons, each of which is independently a type j coupon with probability p_j, $\sum_{j=1}^{n} p_j = 1$, is collected. Find the probability that the set contains a type j coupon given that it contains a type i, $i \neq j$.

SOLUTION   Let A_r be the event that the set contains a type r coupon. Then

$$P(A_j | A_i) = \frac{P(A_j A_i)}{P(A_i)}$$

To compute $P(A_i)$ and $P(A_j A_i)$, consider the probability of their complements:

$$P(A_i) = 1 - P(A_i^c)$$
$$= 1 - P\{\text{no coupon is type } i\}$$
$$= 1 - (1 - p_i)^k$$
$$P(A_i A_j) = 1 - P(A_i^c \cup A_j^c)$$
$$= 1 - [P(A_i^c) + P(A_j^c) - P(A_i^c A_j^c)]$$
$$= 1 - (1 - p_i)^k - (1 - p_j)^k + P\{\text{no coupon is type } i \text{ or type } j\}$$
$$= 1 - (1 - p_i)^k - (1 - p_j)^k + (1 - p_i - p_j)^k$$

where the final equality follows because each of the k coupons is, independently, neither of type i or of type j with probability $1 - p_i - p_j$. Consequently,

$$P(A_j|A_i) = \frac{1 - (1-p_i)^k - (1-p_j)^k + (1-p_i-p_j)^k}{1 - (1-p_i)^k} \quad \blacksquare$$

Problems

1. A box contains three marbles — one red, one green, and one blue. Consider an experiment that consists of taking one marble from the box, then replacing it in the box and drawing a second marble from the box. Describe the sample space. Repeat for the case in which the second marble is drawn without first replacing the first marble.

2. An experiment consists of tossing a coin three times. What is the sample space of this experiment? Which event corresponds to the experiment resulting in more heads than tails?

3. Let $S = \{1, 2, 3, 4, 5, 6, 7\}$, $E = \{1, 3, 5, 7\}$, $F = \{7, 4, 6\}$, $G = \{1, 4\}$. Find
   (a) EF;   (c) EG^c;   (e) $E^c(F \cup G)$;
   (b) $E \cup FG$; (d) $EF^c \cup G$; (f) $EG \cup FG$.

4. Two dice are thrown. Let E be the event that the sum of the dice is odd, let F be the event that the first die lands on 1, and let G be the event that the sum is 5. Describe the events EF, $E \cup F$, FG, EF^c, EFG.

5. A system is composed of four components, each of which is either working or failed. Consider an experiment that consists of observing the status of each component, and let the outcome of the experiment be given by the vector (x_1, x_2, x_3, x_4) where x_i is equal to 1 if component i is working and is equal to 0 if component i is failed.

   (a) How many outcomes are in the sample space of this experiment?
   (b) Suppose that the system will work if components 1 and 2 are both working, or if components 3 and 4 are both working. Specify all the outcomes in the event that the system works.
   (c) Let E be the event that components 1 and 3 are both failed. How many outcomes are contained in event E?

6. Let E, F, G be three events. Find expressions for the events that of E, F, G

   (a) only E occurs;
   (b) both E and G but not F occur;
   (c) at least one of the events occurs;

    (d) at least two of the events occur;

    (e) all three occur;

    (f) none of the events occurs;

    (g) at most one of them occurs;

    (h) at most two of them occur;

    (i) exactly two of them occur;

    (j) at most three of them occur.

7. Find simple expressions for the events

    (a) $E \cup E^c$;

    (b) EE^c;

    (c) $(E \cup F)(E \cup F^c)$;

    (d) $(E \cup F)(E^c \cup F)(E \cup F^c)$;

    (e) $(E \cup F)(F \cup G)$.

8. Use Venn diagrams (or any other method) to show that

    (a) $EF \subset E, E \subset E \cup F$;

    (b) if $E \subset F$ then $F^c \subset E^c$;

    (c) the commutative laws are valid;

    (d) the associative laws are valid;

    (e) $F = FE \cup FE^c$;

    (f) $E \cup F = E \cup E^c F$;

    (g) DeMorgan's laws are valid.

9. For the following Venn diagram, describe in terms of E, F, and G the events denoted in the diagram by the Roman numerals I through VII.

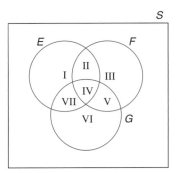

10. Show that if $E \subset F$ then $P(E) \le P(F)$. (*Hint:* Write F as the union of two mutually exclusive events, one of them being E.)

11. Prove Boole's inequality, namely that

$$P\left(\bigcup_{i=1}^{n} E_i\right) \le \sum_{i=1}^{n} P(E_i)$$

12. If $P(E) = .9$ and $P(F) = .9$, show that $P(EF) \geq .8$. In general, prove Bonferroni's inequality, namely that

$$P(EF) \geq P(E) + P(F) - 1$$

13. Prove that

   (a) $P(EF^c) = P(E) - P(EF)$
   (b) $P(E^c F^c) = 1 - P(E) - P(F) + P(EF)$

14. Show that the probability that exactly one of the events E or F occurs is equal to $P(E) + P(F) - 2P(EF)$.

15. Calculate $\binom{9}{3}$, $\binom{9}{6}$, $\binom{7}{2}$, $\binom{7}{5}$, $\binom{10}{7}$.

16. Show that

$$\binom{n}{r} = \binom{n}{n-r}$$

   Now present a combinatorial argument for the foregoing by explaining why a choice of r items from a set of size n is equivalent to a choice of $n - r$ items from that set.

17. Show that

$$\binom{n}{r} = \binom{n-1}{r-1} + \binom{n-1}{r}$$

   For a combinatorial argument, consider a set of n items and fix attention on one of these items. How many different sets of size r contain this item, and how many do not?

18. A group of 5 boys and 10 girls is lined up in random order — that is, each of the 15! permutations is assumed to be equally likely.

   (a) What is the probability that the person in the 4th position is a boy?
   (b) What about the person in the 12th position?
   (c) What is the probability that a particular boy is in the 3rd position?

19. Consider a set of 23 unrelated people. Because each pair of people shares the same birthday with probability 1/365, and there are $\binom{23}{2} = 253$ pairs, why isn't the probability that at least two people have the same birthday equal to 253/365?

20. Suppose that distinct integer values are written on each of 3 cards. These cards are then randomly given the designations A, B, and C. The values on cards A and B are then compared. If the smaller of these values is then compared with the value on card C, what is the probability that it is also smaller than the value on card C?

21. There is a 60 percent chance that the event A will occur. If A does not occur, then there is a 10 percent chance that B will occur.

(a) What is the probability that at least one of the events A or B occurs?

(b) If A is the event that the democratic candidate wins the presidential election in 2012 and B is the event that there is a 6.2 or higher earthquake in Los Angeles sometime in 2013, what would you take as the probability that both A and B occur? What assumption are you making?

22. The sample mean of the annual salaries of a group of 100 accountants who work at a large accounting firm is $130,000 with a sample standard deviation of $20,000. If a member of this group is randomly chosen, what can we say about

(a) the probability that his or her salary is between $90,000 and $170,000?

(b) the probability that his or her salary exceeds $150,000?

Hint: Use the Chebyshev inequality.

23. Of three cards, one is painted red on both sides; one is painted black on both sides; and one is painted red on one side and black on the other. A card is randomly chosen and placed on a table. If the side facing up is red, what is the probability that the other side is also red?

24. A couple has 2 children. What is the probability that both are girls if the eldest is a girl?

25. Fifty-two percent of the students at a certain college are females. Five percent of the students in this college are majoring in computer science. Two percent of the students are women majoring in computer science. If a student is selected at random, find the conditional probability that

(a) this student is female, given that the student is majoring in computer science;

(b) this student is majoring in computer science, given that the student is female.

26. A total of 500 married working couples were polled about their annual salaries, with the following information resulting.

Wife	Husband	
	Less than $50,000	More than $50,000
Less than $50,000	212	198
More than $50,000	36	54

Thus, for instance, in 36 of the couples the wife earned more and the husband earned less than $50,000. If one of the couples is randomly chosen, what is

(a) the probability that the husband earns less than $50,000;

(b) the conditional probability that the wife earns more than $50,000 given that the husband earns more than this amount;

(c) the conditional probability that the wife earns more than $50,000 given that the husband earns less than this amount?

27. There are two local factories that produce microwaves. Each microwave produced at factory A is defective with probability .05, whereas each one produced at factory B is defective with probability .01. Suppose you purchase two microwaves that were produced at the same factory, which is equally likely to have been either factory A or factory B. If the first microwave that you check is defective, what is the conditional probability that the other one is also defective?

28. A red die, a blue die, and a yellow die (all six-sided) are rolled. We are interested in the probability that the number appearing on the blue die is less than that appearing on the yellow die which is less than that appearing on the red die. (That is, if B (R) $[Y]$ is the number appearing on the blue (red) [yellow] die, then we are interested in $P(B < Y < R)$.)

   (a) What is the probability that no two of the dice land on the same number?

   (b) Given that no two of the dice land on the same number, what is the conditional probability that $B < Y < R$?

   (c) What is $P(B < Y < R)$?

   (d) If we regard the outcome of the experiment as the vector B, R, Y, how many outcomes are there in the sample space?

   (e) Without using the answer to (c), determine the number of outcomes that result in $B < Y < R$.

   (f) Use the results of parts (d) and (e) to verify your answer to part (c).

29. You ask your neighbor to water a sickly plant while you are on vacation. Without water it will die with probability .8; with water it will die with probability .15. You are 90 percent certain that your neighbor will remember to water the plant.

   (a) What is the probability that the plant will be alive when you return?

   (b) If it is dead, what is the probability your neighbor forgot to water it?

30. Two balls, each equally likely to be colored either red or blue, are put in an urn. At each stage one of the balls is randomly chosen, its color is noted, and it is then returned to the urn. If the first two balls chosen are colored red, what is the probability that

   (a) both balls in the urn are colored red;

   (b) the next ball chosen will be red?

31. A total of 600 of the 1,000 people in a retirement community classify themselves as Republicans, while the others classify themselves as Democrats. In a local election in which everyone voted, 60 Republicans voted for the Democratic candidate, and 50 Democrats voted for the Republican candidate. If a randomly chosen community member voted for the Republican, what is the probability that she or he is a Democrat?

32. Each of 2 balls is painted black or gold and then placed in an urn. Suppose that each ball is colored black with probability $\frac{1}{2}$, and that these events are independent.

  (a) Suppose that you obtain information that the gold paint has been used (and thus at least one of the balls is painted gold). Compute the conditional probability that both balls are painted gold.

  (b) Suppose, now, that the urn tips over and 1 ball falls out. It is painted gold. What is the probability that both balls are gold in this case? Explain.

33. Each of 2 cabinets identical in appearance has 2 drawers. Cabinet A contains a silver coin in each drawer, and cabinet B contains a silver coin in one of its drawers and a gold coin in the other. A cabinet is randomly selected, one of its drawers is opened, and a silver coin is found. What is the probability that there is a silver coin in the other drawer?

34. Prostate cancer is the most common type of cancer found in males. As an indicator of whether a male has prostate cancer, doctors often perform a test that measures the level of the PSA protein (prostate specific antigen) that is produced only by the prostate gland. Although higher PSA levels are indicative of cancer, the test is notoriously unreliable. Indeed, the probability that a noncancerous man will have an elevated PSA level is approximately .135, with this probability increasing to approximately .268 if the man does have cancer. If, based on other factors, a physician is 70 percent certain that a male has prostate cancer, what is the conditional probability that he has the cancer given that

  (a) the test indicates an elevated PSA level;

  (b) the test does not indicate an elevated PSA level?

  Repeat the preceding, this time assuming that the physician initially believes there is a 30 percent chance the man has prostate cancer.

35. Suppose that an insurance company classifies people into one of three classes — good risks, average risks, and bad risks. Their records indicate that the probabilities that good, average, and bad risk persons will be involved in an accident over a 1-year span are, respectively, .05, .15, and .30. If 20 percent of the population are "good risks," 50 percent are "average risks," and 30 percent are "bad risks," what proportion of people have accidents in a fixed year? If policy holder A had no accidents in 1987, what is the probability that he or she is a good (average) risk?

36. A pair of fair dice is rolled. Let E denote the event that the sum of the dice is equal to 7.

  (a) Show that E is independent of the event that the first die lands on 4.

  (b) Show that E is independent of the event that the second die lands on 3.

37. The probability of the closing of the ith relay in the circuits shown is given by $p_i, i = 1, 2, 3, 4, 5$. If all relays function independently, what is the probability that a current flows between A and B for the respective circuits?

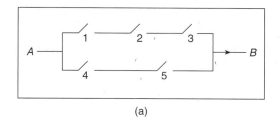

(a)

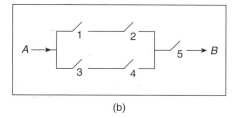

(b)

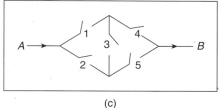

(c)

38. An engineering system consisting of n components is said to be a k-out-of-n system ($k \leq n$) if the system functions if and only if at least k of the n components function. Suppose that all components function independently of each other.

    (a) If the ith component functions with probability $P_i, i = 1, 2, 3, 4$, compute the probability that a 2-out-of-4 system functions.

    (b) Repeat (a) for a 3-out-of-5 system.

39. Five independent flips of a fair coin are made. Find the probability that

    (a) the first three flips are the same;

    (b) either the first three flips are the same, or the last three flips are the same;

    (c) there are at least two heads among the first three flips, and at least two tails among the last three flips.

40. Suppose that n independent trials, each of which results in any of the outcomes 0, 1, or 2, with respective probabilities .3, .5, and .2, are performed. Find the probability that both outcome 1 and outcome 2 occur at least once. (*Hint:* Consider the complementary probability.)

41. A parallel system functions whenever at least one of its components works. Consider a parallel system of n components, and suppose that each component independently works with probability $\frac{1}{2}$. Find the conditional probability that component 1 works, given that the system is functioning.

42. A certain organism possesses a pair of each of 5 different genes (which we will designate by the first 5 letters of the English alphabet). Each gene appears in 2 forms (which we designate by lowercase and capital letters). The capital letter will be assumed to be the dominant gene in the sense that if an organism possesses the gene pair xX, then it will outwardly have the appearance of the X gene. For instance, if X stands for brown eyes and x for blue eyes, then an individual having either gene pair XX or xX will have brown eyes, whereas one having gene pair xx will be blue-eyed. The characteristic appearance of an organism is called its *phenotype*, whereas its genetic constitution is called its *genotype*. (Thus 2 organisms with respective genotypes aA, bB, cc, dD, ee and AA, BB, cc, DD, ee would have different genotypes but the same phenotype.) In a mating between 2 organisms each one contributes, at random, one of its gene pairs of each type. The 5 contributions of an organism (one of each of the 5 types) are assumed to be independent and are also independent of the contributions of its mate. In a mating between organisms having genotypes aA, bB, cC, dD, eE, and aa, bB, cc, Dd, ee, what is the probability that the progeny will (1) phenotypically, (2) genotypically resemble

(a) the first parent;
(b) the second parent;
(c) either parent;
(d) neither parent?

43. Three prisoners are informed by their jailer that one of them has been chosen at random to be executed, and the other two are to be freed. Prisoner A asks the jailer to tell him privately which of his fellow prisoners will be set free, claiming that there would be no harm in divulging this information because he already knows that at least one of the two will go free. The jailer refuses to answer this question, pointing out that if A knew which of his fellow prisoners were to be set free, then his own probability of being executed would rise from $\frac{1}{3}$ to $\frac{1}{2}$ because he would then be one of two prisoners. What do you think of the jailer's reasoning?

44. Although both my parents have brown eyes, I have blue eyes. What is the probability that my sister has blue eyes? (As stated in Problem 42, an individual who receives a blue-eyed gene from each parent will have blue eyes, whereas one who receives one blue-eyed and one brown-eyed gene will have brown eyes.)

45. In a 7 game series played with two teams, the first team to win a total of 4 games is the winner. Suppose that each game played is independently won by team A with probability p.

(a) Given that one team leads 3 to 0, what is the probability that it is team A that is leading?
(b) Given that one team leads 3 to 0, what is the probability that team wins the series?
(c) If $p = \frac{1}{2}$, what is the probability that the team that wins the first game wins the series?

46. Suppose that distinct integer values are written on each of 3 cards. Suppose you are to be offered these cards in a random order. When you are offered a card you

must immediately either accept it or reject it. If you accept a card, the process ends. If you reject a card, then the next card (if a card remains) is offered. If you reject the first two cards offered, then you must accept the final card.

(a) If you plan to accept the first card offered, what is the probability that you will accept the highest valued card?

(b) If you plan to reject the first card offered, and to then accept the second card if and only if its value is greater than the value of the first card, what is the probability that you will accept the highest valued card?

47. Let A, B, C be events such that $P(A) = .2$, $P(B) = .3$, $P(C) = .4$. Find the probability that at least one of the events A and B occurs if

(a) A and B are mutually exclusive;

(b) A and B are independent.

Find the probability that all of the events A, B, C occur if

(c) A, B, C are independent;

(d) A, B, C are mutually exclusive.

48. Two percent of woman of age 45 who participate in routine screening have breast cancer. Ninety percent of those with breast cancer have positive mammographies. Ten percent of the women who do not have breast cancer will also have positive mammographies. Given a woman has a positive mammography, what is the probability she has breast cancer?

49. Twelve percent of all US households are in California. A total of 3.3 percent of all US households earn over 250, 000 per year, while a total of 6.3 percent of California households earn over 250, 000 per year. If a randomly chosen US household earns over 250, 000 per year, what is the probability it is from California?

50. There is a 60 percent chance that the event A will occur. If A does not occur, there is a 10 percent chance that B will occur. What is the probability that at least one of the events A or B occur?

51. Suppose distinct values are written on each of three cards, which are then randomly given the designations A, B, and C. The values on cards A and B are then compared. What is the probability that the smaller of these values is also smaller than the value on card C?

Chapter 4

RANDOM VARIABLES AND EXPECTATION

4.1 RANDOM VARIABLES

When a random experiment is performed, we are often not interested in all of the details of the experimental result but only in the value of some numerical quantity determined by the result. For instance, in tossing dice we are often interested in the sum of the two dice and are not really concerned about the values of the individual dice. That is, we may be interested in knowing that the sum is 7 and not be concerned over whether the actual outcome was (1, 6) or (2, 5) or (3, 4) or (4, 3) or (5, 2) or (6, 1). Also, a civil engineer may not be directly concerned with the daily risings and declines of the water level of a reservoir (which we can take as the experimental result) but may only care about the level at the end of a rainy season. These quantities of interest that are determined by the result of the experiment are known as *random variables*.

Since the value of a random variable is determined by the outcome of the experiment, we may assign probabilities of its possible values.

EXAMPLE 4.1a Letting X denote the random variable that is defined as the sum of two fair dice, then

$$P\{X = 2\} = P\{(1, 1)\} = \tfrac{1}{36} \tag{4.1.1}$$
$$P\{X = 3\} = P\{(1, 2), (2, 1)\} = \tfrac{2}{36}$$
$$P\{X = 4\} = P\{(1, 3), (2, 2), (3, 1)\} = \tfrac{3}{36}$$
$$P\{X = 5\} = P\{(1, 4), (2, 3), (3, 2), (4, 1)\} = \tfrac{4}{36}$$
$$P\{X = 6\} = P\{(1, 5), (2, 4), (3, 3), (4, 2), (5, 1)\} = \tfrac{5}{36}$$
$$P\{X = 7\} = P\{(1, 6), (2, 5), (3, 4), (4, 3), (5, 2), (6, 1)\} = \tfrac{6}{36}$$
$$P\{X = 8\} = P\{(2, 6), (3, 5), (4, 4), (5, 3), (6, 2)\} = \tfrac{5}{36}$$

$$P\{X = 9\} = P\{(3,6),(4,5),(5,4),(6,3)\} = \tfrac{4}{36}$$

$$P\{X = 10\} = P\{(4,6),(5,5),(6,4)\} = \tfrac{3}{36}$$

$$P\{X = 11\} = P\{(5,6),(6,5)\} = \tfrac{2}{36}$$

$$P\{X = 12\} = P\{(6,6)\} = \tfrac{1}{36}$$

In other words, the random variable X can take on any integral value between 2 and 12 and the probability that it takes on each value is given by Equation 4.1.1. Since X must take on some value, we must have

$$1 = P(S) = P\left(\bigcup_{i=2}^{12}\{X = i\}\right) = \sum_{i=2}^{12} P\{X = i\}$$

which is easily verified from Equation 4.1.1.

Another random variable of possible interest in this experiment is the value of the first die. Letting Y denote this random variable, then Y is equally likely to take on any of the values 1 through 6. That is,

$$P\{Y = i\} = 1/6, \qquad i = 1,2,3,4,5,6 \quad \blacksquare$$

EXAMPLE 4.1b Suppose that an individual purchases two electronic components, each of which may be either defective or acceptable. In addition, suppose that the four possible results — (d, d), (d, a), (a, d), (a, a) — have respective probabilities .09, .21, .21, .49 [where (d, d) means that both components are defective, (d, a) that the first component is defective and the second acceptable, and so on]. If we let X denote the number of acceptable components obtained in the purchase, then X is a random variable taking on one of the values 0, 1, 2 with respective probabilities

$$P\{X = 0\} = .09$$

$$P\{X = 1\} = .42$$

$$P\{X = 2\} = .49$$

If we were mainly concerned with whether there was at least one acceptable component, we could define the random variable I by

$$I = \begin{cases} 1 & \text{if } X = 1 \text{ or } 2 \\ 0 & \text{if } X = 0 \end{cases}$$

If A denotes the event that at least one acceptable component is obtained, then the random variable I is called the *indicator* random variable for the event A, since I will equal 1 or 0 depending upon whether A occurs. The probabilities attached to the possible values of I are

$$P\{I=1\} = .91$$
$$P\{I=0\} = .09 \quad \blacksquare$$

In the two foregoing examples, the random variables of interest took on a finite number of possible values. Random variables whose set of possible values can be written either as a finite sequence $x_1, \ldots, x_n$, or as an infinite sequence $x_1, \ldots$ are said to be *discrete*. For instance, a random variable whose set of possible values is the set of nonnegative integers is a discrete random variable. However, there also exist random variables that take on a continuum of possible values. These are known as *continuous* random variables. One example is the random variable denoting the lifetime of a car, when the car's lifetime is assumed to take on any value in some interval (a, b).

The *cumulative distribution function*, or more simply the *distribution function*, F of the random variable X is defined for any real number x by

$$F(x) = P\{X \leq x\}$$

That is, $F(x)$ is the probability that the random variable X takes on a value that is less than or equal to x.

Notation: We will use the notation $X \sim F$ to signify that F is the distribution function of X.

All probability questions about X can be answered in terms of its distribution function F. For example, suppose we wanted to compute $P\{a < X \leq b\}$. This can be accomplished by first noting that the event $\{X \leq b\}$ can be expressed as the union of the two mutually exclusive events $\{X \leq a\}$ and $\{a < X \leq b\}$. Therefore, applying Axiom 3, we obtain that

$$P\{X \leq b\} = P\{X \leq a\} + P\{a < X \leq b\}$$

or

$$P\{a < X \leq b\} = F(b) - F(a)$$

EXAMPLE 4.1c Suppose the random variable X has distribution function

$$F(x) = \begin{cases} 0 & x \leq 0 \\ 1 - \exp\{-x^2\} & x > 0 \end{cases}$$

What is the probability that X exceeds 1?

SOLUTION   The desired probability is computed as follows:

$$P\{X > 1\} = 1 - P\{X \leq 1\}$$
$$= 1 - F(1)$$
$$= e^{-1}$$
$$= .368 \quad \blacksquare$$

4.2   TYPES OF RANDOM VARIABLES

As was previously mentioned, a random variable whose set of possible values is a sequence is said to be *discrete*. For a discrete random variable X, we define the *probability mass function* $p(a)$ of X by

$$p(a) = P\{X = a\}$$

The probability mass function $p(a)$ is positive for at most a countable number of values of a. That is, if X must assume one of the values $x_1, x_2, \ldots$, then

$$p(x_i) > 0, \qquad i = 1, 2, \ldots$$
$$p(x) = 0, \qquad \text{all other values of } x$$

Since X must take on one of the values x_i, we have

$$\sum_{i=1}^{\infty} p(x_i) = 1$$

EXAMPLE 4.2a   Consider a random variable X that is equal to 1, 2, or 3. If we know that

$$p(1) = \tfrac{1}{2} \qquad \text{and} \qquad p(2) = \tfrac{1}{3}$$

then it follows (since $p(1) + p(2) + p(3) = 1$) that

$$p(3) = \tfrac{1}{6}$$

A graph of $p(x)$ is presented in Figure 4.1. $\blacksquare$

The cumulative distribution function F can be expressed in terms of $p(x)$ by

$$F(a) = \sum_{\text{all } x \leq a} p(x)$$

If X is a discrete random variable whose set of possible values are $x_1, x_2, x_3, \ldots$, where $x_1 < x_2 < x_3 < \cdots$, then its distribution function F is a step function. That is, the value of F is constant in the intervals $[x_{i-1}, x_i)$ and then takes a step (or jump) of size $p(x_i)$ at x_i.

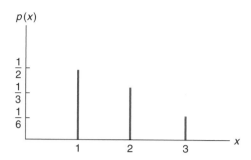

FIGURE 4.1   *Graph of p(x), Example 4.2a.*

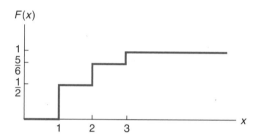

FIGURE 4.2   *Graph of F(x).*

For instance, suppose X has a probability mass function given (as in Example 4.2a) by

$$p(1) = \tfrac{1}{2}, \qquad p(2) = \tfrac{1}{3}, \qquad p(3) = \tfrac{1}{6}$$

Then the cumulative distribution function F of X is given by

$$F(a) = \begin{cases} 0 & a < 1 \\ \tfrac{1}{2} & 1 \le a < 2 \\ \tfrac{5}{6} & 2 \le a < 3 \\ 1 & 3 \le a \end{cases}$$

This is graphically presented in Figure 4.2.

Whereas the set of possible values of a discrete random variable is a sequence, we often must consider random variables whose set of possible values is an interval. Let X be such a random variable. We say that X is a *continuous* random variable if there exists a nonnegative function $f(x)$, defined for all real $x \in (-\infty, \infty)$, having the property that for any set B of real numbers

$$P\{X \in B\} = \int_B f(x)\,dx \qquad (4.2.1)$$

The function $f(x)$ is called the *probability density function* of the random variable X.

In words, Equation 4.2.1 states that the probability that X will be in B may be obtained by integrating the probability density function over the set B. Since X must assume some value, $f(x)$ must satisfy

$$1 = P\{X \in (-\infty, \infty)\} = \int_{-\infty}^{\infty} f(x)\, dx$$

All probability statements about X can be answered in terms of $f(x)$. For instance, letting $B = [a, b]$, we obtain from Equation 4.2.1 that

$$P\{a \leq X \leq b\} = \int_{a}^{b} f(x)\, dx \tag{4.2.2}$$

If we let $a = b$ in the above, then

$$P\{X = a\} = \int_{a}^{a} f(x)\, dx = 0$$

In words, this equation states that the probability that a continuous random variable will assume any *particular* value is zero. (See Figure 4.3.)

The relationship between the cumulative distribution $F(\cdot)$ and the probability density $f(\cdot)$ is expressed by

$$F(a) = P\{X \in (-\infty, a]\} = \int_{-\infty}^{a} f(x)\, dx$$

Differentiating both sides yields

$$\frac{d}{da}F(a) = f(a)$$

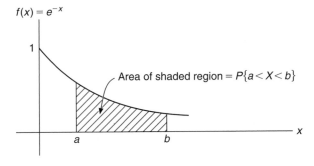

FIGURE 4.3   *The probability density function* $f(x) = \begin{cases} e^{-x} & x \geq 0 \\ 0 & x < 0 \end{cases}$.

That is, the density is the derivative of the cumulative distribution function. A somewhat more intuitive interpretation of the density function may be obtained from Equation 4.2.2 as follows:

$$P\left\{a - \frac{\varepsilon}{2} \le X \le a + \frac{\varepsilon}{2}\right\} = \int_{a-\varepsilon/2}^{a+\varepsilon/2} f(x)\, dx \approx \varepsilon f(a)$$

when ε is small. In other words, the probability that X will be contained in an interval of length ε around the point a is approximately $\varepsilon f(a)$. From this, we see that $f(a)$ is a measure of how likely it is that the random variable will be near a.

EXAMPLE 4.2b Suppose that X is a continuous random variable whose probability density function is given by

$$f(x) = \begin{cases} C(4x - 2x^2) & 0 < x < 2 \\ 0 & \text{otherwise} \end{cases}$$

(a) What is the value of C?
(b) Find $P\{X > 1\}$.

SOLUTION (a) Since f is a probability density function, we must have that $\int_{-\infty}^{\infty} f(x)\, dx = 1$, implying that

$$C \int_0^2 (4x - 2x^2)\, dx = 1$$

or

$$C \left[2x^2 - \frac{2x^3}{3} \right] \Big|_{x=0}^{x=2} = 1$$

or

$$C = \tfrac{3}{8}$$

(b) Hence

$$P\{X > 1\} = \int_1^{\infty} f(x)\, dx = \tfrac{3}{8} \int_1^2 (4x - 2x^2)\, dx = \tfrac{1}{2} \quad \blacksquare$$

4.3 JOINTLY DISTRIBUTED RANDOM VARIABLES

For a given experiment, we are often interested not only in probability distribution functions of individual random variables but also in the relationships between two or more random variables. For instance, in an experiment into the possible causes of cancer, we might be interested in the relationship between the average number of cigarettes smoked daily and the age at which an individual contracts cancer. Similarly, an engineer might be interested in the relationship between the shear strength and the diameter of a spot weld in a fabricated sheet steel specimen.

To specify the relationship between two random variables, we define the joint cumulative probability distribution function of X and Y by

$$F(x, y) = P\{X \le x, Y \le y\}$$

A knowledge of the joint probability distribution function enables one, at least in theory, to compute the probability of any statement concerning the values of X and Y. For instance, the distribution function of X — call it F_X — can be obtained from the joint distribution function F of X and Y as follows:

$$\begin{aligned} F_X(x) &= P\{X \le x\} \\ &= P\{X \le x, Y < \infty\} \\ &= F(x, \infty) \end{aligned}$$

Similarly, the cumulative distribution function of Y is given by

$$F_Y(y) = F(\infty, y)$$

In the case where X and Y are both discrete random variables whose possible values are, respectively, $x_1, x_2, \ldots$, and $y_1, y_2, \ldots$, we define the *joint probability mass function* of X and Y, $p(x_i, y_j)$, by

$$p(x_i, y_j) = P\{X = x_i, Y = y_j\}$$

The individual probability mass functions of X and Y are easily obtained from the joint probability mass function by the following reasoning. Since Y must take on some value y_j, it follows that the event $\{X = x_i\}$ can be written as the union, over all j, of the mutually exclusive events $\{X = x_i, Y = y_j\}$. That is,

$$\{X = x_i\} = \bigcup_j \{X = x_i, Y = y_j\}$$

and so, using Axiom 3 of the probability function, we see that

$$\begin{aligned} P\{X = x_i\} &= P\left(\bigcup_j \{X = x_i, Y = y_j\} \right) &&\text{(4.3.1)} \\ &= \sum_j P\{X = x_i, Y = y_j\} \\ &= \sum_j p(x_i, y_j) \end{aligned}$$

Similarly, we can obtain $P\{Y = y_j\}$ by summing $p(x_i, y_j)$ over all possible values of x_i, that is,

$$P\{Y = y_j\} = \sum_i P\{X = x_i, Y = y_j\} \qquad (4.3.2)$$

$$= \sum_i p(x_i, y_j)$$

Hence, specifying the joint probability mass function always determines the individual mass functions. However, it should be noted that the reverse is not true. Namely, knowledge of $P\{X = x_i\}$ and $P\{Y = y_j\}$ does not determine the value of $P\{X = x_i, Y = y_j\}$.

EXAMPLE 4.3a Suppose that 3 batteries are randomly chosen from a group of 3 new, 4 used but still working, and 5 defective batteries. If we let X and Y denote, respectively, the number of new and used but still working batteries that are chosen, then the joint probability mass function of X and Y, $p(i, j) = P\{X = i, Y = j\}$, is given by

$$p(i, j) = \frac{\binom{3}{i}\binom{4}{j}\binom{5}{3-i-j}}{\binom{12}{3}}$$

where the preceding follows because of the $\binom{12}{3}$ equally likely outcomes, there are, by the basic principle of counting, $\binom{3}{i}\binom{4}{j}\binom{5}{3-i-j}$ possible choices that contain exactly i new, j used, and $3 - i - j$ defective batteries. Consequently,

$$p(0, 0) = \binom{5}{3} \Big/ \binom{12}{3} = 10/220$$

$$p(0, 1) = \binom{4}{1}\binom{5}{2} \Big/ \binom{12}{3} = 40/220$$

$$p(0, 2) = \binom{4}{2}\binom{5}{1} \Big/ \binom{12}{3} = 30/220$$

$$p(0, 3) = \binom{4}{3} \Big/ \binom{12}{3} = 4/220$$

$$p(1, 0) = \binom{3}{1}\binom{5}{2} \Big/ \binom{12}{3} = 30/220$$

$$p(1, 1) = \binom{3}{1}\binom{4}{1}\binom{5}{1} \Big/ \binom{12}{3} = 60/220$$

$$p(1, 2) = \binom{3}{1}\binom{4}{2} \Big/ \binom{12}{3} = 18/220$$

$$p(2, 0) = \binom{3}{2}\binom{5}{1} \Big/ \binom{12}{3} = 15/220$$

TABLE 4.1 $P\{X = i, Y = j\}$

i \ j	0	1	2	3	Row Sum $= P\{X = i\}$
0	$\frac{10}{220}$	$\frac{40}{220}$	$\frac{30}{220}$	$\frac{4}{220}$	$\frac{84}{220}$
1	$\frac{30}{220}$	$\frac{60}{220}$	$\frac{18}{220}$	0	$\frac{108}{220}$
2	$\frac{15}{220}$	$\frac{12}{220}$	0	0	$\frac{27}{220}$
3	$\frac{1}{220}$	0	0	0	$\frac{1}{220}$
Column Sums = $P\{Y=j\}$	$\frac{56}{220}$	$\frac{112}{220}$	$\frac{48}{220}$	$\frac{4}{220}$	

$$p(2, 1) = \binom{3}{2}\binom{4}{1}\bigg/\binom{12}{3} = 12/220$$

$$p(3, 0) = \binom{3}{3}\bigg/\binom{12}{3} = 1/220$$

These probabilities can most easily be expressed in tabular form as shown in Table 4.1.

The reader should note that the probability mass function of X is obtained by computing the row sums, in accordance with the Equation 4.3.1, whereas the probability mass function of Y is obtained by computing the column sums, in accordance with Equation 4.3.2. Because the individual probability mass functions of X and Y thus appear in the margin of such a table, they are often referred to as being the marginal probability mass functions of X and Y, respectively. It should be noted that to check the correctness of such a table we could sum the marginal row (or the marginal column) and verify that its sum is 1. (Why must the sum of the entries in the marginal row (or column) equal 1?)   ■

EXAMPLE 4.3b   Suppose that 15 percent of the families in a certain community have no children, 20 percent have 1, 35 percent have 2, and 30 percent have 3 children; suppose further that each child is equally likely (and independently) to be a boy or a girl. If a family is chosen at random from this community, then B, the number of boys, and G, the number of girls, in this family will have the joint probability mass function shown in Table 4.2.

These probabilities are obtained as follows:

$$P\{B = 0, G = 0\} = P\{\text{no children}\}$$
$$= .15$$
$$P\{B = 0, G = 1\} = P\{1 \text{ girl and total of 1 child}\}$$
$$= P\{1 \text{ child}\}P\{1 \text{ girl}|1 \text{ child}\}$$
$$= (.20)\left(\tfrac{1}{2}\right) = .1$$

TABLE 4.2 $P\{B = i, G = j\}$

i \ j	0	1	2	3	Row Sum = $P\{B = i\}$
0	.15	.10	.0875	.0375	.3750
1	.10	.175	.1125	0	.3875
2	.0875	.1125	0	0	.2000
3	.0375	0	0	0	.0375
Column Sum = $P\{G=j\}$	.3750	.3875	.2000	.0375	

$$P\{B = 0, G = 2\} = P\{2 \text{ girls and total of 2 children}\}$$
$$= P\{2 \text{ children}\}P\{2 \text{ girls}|2 \text{ children}\}$$
$$= (.35) \left(\tfrac{1}{2}\right)^2 = .0875$$
$$P\{B = 0, G = 3\} = P\{3 \text{ girls and total of 3 children}\}$$
$$= P\{3 \text{ children}\}P\{3 \text{ girls}|3 \text{ children}\}$$
$$= (.30) \left(\tfrac{1}{2}\right)^3 = .0375$$

We leave it to the reader to verify the remainder of Table 4.2, which tells us, among other things, that the family chosen will have at least 1 girl with probability .625. ■

We say that X and Y are *jointly continuous* if there exists a function $f(x, y)$ defined for all real x and y, having the property that for every set C of pairs of real numbers (that is, C is a set in the two-dimensional plane)

$$P\{(X, Y) \in C\} = \iint\limits_{(x,y)\in C} f(x, y) \, dx \, dy \qquad (4.3.3)$$

The function $f(x, y)$ is called the *joint probability density function* of X and Y. If A and B are any sets of real numbers, then by defining $C = \{(x, y) : x \in A, y \in B\}$, we see from Equation 4.3.3 that

$$P\{X \in A, Y \in B\} = \int_B \int_A f(x, y) \, dx \, dy \qquad (4.3.4)$$

Because

$$F(a, b) = P\{X \in (-\infty, a], Y \in (-\infty, b]\}$$
$$= \int_{-\infty}^{b} \int_{-\infty}^{a} f(x, y) \, dx \, dy$$

it follows, upon differentiation, that

$$f(a, b) = \frac{\partial^2}{\partial a\, \partial b} F(a, b)$$

wherever the partial derivatives are defined. Another interpretation of the joint density function is obtained from Equation 4.3.4 as follows:

$$P\{a < X < a + da, b < Y < b + db\} = \int_b^{d+db} \int_a^{a+da} f(x, y)\, dx\, dy$$

$$\approx f(a, b)\, da\, db$$

when da and db are small and $f(x, y)$ is continuous at a, b. Hence $f(a, b)$ is a measure of how likely it is that the random vector (X, Y) will be near (a, b).

If X and Y are jointly continuous, they are individually continuous, and their probability density functions can be obtained as follows:

$$P\{X \in A\} = P\{X \in A, Y \in (-\infty, \infty)\} \tag{4.3.5}$$

$$= \int_A \int_{-\infty}^{\infty} f(x, y)\, dy\, dx$$

$$= \int_A f_X(x)\, dx$$

where

$$f_X(x) = \int_{-\infty}^{\infty} f(x, y)\, dy$$

is thus the probability density function of X. Similarly, the probability density function of Y is given by

$$f_Y(y) = \int_{-\infty}^{\infty} f(x, y)\, dx \tag{4.3.6}$$

EXAMPLE 4.3c The joint density function of X and Y is given by

$$f(x, y) = \begin{cases} 2e^{-x}e^{-2y} & 0 < x < \infty, 0 < y < \infty \\ 0 & \text{otherwise} \end{cases}$$

Compute **(a)** $P\{X > 1, Y < 1\}$; **(b)** $P\{X < Y\}$; and **(c)** $P\{X < a\}$.

SOLUTION

(a)
$$P\{X > 1, Y < 1\} = \int_0^1 \int_1^\infty 2e^{-x} e^{-2y}\, dx\, dy$$

$$= \int_0^1 2e^{-2y}(-e^{-x}|_1^\infty)\, dy$$

$$= e^{-1} \int_0^1 2e^{-2y}\, dy$$

$$= e^{-1}(1 - e^{-2})$$

(b)
$$P\{X < Y\} = \iint\limits_{(x,y):x<y} 2e^{-x} e^{-2y}\, dx\, dy$$

$$= \int_0^\infty \int_0^y 2e^{-x} e^{-2y}\, dx\, dy$$

$$= \int_0^\infty 2e^{-2y}(1 - e^{-y})\, dy$$

$$= \int_0^\infty 2e^{-2y}\, dy - \int_0^\infty 2e^{-3y}\, dy$$

$$= 1 - \tfrac{2}{3}$$

$$= \tfrac{1}{3}$$

(c)
$$P\{X < a\} = \int_0^a \int_0^\infty 2e^{-2y} e^{-x}\, dy\, dx$$

$$= \int_0^a e^{-x}\, dx$$

$$= 1 - e^{-a} \quad \blacksquare$$

4.3.1 INDEPENDENT RANDOM VARIABLES

The random variables X and Y are said to be independent if for any two sets of real numbers A and B

$$P\{X \in A, Y \in B\} = P\{X \in A\}P\{Y \in B\} \qquad (4.3.7)$$

In other words, X and Y are independent if, for all A and B, the events $E_A = \{X \in A\}$ and $F_B = \{Y \in B\}$ are independent.

It can be shown by using the three axioms of probability that Equation 4.3.7 will follow if and only if for all a, b

$$P\{X \le a, Y \le b\} = P\{X \le a\}P\{Y \le b\}$$

Hence, in terms of the joint distribution function F of X and Y, we have that X and Y are independent if

$$F(a, b) = F_X(a)F_Y(b) \qquad \text{for all } a, b$$

When X and Y are discrete random variables, the condition of independence Equation 4.3.7 is equivalent to

$$p(x, y) = p_X(x)p_Y(y) \qquad \text{for all } x, y \qquad (4.3.8)$$

where p_X and p_Y are the probability mass functions of X and Y. The equivalence follows because, if Equation 4.3.7 is satisfied, then we obtain Equation 4.3.8 by letting A and B be, respectively, the one-point sets $A = \{x\}, B = \{y\}$. Furthermore, if Equation 4.3.8 is valid, then for any sets A, B

$$
\begin{aligned}
P\{X \in A, Y \in B\} &= \sum_{y \in B} \sum_{x \in A} p(x, y) \\
&= \sum_{y \in B} \sum_{x \in A} p_X(x)p_Y(y) \\
&= \sum_{y \in B} p_Y(y) \sum_{x \in A} p_X(x) \\
&= P\{Y \in B\}P\{X \in A\}
\end{aligned}
$$

and thus Equation 4.3.7 is established.

In the jointly continuous case, the condition of independence is equivalent to

$$f(x, y) = f_X(x)f_Y(y) \qquad \text{for all } x, y$$

Loosely speaking, X and Y are independent if knowing the value of one does not change the distribution of the other. Random variables that are not independent are said to be dependent.

EXAMPLE 4.3d Suppose that X and Y are independent random variables having the common density function

$$
f(x) = \begin{cases} e^{-x} & x > 0 \\ 0 & \text{otherwise} \end{cases}
$$

Find the density function of the random variable X/Y.

SOLUTION We start by determining the distribution function of X/Y. For $a > 0$

$$F_{X/Y}(a) = P\{X/Y \leq a\}$$

$$= \iint\limits_{x/y \leq a} f(x, y)\, dx\, dy$$

$$= \iint\limits_{x/y \leq a} e^{-x} e^{-y}\, dx\, dy$$

$$= \int_0^\infty \int_0^{ay} e^{-x} e^{-y}\, dx\, dy$$

$$= \int_0^\infty (1 - e^{-ay}) e^{-y}\, dy$$

$$= \left[-e^{-y} + \frac{e^{-(a+1)y}}{a+1} \right] \Big|_0^\infty$$

$$= 1 - \frac{1}{a+1}$$

Differentiation yields that the density function of X/Y is given by

$$f_{X/Y}(a) = 1/(a+1)^2, \qquad 0 < a < \infty \quad \blacksquare$$

We can also define joint probability distributions for n random variables in exactly the same manner as we did for $n = 2$. For instance, the joint cumulative probability distribution function $F(a_1, a_2, \ldots, a_n)$ of the n random variables $X_1, X_2, \ldots, X_n$ is defined by

$$F(a_1, a_2, \ldots, a_n) = P\{X_1 \leq a_1, X_2 \leq a_2, \ldots, X_n \leq a_n\}$$

If these random variables are discrete, we define their joint probability mass function $p(x_1, x_2, \ldots, x_n)$ by

$$p(x_1, x_2, \ldots, x_n) = P\{X_1 = x_1, X_2 = x_2, \ldots, X_n = x_n\}$$

Further, the n random variables are said to be jointly continuous if there exists a function $f(x_1, x_2, \ldots, x_n)$, called the joint probability density function, such that for any set C in n-space

$$P\{(X_1, X_2, \ldots, X_n) \in C\} = \underset{(x_1, \ldots, x_n) \in C}{\int \int} \cdots \int f(x_1, \ldots, x_n)\, dx_1\, dx_2 \cdots dx_n$$

In particular, for any n sets of real numbers $A_1, A_2, \ldots, A_n$

$$P\{X_1 \in A_1, X_2 \in A_2, \ldots, X_n \in A_n\}$$
$$= \int_{A_n} \int_{A_{n-1}} \cdots \int_{A_1} f(x_1, \ldots, x_n) \, dx_1 \, dx_2 \ldots dx_n$$

The concept of independence may, of course, also be defined for more than two random variables. In general, the n random variables $X_1, X_2, \ldots, X_n$ are said to be independent if, for all sets of real numbers $A_1, A_2, \ldots, A_n$,

$$P\{X_1 \in A_1, X_2 \in A_2, \ldots, X_n \in A_n\} = \prod_{i=1}^{n} P\{X_i \in A_i\}$$

As before, it can be shown that this condition is equivalent to

$$P\{X_1 \leq a_1, X_2 \leq a_2, \ldots, X_n \leq a_n\}$$
$$= \prod_{i=1}^{n} P\{X_1 \leq a_i\} \qquad \text{for all } a_1, a_2, \ldots, a_n$$

Finally, we say that an infinite collection of random variables is independent if every finite subcollection of them is independent.

EXAMPLE 4.3e Suppose that the successive daily changes of the price of a given stock are assumed to be independent and identically distributed random variables with probability mass function given by

$$P\{\text{daily change is } i\} = \begin{cases} -3 & \text{with probability .05} \\ -2 & \text{with probability .10} \\ -1 & \text{with probability .20} \\ 0 & \text{with probability .30} \\ 1 & \text{with probability .20} \\ 2 & \text{with probability .10} \\ 3 & \text{with probability .05} \end{cases}$$

Then the probability that the stock's price will increase successively by 1, 2, and 0 points in the next three days is

$$P\{X_1 = 1, X_2 = 2, X_3 = 0\} = (.20)(.10)(.30) = .006$$

where we have let X_i denote the change on the ith day. ∎

*4.3.2 Conditional Distributions

The relationship between two random variables can often be clarified by consideration of the conditional distribution of one given the value of the other.

Recall that for any two events E and F, the conditional probability of E given F is defined, provided that $P(F) > 0$, by

$$P(E|F) = \frac{P(EF)}{P(F)}$$

Hence, if X and Y are discrete random variables, it is natural to define the conditional probability mass function of X given that $Y = y$, by

$$\begin{aligned} p_{X|Y}(x|y) &= P\{X = x|Y = y\} \\ &= \frac{P\{X = x, Y = y\}}{P\{Y = y\}} \\ &= \frac{p(x, y)}{p_Y(y)} \end{aligned}$$

for all values of y such that $p_Y(y) > 0$.

EXAMPLE 4.3f If we know, in Example 4.3b, that the family chosen has one girl, compute the conditional probability mass function of the number of boys in the family.

SOLUTION We first note from Table 4.2 that

$$P\{G = 1\} = .3875$$

Hence,

$$P\{B = 0|G = 1\} = \frac{P\{B = 0, G = 1\}}{P\{G = 1\}} = \frac{.10}{.3875} = 8/31$$

$$P\{B = 1|G = 1\} = \frac{P\{B = 1, G = 1\}}{P\{G = 1\}} = \frac{.175}{.3875} = 14/31$$

$$P\{B = 2|G = 1\} = \frac{P\{B = 2, G = 1\}}{P\{G = 1\}} = \frac{.1125}{.3875} = 9/31$$

$$P\{B = 3|G = 1\} = \frac{P\{B = 3, G = 1\}}{P\{G = 1\}} = 0$$

Thus, for instance, given 1 girl, there are 23 chances out of 31 that there will also be at least 1 boy. ∎

---

* Optional section.

EXAMPLE 4.3g Suppose that $p(x, y)$, the joint probability mass function of X and Y, is given by

$$p(0, 0) = .4, \quad p(0, 1) = .2, \quad p(1, 0) = .1, \quad p(1, 1) = .3$$

Calculate the conditional probability mass function of X given that $Y = 1$.

SOLUTION We first note that

$$P\{Y = 1\} = \sum_x p(x, 1) = p(0, 1) + p(1, 1) = .5$$

Hence,

$$P\{X = 0 | Y = 1\} = \frac{p(0, 1)}{P\{Y = 1\}} = 2/5$$

$$P\{X = 1 | Y = 1\} = \frac{p(1, 1)}{P\{Y = 1\}} = 3/5 \quad \blacksquare$$

If X and Y have a joint probability density function $f(x, y)$, then the conditional probability density function of X, given that $Y = y$, is defined for all values of y such that $f_Y(y) > 0$, by

$$f_{X|Y}(x | y) = \frac{f(x, y)}{f_Y(y)}$$

To motivate this definition, multiply the left-hand side by dx and the right-hand side by $(dx\, dy)/dy$ to obtain

$$
\begin{aligned}
f_{X|Y}(x | y)\, dx &= \frac{f(x, y)\, dx\, dy}{f_Y(y)\, dy} \\
&\approx \frac{P\{x \leq X \leq x + dx, y \leq Y \leq y + dy\}}{P\{y \leq Y \leq y + dy\}} \\
&= P\{x \leq X \leq x + dy | y \leq Y \leq y + dy\}
\end{aligned}
$$

In other words, for small values of dx and dy, $f_{X|Y}(x | y)\, dx$ represents the conditional probability that X is between x and $x + dx$, given that Y is between y and $y + dy$.

The use of conditional densities allows us to define conditional probabilities of events associated with one random variable when we are given the value of a second random variable. That is, if X and Y are jointly continuous, then, for any set A,

$$P\{X \in A | Y = y\} = \int_A f_{X|Y}(x | y)\, dx$$

EXAMPLE 4.3h The joint density of X and Y is given by

$$f(x,y) = \begin{cases} \frac{12}{5}x(2-x-y) & 0 < x < 1, 0 < y < 1 \\ 0 & \text{otherwise} \end{cases}$$

Compute the conditional density of X, given that $Y = y$, where $0 < y < 1$.

SOLUTION For $0 < x < 1, 0 < y < 1$, we have

$$f_{X|Y}(x|y) = \frac{f(x,y)}{f_Y(y)}$$

$$= \frac{f(x,y)}{\int_{-\infty}^{\infty} f(x,y)\,dx}$$

$$= \frac{x(2-x-y)}{\int_0^1 x(2-x-y)\,dx}$$

$$= \frac{x(2-x-y)}{\frac{2}{3} - y/2}$$

$$= \frac{6x(2-x-y)}{4 - 3y} \quad \blacksquare$$

4.4 EXPECTATION

One of the most important concepts in probability theory is that of the expectation of a random variable. If X is a discrete random variable taking on the possible values $x_1, x_2, \ldots$, then the *expectation* or *expected value* of X, denoted by $E[X]$, is defined by

$$E[X] = \sum_i x_i P\{X = x_i\}$$

In words, the expected value of X is a weighted average of the possible values that X can take on, each value being weighted by the probability that X assumes it. For instance, if the probability mass function of X is given by

$$p(0) = \tfrac{1}{2} = p(1)$$

then

$$E[X] = 0\left(\tfrac{1}{2}\right) + 1\left(\tfrac{1}{2}\right) = \tfrac{1}{2}$$

is just the ordinary average of the two possible values 0 and 1 that X can assume. On the other hand, if

$$p(0) = \tfrac{1}{3}, \quad p(1) = \tfrac{2}{3}$$

then

$$E[X] = 0 \left(\tfrac{1}{3}\right) + 1 \left(\tfrac{2}{3}\right) = \tfrac{2}{3}$$

is a weighted average of the two possible values 0 and 1 where the value 1 is given twice as much weight as the value 0 since $p(1) = 2p(0)$.

Another motivation of the definition of expectation is provided by the frequency interpretation of probabilities. This interpretation assumes that if an infinite sequence of independent replications of an experiment is performed, then for any event E, the proportion of time that E occurs will be $P(E)$. Now, consider a random variable X that must take on one of the values $x_1, x_2, \ldots, x_n$ with respective probabilities $p(x_1), p(x_2), \ldots, p(x_n)$; and think of X as representing our winnings in a single game of chance. That is, with probability $p(x_i)$ we shall win x_i units $i = 1, 2, \ldots, n$. Now by the frequency interpretation, it follows that if we continually play this game, then the proportion of time that we win x_i will be $p(x_i)$. Since this is true for all i, $i = 1, 2, \ldots, n$, it follows that our average winnings per game will be

$$\sum_{i=1}^{n} x_i p(x_i) = E[X]$$

To see this argument more clearly, suppose that we play N games where N is very large. Then in approximately $Np(x_i)$ of these games, we shall win x_i, and thus our total winnings in the N games will be

$$\sum_{i=1}^{n} x_i N\, p(x_i)$$

implying that our average winnings per game are

$$\sum_{i=1}^{n} \frac{x_i Np(x_i)}{N} = \sum_{i=1}^{n} x_i p(x_i) = E[X]$$

EXAMPLE 4.4a Find $E[X]$ where X is the outcome when we roll a fair die.

SOLUTION Since $p(1) = p(2) = p(3) = p(4) = p(5) = p(6) = \tfrac{1}{6}$, we obtain that

$$E[X] = 1 \left(\tfrac{1}{6}\right) + 2 \left(\tfrac{1}{6}\right) + 3 \left(\tfrac{1}{6}\right) + 4 \left(\tfrac{1}{6}\right) + 5 \left(\tfrac{1}{6}\right) + 6 \left(\tfrac{1}{6}\right) = \tfrac{7}{2}$$

The reader should note that, for this example, the expected value of X is not a value that X could possibly assume. (That is, rolling a die cannot possibly lead to an outcome of 7/2.) Thus, even though we call $E[X]$ the *expectation* of X, it should not be interpreted as the value that we *expect* X to have but rather as the average value of X in a large number of repetitions of the experiment. That is, if we continually roll a fair die, then after a large number of rolls the average of all the outcomes will be approximately 7/2. (The interested reader should try this as an experiment.) ■

EXAMPLE 4.4b If I is an indicator random variable for the event A, that is, if

$$I = \begin{cases} 1 & \text{if } A \text{ occurs} \\ 0 & \text{if } A \text{ does not occur} \end{cases}$$

then

$$E[I] = 1P(A) + 0P(A^c) = P(A)$$

Hence, the expectation of the indicator random variable for the event A is just the probability that A occurs. ■

EXAMPLE 4.4c (Entropy) For a given random variable X, how much information is conveyed in the message that $X = x$? Let us begin our attempts at quantifying this statement by agreeing that the amount of information in the message that $X = x$ should depend on how likely it was that X would equal x. In addition, it seems reasonable that the more unlikely it was that X would equal x, the more informative would be the message. For instance, if X represents the sum of two fair dice, then there seems to be more information in the message that X equals 12 than there would be in the message that X equals 7, since the former event has probability $\frac{1}{36}$ and the latter $\frac{1}{6}$.

Let us denote by $I(p)$ the amount of information contained in the message that an event, whose probability is p, has occurred. Clearly $I(p)$ should be a nonnegative, decreasing function of p. To determine its form, let X and Y be independent random variables, and suppose that $P\{X = x\} = p$ and $P\{Y = y\} = q$. How much information is contained in the message that X equals x and Y equals y? To answer this, note first that the amount of information in the statement that X equals x is $I(p)$. Also, since knowledge of the fact that X is equal to x does not affect the probability that Y will equal y (since X and Y are independent), it seems reasonable that the additional amount of information contained in the statement that $Y = y$ should equal $I(q)$. Thus, it seems that the amount of information in the message that X equals x and Y equals y is $I(p) + I(q)$. On the other hand, however, we have that

$$P\{X = x, Y = y\} = P\{X = x\}P\{Y = y\} = pq$$

which implies that the amount of information in the message that X equals x and Y equals y is $I(pq)$. Therefore, it seems that the function I should satisfy the identity

$$I(pq) = I(p) + I(q)$$

However, if we define the function G by

$$G(p) = I(2^{-p})$$

then we see from the above that

$$
\begin{aligned}
G(p+q) &= I(2^{-(p+q)}) \\
&= I(2^{-p}2^{-q}) \\
&= I(2^{-p}) + I(2^{-q}) \\
&= G(p) + G(q)
\end{aligned}
$$

However, it can be shown that the only (monotone) functions G that satisfy the foregoing functional relationship are those of the form

$$G(p) = cp$$

for some constant c. Therefore, we must have that

$$I(2^{-p}) = cp$$

or, letting $q = 2^{-p}$

$$I(q) = -c\log_2(q)$$

for some positive constant c. It is traditional to let $c = 1$ and to say that the information is measured in units of *bits* (short for binary digits).

Consider now a random variable X, which must take on one of the values $x_1, \ldots, x_n$ with respective probabilities $p_1, \ldots, p_n$. As $-\log_2(p_i)$ represents the information conveyed by the message that X is equal to x_i, it follows that the expected amount of information that will be conveyed when the value of X is transmitted is given by

$$H(X) = -\sum_{i=1}^{n} p_i \log_2(p_i)$$

The quantity $H(X)$ is known in information theory as the *entropy* of the random variable X. ∎

We can also define the expectation of a continuous random variable. Suppose that X is a continuous random variable with probability density function f. Since, for dx small

$$f(x)\,dx \approx P\{x < X < x + dx\}$$

it follows that a weighted average of all possible values of X, with the weight given to x equal to the probability that X is near x, is just the integral over all x of $xf(x)\,dx$. Hence,

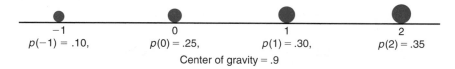

Center of gravity = .9

FIGURE 4.4

it is natural to define the expected value of X by

$$E[X] = \int_{-\infty}^{\infty} xf(x)\, dx$$

EXAMPLE 4.4d Suppose that you are expecting a message at some time past 5 P.M. From experience you know that X, the number of hours after 5 P.M. until the message arrives, is a random variable with the following probability density function:

$$f(x) = \begin{cases} \dfrac{1}{1.5} & \text{if } 0 < x < 1.5 \\ 0 & \text{otherwise} \end{cases}$$

The expected amount of time past 5 P.M. until the message arrives is given by

$$E[X] = \int_{0}^{1.5} \frac{x}{1.5}\, dx = .75$$

Hence, on average, you would have to wait three-fourths of an hour. ■

REMARKS

(a) The concept of expectation is analogous to the physical concept of the center of gravity of a distribution of mass. Consider a discrete random variable X having probability mass function $p(x_i), i \geq 1$. If we now imagine a weightless rod in which weights with mass $p(x_i), i \geq 1$ are located at the points $x_i, i \geq 1$ (see Figure 4.4), then the point at which the rod would be in balance is known as the center of gravity. For those readers acquainted with elementary statics, it is now a simple matter to show that this point is at $E[X]$.[*]
(b) $E[X]$ has the same units of measurement as does X.

4.5 PROPERTIES OF THE EXPECTED VALUE

Suppose now that we are given a random variable X and its probability distribution (that is, its probability mass function in the discrete case or its probability density function in the continuous case). Suppose also that we are interested in calculating, not the expected

---

[*] To prove this, we must show that the sum of the torques tending to turn the point around $E[X]$ is equal to 0. That is, we must show that $0 = \sum_i (x_i - E[X])p(x_i)$, which is immediate.

value of X, but the expected value of some function of X, say $g(X)$. How do we go about doing this? One way is as follows. Since $g(X)$ is itself a random variable, it must have a probability distribution, which should be computable from a knowledge of the distribution of X. Once we have obtained the distribution of $g(X)$, we can then compute $E[g(X)]$ by the definition of the expectation.

EXAMPLE 4.5a Suppose X has the following probability mass function

$$p(0) = .2, \quad p(1) = .5, \quad p(2) = .3$$

Calculate $E[X^2]$.

SOLUTION Letting $Y = X^2$, we have that Y is a random variable that can take on one of the values $0^2, 1^2, 2^2$ with respective probabilities

$$p_Y(0) = P\{Y = 0^2\} = .2$$

$$p_Y(1) = P\{Y = 1^2\} = .5$$

$$p_Y(4) = P\{Y = 2^2\} = .3$$

Hence,

$$E[X^2] = E[Y] = 0(.2) + 1(.5) + 4(.3) = 1.7 \quad \blacksquare$$

EXAMPLE 4.5b The time, in hours, it takes to locate and repair an electrical break-down in a certain factory is a random variable — call it X — whose density function is given by

$$f_X(x) = \begin{cases} 1 & \text{if } 0 < x < 1 \\ 0 & \text{otherwise} \end{cases}$$

If the cost involved in a breakdown of duration x is x^3, what is the expected cost of such a breakdown?

SOLUTION Letting $Y = X^3$ denote the cost, we first calculate its distribution function as follows. For $0 \le a \le 1$,

$$F_Y(a) = P\{Y \le a\}$$

$$= P\{X^3 \le a\}$$

$$= P\{X \le a^{1/3}\}$$

$$= \int_0^{a^{1/3}} dx$$

$$= a^{1/3}$$

By differentiating $F_Y(a)$, we obtain the density of Y,

$$f_Y(a) = \frac{1}{3}a^{-2/3}, \quad 0 \le a < 1$$

Hence,

$$E[X^3] = E[Y] = \int_{-\infty}^{\infty} af_Y(a)\,da$$

$$= \int_0^1 a\frac{1}{3}a^{-2/3}\,da$$

$$= \frac{1}{3}\int_0^1 a^{1/3}\,da$$

$$= \frac{1}{3}\frac{3}{4}a^{4/3}\Big|_0^1$$

$$= \frac{1}{4} \quad \blacksquare$$

While the foregoing procedure will, in theory, always enable us to compute the expectation of any function of X from a knowledge of the distribution of X, there is an easier way of doing this. Suppose, for instance, that we wanted to compute the expected value of $g(X)$. Since $g(X)$ takes on the value $g(x)$ when $X = x$, it seems intuitive that $E[g(X)]$ should be a weighted average of the possible values $g(x)$ with, for a given x, the weight given to $g(x)$ being equal to the probability (or probability density in the continuous case) that X will equal x. Indeed, the foregoing can be shown to be true and we thus have the following proposition.

PROPOSITION 4.5.1 EXPECTATION OF A FUNCTION OF A RANDOM VARIABLE

(a) If X is a discrete random variable with probability mass function $p(x)$, then for any real-valued function g,

$$E[g(X)] = \sum_x g(x)p(x)$$

(b) If X is a continuous random variable with probability density function $f(x)$, then for any real-valued function g,

$$E[g(X)] = \int_{-\infty}^{\infty} g(x)f(x)\,dx$$

EXAMPLE 4.5c Applying Proposition 4.5.1 to Example 4.5a yields

$$E[X^2] = 0^2(0.2) + (1^2)(0.5) + (2^2)(0.3) = 1.7$$

which, of course, checks with the result derived in Example 4.5a. ■

EXAMPLE 4.5d Applying the proposition to Example 4.5b yields

$$E[X^3] = \int_0^1 x^3 dx \quad (\text{since } f(x) = 1, 0 < x < 1)$$

$$= \frac{1}{4} \quad ■$$

An immediate corollary of Proposition 4.5.1 is the following.

Corollary 4.5.2

If a and b are constants, then

$$E[aX + b] = aE[X] + b$$

Proof

In the discrete case,

$$E[aX + b] = \sum_x (ax + b)p(x)$$

$$= a \sum_x x p(x) + b \sum_x p(x)$$

$$= aE[X] + b$$

In the continuous case,

$$E[aX + b] = \int_{-\infty}^{\infty} (ax + b)f(x)\, dx$$

$$= a \int_{-\infty}^{\infty} x f(x)\, dx + b \int_{-\infty}^{\infty} f(x)\, dx$$

$$= aE[X] + b \quad ■$$

If we take $a = 0$ in Corollary 4.5.2, we see that

$$E[b] = b$$

That is, the expected value of a constant is just its value. (Is this intuitive?) Also, if we take $b = 0$, then we obtain

$$E[aX] = aE[X]$$

or, in words, the expected value of a constant multiplied by a random variable is just the constant times the expected value of the random variable. The expected value of a random variable X, $E[X]$, is also referred to as the *mean* or the *first moment* of X. The quantity $E[X^n]$, $n \geq 1$, is called the nth moment of X. By Proposition 4.5.1, we note that

$$E[X^n] = \begin{cases} \displaystyle\sum_x x^n p(x) & \text{if } X \text{ is discrete} \\ \displaystyle\int_{-\infty}^{\infty} x^n f(x)\, dx & \text{if } X \text{ is continuous} \end{cases}$$

4.5.1 EXPECTED VALUE OF SUMS OF RANDOM VARIABLES

The two-dimensional version of Proposition 4.5.1 states that if X and Y are random variables and g is a function of two variables, then

$$E[g(X, Y)] = \sum_y \sum_x g(x, y) p(x, y) \qquad \text{in the discrete case}$$

$$= \int_{-\infty}^{\infty} \int_{-\infty}^{\infty} g(x, y) f(x, y)\, dx\, dy \quad \text{in the continuous case}$$

For example, if $g(X, Y) = X + Y$, then, in the continuous case,

$$E[X + Y] = \int_{-\infty}^{\infty} \int_{-\infty}^{\infty} (x + y) f(x, y)\, dx\, dy$$

$$= \int_{-\infty}^{\infty} \int_{-\infty}^{\infty} x f(x, y)\, dx\, dy + \int_{-\infty}^{\infty} \int_{-\infty}^{\infty} y f(x, y)\, dx\, dy$$

$$= E[X] + E[Y]$$

where the final equality followed by applying the identity

$$E[g(X, Y)] = \int_{-\infty}^{\infty} \int_{-\infty}^{\infty} g(x, y) f(x, y)\, dx\, dy$$

first to the function $g(x, y) = x$ and then to the function $g(x, y) = y$.

A similar result can be shown in the discrete case and indeed, for any random variables X and Y,

$$E[X + Y] = E[X] + E[Y] \qquad\qquad (4.5.1)$$

By repeatedly applying Equation 4.5.1 we can show that the expected value of the sum of any number of random variables equals the sum of their individual expectations. For instance,

$$E[X + Y + Z] = E[(X + Y) + Z]$$
$$= E[X + Y] + E[Z] \qquad \text{by Equation 4.5.1}$$
$$= E[X] + E[Y] + E[Z] \qquad \text{again by Equation 4.5.1}$$

And in general, for any n,

$$E[X_1 + X_2 \cdots + X_n] = E[X_1] + E[X_2] + \cdots + E[X_n] \qquad (4.5.2)$$

Equation 4.5.2 is an extremely useful formula whose utility will now be illustrated by a series of examples.

EXAMPLE 4.5e Find the expected value of the sum obtained when two fair dice are rolled.

SOLUTION If X is the sum, then $E[X]$ can be obtained from the formula

$$E[X] = \sum_{i=2}^{12} i \, P(X = i)$$

However, it is simpler to name the dice, and let X_i be the value on dice $i, i = 1, 2$. As, $X = X_1 + X_2$, this yields that

$$E[X] = E[X_1] + E[X_2]$$

Thus, from Example 4.4a, we see that $E[X] = 7$. ∎

EXAMPLE 4.5f A construction firm has recently sent in bids for 3 jobs worth (in profits) 10, 20, and 40 (thousand) dollars. If its probabilities of winning the jobs are respectively .2, .8, and .3, what is the firm's expected total profit?

SOLUTION Letting $X_i, i = 1, 2, 3$ denote the firm's profit from job i, then

$$\text{total profit} = X_1 + X_2 + X_3$$

and so

$$E[\text{total profit}] = E[X_1] + E[X_2] + E[X_3]$$

Now

$$E[X_1] = 10(.2) + 0(.8) = 2$$
$$E[X_2] = 20(.8) + 0(.2) = 16$$
$$E[X_3] = 40(.3) + 0(.7) = 12$$

and thus the firm's expected total profit is 30 thousand dollars. ∎

EXAMPLE 4.5g A secretary has typed N letters along with their respective envelopes. The envelopes get mixed up when they fall on the floor. If the letters are placed in the mixed-up envelopes in a completely random manner (that is, each letter is equally likely to end up in any of the envelopes), what is the expected number of letters that are placed in the correct envelopes?

SOLUTION Letting X denote the number of letters that are placed in the correct envelope, we can most easily compute $E[X]$ by noting that

$$X = X_1 + X_2 + \cdots + X_N$$

where

$$X_i = \begin{cases} 1 & \text{if the } i\text{th letter is placed in its proper envelope} \\ 0 & \text{otherwise} \end{cases}$$

Now, since the ith letter is equally likely to be put in any of the N envelopes, it follows that

$$P\{X_i = 1\} = P\{i\text{th letter is in its proper envelope}\} = 1/N$$

and so

$$E[X_i] = 1P\{X_i = 1\} + 0P\{X_i = 0\} = 1/N$$

Hence, from Equation 4.5.2 we obtain that

$$E[X] = E[X_1] + \cdots + E[X_N] = \left(\frac{1}{N}\right)N = 1$$

Hence, no matter how many letters there are, on the average, exactly one of the letters will be in its own envelope. ■

EXAMPLE 4.5h Suppose there are 20 different types of coupons and suppose that each time one obtains a coupon it is equally likely to be any one of the types. Compute the expected number of different types that are contained in a set for 10 coupons.

SOLUTION Let X denote the number of different types in the set of 10 coupons. We compute $E[X]$ by using the representation

$$X = X_1 + \cdots + X_{20}$$

where

$$X_i = \begin{cases} 1 & \text{if at least one type } i \text{ coupon is contained in the set of 10} \\ 0 & \text{otherwise} \end{cases}$$

Now

$$E[X_i] = P\{X_i = 1\}$$

$$= P\{\text{at least one type } i \text{ coupon is in the set of } 10\}$$

$$= 1 - P\{\text{no type } i \text{ coupons are contained in the set of } 10\}$$

$$= 1 - \left(\tfrac{19}{20}\right)^{10}$$

when the last equality follows since each of the 10 coupons will (independently) not be a type i with probability $\tfrac{19}{20}$. Hence,

$$E[X] = E[X_1] + \cdots + E[X_{20}] = 20\left[1 - \left(\tfrac{19}{20}\right)^{10}\right] = 8.025 \quad \blacksquare$$

An important property of the mean arises when one must predict the value of a random variable. That is, suppose that the value of a random variable X is to be predicted. If we predict that X will equal c, then the square of the "error" involved will be $(X - c)^2$. We will now show that the average squared error is minimized when we predict that X will equal its mean μ. To see this, note that for any constant c

$$E[(X - c)^2] = E[(X - \mu + \mu - c)^2]$$

$$= E[(X - \mu)^2 + 2(\mu - c)(X - \mu) + (\mu - c)^2]$$

$$= E[(X - \mu)^2] + 2(\mu - c)E[X - \mu] + (\mu - c)^2$$

$$= E[(X - \mu)^2] + (\mu - c)^2 \quad \text{since} \quad E[X - \mu] = E[X] - \mu = 0$$

$$\geq E[(X - \mu)^2]$$

Hence, the best predictor of a random variable, in terms of minimizing the expected square of its error, is just its mean.

4.6 VARIANCE

Given a random variable X along with its probability distribution function, it would be extremely useful if we were able to summarize the essential properties of the mass function by certain suitably defined measures. One such measure would be $E[X]$, the expected value of X. However, while $E[X]$ yields the weighted average of the possible values of X, it does not tell us anything about the variation, or spread, of these values. For instance, while the following random variables W, Y, and Z having probability mass functions determined by

$$W = 0 \quad \text{with probability } 1$$

$$Y = \begin{cases} -1 & \text{with probability } \tfrac{1}{2} \\ 1 & \text{with probability } \tfrac{1}{2} \end{cases}$$

$$Z = \begin{cases} -100 & \text{with probability } \frac{1}{2} \\ 100 & \text{with probability } \frac{1}{2} \end{cases}$$

all have the same expectation — namely, 0 — there is much greater spread in the possible values of Y than in those of W (which is a constant) and in the possible values of Z than in those of Y.

Because we expect X to take on values around its mean $E[X]$, it would appear that a reasonable way of measuring the possible variation of X would be to look at how far apart X would be from its mean on the average. One possible way to measure this would be to consider the quantity $E[|X - \mu|]$, where $\mu = E[X]$, and $|X - \mu|$ represents the absolute value of $X - \mu$. However, it turns out to be mathematically inconvenient to deal with this quantity and so a more tractable quantity is usually considered — namely, the expectation of the square of the difference between X and its mean. We thus have the following definition.

Definition

If X is a random variable with mean μ, then the *variance* of X, denoted by $\text{Var}(X)$, is defined by

$$\text{Var}(X) = E[(X - \mu)^2]$$

An alternative formula for $\text{Var}(X)$ can be derived as follows:

$$\begin{aligned} \text{Var}(X) &= E[(X - \mu)^2] \\ &= E[X^2 - 2\mu X + \mu^2] \\ &= E[X^2] - E[2\mu X] + E[\mu^2] \\ &= E[X^2] - 2\mu E[X] + \mu^2 \\ &= E[X^2] - \mu^2 \end{aligned}$$

That is,

$$\text{Var}(X) = E[X^2] - (E[X])^2 \tag{4.6.1}$$

or, in words, the variance of X is equal to the expected value of the square of X minus the square of the expected value of X. This is, in practice, often the easiest way to compute $\text{Var}(X)$.

EXAMPLE 4.6a Compute Var(X) when X represents the outcome when we roll a fair die.

SOLUTION Since $P\{X = i\} = \frac{1}{6}, i = 1, 2, 3, 4, 5, 6,$ we obtain

$$E[X^2] = \sum_{i-1}^{6} i^2 P\{X = i\}$$
$$= 1^2 \left(\tfrac{1}{6}\right) + 2^2 \left(\tfrac{1}{6}\right) + 3^2 \left(\tfrac{1}{6}\right) + 4^2 \left(\tfrac{1}{6}\right) + 5^2 \left(\tfrac{1}{6}\right) + 6^2 \left(\tfrac{1}{6}\right)$$
$$= \tfrac{91}{6}$$

Hence, since it was shown in Example 4.4a that $E[X] = \frac{7}{2}$, we obtain from Equation 4.6.1 that

$$\text{Var}(X) = E[X^2] - (E[X])^2$$
$$= \tfrac{91}{6} - \left(\tfrac{7}{2}\right)^2 = \tfrac{35}{12} \quad \blacksquare$$

EXAMPLE 4.6b Variance of an Indicator Random Variable. If, for some event A,

$$I = \begin{cases} 1 & \text{if event } A \text{ occurs} \\ 0 & \text{if event } A \text{ does not occur} \end{cases}$$

then

$$\text{Var}(I) = E[I^2] - (E[I])^2$$
$$= E[I] - (E[I])^2 \quad \text{since } I^2 = I \text{ (as } 1^2 = 1 \text{ and } 0^2 = 0)$$
$$= E[I](1 - E[I])$$
$$= P(A)[1 - P(A)] \quad \text{since } E[I] = P(A) \text{ from Example 4.4b} \quad \blacksquare$$

A useful identity concerning variances is that for any constants a and b,

$$\text{Var}(aX + b) = a^2 \text{Var}(X) \tag{4.6.2}$$

To prove Equation 4.6.2, let $\mu = E[X]$ and recall that $E[aX + b] = a\mu + b$. Thus, by the definition of variance, we have

$$\text{Var}(aX + b) = E[(aX + b - E[aX + b])^2]$$
$$= E[(aX + b - a\mu - b)^2]$$
$$= E[(aX - a\mu)^2]$$
$$= E[a^2(X - \mu)^2]$$
$$= a^2 E[(X - \mu)^2]$$
$$= a^2 \text{Var}(X)$$

Specifying particular values for a and b in Equation 4.6.2 leads to some interesting corollaries. For instance, by setting $a = 0$ in Equation 4.6.2 we obtain that

$$\text{Var}(b) = 0$$

That is, the variance of a constant is 0. (Is this intuitive?) Similarly, by setting $a = 1$ we obtain

$$\text{Var}(X + b) = \text{Var}(X)$$

That is, the variance of a constant plus a random variable is equal to the variance of the random variable. (Is this intuitive? Think about it.) Finally, setting $b = 0$ yields

$$\text{Var}(aX) = a^2 \text{Var}(X)$$

The quantity $\sqrt{\text{Var}(X)}$ is called the *standard deviation* of X. The standard deviation has the same units as does the mean.

REMARK

Analogous to the mean's being the center of gravity of a distribution of mass, the variance represents, in the terminology of mechanics, the moment of inertia.

4.7 COVARIANCE AND VARIANCE OF SUMS OF RANDOM VARIABLES

We showed in Section 4.5 that the expectation of a sum of random variables is equal to the sum of their expectations. The corresponding result for variances is, however, not generally valid. Consider

$$\begin{aligned}
\text{Var}(X + X) &= \text{Var}(2X) \\
&= 2^2 \text{Var}(X) \\
&= 4 \text{Var}(X) \\
&\neq \text{Var}(X) + \text{Var}(X)
\end{aligned}$$

There is, however, an important case in which the variance of a sum of random variables is equal to the sum of the variances; and this is when the random variables are independent. Before proving this, however, let us define the concept of the covariance of two random variables.

Definition

The *covariance* of two random variables X and Y, written $\text{Cov}(X, Y)$, is defined by

$$\text{Cov}(X, Y) = E[(X - \mu_x)(Y - \mu_y)]$$

where μ_x and μ_y are the means of X and Y, respectively.

A useful expression for $\text{Cov}(X, Y)$ can be obtained by expanding the right side of the definition. This yields

$$
\begin{aligned}
\text{Cov}(X, Y) &= E[XY - \mu_x Y - \mu_y X + \mu_x \mu_y] \\
&= E[XY] - \mu_x E[Y] - \mu_y E[X] + \mu_x \mu_y \\
&= E[XY] - \mu_x \mu_y - \mu_y \mu_x + \mu_x \mu_y \\
&= E[XY] - E[X]E[Y]
\end{aligned}
\tag{4.7.1}
$$

From its definition we see that covariance satisfies the following properties:

$$
\text{Cov}(X, Y) = \text{Cov}(Y, X)
\tag{4.7.2}
$$

and

$$
\text{Cov}(X, X) = \text{Var}(X)
\tag{4.7.3}
$$

Another property of covariance, which immediately follows from its definition, is that, for any constant a,

$$
\text{Cov}(aX, Y) = a\,\text{Cov}(X, Y)
\tag{4.7.4}
$$

The proof of Equation 4.7.4 is left as an exercise.

Covariance, like expectation, possesses an additive property.

Lemma 4.7.1

$$
\text{Cov}(X_1 + X_2, Y) = \text{Cov}(X_1, Y) + \text{Cov}(X_2, Y)
$$

Proof

$$
\begin{aligned}
\text{Cov}(X_1 &+ X_2, Y) \\
&= E[(X_1 + X_2)Y] - E[X_1 + X_2]E[Y] \quad \text{from Equation 4.7.1} \\
&= E[X_1 Y] + E[X_2 Y] - (E[X_1] + E[X_2])E[Y] \\
&= E[X_1 Y] - E[X_1]E[Y] + E[X_2 Y] - E[X_2]E[Y] \\
&= \text{Cov}(X_1, Y) + \text{Cov}(X_2, Y) \quad \blacksquare
\end{aligned}
$$

Lemma 4.7.1 can be easily generalized (see Problem 48) to show that

$$
\text{Cov}\left(\sum_{i=1}^{n} X_i, Y\right) = \sum_{i=1}^{n} \text{Cov}(X_i, Y)
\tag{4.7.5}
$$

which gives rise to the following.

PROPOSITION 4.7.2

$$\text{Cov}\left(\sum_{i=1}^{n} X_i, \sum_{j=1}^{m} Y_j\right) = \sum_{i=1}^{n}\sum_{j=1}^{m}\text{Cov}(X_i, Y_j)$$

Proof

$$\text{Cov}\left(\sum_{i=1}^{n} X_i, \sum_{j=1}^{m} Y_j\right)$$

$$= \sum_{i=1}^{n}\text{Cov}\left(X_i, \sum_{j=1}^{m} Y_j\right) \quad \text{from Equation 4.7.5}$$

$$= \sum_{i=1}^{n}\text{Cov}\left(\sum_{j=1}^{m} Y_j, X_i\right) \quad \text{by the symmetry property Equation 4.7.2}$$

$$= \sum_{i=1}^{n}\sum_{j=1}^{m}\text{Cov}(Y_j, X_i) \quad \text{again from Equation 4.7.5}$$

and the result now follows by again applying the symmetry property Equation 4.7.2. ∎

Using Equation 4.7.3 gives rise to the following formula for the variance of a sum of random variables.

Corollary 4.7.3

$$\text{Var}\left(\sum_{i=1}^{n} X_i\right) = \sum_{i=1}^{n}\text{Var}(X_i) + \sum_{i=1}^{n}\sum_{\substack{j=1 \\ j\neq i}}^{n}\text{Cov}(X_i, X_j)$$

Proof

Because $\text{Cov}(X, X) = \text{Var}(X)$, for any random variable X, we obtain from Proposition 4.7.2 that

$$\text{Var}\left(\sum_{i=1}^{n} X_i\right) = \text{Cov}\left(\sum_{i=1}^{n} X_i, \sum_{j=1}^{n} X_j\right)$$

$$= \sum_{i=1}^{n}\sum_{j=1}^{n}\text{Cov}(X_i, X_j)$$

$$= \sum_{i=1}^{n} \left[\sum_{j\neq i} \mathrm{Cov}(X_i, X_j) + \mathrm{Cov}(X_i, X_i) \right]$$

$$= \sum_{i=1}^{n} \sum_{j\neq i} \mathrm{Cov}(X_i, X_j) + \sum_{i=1}^{n} \mathrm{Cov}(X_i, X_i)$$

$$= \sum_{i=1}^{n} \sum_{j\neq i} \mathrm{Cov}(X_i, X_j) + \sum_{i=1}^{n} \mathrm{Var}(X_i) \qquad \blacksquare$$

In the case of $n = 2$, Corollary 4.7.3 yields that

$$\mathrm{Var}(X + Y) = \mathrm{Var}(X) + \mathrm{Var}(Y) + \mathrm{Cov}(X, Y) + \mathrm{Cov}(Y, X)$$

or, using Equation 4.7.2,

$$\mathrm{Var}(X + Y) = \mathrm{Var}(X) + \mathrm{Var}(Y) + 2\,\mathrm{Cov}(X, Y) \qquad (4.7.6)$$

Theorem 4.7.4

If X and Y are independent random variables, then

$$\mathrm{Cov}(X, Y) = 0$$

and so for independent $X_1, \ldots, X_n$,

$$\mathrm{Var}\left(\sum_{i=1}^{n} X_i \right) = \sum_{i=1}^{n} \mathrm{Var}(X_i)$$

Proof

We need to prove that $E[XY] = E[X]E[Y]$. Now, in the discrete case,

$$E[XY] = \sum_{j} \sum_{i} x_i y_j P\{X = x_i, Y = y_j\}$$

$$= \sum_{j} \sum_{i} x_i y_j P\{X = x_i\} P\{Y = y_j\} \quad \text{by independence}$$

$$= \sum_{y} y_j P\{Y = y_j\} \sum_{i} x_i P\{X = x_i\}$$

$$= E[Y]E[X]$$

Because a similar argument holds in all other cases, the result is proven. $\blacksquare$

EXAMPLE 4.7a Compute the variance of the sum obtained when 10 independent rolls of a fair die are made.

SOLUTION Letting X_i denote the outcome of the ith roll, we have that

$$\text{Var}\left(\sum_1^{10} X_i\right) = \sum_1^{10} \text{Var}(X_i)$$

$$= 10\frac{35}{12} \quad \text{from Example 4.6a}$$

$$= \frac{175}{6} \quad \blacksquare$$

EXAMPLE 4.7b Compute the variance of the number of heads resulting from 10 independent tosses of a fair coin.

SOLUTION Letting

$$I_j = \begin{cases} 1 & \text{if the } j\text{th toss lands heads} \\ 0 & \text{if the } j\text{th toss lands tails} \end{cases}$$

then the total number of heads is equal to

$$\sum_{j=1}^{10} I_j$$

Hence, from Theorem 4.7.4,

$$\text{Var}\left(\sum_{j=1}^{10} I_j\right) = \sum_{j=1}^{10} \text{Var}(I_j)$$

Now, since I_j is an indicator random variable for an event having probability $\frac{1}{2}$, it follows from Example 4.6b that

$$\text{Var}(I_j) = \frac{1}{2}\left(1 - \frac{1}{2}\right) = \frac{1}{4}$$

and thus

$$\text{Var}\left(\sum_{j=1}^{10} I_j\right) = \frac{10}{4} \quad \blacksquare$$

The covariance of two random variables is important as an indicator of the relationship between them. For instance, consider the situation where X and Y are indicator variables for whether or not the events A and B occur. That is, for events A and B, define

$$X = \begin{cases} 1 & \text{if } A \text{ occurs} \\ 0 & \text{otherwise} \end{cases}, \qquad Y = \begin{cases} 1 & \text{if } B \text{ occurs} \\ 0 & \text{otherwise} \end{cases}$$

and note that

$$XY = \begin{cases} 1 & \text{if } X = 1, Y = 1 \\ 0 & \text{otherwise} \end{cases}$$

Thus,

$$\begin{aligned} \text{Cov}(X, Y) &= E[XY] - E[X]E[Y] \\ &= P\{X = 1, Y = 1\} - P\{X = 1\}P\{Y = 1\} \end{aligned}$$

From this we see that

$$\begin{aligned} \text{Cov}(X, Y) > 0 &\Leftrightarrow P\{X = 1, Y = 1\} > P\{X = 1\}P\{Y = 1\} \\ &\Leftrightarrow \frac{P\{X = 1, Y = 1\}}{P\{X = 1\}} > P\{Y = 1\} \\ &\Leftrightarrow P\{Y = 1 | X = 1\} > P\{Y = 1\} \end{aligned}$$

that $Y = 1$; whereas the covariance of X and Y is negative if the outcome $X = 1$ makes it less likely that $Y = 1$, and so makes it more likely that $Y = 0$. (By the symmetry of the covariance, the preceding remains true when X and Y are interchanged.)

In general, it can be shown that a positive value of $\text{Cov}(X, Y)$ is an indication that Y tends to increase as X does, whereas a negative value indicates that Y tends to decrease as X increases. The strength of the relationship between X and Y is indicated by the correlation between X and Y, a dimensionless quantity obtained by dividing the covariance by the product of the standard deviations of X and Y. That is,

$$\text{Corr}(X, Y) = \frac{\text{Cov}(X, Y)}{\sqrt{\text{Var}(X)\text{Var}(Y)}}$$

It can be shown (see Problem 49) that this quantity always has a value between -1 and $+1$.

4.8 MOMENT GENERATING FUNCTIONS

The moment generating function $\phi(t)$ of the random variable X is defined for all values t by

$$\phi(t) = E[e^{tX}] = \begin{cases} \displaystyle\sum_x e^{tx} p(x) & \text{if } X \text{ is discrete} \\ \displaystyle\int_{-\infty}^{\infty} e^{tx} f(x)\, dx & \text{if } X \text{ is continuous} \end{cases}$$

We call $\phi(t)$ the moment generating function because all of the moments of X can be obtained by successively differentiating $\phi(t)$. For example,

$$\phi'(t) = \frac{d}{dt} E[e^{tX}]$$

$$= E\left[\frac{d}{dt}(e^{tX})\right]$$

$$= E[Xe^{tX}]$$

Hence,

$$\phi'(0) = E[X]$$

Similarly,

$$\phi''(t) = \frac{d}{dt}\phi'(t)$$

$$= \frac{d}{dt} E[Xe^{tX}]$$

$$= E\left[\frac{d}{dt}(Xe^{tX})\right]$$

$$= E[X^2 e^{tX}]$$

and so

$$\phi''(0) = E[X^2]$$

In general, the nth derivative of $\phi(t)$ evaluated at $t = 0$ equals $E[X^n]$; that is,

$$\phi^n(0) = E[X^n], \quad n \geq 1$$

An important property of moment generating functions is that the *moment generating function of the sum of independent random variables is just the product of the individual moment generating functions*. To see this, suppose that X and Y are independent and have moment generating functions $\phi_X(t)$ and $\phi_Y(t)$, respectively. Then $\phi_{X+Y}(t)$, the moment generating function of $X + Y$, is given by

$$\phi_{X+Y}(t) = E[e^{t(X+Y)}]$$
$$= E[e^{tX} e^{tY}]$$
$$= E[e^{tX}]E[e^{tY}]$$
$$= \phi_X(t)\phi_Y(t)$$

where the next to the last equality follows from Theorem 4.7.4 since X and Y, and thus e^{tX} and e^{tY}, are independent.

Another important result is that the *moment generating function uniquely determines the distribution*. That is, there exists a one-to-one correspondence between the moment generating function and the distribution function of a random variable.

4.9 CHEBYSHEV'S INEQUALITY AND THE WEAK LAW OF LARGE NUMBERS

We start this section by proving a result known as Markov's inequality.

PROPOSITION 4.9.1 MARKOV'S INEQUALITY

If X is a random variable that takes only nonnegative values, then for any value $a > 0$

$$P\{X \geq a\} \leq \frac{E[X]}{a}$$

Proof

We give a proof for the case where X is continuous with density f.

$$E[X] = \int_0^\infty x f(x)\, dx$$
$$= \int_0^a x f(x)\, dx + \int_a^\infty x f(x)\, dx$$
$$\geq \int_a^\infty x f(x)\, dx$$
$$\geq \int_a^\infty a f(x)\, dx$$
$$= a \int_a^\infty f(x)\, dx$$
$$= a P\{X \geq a\}$$

and the result is proved.   ■

As a corollary, we obtain Proposition 4.9.2.

PROPOSITION 4.9.2 CHEBYSHEV'S INEQUALITY

If X is a random variable with mean μ and variance σ^2, then for any value $k > 0$

$$P\{|X - \mu| \geq k\} \leq \frac{\sigma^2}{k^2}$$

Proof

Since $(X - \mu)^2$ is a nonnegative random variable, we can apply Markov's inequality (with $a = k^2$) to obtain

$$P\{(X - \mu)^2 \geq k^2\} \leq \frac{E[(X - \mu)^2]}{k^2} \qquad (4.9.1)$$

But since $(X - \mu) \geq k^2$ if and only if $|X - \mu| \geq k$, Equation 4.9.1 is equivalent to

$$P\{|X - \mu| \geq k\} \leq \frac{E[(X - \mu)^2]}{k^2} = \frac{\sigma^2}{k^2}$$

and the proof is complete. ■

The importance of Markov's and Chebyshev's inequalities is that they enable us to derive bounds on probabilities when only the mean, or both the mean and the variance, of the probability distribution are known. Of course, if the actual distribution were known, then the desired probabilities could be exactly computed and we would not need to resort to bounds.

EXAMPLE 4.9a Suppose that it is known that the number of items produced in a factory during a week is a random variable with mean 50.

(a) What can be said about the probability that this week's production will exceed 75?

(b) If the variance of a week's production is known to equal 25, then what can be said about the probability that this week's production will be between 40 and 60?

SOLUTION Let X be the number of items that will be produced in a week:

(a) By Markov's inequality

$$P\{X > 75\} \leq \frac{E[X]}{75} = \frac{50}{75} = \frac{2}{3}$$

(b) By Chebyshev's inequality

$$P\{|X - 50| \geq 10\} \leq \frac{\sigma^2}{10^2} = \frac{1}{4}$$

Hence

$$P\{|X - 50| < 10\} \geq 1 - \frac{1}{4} = \frac{3}{4}$$

and so the probability that this week's production will be between 40 and 60 is at least .75. ■

By replacing k by $k\sigma$ in Equation 4.9.1, we can write Chebyshev's inequality as

$$P\{|X - \mu| > k\sigma\} \leq 1/k^2$$

Thus it states that the probability a random variable differs from its mean by more than k standard deviations is bounded by $1/k^2$.

We will end this section by using Chebyshev's inequality to prove the weak law of large numbers, which states that the probability that the average of the first n terms in a sequence of independent and identically distributed random variables differs by its mean by more than ε goes to 0 as n goes to infinity.

Theorem 4.9.3 The Weak Law of Large Numbers

Let $X_1, X_2, \ldots,$ be a sequence of independent and identically distributed random variables, each having mean $E[X_i] = \mu$. Then, for any $\varepsilon > 0$,

$$P\left\{\left|\frac{X_1 + \cdots + X_n}{n} - \mu\right| > \varepsilon\right\} \to 0 \quad \text{as } n \to \infty$$

Proof

We shall prove the result only under the additional assumption that the random variables have a finite variance σ^2. Now, as

$$E\left[\frac{X_1 + \cdots + X_n}{n}\right] = \mu \quad \text{and} \quad \text{Var}\left(\frac{X_1 + \cdots + X_n}{n}\right) = \frac{\sigma^2}{n}$$

it follows from Chebyshev's inequality that

$$P\left\{\left|\frac{X_1 + \cdots + X_n}{n} - \mu\right| > \epsilon\right\} \leq \frac{\sigma^2}{n\epsilon^2}$$

and the result is proved. ■

For an application of the above, suppose that a sequence of independent trials is performed. Let E be a fixed event and denote by $P(E)$ the probability that E occurs on a given trial. Letting

$$X_i = \begin{cases} 1 & \text{if } E \text{ occurs on trial } i \\ 0 & \text{if } E \text{ does not occur on trial } i \end{cases}$$

it follows that $X_1 + X_2 + \cdots + X_n$ represents the number of times that E occurs in the first n trials. Because $E[X_i] = P(E)$, it thus follows from the weak law of large numbers that for any positive number ε, no matter how small, the probability that the proportion of the first n trials in which E occurs differs from $P(E)$ by more than ε goes to 0 as n increases.

Problems

1. Five men and 5 women are ranked according to their scores on an examination. Assume that no two scores are alike and all 10! possible rankings are equally likely. Let X denote the highest ranking achieved by a woman (for instance, $X = 2$ if the top-ranked person was male and the next-ranked person was female). Find $P\{X = i\}, i = 1, 2, 3, \ldots, 8, 9, 10$.

2. Let X represent the difference between the number of heads and the number of tails obtained when a coin is tossed n times. What are the possible values of X?

3. In Problem 2, if the coin is assumed fair, for $n = 3$, what are the probabilities associated with the values that X can take on?

4. The distribution function of the random variable X is given

$$F(x) = \begin{cases} 0 & x < 0 \\ \dfrac{x}{2} & 0 \leq x < 1 \\ \dfrac{2}{3} & 1 \leq x < 2 \\ \dfrac{11}{12} & 2 \leq x < 3 \\ 1 & 3 \leq x \end{cases}$$

   (a) Plot this distribution function.
   (b) What is $P\{X > \frac{1}{2}\}$?
   (c) What is $P\{2 < X \leq 4\}$?
   (d) What is $P\{X < 3\}$?
   (e) What is $P\{X = 1\}$?

5. Suppose the random variable X has probability density function

$$f(x) = \begin{cases} cx^3, & \text{if } 0 \leq x \leq 1 \\ 0, & \text{otherwise} \end{cases}$$

   (a) Find the value of c.
   (b) Find $P\{.4 < X < .8\}$.

6. The amount of time, in hours, that a computer functions before breaking down is a continuous random variable with probability density function given by

$$f(x) = \begin{cases} \lambda e^{-x/100} & x \geq 0 \\ 0 & x < 0 \end{cases}$$

What is the probability that a computer will function between 50 and 150 hours before breaking down? What is the probability that it will function less than 100 hours?

7. The lifetime in hours of a certain kind of radio tube is a random variable having a probability density function given by

$$f(x) = \begin{cases} 0 & x \leq 100 \\ \dfrac{100}{x^2} & x > 100 \end{cases}$$

What is the probability that exactly 2 of 5 such tubes in a radio set will have to be replaced within the first 150 hours of operation? Assume that the events $E_i, i = 1, 2, 3, 4, 5$, that the ith such tube will have to be replaced within this time are independent.

8. If the density function of X equals

$$f(x) = \begin{cases} c e^{-2x} & 0 < x < \infty \\ 0 & x < 0 \end{cases}$$

find c. What is $P\{X > 2\}$?

9. A set of five transistors are to be tested, one at a time in a random order, to see which of them are defective. Suppose that three of the five transistors are defective, and let N_1 denote the number of tests made until the first defective is spotted, and let N_2 denote the number of additional tests until the second defective is spotted. Find the joint probability mass function of N_1 and N_2.

10. The joint probability density function of X and Y is given by

$$f(x, y) = \frac{6}{7}\left(x^2 + \frac{xy}{2}\right), \quad 0 < x < 1, \quad 0 < y < 2$$

(a) Verify that this is indeed a joint density function.
(b) Compute the density function of X.
(c) Find $P\{X > Y\}$.

11. Let $X_1, X_2, \ldots, X_n$ be independent random variables, each having a uniform distribution over $(0, 1)$. Let $M = \text{maximum } (X_1, X_2, \ldots, X_n)$. Show that the distribution function of M is given by

$$F_M(x) = x^n, \quad 0 \leq x \leq 1$$

What is the probability density function of M?

12. The joint density of X and Y is given by

$$f(x, y) = \begin{cases} xe^{-(x+y)} & x > 0, y > 0 \\ 0 & \text{otherwise} \end{cases}$$

(a) Compute the density of X.
(b) Compute the density of Y.
(c) Are X and Y independent?

13. The joint density of X and Y is

$$f(x, y) = \begin{cases} 2 & 0 < x < y, 0 < y < 1 \\ 0 & \text{otherwise} \end{cases}$$

(a) Compute the density of X.
(b) Compute the density of Y.
(c) Are X and Y independent?

14. If the joint density function of X and Y factors into one part depending only on x and one depending only on y, show that X and Y are independent. That is, if

$$f(x, y) = k(x) h(y), \quad -\infty < x < \infty, \quad -\infty < y < \infty$$

show that X and Y are independent.

15. Is Problem 14 consistent with the results of Problems 12 and 13?

16. Suppose that X and Y are independent continuous random variables. Show that

(a) $P\{X + Y \leq a\} = \displaystyle\int_{-\infty}^{\infty} F_X(a - y) f_Y(y) \, dy$

(b) $P\{X \leq Y\} = \displaystyle\int_{-\infty}^{\infty} F_X(y) f_Y(y) \, dy$

where f_Y is the density function of Y, and F_X is the distribution function of X.

17. When a current I (measured in amperes) flows through a resistance R (measured in ohms), the power generated (measured in watts) is given by $W = I^2 R$. Suppose that I and R are independent random variables with densities

$$f_I(x) = 6x(1 - x) \quad 0 \leq x \leq 1$$
$$f_R(x) = 2x \quad 0 \leq x \leq 1$$

Determine the density function of W.

18. In Example 4.3b, determine the conditional probability mass function of the size of a randomly chosen family containing 2 girls.

19. Compute the conditional density function of X given $Y = y$ in **(a)** Problem 10 and **(b)** Problem 13.

20. Show that X and Y are independent if and only if

    (a) $p_{X|Y}(x|y) = p_X(x)$ in the discrete case

    (b) $f_{X|Y}(x|y) = f_X(x)$ in the continuous case

21. Compute the expected value of the random variable in Problem 1.

22. Compute the expected value of the random variable in Problem 3.

23. Each night different meteorologists give us the "probability" that it will rain the next day. To judge how well these people predict, we will score each of them as follows: If a meteorologist says that it will rain with probability p, then he or she will receive a score of

$$1 - (1 - p)^2 \quad \text{if it does rain}$$
$$1 - p^2 \qquad\quad \text{if it does not rain}$$

We will then keep track of scores over a certain time span and conclude that the meteorologist with the highest average score is the best predictor of weather. Suppose now that a given meteorologist is aware of this and so wants to maximize his or her expected score. If this individual truly believes that it will rain tomorrow with probability p^*, what value of p should he or she assert so as to maximize the expected score?

24. An insurance company writes a policy to the effect that an amount of money A must be paid if some event E occurs within a year. If the company estimates that E will occur within a year with probability p, what should it charge the customer so that its expected profit will be 10 percent of A?

25. A total of 4 buses carrying 148 students from the same school arrive at a football stadium. The buses carry, respectively, 40, 33, 25, and 50 students. One of the students is randomly selected. Let X denote the number of students that were on the bus carrying this randomly selected student. One of the 4 bus drivers is also randomly selected. Let Y denote the number of students on her bus.

    (a) Which of $E[X]$ or $E[Y]$ do you think is larger? Why?

    (b) Compute $E[X]$ and $E[Y]$.

26. Suppose that two teams play a series of games that end when one of them has won i games. Suppose that each game played is, independently, won by team A with probability p. Find the expected number of games that are played when $i = 2$. Also show that this number is maximized when $p = \frac{1}{2}$.

27. The density function of X is given by

$$f(x) = \begin{cases} a + bx^2 & 0 \le x \le 1 \\ 0 & \text{otherwise} \end{cases}$$

If $E[X] = \frac{3}{5}$, find a, b.

28. The lifetime in hours of electronic tubes is a random variable having a probability density function given by

$$f(x) = a^2 x e^{-ax}, \quad x \ge 0$$

Compute the expected lifetime of such a tube.

29. Let $X_1, X_2, \ldots, X_n$ be independent random variables having the common density function

$$f(x) = \begin{cases} 1 & 0 < x < 1 \\ 0 & \text{otherwise} \end{cases}$$

Find (a) $E[\text{Max}(X_1, \ldots, X_n)]$ and (b) $E[\text{Min}(X_1, \ldots, X_n)]$.

30. Suppose that X has density function

$$f(x) = \begin{cases} 1 & 0 < x < 1 \\ 0 & \text{otherwise} \end{cases}$$

Compute $E[X^n]$ (a) by computing the density of X^n and then using the definition of expectation and (b) by using Proposition 4.5.1.

31. The time it takes to repair a personal computer is a random variable whose density, in hours, is given by

$$f(x) = \begin{cases} \frac{1}{2} & 0 < x < 2 \\ 0 & \text{otherwise} \end{cases}$$

The cost of the repair depends on the time it takes and is equal to $40 + 30\sqrt{x}$ when the time is x. Compute the expected cost to repair a personal computer.

32. If $E[X] = 2$ and $E[X^2] = 8$, calculate (a) $E[(2+4X)^2]$ and (b) $E[X^2+(X+1)^2]$.

33. Ten balls are randomly chosen from an urn containing 17 white and 23 black balls. Let X denote the number of white balls chosen. Compute $E[X]$

(a) by defining appropriate indicator variables X_i, $i = 1, \ldots, 10$ so that

$$X = \sum_{i=1}^{10} X_i$$

(b) by defining appropriate indicator variables $Y_i = 1, \ldots, 17$ so that

$$X = \sum_{i=1}^{17} Y_i$$

34. If X is a continuous random variable having distribution function F, then its *median* is defined as that value of m for which

$$F(m) = 1/2$$

Find the median of the random variables with density function

(a) $f(x) = e^{-x}, \quad x \geq 0$;

(b) $f(x) = 1, \quad 0 \leq x \leq 1$.

35. The median, like the mean, is important in predicting the value of a random variable. Whereas it was shown in the text that the mean of a random variable is the best predictor from the point of view of minimizing the expected value of the square of the error, the median is the best predictor if one wants to minimize the expected value of the absolute error. That is, $E[|X - c|]$ is minimized when c is the median of the distribution function of X. Prove this result when X is continuous with distribution function F and density function f. *Hint:* Write

$$E[|X - c|] = \int_{-\infty}^{\infty} |x - c| f(x) \, dx$$

$$= \int_{-\infty}^{c} |x - c| f(x) \, dx + \int_{c}^{\infty} |x - c| f(x) \, dx$$

$$= \int_{-\infty}^{c} (c - x) f(x) \, dx + \int_{c}^{\infty} (x - c) f(x) \, dx$$

$$= c F(c) - \int_{-\infty}^{c} x f(x) \, dx + \int_{c}^{\infty} x f(x) \, dx - c[1 - F(c)]$$

Now, use calculus to find the minimizing value of c.

36. We say that m_p is the *100p percentile* of the distribution function F if

$$F(m_p) = p$$

Find m_p for the distribution having density function

$$f(x) = 2e^{-2x}, \quad x \geq 0$$

37. A community consists of 100 married couples. If 50 members of the community die, what is the expected number of marriages that remain intact? Assume that the

set of people who die is equally likely to be any of the $\binom{200}{50}$ groups of size 50.
Hint: For $i = 1, \ldots, 100$ let

$$X_i = \begin{cases} 1 & \text{if neither member of couple } i \text{ dies} \\ 0 & \text{otherwise} \end{cases}$$

38. Compute the expectation and variance of the number of successes in n independent trials, each of which results in a success with probability p. Is independence necessary?

39. Suppose that X is equally likely to take on any of the values 1, 2, 3, 4. Compute **(a)** $E[X]$ and **(b)** $\text{Var}(X)$.

40. Let $p_i = P\{X = i\}$ and suppose that $p_1 + p_2 + p_3 = 1$. If $E[X] = 2$, what values of p_1, p_2, p_3 **(a)** maximize and **(b)** minimize $\text{Var}(X)$?

41. Compute the mean and variance of the number of heads that appear in 3 flips of a fair coin.

42. Argue that for any random variable X

$$E[X^2] \geq (E[X])^2$$

When does one have equality?

43. A random variable X, which represents the weight (in ounces) of an article, has density function,

$$f(z) = \begin{cases} z - 8 & \text{for } 8 \leq z \leq 9 \\ 10 - z & \text{for } 9 < z \leq 10 \\ 0 & \text{otherwise} \end{cases}$$

   (a) Calculate the mean and variance of the random variable X.
   (b) The manufacturer sells the article for a fixed price of \$2.00. He guarantees to refund the purchase money to any customer who finds the weight of his article to be less than 8.25 oz. His cost of production is related to the weight of the article by the relation $x/15 + .35$. Find the expected profit per article.

44. Let X_i denote the percentage of votes cast in a given election that are for candidate i, and suppose that X_1 and X_2 have a joint density function

$$f_{X_1,X_2}(x,y) = \begin{cases} 3(x+y), & \text{if } x \geq 0, y \geq 0, 0 \leq x + y \leq 1 \\ 0, & \text{if otherwise} \end{cases}$$

   (a) Find the marginal densities of X_1 and X_2.
   (b) Find $E[X_i]$ and $\text{Var}(X_i)$ for $i = 1, 2$.

45. A product is classified according to the number of defects it contains and the factory that produces it. Let X_1 and X_2 be the random variables that represent the number of defects per unit (taking on possible values of 0, 1, 2, or 3) and the factory number (taking on possible values 1 or 2), respectively. The entries in the table represent the joint possibility mass function of a randomly chosen product.

X_1 \ X_2	1	2
0	$\frac{1}{8}$	$\frac{1}{16}$
1	$\frac{1}{16}$	$\frac{1}{16}$
2	$\frac{3}{16}$	$\frac{1}{8}$
3	$\frac{1}{8}$	$\frac{1}{4}$

(a) Find the marginal probability distributions of X_1 and X_2.
(b) Find $E[X_1]$, $E[X_2]$, $\text{Var}(X_1)$, $\text{Var}(X_2)$, and $\text{Cov}(X_1, X_2)$.

46. Find $\text{Corr}(X_1, X_2)$ for the random variables of Problem 44.
47. Verify Equation 4.7.4.
48. Prove Equation 4.7.5 by using mathematical induction.
49. Let X have variance σ_x^2 and let Y have variance σ_y^2. Starting with

$$0 \leq \text{Var}(X/\sigma_x + Y/\sigma_y)$$

show that

$$-1 \leq \text{Corr}(X, Y)$$

Now using that

$$0 \leq \text{Var}(X/\sigma_x - Y/\sigma_y)$$

conclude that

$$-1 \leq \text{Corr}(X, Y) \leq 1$$

Using the result that $\text{Var}(Z) = 0$ implies that Z is constant, argue that, if $\text{Corr}(X, Y) = 1$ or -1, then X and Y are related by

$$Y = a + bx$$

where the sign of b is positive when the correlation is 1 and negative when it is -1.

50. Consider n independent trials, each of which results in any of the outcomes $i, i = 1, 2, 3$, with respective probabilities $p_1, p_2, p_3, \sum_{i=1}^{3} p_i = 1$. Let N_i denote the number of trials that result in outcome i, and show that $\text{Cov}(N_1, N_2) = -np_1p_2$. Also explain why it is intuitive that this covariance is negative. (*Hint:* For $i = 1, \ldots, n$, let

$$ X_i = \begin{cases} 1 & \text{if trial } i \text{ results in outcome 1} \\ 0 & \text{if trial } i \text{ does not result in outcome 1} \end{cases} $$

Similarly, for $j = 1, \ldots, n$, let

$$ Y_j = \begin{cases} 1 & \text{if trial } j \text{ results in outcome 2} \\ 0 & \text{if trial } j \text{ does not result in outcome 2} \end{cases} $$

Argue that

$$ N_1 = \sum_{i=1}^{n} X_i, \quad N_2 = \sum_{j=1}^{n} Y_j $$

Then use Proposition 4.7.2 and Theorem 4.7.4.)

51. In Example 4.5f, compute $\text{Cov}(X_i, X_j)$ and use this result to show that $\text{Var}(X) = 1$.

52. If X_1 and X_2 have the same probability distribution function, show that

$$ \text{Cov}(X_1 - X_2, \ X_1 + X_2) = 0 $$

Note that independence is not being assumed.

53. Suppose that X has density function

$$ f(x) = e^{-x}, \quad x > 0 $$

Compute the moment generating function of X and use your result to determine its mean and variance. Check your answer for the mean by a direct calculation.

54. If the density function of X is

$$ f(x) = 1, \quad 0 < x < 1 $$

determine $E[e^{tX}]$. Differentiate to obtain $E[X^n]$ and then check your answer.

55. Suppose that X is a random variable with mean and variance both equal to 20. What can be said about $P\{0 \le X \le 40\}$?

56. From past experience, a professor knows that the test score of a student taking her final examination is a random variable with mean 75.

    (a) Give an upper bound to the probability that a student's test score will exceed 85.
    Suppose in addition the professor knows that the variance of a student's test score is equal to 25.

    (b) What can be said about the probability that a student will score between 65 and 85?

    (c) How many students would have to take the examination so as to ensure, with probability at least .9, that the class average would be within 5 of 75?

57. Let X and Y have respective distribution functions F_X and F_Y, and suppose that for some constants a and $b > 0$,

$$F_X(x) = F_Y\left(\frac{x-a}{b}\right)$$

    (a) Determine $E[X]$ in terms of $E[Y]$.
    (b) Determine $\text{Var}(X)$ in terms of $\text{Var}(Y)$.

    Hint: X has the same distribution as what other random variable?

Chapter 5

SPECIAL RANDOM VARIABLES

Certain types of random variables occur over and over again in applications. In this chapter, we will study a variety of them.

5.1 THE BERNOULLI AND BINOMIAL RANDOM VARIABLES

Suppose that a trial, or an experiment, whose outcome can be classified as either a "success" or as a "failure" is performed. If we let $X = 1$ when the outcome is a success and $X = 0$ when it is a failure, then the probability mass function of X is given by

$$P\{X = 0\} = 1 - p \qquad \qquad (5.1.1)$$
$$P\{X = 1\} = p$$

where $p, 0 \leq p \leq 1$, is the probability that the trial is a "success."

A random variable X is said to be a Bernoulli random variable (after the Swiss mathematician James Bernoulli) if its probability mass function is given by Equations 5.1.1 for some $p \in (0, 1)$. Its expected value is

$$E[X] = 1 \cdot P\{X = 1\} + 0 \cdot P\{X = 0\} = p$$

That is, the expectation of a Bernoulli random variable is the probability that the random variable equals 1.

Suppose now that n independent trials, each of which results in a "success" with probability p and in a "failure" with probability $1 - p$, are to be performed. If X represents the number of successes that occur in the n trials, then X is said to be a *binomial* random variable with parameters (n, p).

The probability mass function of a binomial random variable with parameters n and p is given by

$$P\{X = i\} = \binom{n}{i} p^i (1 - p)^{n-i}, \quad i = 0, 1, \ldots, n \qquad (5.1.2)$$

where $\binom{n}{i} = n!/[i!(n - i)!]$ is the number of different groups of i objects that can be chosen from a set of n objects. The validity of Equation 5.1.2 may be verified by first noting that the probability of any particular sequence of the n outcomes containing i successes and $n - i$ failures is, by the assumed independence of trials, $p^i (1 - p)^{n-i}$. Equation 5.1.2 then follows since there are $\binom{n}{i}$ different sequences of the n outcomes leading to i successes and $n - i$ failures — which can perhaps most easily be seen by noting that there are $\binom{n}{i}$ different selections of the i trials that result in successes. For instance, if $n = 5$, $i = 2$, then there are $\binom{5}{2}$ choices of the two trials that are to result in successes — namely, any of the outcomes

$$
\begin{array}{lll}
(s,s,f,f,f) & (f,s,s,f,f) & (f,f,s,f,s) \\
(s,f,s,f,f) & (f,s,f,s,f) & \\
(s,f,f,s,f) & (f,s,f,f,s) & (f,f,f,s,s) \\
(s,f,f,f,s) & (f,f,s,s,f) &
\end{array}
$$

where the outcome (f,s,f,s,f) means, for instance, that the two successes appeared on trials 2 and 4. Since each of the $\binom{5}{2}$ outcomes has probability $p^2(1 - p)^3$, we see that the probability of a total of 2 successes in 5 independent trials is $\binom{5}{2}p^2(1 - p)^3$. As a check, note that, by the binomial theorem, the probabilities sum to 1; that is,

$$\sum_{i=0}^{\infty} p(i) = \sum_{i=0}^{n} \binom{n}{i} p^i (1 - p)^{n-i} = [p + (1 - p)]^n = 1$$

The probability mass function of three binomial random variables with respective parameters (10, .5), (10, .3), and (10, .6) are presented in Figure 5.1. The first of these is symmetric about the value .5, whereas the second is somewhat weighted, or *skewed*, to lower values and the third to higher values.

EXAMPLE 5.1a It is known that disks produced by a certain company will be defective with probability .01 independently of each other. The company sells the disks in packages of 10 and offers a money-back guarantee that at most 1 of the 10 disks is defective. What proportion of packages is returned? If someone buys three packages, what is the probability that exactly one of them will be returned?

SOLUTION If X is the number of defective disks in a package, then assuming that customers always take advantage of the guarantee, it follows that X is a binomial random variable

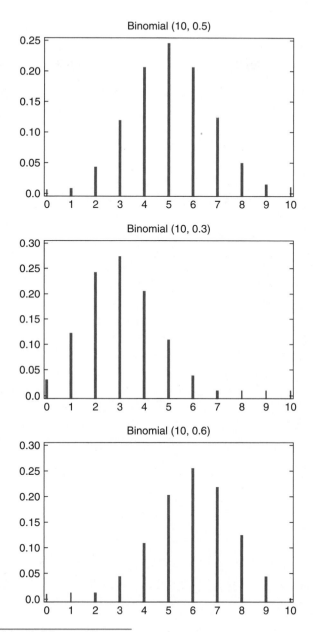

FIGURE 5.1   *Binomial probability mass functions.*

with parameters $(10, .01)$. Hence the probability that a package will have to be replaced is

$$P\{X > 1\} = 1 - P\{X = 0\} - P\{X = 1\}$$

$$= 1 - \binom{10}{0}(.01)^0(.99)^{10} - \binom{10}{1}(.01)^1(.99)^9 \approx .005$$

Because each package will, independently, have to be replaced with probability .005, it follows from the law of large numbers that in the long run .5 percent of the packages will have to be replaced.

It follows from the foregoing that the number of packages that will be returned by a buyer of three packages is a binomial random variable with parameters $n = 3$ and $p = .005$. Therefore, the probability that exactly one of the three packages will be returned is $\binom{3}{1}(.005)(.995)^2 = .015.$ ∎

EXAMPLE 5.1b The color of one's eyes is determined by a single pair of genes, with the gene for brown eyes being dominant over the one for blue eyes. This means that an individual having two blue-eyed genes will have blue eyes, while one having either two brown-eyed genes or one brown-eyed and one blue-eyed gene will have brown eyes. When two people mate, the resulting offspring receives one randomly chosen gene from each of its parents' gene pair. If the eldest child of a pair of brown-eyed parents has blue eyes, what is the probability that exactly two of the four other children (none of whom is a twin) of this couple also have blue eyes?

SOLUTION To begin, note that since the eldest child has blue eyes, it follows that both parents must have one blue-eyed and one brown-eyed gene. (For if either had two brown-eyed genes, then each child would receive at least one brown-eyed gene and would thus have brown eyes.) The probability that an offspring of this couple will have blue eyes is equal to the probability that it receives the blue-eyed gene from both parents, which is $\left(\frac{1}{2}\right)\left(\frac{1}{2}\right) = \frac{1}{4}$. Hence, because each of the other four children will have blue eyes with probability $\frac{1}{4}$, it follows that the probability that exactly two of them have this eye color is

$$\binom{4}{2}(1/4)^2(3/4)^2 = 27/128$$ ∎

EXAMPLE 5.1c A communications system consists of n components, each of which will, independently, function with probability p. The total system will be able to operate effectively if at least one-half of its components function.

(a) For what values of p is a 5-component system more likely to operate effectively than a 3-component system?

(b) In general, when is a $2k + 1$ component system better than a $2k - 1$ component system?

SOLUTION

(a) Because the number of functioning components is a binomial random variable with parameters (n, p), it follows that the probability that a 5-component system will be effective is

$$\binom{5}{3}p^3(1 - p)^2 + \binom{5}{4}p^4(1 - p) + p^5$$

whereas the corresponding probability for a 3-component system is

$$\binom{3}{2} p^2 (1 - p) + p^3$$

Hence, the 5-component system is better if

$$10 p^3 (1 - p)^2 + 5 p^4 (1 - p) + p^5 \geq 3 p^2 (1 - p) + p^3$$

which reduces to

$$3(p - 1)^2 (2p - 1) \geq 0$$

or

$$p \geq \frac{1}{2}$$

(b) In general, a system with $2k + 1$ components will be better than one with $2k - 1$ components if (and only if) $p \geq \frac{1}{2}$. To prove this, consider a system of $2k + 1$ components and let X denote the number of the first $2k - 1$ that function. Then

$$P_{2k+1}(\text{effective}) = P\{X \geq k + 1\} + P\{X = k\}(1 - (1 - p)^2) + P\{X = k - 1\} p^2$$

which follows since the $2k + 1$ component system will be effective if either

(1) $X \geq k + 1$;
(2) $X = k$ and at least one of the remaining 2 components function; or
(3) $X = k - 1$ and both of the next 2 function.

Because

$$P_{2k-1}(\text{effective}) = P\{X \geq k\}$$
$$= P\{X = k\} + P\{X \geq k + 1\}$$

we obtain that

$$P_{2k+1}(\text{effective}) - P_{2k-1}(\text{effective})$$
$$= P\{X = k - 1\} p^2 - (1 - p)^2 P\{X = k\}$$
$$= \binom{2k - 1}{k - 1} p^{k-1}(1 - p)^k p^2 - (1 - p)^2 \binom{2k - 1}{k} p^k (1 - p)^{k-1}$$
$$= \binom{2k - 1}{k} p^k (1 - p)^k [p - (1 - p)] \qquad \text{since } \binom{2k - 1}{k - 1} = \binom{2k - 1}{k}$$
$$\geq 0 \Leftrightarrow p \geq \frac{1}{2} \quad \blacksquare$$

EXAMPLE 5.1d Suppose that 10 percent of the chips produced by a computer hardware manufacturer are defective. If we order 100 such chips, will X, the number of defective ones we receive, be a binomial random variable?

SOLUTION The random variable X will be a binomial random variable with parameters $(100, .1)$ if each chip has probability .9 of being functional and if the functioning of successive chips is independent. Whether this is a reasonable assumption when we know that 10 percent of the chips produced are defective depends on additional factors. For instance, suppose that all the chips produced on a given day are always either functional or defective (with 90 percent of the days resulting in functional chips). In this case, if we know that all of our 100 chips were manufactured on the same day, then X will not be a binomial random variable. This is so since the independence of successive chips is not valid. In fact, in this case, we would have

$$P\{X = 100\} = .1$$
$$P\{X = 0\} = .9 \quad \blacksquare$$

Since a binomial random variable X, with parameters n and p, represents the number of successes in n independent trials, each having success probability p, we can represent X as follows:

$$X = \sum_{i=1}^{n} X_i \qquad (5.1.3)$$

where

$$X_i = \begin{cases} 1 & \text{if the } i\text{th trial is a success} \\ 0 & \text{otherwise} \end{cases}$$

Because the $X_i, i = 1, \ldots, n$ are independent Bernoulli random variables, we have that

$$E[X_i] = P\{X_i = 1\} = p$$
$$\text{Var}(X_i) = E[X_i^2] - p^2$$
$$= p(1 - p)$$

where the last equality follows since $X_i^2 = X_i$, and so $E[X_i^2] = E[X_i] = p$.

Using the representation Equation 5.1.3, it is now an easy matter to compute the mean and variance of X:

$$E[X] = \sum_{i=1}^{n} E[X_i]$$
$$= np$$

$$\text{Var}(X) = \sum_{i=1}^{n} \text{Var}(X_i) \quad \text{since the } X_i \text{ are independent}$$

$$= np(1-p)$$

If X_1 and X_2 are independent binomial random variables having respective parameters (n_i, p), $i = 1, 2$, then their sum is binomial with parameters $(n_1 + n_2, p)$. This can most easily be seen by noting that because X_i, $i = 1, 2$, represents the number of successes in n_i independent trials each of which is a success with probability p, then $X_1 + X_2$ represents the number of successes in $n_1 + n_2$ independent trials each of which is a success with probability p. Therefore, $X_1 + X_2$ is binomial with parameters $(n_1 + n_2, p)$.

5.1.1 COMPUTING THE BINOMIAL DISTRIBUTION FUNCTION

Suppose that X is binomial with parameters (n, p). The key to computing its distribution function

$$P\{X \le i\} = \sum_{k=0}^{i} \binom{n}{k} p^k (1-p)^{n-k}, \qquad i = 0, 1, \dots, n$$

is to utilize the following relationship between $P\{X = k+1\}$ and $P\{X = k\}$:

$$P\{X = k+1\} = \frac{p}{1-p} \frac{n-k}{k+1} P\{X = k\} \qquad (5.1.4)$$

The proof of this equation is left as an exercise.

EXAMPLE 5.1e Let X be a binomial random variable with parameters $n = 6, p = .4$. Then, starting with $P\{X = 0\} = (.6)^6$ and recursively employing Equation 5.1.4, we obtain

$$P\{X = 0\} = (.6)^6 = .0467$$

$$P\{X = 1\} = \frac{4}{6}\frac{6}{1} P\{X = 0\} = .1866$$

$$P\{X = 2\} = \frac{4}{6}\frac{5}{2} P\{X = 1\} = .3110$$

$$P\{X = 3\} = \frac{4}{6}\frac{4}{3} P\{X = 2\} = .2765$$

$$P\{X = 4\} = \frac{4}{6}\frac{3}{4} P\{X = 3\} = .1382$$

$$P\{X = 5\} = \frac{4}{6}\frac{2}{5} P\{X = 4\} = .0369$$

$$P\{X = 6\} = \frac{4}{6}\frac{1}{6} P\{X = 5\} = .0041 \quad \blacksquare$$

The text disk uses Equation 5.1.4 to compute binomial probabilities. In using it, one enters the binomial parameters n and p and a value i and the program computes the probabilities that a binomial (n, p) random variable is equal to and is less than or equal to i.

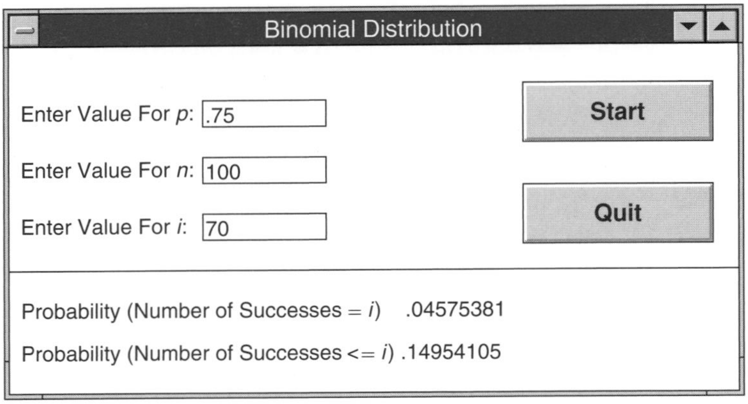

```
┌─────────────────────────────────────────────────────────────┐
│ ─                    Binomial Distribution              ▼ ▲ │
├─────────────────────────────────────────────────────────────┤
│                                                               │
│   Enter Value For p: │.75        │      ┌──────────────┐      │
│                                         │    Start     │      │
│   Enter Value For n: │100        │      └──────────────┘      │
│                                                               │
│   Enter Value For i: │70         │      ┌──────────────┐      │
│                                         │    Quit      │      │
│                                         └──────────────┘      │
├─────────────────────────────────────────────────────────────┤
│  Probability (Number of Successes = i)   .04575381           │
│  Probability (Number of Successes <= i) .14954105            │
└─────────────────────────────────────────────────────────────┘
```

FIGURE 5.2

EXAMPLE 5.1f If X is a binomial random variable with parameters $n = 100$ and $p = .75$, find $P\{X = 70\}$ and $P\{X \leq 70\}$.

SOLUTION The text disk gives the answers shown in Figure 5.2. ■

5.2 THE POISSON RANDOM VARIABLE

A random variable X, taking on one of the values 0, 1, 2, ..., is said to be a Poisson random variable with parameter $\lambda, \lambda > 0$, if its probability mass function is given by

$$P\{X = i\} = e^{-\lambda}\frac{\lambda^i}{i!}, \qquad i = 0, 1, \ldots \tag{5.2.1}$$

The symbol e stands for a constant approximately equal to 2.7183. It is a famous constant in mathematics, named after the Swiss mathematician L. Euler, and it is also the base of the so-called natural logarithm.

Equation 5.2.1 defines a probability mass function, since

$$\sum_{i=0}^{\infty} p(i) = e^{-\lambda}\sum_{i=0}^{\infty}\lambda^i/i! = e^{-\lambda}e^{\lambda} = 1$$

A graph of this mass function when $\lambda = 4$ is given in Figure 5.3.

The Poisson probability distribution was introduced by S. D. Poisson in a book he wrote dealing with the application of probability theory to lawsuits, criminal trials, and the like. This book, published in 1837, was entitled *Recherches sur la probabilité des jugements en matière criminelle et en matière civile.*

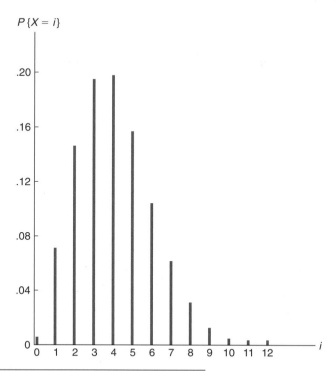

FIGURE 5.3   *The Poisson probability mass function with $\lambda = 4$.*

As a prelude to determining the mean and variance of a Poisson random variable, let us first determine its moment generating function.

$$\phi(t) = E[e^{tX}]$$

$$= \sum_{i=0}^{\infty} e^{ti} e^{-\lambda} \lambda^i / i!$$

$$= e^{-\lambda} \sum_{i=0}^{\infty} (\lambda e^t)^i / i!$$

$$= e^{-\lambda} e^{\lambda e^t}$$

$$= \exp\{\lambda(e^t - 1)\}$$

Differentiation yields

$$\phi'(t) = \lambda e^t \exp\{\lambda(e^t - 1)\}$$

$$\phi''(t) = (\lambda e^t)^2 \exp\{\lambda(e^t - 1)\} + \lambda e^t \exp\{\lambda(e^t - 1)\}$$

Evaluating at $t = 0$ gives that

$$E[X] = \phi'(0) = \lambda$$
$$\text{Var}(X) = \phi''(0) - (E[X])^2$$
$$= \lambda^2 + \lambda - \lambda^2 = \lambda$$

Thus both the mean and the variance of a Poisson random variable are equal to the parameter λ.

The Poisson random variable has a wide range of applications in a variety of areas because it may be used as an approximation for a binomial random variable with parameters (n, p) when n is large and p is small. To see this, suppose that X is a binomial random variable with parameters (n, p) and let $\lambda = np$. Then

$$P\{X = i\} = \frac{n!}{(n-1)!i!}p^i(1-p)^{n-i}$$

$$= \frac{n!}{(n-1)!i!}\left(\frac{\lambda}{n}\right)^i\left(1 - \frac{\lambda}{n}\right)^{n-i}$$

$$= \frac{n(n-1)\ldots(n-i+1)}{n^i}\frac{\lambda^i}{i!}\frac{(1-\lambda/n)^n}{(1-\lambda/n)^i}$$

Now, for n large and p small,

$$\left(1 - \frac{\lambda}{n}\right)^n \approx e^{-\lambda} \qquad \frac{n(n-1)\ldots(n-i+1)}{n^i} \approx 1 \qquad \left(1 - \frac{\lambda}{n}\right)^i \approx 1$$

Hence, for n large and p small,

$$P\{X = i\} \approx e^{-\lambda}\frac{\lambda^i}{i!}$$

In other words, if n independent trials, each of which results in a "success" with probability p, are performed, then when n is large and p small, the number of successes occurring is approximately a Poisson random variable with mean $\lambda = np$.

Some examples of random variables that usually obey, to a good approximation, the Poisson probability law (that is, they usually obey Equation 5.2.1 for some value of λ) are:

1. The number of misprints on a page (or a group of pages) of a book.
2. The number of people in a community living to 100 years of age.
3. The number of wrong telephone numbers that are dialed in a day.
4. The number of transistors that fail on their first day of use.
5. The number of customers entering a post office on a given day.

6. The number of α-particles discharged in a fixed period of time from some radioactive particle.

Each of the foregoing, and numerous other random variables, is approximately Poisson for the same reason — namely, because of the Poisson approximation to the binomial. For instance, we can suppose that there is a small probability p that each letter typed on a page will be misprinted, and so the number of misprints on a given page will be approximately Poisson with mean $\lambda = np$ where n is the (presumably) large number of letters on that page. Similarly, we can suppose that each person in a given community, independently, has a small probability p of reaching the age 100, and so the number of people that do will have approximately a Poisson distribution with mean np where n is the large number of people in the community. We leave it for the reader to reason out why the remaining random variables in examples 3 through 6 should have approximately a Poisson distribution.

EXAMPLE 5.2a Suppose that the average number of accidents occurring weekly on a particular stretch of a highway equals 3. Calculate the probability that there is at least one accident this week.

SOLUTION Let X denote the number of accidents occurring on the stretch of highway in question during this week. Because it is reasonable to suppose that there are a large number of cars passing along that stretch, each having a small probability of being involved in an accident, the number of such accidents should be approximately Poisson distributed. Hence,

$$P\{X \geq 1\} = 1 - P\{X = 0\}$$

$$= 1 - e^{-3}\frac{3^0}{0!}$$

$$= 1 - e^{-3}$$

$$\approx .9502 \quad \blacksquare$$

EXAMPLE 5.2b Suppose the probability that an item produced by a certain machine will be defective is .1. Find the probability that a sample of 10 items will contain at most one defective item. Assume that the quality of successive items is independent.

SOLUTION The desired probability is $\binom{10}{0}(.1)^0(.9)^{10} + \binom{10}{1}(.1)^1(.9)^9 = .7361$, whereas the Poisson approximation yields the value

$$e^{-1}\frac{1^0}{0!} + e^{-1}\frac{1^1}{1!} = 2e^{-1} \approx .7358 \quad \blacksquare$$

EXAMPLE 5.2c Consider an experiment that consists of counting the number of α particles given off in a 1-second interval by 1 gram of radioactive material. If we know from past experience that, on the average, 3.2 such α-particles are given off, what is a good approximation to the probability that no more than 2 α-particles will appear?

SOLUTION If we think of the gram of radioactive material as consisting of a large number n of atoms each of which has probability $3.2/n$ of disintegrating and sending off an α-particle during the second considered, then we see that, to a very close approximation, the number of α-particles given off will be a Poisson random variable with parameter $\lambda = 3.2$. Hence the desired probability is

$$P\{X \le 2\} = e^{-3.2} + 3.2e^{-3.2} + \frac{(3.2)^2}{2}e^{-3.2}$$

$$= .382 \quad \blacksquare$$

EXAMPLE 5.2d If the average number of claims handled daily by an insurance company is 5, what proportion of days have less than 3 claims? What is the probability that there will be 4 claims in exactly 3 of the next 5 days? Assume that the number of claims on different days is independent.

SOLUTION Because the company probably insures a large number of clients, each having a small probability of making a claim on any given day, it is reasonable to suppose that the number of claims handled daily, call it X, is a Poisson random variable. Since $E(X) = 5$, the probability that there will be fewer than 3 claims on any given day is

$$P\{X \le 3\} = P\{X = 0\} + P\{X = 1\} + P\{X = 2\}$$

$$= e^{-5} + e^{-5}\frac{5^1}{1!} + e^{-5}\frac{5^2}{2!}$$

$$= \frac{37}{2}e^{-5}$$

$$\approx .1247$$

Since any given day will have fewer than 3 claims with probability .125, it follows, from the law of large numbers, that over the long run 12.5 percent of days will have fewer than 3 claims.

It follows from the assumed independence of the number of claims over successive days that the number of days in a 5-day span that have exactly 4 claims is a binomial random variable with parameters 5 and $P\{X = 4\}$. Because

$$P\{X = 4\} = e^{-5}\frac{5^4}{4!} \approx .1755$$

it follows that the probability that 3 of the next 5 days will have 4 claims is equal to

$$\binom{5}{3}(.1755)^3(.8245)^2 \approx .0367 \quad \blacksquare$$

The Poisson approximation result can be shown to be valid under even more general conditions than those so far mentioned. For instance, suppose that n independent trials are to be performed, with the ith trial resulting in a success with probability p_i, $i = 1, \ldots, n$. Then it can be shown that if n is large and each p_i is small, then the number of successful trials is approximately Poisson distributed with mean equal to $\sum_{i=1}^{n} p_i$. In fact, this result will sometimes remain true even when the trials are not independent, provided that their dependence is "weak." For instance, consider the following example.

EXAMPLE 5.2e   At a party n people put their hats in the center of a room, where the hats are mixed together. Each person then randomly chooses a hat. If X denotes the number of people who select their own hat, then, for large n, it can be shown that X has approximately a Poisson distribution with mean 1. To see why this might be true, let

$$X_i = \begin{cases} 1 & \text{if the } i\text{th person selects his or her own hat} \\ 0 & \text{otherwise} \end{cases}$$

Then we can express X as

$$X = X_1 + \cdots + X_n$$

and so X can be regarded as representing the number of "successes" in n "trials" where trial i is said to be a success if the ith person chooses her own hat. Now, since the ith person is equally likely to end up with any of the n hats, one of which is her own, it follows that

$$P\{X_i = 1\} = \frac{1}{n} \tag{5.2.2}$$

Suppose now that $i \neq j$ and consider the conditional probability that the ith person chooses her own hat given that the jth person does — that is, consider $P\{X_i = 1 | X_j = 1\}$. Now given that the jth person indeed selects her own hat, it follows that the ith individual is equally likely to end up with any of the remaining $n - 1$, one of which is her own. Hence, it follows that

$$P\{X_i = 1 | X_j = 1\} = \frac{1}{n - 1} \tag{5.2.3}$$

Thus, we see from Equations 5.2.2 and 5.2.3 that whereas the trials are not independent, their dependence is rather weak [since, if the above conditional probability were equal to $1/n$ rather than $1/(n-1)$, then trials i and j would be independent]; and thus it is not at all surprising that X has approximately a Poisson distribution. The fact that $E[X] = 1$ follows since

$$E[X] = E[X_1 + \cdots + X_n]$$
$$= E[X_1] + \cdots + E[X_n]$$
$$= n\left(\frac{1}{n}\right) = 1$$

The last equality follows since, from Equation 5.2.2,

$$E[X_i] = P\{X_i = 1\} = \frac{1}{n} \quad \blacksquare$$

The Poisson distribution possesses the reproductive property that the sum of independent Poisson random variables is also a Poisson random variable. To see this, suppose that X_1 and X_2 are independent Poisson random variables having respective means λ_1 and λ_2. Then the moment generating function of $X_1 + X_2$ is as follows:

$$\begin{aligned}
E[e^{t(X_1+X_2)}] &= E[e^{tX_1} e^{tX_2}]\\
&= E[e^{tX_1}]E[e^{tX_2}] \quad \text{by independence}\\
&= \exp\{\lambda_1(e^t - 1)\} \ \exp\{\lambda_2(e^t - 1)\}\\
&= \exp\{(\lambda_1 + \lambda_2)(e^t - 1)\}
\end{aligned}$$

Because $\exp\{(\lambda_1 + \lambda_2)(e^t - 1)\}$ is the moment generating function of a Poisson random variable having mean $\lambda_1 + \lambda_2$, we may conclude, from the fact that the moment generating function uniquely specifies the distribution, that $X_1 + X_2$ is Poisson with mean $\lambda_1 + \lambda_2$.

EXAMPLE 5.2f It has been established that the number of defective stereos produced daily at a certain plant is Poisson distributed with mean 4. Over a 2-day span, what is the probability that the number of defective stereos does not exceed 3?

SOLUTION Assuming that X_1, the number of defectives produced during the first day, is independent of X_2, the number produced during the second day, then $X_1 + X_2$ is Poisson with mean 8. Hence,

$$P\{X_1 + X_2 \le 3\} = \sum_{i=0}^{3} e^{-8}\frac{8^i}{i!} = .04238 \quad \blacksquare$$

Consider now a situation in which a random number, call it N, of events will occur, and suppose that each of these events will independently be a type 1 event with probability p or a type 2 event with probability $1 - p$. Let N_1 and N_2 denote, respectively, the numbers of type 1 and type 2 events that occur. (So $N = N_1 + N_2$.) If N is Poisson distributed with mean λ, then the joint probability mass function of N_1 and N_2 is obtained as follows.

$$\begin{aligned}
P\{N_1 = n, N_2 = m\} &= P\{N_1 = n, N_2 = m, N = n + m\}\\
&= P\{N_1 = n, N_2 = m | N = n + m\}P\{N = n + m\}\\
&= P\{N_1 = n, N_2 = m | N = n + m\}e^{-\lambda}\frac{\lambda^{n+m}}{(n + m)!}
\end{aligned}$$

Now, given a total of $n + m$ events, because each one of these events is independently type 1 with probability p, it follows that the conditional probability that there are exactly n type 1 events (and m type 2 events) is the probability that a binomial $(n+m, p)$ random variable is equal to n. Consequently,

$$P\{N_1 = n, N_2 = m\} = \frac{(n+m)!}{n!m!}p^n(1-p)^m e^{-\lambda}\frac{\lambda^{n+m}}{(n+m)!}$$

$$= e^{-\lambda p}\frac{(\lambda p)^n}{n!}e^{-\lambda(1-p)}\frac{(\lambda(1-p))^m}{m!} \qquad (5.2.4)$$

The probability mass function of N_1 is thus

$$P\{N_1 = n\} = \sum_{m=0}^{\infty} P\{N_1 = n, N_2 = m\}$$

$$= e^{-\lambda p}\frac{(\lambda p)^n}{n!}\sum_{m=0}^{\infty} e^{-\lambda(1-p)}\frac{(\lambda(1-p))^m}{m!}$$

$$= e^{-\lambda p}\frac{(\lambda p)^n}{n!} \qquad (5.2.5)$$

Similarly,

$$P\{N_2 = m\} = \sum_{n=0}^{\infty} P\{N_1 = n, N_2 = m\} = e^{-\lambda(1-p)}\frac{(\lambda(1-p))^m}{m!} \qquad (5.2.6)$$

It now follows from Equations 5.2.4, 5.2.5, and 5.2.6, that N_1 and N_2 are independent Poisson random variables with respective means λp and $\lambda(1-p)$.

The preceding result generalizes when each of the Poisson number of events can be classified into any of r categories, to yield the following important property of the Poisson distribution: *If each of a Poisson number of events having mean λ is independently classified as being of one of the types $1, \ldots, r$, with respective probabilities $p_1, \ldots, p_r$, $\sum_{i=1}^{r} p_i = 1$, then the numbers of type $1, \ldots, r$ events are independent Poisson random variables with respective means $\lambda p_1, \ldots, \lambda p_r$.*

5.2.1 COMPUTING THE POISSON DISTRIBUTION FUNCTION

If X is Poisson with mean λ, then

$$\frac{P\{X = i+1\}}{P\{X = i\}} = \frac{e^{-\lambda}\lambda^{i+1}/(i+1)!}{e^{-\lambda}\lambda^i/i!} = \frac{\lambda}{i+1} \qquad (5.2.7)$$

Starting with $P\{X = 0\} = e^{-\lambda}$, we can use Equation 5.2.7 to successively compute

$$P\{X = 1\} = \lambda P\{X = 0\}$$

$$P\{X = 2\} = \frac{\lambda}{2}P\{X = 1\}$$

$$\vdots$$

$$P\{X = i + 1\} = \frac{\lambda}{i + 1}P\{X = i\}$$

The text disk includes a program that uses Equation 5.2.7 to compute Poisson probabilities.

5.3 THE HYPERGEOMETRIC RANDOM VARIABLE

A bin contains $N + M$ batteries, of which N are of acceptable quality and the other M are defective. A sample of size n is to be randomly chosen (without replacements) in the sense that the set of sampled batteries is equally likely to be any of the $\binom{N+M}{n}$ subsets of size n. If we let X denote the number of acceptable batteries in the sample, then

$$P\{X = i\} = \frac{\binom{N}{i}\binom{M}{n-i}}{\binom{N+M}{n}}, \qquad i = 0, 1, \ldots, \min(N, n)^* \qquad (5.3.1)$$

Any random variable X whose probability mass function is given by Equation 5.3.1 is said to be a *hypergeometric* random variable with parameters N, M, n.

EXAMPLE 5.3a   The components of a 6-component system are to be randomly chosen from a bin of 20 used components. The resulting system will be functional if at least 4 of its 6 components are in working condition. If 15 of the 20 components in the bin are in working condition, what is the probability that the resulting system will be functional?

SOLUTION   If X is the number of working components chosen, then X is hypergeometric with parameters 15, 5, 6. The probability that the system will be functional is

$$P\{X \geq 4\} = \sum_{i=4}^{6} P\{X = i\}$$

$$= \frac{\binom{15}{4}\binom{5}{2} + \binom{15}{5}\binom{5}{1} + \binom{15}{6}\binom{5}{0}}{\binom{20}{6}}$$

$$\approx .8687 \quad \blacksquare$$

---

* We are following the convention that $\binom{m}{r} = 0$ if $r > m$ or if $r < 0$.

To compute the mean and variance of a hypergeometric random variable whose probability mass function is given by Equation 5.3.1, imagine that the batteries are drawn sequentially and let

$$X_i = \begin{cases} 1 & \text{if the } i\text{th selection is acceptable} \\ 0 & \text{otherwise} \end{cases}$$

Now, since the ith selection is equally likely to be any of the $N + M$ batteries, of which N are acceptable, it follows that

$$P\{X_i = 1\} = \frac{N}{N + M} \qquad (5.3.2)$$

Also, for $i \neq j$,

$$
\begin{aligned}
P\{X_i = 1, X_j = 1\} &= P\{X_i = 1\}P\{X_j = 1 | X_i = 1\} \\
&= \frac{N}{N + M} \frac{N - 1}{N + M - 1}
\end{aligned}
\qquad (5.3.3)
$$

which follows since, given that the ith selection is acceptable, the jth selection is equally likely to be any of the other $N + M - 1$ batteries of which $N - 1$ are acceptable.

To compute the mean and variance of X, the number of acceptable batteries in the sample of size n, use the representation

$$X = \sum_{i=1}^{n} X_i$$

This gives

$$E[X] = \sum_{i=1}^{n} E[X_i] = \sum_{i=1}^{n} P\{X_i = 1\} = \frac{nN}{N + M} \qquad (5.3.4)$$

Also, Corollary 4.7.3 for the variance of a sum of random variables gives

$$\text{Var}(X) = \sum_{i=1}^{n} \text{Var}(X_i) + 2 \sum\sum_{1 \le i < j \le n} \text{Cov}(X_i, X_j) \qquad (5.3.5)$$

Now, X_i is a Bernoulli random variable and so

$$\text{Var}(X_i) = P\{X_i = 1\}(1 - P\{X_i = 1\}) = \frac{N}{N + M} \frac{M}{N + M} \qquad (5.3.6)$$

Also, for $i < j$,

$$\text{Cov}(X_i, X_j) = E[X_i X_j] - E[X_i]E[X_j]$$

Now, because both X_i and X_j are Bernoulli (that is, $0 - 1$) random variables, it follows that $X_i X_j$ is a Bernoulli random variable, and so

$$
\begin{aligned}
E[X_i X_j] &= P\{X_i X_j = 1\} \\
&= P\{X_i = 1, X_j = 1\} \\
&= \frac{N(N-1)}{(N+M)(N+M-1)} \quad \text{from Equation 5.3.3} \qquad (5.3.7)
\end{aligned}
$$

So from Equation 5.3.2 and the foregoing we see that for $i \neq j$,

$$
\begin{aligned}
\text{Cov}(X_i, X_j) &= \frac{N(N-1)}{(N+M)(N+M-1)} - \left(\frac{N}{N+M}\right)^2 \\
&= \frac{-NM}{(N+M)^2(N+M-1)}
\end{aligned}
$$

Hence, since there are $\binom{n}{2}$ terms in the second sum on the right side of Equation 5.3.5, we obtain from Equation 5.3.6

$$
\begin{aligned}
\text{Var}(X) &= \frac{nNM}{(N+M)^2} - \frac{n(n-1)NM}{(N+M)^2(N+M-1)} \\
&= \frac{nNM}{(N+M)^2}\left(1 - \frac{n-1}{N+M-1}\right) \qquad (5.3.8)
\end{aligned}
$$

If we let $p = N/(N+M)$ denote the proportion of batteries in the bin that are acceptable, we can rewrite Equations 5.3.4 and 5.3.8 as follows.

$$E(X) = np$$

$$\text{Var}(X) = np(1-p)\left[1 - \frac{n-1}{N+M-1}\right]$$

It should be noted that, for fixed p, as $N + M$ increases to ∞, $\text{Var}(X)$ converges to $np(1-p)$, which is the variance of a binomial random variable with parameters (n, p). (Why was this to be expected?)

EXAMPLE 5.3b An unknown number, say N, of animals inhabit a certain region. To obtain some information about the population size, ecologists often perform the following experiment: They first catch a number, say r, of these animals, mark them in some manner, and release them. After allowing the marked animals time to disperse throughout the region, a new catch of size, say, n is made. Let X denote the number of marked animals in this second capture. If we assume that the population of animals in the region remained fixed between the time of the two catches and that each time an animal was caught it was equally likely to be any of the remaining uncaught animals, it follows that X is a hypergeometric random variable such that

$$P\{X = i\} = \frac{\binom{r}{i}\binom{N-r}{n-i}}{\binom{N}{n}} \equiv P_i(N)$$

Suppose now that X is observed to equal i. That is, the fraction i/n of the animals in the second catch were marked. By taking this as an approximation of r/N, the proportion of animals in the region that are marked, we obtain the estimate rn/i of the number of animals in the region. For instance, if $r = 50$ animals are initially caught, marked, and then released, and a subsequent catch of $n = 100$ animals revealed $X = 25$ of them that were marked, then we would estimate the number of animals in the region to be about 200. ■

There is a relationship between binomial random variables and the hypergeometric distribution that will be useful to us in developing a statistical test concerning two binomial populations.

EXAMPLE 5.3c Let X and Y be independent binomial random variables having respective parameters (n, p) and (m, p). The conditional probability mass function of X given that $X + Y = k$ is as follows.

$$P\{X = i | X + Y = k\} = \frac{P\{X = i, X + Y = k\}}{P\{X + Y = k\}}$$

$$= \frac{P\{X = i, Y = k - i\}}{P\{X + Y = k\}}$$

$$= \frac{P\{X = i\}P\{Y = k - i\}}{P\{X + Y = k\}}$$

$$= \frac{\binom{n}{i}p^i(1-p)^{n-i}\binom{m}{k-i}p^{k-i}(1-p)^{m-(k-i)}}{\binom{n+m}{k}p^k(1-p)^{n+m-k}}$$

$$= \frac{\binom{n}{i}\binom{m}{k-i}}{\binom{n+m}{k}}$$

where the next-to-last equality used the fact that $X + Y$ is binomial with parameters $(n + m, p)$. Hence, we see that the conditional distribution of X given the value of $X + Y$ is hypergeometric.

It is worth noting that the preceding is quite intuitive. For suppose that $n + m$ independent trials, each of which has the same probability of being a success, are performed; let X be the number of successes in the first n trials, and let Y be the number of successes in the final m trials. Given a total of k successes in the $n + m$ trials, it is quite intuitive that each subgroup of k trials is equally likely to consist of those trials that resulted in successes. That is, the k success trials are distributed as a random selection of k of the $n + m$ trials, and so the number that are from the first n trials is hypergeometric. ■

5.4 THE UNIFORM RANDOM VARIABLE

A random variable X is said to be uniformly distributed over the interval $[\alpha, \beta]$ if its probability density function is given by

$$f(x) = \begin{cases} \dfrac{1}{\beta - \alpha} & \text{if } \alpha \le x \le \beta \\ 0 & \text{otherwise} \end{cases}$$

A graph of this function is given in Figure 5.4. Note that the foregoing meets the requirements of being a probability density function since

$$\frac{1}{\beta - \alpha} \int_\alpha^\beta dx = 1$$

The uniform distribution arises in practice when we suppose a certain random variable is equally likely to be near any value in the interval $[\alpha, \beta]$.

The probability that X lies in any subinterval of $[\alpha, \beta]$ is equal to the length of that subinterval divided by the length of the interval $[\alpha, \beta]$. This follows since when $[a, b]$ is a subinterval of $[\alpha, \beta]$ (see Figure 5.5),

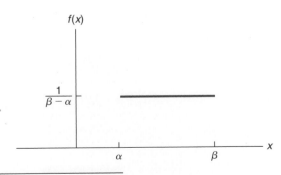

FIGURE 5.4   *Graph of $f(x)$ for a uniform $[\alpha, \beta]$.*

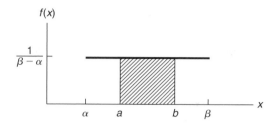

FIGURE 5.5   *Probabilities of a uniform random variable.*

$$P\{a < X < b\} = \frac{1}{\beta - \alpha} \int_a^b dx$$
$$= \frac{b - a}{\beta - \alpha}$$

EXAMPLE 5.4a   If X is uniformly distributed over the interval $[0, 10]$, compute the probability that **(a)** $2 < X < 9$, **(b)** $1 < X < 4$, **(c)** $X < 5$, **(d)** $X > 6$.

SOLUTION   The respective answers are **(a)** 7/10, **(b)** 3/10, **(c)** 5/10, **(d)** 4/10.   ■

EXAMPLE 5.4b   Buses arrive at a specified stop at 15-minute intervals starting at 7 A.M. That is, they arrive at 7, 7:15, 7:30, 7:45, and so on. If a passenger arrives at the stop at a time that is uniformly distributed between 7 and 7:30, find the probability that he waits

(a)   less than 5 minutes for a bus;
(b)   at least 12 minutes for a bus.

SOLUTION   Let X denote the time in minutes past 7 A.M. that the passenger arrives at the stop. Since X is a uniform random variable over the interval $(0, 30)$, it follows that the passenger will have to wait less than 5 minutes if he arrives between 7:10 and 7:15 or between 7:25 and 7:30. Hence, the desired probability for **(a)** is

$$P\{10 < X < 15\} + P\{25 < X < 30\} = \frac{5}{30} + \frac{5}{30} = \frac{1}{3}$$

Similarly, he would have to wait at least 12 minutes if he arrives between 7 and 7:03 or between 7:15 and 7:18, and so the probability for **(b)** is

$$P\{0 < X < 3\} + P\{15 < X < 18\} = \frac{3}{30} + \frac{3}{30} = \frac{1}{5} \quad ■$$

The mean of a uniform $[\alpha, \beta]$ random variable is

$$
\begin{aligned}
E[X] &= \int_\alpha^\beta \frac{x}{\beta - \alpha}\, dx \\
&= \frac{\beta^2 - \alpha^2}{2(\beta - \alpha)} \\
&= \frac{(\beta - \alpha)(\beta + \alpha)}{2(\beta - \alpha)}
\end{aligned}
$$

or

$$
E[X] = \frac{\alpha + \beta}{2}
$$

Or, in other words, the expected value of a uniform $[\alpha, \beta]$ random variable is equal to the midpoint of the interval $[\alpha, \beta]$, which is clearly what one would expect. (Why?)

The variance is computed as follows.

$$
\begin{aligned}
E[X^2] &= \frac{1}{\beta - \alpha} \int_\alpha^\beta x^2\, dx \\
&= \frac{\beta^3 - \alpha^3}{3(\beta - \alpha)} \\
&= \frac{\beta^2 + \alpha\beta + \alpha^2}{3}
\end{aligned}
$$

where the final equation used that

$$
\beta^3 - \alpha^3 = (\beta^2 + \alpha\beta + \alpha^2)(\beta - \alpha)
$$

Hence,

$$
\begin{aligned}
\mathrm{Var}(X) &= \frac{\beta^2 + \alpha\beta + \alpha^2}{3} - \left(\frac{\alpha + \beta}{2}\right)^2 \\
&= \frac{4(\beta^2 + \alpha\beta + \alpha^2) - 3(\alpha^2 + 2\alpha\beta + \beta^2)}{12} \\
&= \frac{\alpha^2 + \beta^2 - 2\alpha\beta}{12} \\
&= \frac{(\beta - \alpha)^2}{12}
\end{aligned}
$$

EXAMPLE 5.4c   The current in a semiconductor diode is often measured by the Shockley equation

$$
I = I_0(e^{aV} - 1)
$$

where V is the voltage across the diode; I_0 is the reverse current; a is a constant; and I is the resulting diode current. Find $E[I]$ if $a = 5$, $I_0 = 10^{-6}$, and V is uniformly distributed over $(1, 3)$.

SOLUTION

$$
\begin{aligned}
E[I] &= E[I_0(e^{aV} - 1)] \\
&= I_0 E[e^{aV} - 1] \\
&= I_0(E[e^{aV}] - 1) \\
&= 10^{-6} \int_1^3 e^{5x} \frac{1}{2} dx - 10^{-6} \\
&= 10^{-7}(e^{15} - e^5) - 10^{-6} \\
&\approx .3269 \quad \blacksquare
\end{aligned}
$$

The value of a uniform $(0, 1)$ random variable is called a *random number*. Most computer systems have a built-in subroutine for generating (to a high level of approximation) sequences of independent random numbers — for instance, Table 5.1 presents a set of independent random numbers. Random numbers are quite useful in probability and statistics because their use enables one to empirically estimate various probabilities and expectations.

TABLE 5.1   *A Random Number Table*

.68587	.25848	.85227	.78724	.05302	.70712	.76552	.70326	.80402	.49479
.73253	.41629	.37913	.00236	.60196	.59048	.59946	.75657	.61849	.90181
.84448	.42477	.94829	.86678	.14030	.04072	.45580	.36833	.10783	.33199
.49564	.98590	.92880	.69970	.83898	.21077	.71374	.85967	.20857	.51433
.68304	.46922	.14218	.63014	.50116	.33569	.97793	.84637	.27681	.04354
.76992	.70179	.75568	.21792	.50646	.07744	.38064	.06107	.41481	.93919
.37604	.27772	.75615	.51157	.73821	.29928	.62603	.06259	.21552	.72977
.43898	.06592	.44474	.07517	.44831	.01337	.04538	.15198	.50345	.65288
.86039	.28645	.44931	.59203	.98254	.56697	.55897	.25109	.47585	.59524
.28877	.84966	.97319	.66633	.71350	.28403	.28265	.61379	.13886	.78325
.44973	.12332	.16649	.88908	.31019	.33358	.68401	.10177	.92873	.13065
.42529	.37593	.90208	.50331	.37531	.72208	.42884	.07435	.58647	.84972
.82004	.74696	.10136	.35971	.72014	.08345	.49366	.68501	.14135	.15718
.67090	.08493	.47151	.06464	.14425	.28381	.40455	.87302	.07135	.04507
.62825	.83809	.37425	.17693	.69327	.04144	.00924	.68246	.48573	.24647
.10720	.89919	.90448	.80838	.70997	.98438	.51651	.71379	.10830	.69984
.69854	.89270	.54348	.22658	.94233	.08889	.52655	.83351	.73627	.39018
.71460	.25022	.06988	.64146	.69407	.39125	.10090	.08415	.07094	.14244
.69040	.33461	.79399	.22664	.68810	.56303	.65947	.88951	.40180	.87943
.13452	.36642	.98785	.62929	.88509	.64690	.38981	.99092	.91137	.02411
.94232	.91117	.98610	.71605	.89560	.92921	.51481	.20016	.56769	.60462
.99269	.98876	.47254	.93637	.83954	.60990	.10353	.13206	.33480	.29440
.75323	.86974	.91355	.12780	.01906	.96412	.61320	.47629	.33890	.22099
.75003	.98538	.63622	.94890	.96744	.73870	.72527	.17745	.01151	.47200

For an illustration of the use of random numbers, suppose that a medical center is planning to test a new drug designed to reduce its users' blood cholesterol levels. To test its effectiveness, the medical center has recruited 1,000 volunteers to be subjects in the test. To take into account the possibility that the subjects' blood cholesterol levels may be affected by factors external to the test (such as changing weather conditions), it has been decided to split the volunteers into 2 groups of size 500 — a *treatment* group that will be given the drug and a *control* group that will be given a placebo. Both the volunteers and the administrators of the drug will not be told who is in each group (such a test is called a *double-blind test*). It remains to determine which of the volunteers should be chosen to constitute the treatment group. Clearly, one would want the treatment group and the control group to be as similar as possible in all respects with the exception that members in the first group are to receive the drug while those in the other group receive a placebo; then it will be possible to conclude that any difference in response between the groups is indeed due to the drug. There is general agreement that the best way to accomplish this is to choose the 500 volunteers to be in the treatment group in a completely random fashion. That is, the choice should be made so that each of the $\binom{1000}{500}$ subsets of 500 volunteers is equally likely to constitute the control group. How can this be accomplished?

*EXAMPLE 5.4.d (Choosing a Random Subset)   From a set of n elements — numbered $1, 2, \ldots, n$ — suppose we want to generate a random subset of size k that is to be chosen in such a manner that each of the $\binom{n}{k}$ subsets is equally likely to be the subset chosen. How can we do this?

To answer this question, let us work backward and suppose that we have indeed randomly generated such a subset of size k. Now for each $j = 1, \ldots, n$, we set

$$I_j = \begin{cases} 1 & \text{if element } j \text{ is in the subset} \\ 0 & \text{otherwise} \end{cases}$$

and compute the conditional distribution of I_j given $I_1, \ldots, I_{j-1}$. To start, note that the probability that element 1 is in the subset of size k is clearly k/n (which can be seen either by noting that there is probability $1/n$ that element 1 would have been the jth element chosen, $j = 1, \ldots, k$; or by noting that the proportion of outcomes of the random selection that results in element 1 being chosen is $\binom{1}{1}\binom{n-1}{k-1}/\binom{n}{k} = k/n$). Therefore, we have that

$$P\{I_1 = 1\} = k/n \qquad (5.4.1)$$

To compute the conditional probability that element 2 is in the subset given I_1, note that if $I_1 = 1$, then aside from element 1 the remaining $k-1$ members of the subset would have been chosen "at random" from the remaining $n-1$ elements (in the sense that each of the subsets of size $k - 1$ of the numbers $2, \ldots, n$ is equally likely to be the other elements of the subset). Hence, we have that

$$P\{I_2 = 1 | I_1 = 1\} = \frac{k - 1}{n - 1} \tag{5.4.2}$$

Similarly, if element 1 is not in the subgroup, then the k members of the subgroup would have been chosen "at random" from the other $n - 1$ elements, and thus

$$P\{I_2 = 1 | I_1 = 0\} = \frac{k}{n - 1} \tag{5.4.3}$$

From Equations 5.4.2 and 5.4.3, we see that

$$P\{I_2 = 1 | I_1\} = \frac{k - I_1}{n - 1}$$

In general, we have that

$$P\{I_j = 1 | I_1, \ldots, I_{j-1}\} = \frac{k - \sum_{i=1}^{j-1} I_i}{n - j + 1}, \quad j = 2, \ldots, n \tag{5.4.4}$$

The preceding formula follows since $\sum_{i=1}^{j-1} I_i$ represents the number of the first $j - 1$ elements that are included in the subset, and so given $I_1, \ldots, I_{j-1}$ there remain $k - \sum_{i=1}^{j-1} I_i$ elements to be selected from the remaining $n - (j - 1)$.

Since $P\{U < a\} = a, 0 \leq a \leq 1$, when U is a uniform $(0, 1)$ random variable, Equations 5.4.1 and 5.4.4 lead to the following method for generating a random subset of size k from a set of n elements: Namely, generate a sequence of (at most n) random numbers $U_1, U_2, \ldots$ and set

$$I_1 = \begin{cases} 1 & \text{if } U_1 < \dfrac{k}{n} \\ 0 & \text{otherwise} \end{cases}$$

$$I_2 = \begin{cases} 1 & \text{if } U_2 < \dfrac{k - I_1}{n - 1} \\ 0 & \text{otherwise} \end{cases}$$

$$\vdots$$

$$I_j = \begin{cases} 1 & \text{if } U_j < \dfrac{k - I_1 - \cdots - I_{j-1}}{n - j + 1} \\ 0 & \text{otherwise} \end{cases}$$

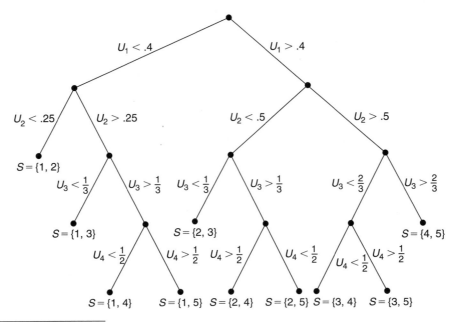

FIGURE 5.6   *Tree diagram.*

This process stops when $I_1 + \cdots + I_j = k$ and the random subset consists of the k elements whose I-value equals 1. That is, $S = \{i : I_i = 1\}$ is the subset.

For instance, if $k = 2$, $n = 5$, then the tree diagram of Figure 5.6 illustrates the foregoing technique. The random subset S is given by the final position on the tree. Note that the probability of ending up in any given final position is equal to $1/10$, which can be seen by multiplying the probabilities of moving through the tree to the desired endpoint. For instance, the probability of ending at the point labeled $S = \{2, 4\}$ is $P\{U_1 > .4\}P\{U_2 < .5\}P\{U_3 > \frac{1}{3}\}P\{U_4 > \frac{1}{2}\} = (.6)(.5)\left(\frac{2}{3}\right)\left(\frac{1}{2}\right) = .1.$

As indicated in the tree diagram (see the rightmost branches that result in $S = \{4, 5\}$), we can stop generating random numbers when the number of remaining places in the subset to be chosen is equal to the remaining number of elements. That is, the general procedure would stop whenever either $\sum_{i=1}^{j} I_i = k$ or $\sum_{i=1}^{j} I_i = k - (n - j)$. In the latter case, $S = \{i \leq j : I_i = 1, j + 1, \ldots, n\}$. ∎

EXAMPLE 5.4e   The random vector X, Y is said to have a *uniform* distribution over the two-dimensional region R if its joint density function is constant for points in R, and is 0 for points outside of R. That is, if

$$f(x, y) = \begin{cases} c & \text{if } (x, y) \in R \\ 0 & \text{if otherwise} \end{cases}$$

Because

$$1 = \int_R f(x, y) \, dx \, dy$$
$$= \int_R c \, dx \, dy$$
$$= c \times \text{Area of } R$$

it follows that

$$c = \frac{1}{\text{Area of } R}$$

For any region $A \subset R$,

$$P\{(X, Y) \in A\} = \int \int_{(x,y) \in A} f(x, y) \, dx \, dy$$
$$= \int \int_{(x,y) \in A} c \, dx \, dy$$
$$= \frac{\text{Area of } A}{\text{Area of } R}$$

Suppose now that X, Y is uniformly distributed over the following rectangular region R:

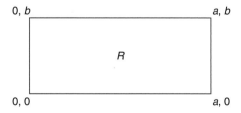

Its joint density function is

$$f(x, y) = \begin{cases} c & \text{if } 0 \leq x \leq a, \ 0 \leq y \leq b \\ 0 & \text{otherwise} \end{cases}$$

where $c = \frac{1}{\text{Area of rectangle}} = \frac{1}{ab}$. In this case, X and Y are independent uniform random variables. To show this, note that for $0 \leq x \leq a, 0 \leq y \leq b$

$$P\{X \leq x, Y \leq y\} = c \int_0^x \int_0^y dy \, dx = \frac{xy}{ab} \qquad (5.4.5)$$

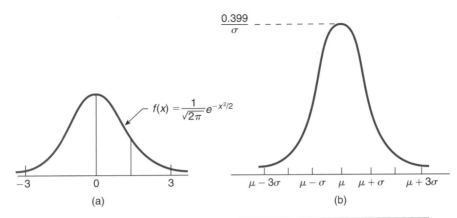

FIGURE 5.7   *The normal density function (a) with $\mu = 0, \sigma = 1$ and (b) with arbitrary μ and σ^2.*

First letting $y = b$, and then letting $x = a$, in the preceding shows that

$$P\{X \leq x\} = \frac{x}{a}, \quad P\{Y \leq y\} = \frac{y}{b} \qquad (5.4.6)$$

Thus, from Equations 5.4.5 and 5.4.6 we can conclude that X and Y are independent, with X being uniform on $(0, a)$ and Y being uniform on $(0, b)$. ■

5.5 NORMAL RANDOM VARIABLES

A random variable is said to be normally distributed with parameters μ and σ^2, and we write $X \sim \mathcal{N}(\mu, \sigma^2)$, if its density is

$$f(x) = \frac{1}{\sqrt{2\pi}\,\sigma} e^{-(x-\mu)^2/2\sigma^2}, \qquad -\infty < x < \infty^*$$

The normal density $f(x)$ is a bell-shaped curve that is symmetric about μ and that attains its maximum value of $\frac{1}{\sqrt{2\pi}\,\sigma} \approx 0.399/\sigma$ at $x = \mu$ (see Figure 5.7).

The normal distribution was introduced by the French mathematician Abraham de Moivre in 1733 and was used by him to approximate probabilities associated with binomial random variables when the binomial parameter n is large. This result was later extended by Laplace and others and is now encompassed in a probability theorem known as the central limit theorem, which gives a theoretical base to the often noted empirical observation that, in practice, many random phenomena obey, at least approximately, a normal probability distribution. Some examples of this behavior are the height of a person, the velocity in any direction of a molecule in gas, and the error made in measuring a physical quantity.

---

* To verify that this is indeed a density function, see Problem 29.

To compute $E[X]$ note that

$$E[X - \mu] = \frac{1}{\sqrt{2\pi}\sigma} \int_{-\infty}^{\infty} (x - \mu) e^{-(x-\mu)^2/2\sigma^2} dx$$

Letting $y = (x - \mu)/\sigma$ gives that

$$E[X - \mu] = \frac{\sigma}{\sqrt{2\pi}} \int_{-\infty}^{\infty} y e^{-y^2/2} dy$$

But

$$\int_{-\infty}^{\infty} y e^{-y^2/2} dy = -e^{-y^2/2}\big|_{-\infty}^{\infty} = 0$$

showing that $E[X - \mu] = 0$, or equivalently that

$$E[X] = \mu$$

Using this, we now compute $\text{Var}(X)$ as follows:

$$\begin{aligned}
\text{Var}(X) &= E[(X - \mu)^2] \\
&= \frac{1}{\sqrt{2\pi}\sigma} \int_{-\infty}^{\infty} (x - \mu)^2 e^{-(x-\mu)^2/2\sigma^2} dx \\
&= \frac{1}{\sqrt{2\pi}} \int_{-\infty}^{\infty} \sigma^2 y^2 e^{-y^2/2} dy
\end{aligned} \qquad (5.5.1)$$

With $u = y$ and $dv = y e^{-y^2/2}$, the integration by parts formula

$$\int u \, dv = uv - \int v \, du$$

yields that

$$\int_{-\infty}^{\infty} y^2 e^{-y^2/2} dy = -y e^{-y^2/2}\big|_{-\infty}^{\infty} + \int_{-\infty}^{\infty} e^{-y^2/2} dy$$

$$= \int_{-\infty}^{\infty} e^{-y^2/2} dy$$

Hence, from (5.5.1)

$$\begin{aligned}
\text{Var}(X) &= \sigma^2 \frac{1}{\sqrt{2\pi}} \int_{-\infty}^{\infty} e^{-y^2/2} dy \\
&= \sigma^2
\end{aligned}$$

where the preceding used that $\frac{1}{\sqrt{2\pi}}e^{-y^2/2}dy$ is the density function of a normal random variable with parameters $\mu = 0$ and $\sigma = 1$, so its integral must equal 1.

Thus μ and σ^2 represent, respectively, the mean and variance of the normal distribution.

A very important property of normal random variables is that if X is normal with mean μ and variance σ^2, then for any constants a and $b, b \neq 0$, the random variable $Y = a + bX$ is also a normal random variable with parameters

$$E[Y] = E[a + bX] = a + bE[X] = a + b\mu$$

and variance

$$\text{Var}(Y) = \text{Var}(a + bX) = b^2\text{Var}(X) = b^2\sigma^2$$

To verify this, let $F_Y(y)$ be the distribution function of Y. Then, for $b > 0$

$$F_Y(y) = P(Y \leq y)$$
$$= P(a + bX \leq y)$$
$$= P\left(X \leq \frac{y-a}{b}\right)$$
$$= F_X\left(\frac{y-a}{b}\right)$$

where F_X is the distribution function of X. Similarly, if $b < 0$, then

$$F_Y(y) = P(a + bX \leq y)$$
$$= P\left(X \geq \frac{y-a}{b}\right)$$
$$= 1 - F_X\left(\frac{y-a}{b}\right)$$

Differentiation yields that the density function of Y is

$$f_Y(y) = \begin{cases} \frac{1}{b}f_X\left(\frac{y-a}{b}\right), & \text{if } b > 0 \\ -\frac{1}{b}f_X\left(\frac{y-a}{b}\right), & \text{if } b < 0 \end{cases}$$

which can be written as

$$f_Y(y) = \frac{1}{|b|}f_X\left(\frac{y-a}{b}\right)$$
$$= \frac{1}{\sqrt{2\pi}\sigma|b|}e^{-\left(\frac{y-a}{b}-\mu\right)^2/2\sigma^2}$$

$$= \frac{1}{\sqrt{2\pi}\,\sigma\,|b|}\,e^{-(y-a-b\mu)^2/2b^2\sigma^2}$$

showing that $Y = a + bX$ is normal with mean $a + b\mu$ and variance $b^2\sigma^2$.

It follows from the foregoing that if $X \sim \mathcal{N}(\mu, \sigma^2)$, then

$$Z = \frac{X - \mu}{\sigma}$$

is a normal random variable with mean 0 and variance 1. Such a random variable Z is said to have a *standard*, or *unit*, normal distribution. Let $\Phi(\cdot)$ denote its distribution function. That is,

$$\Phi(x) = \frac{1}{\sqrt{2\pi}} \int_{-\infty}^{x} e^{-y^2/2}\,dy, \qquad -\infty < x < \infty$$

This result that $Z = (X - \mu)/\sigma$ has a standard normal distribution when X is normal with parameters μ and σ^2 is quite important, for it enables us to write all probability statements about X in terms of probabilities for Z. For instance, to obtain $P\{X < b\}$, we note that X will be less than b if and only if $(X - \mu)/\sigma$ is less than $(b - \mu)/\sigma$, and so

$$P\{X < b\} = P\left\{ \frac{X - \mu}{\sigma} < \frac{b - \mu}{\sigma} \right\}$$

$$= \Phi\left(\frac{b - \mu}{\sigma} \right)$$

Similarly, for any $a < b$,

$$P\{a < X < b\} = P\left\{ \frac{a - \mu}{\sigma} < \frac{X - \mu}{\sigma} < \frac{b - \mu}{\sigma} \right\}$$

$$= P\left\{ \frac{a - \mu}{\sigma} < Z < \frac{b - \mu}{\sigma} \right\}$$

$$= P\left\{ Z < \frac{b - \mu}{\sigma} \right\} - P\left\{ Z < \frac{a - \mu}{\sigma} \right\}$$

$$= \Phi\left(\frac{b - \mu}{\sigma} \right) - \Phi\left(\frac{a - \mu}{\sigma} \right)$$

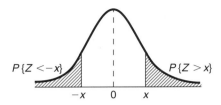

FIGURE 5.8   *Standard normal probabilities.*

It remains for us to compute $\Phi(x)$. This has been accomplished by an approxima-tion and the results are presented in Table A1 of the Appendix, which tabulates $\Phi(x)$ (to a 4-digit level of accuracy) for a wide range of nonnegative values of x. In addition, Program 5.5a of the text disk can be used to obtain $\Phi(x)$.

While Table A1 tabulates $\Phi(x)$ only for nonnegative values of x, we can also obtain $\Phi(-x)$ from the table by making use of the symmetry (about 0) of the standard normal probability density function. That is, for $x > 0$, if Z represents a standard normal random variable, then (see Figure 5.8)

$$\Phi(-x) = P\{Z < -x\}$$
$$= P\{Z > x\} \quad \text{by symmetry}$$
$$= 1 - \Phi(x)$$

Thus, for instance,

$$P\{Z < -1\} = \Phi(-1) = 1 - \Phi(1) = 1 - .8413 = .1587$$

EXAMPLE 5.5a   If X is a normal random variable with mean $\mu = 3$ and variance $\sigma^2 = 16$, find

(a) $P\{X < 11\}$;
(b) $P\{X > -1\}$;
(c) $P\{2 < X < 7\}$.

SOLUTION

(a)
$$P\{X < 11\} = P\left\{\frac{X - 3}{4} < \frac{11 - 3}{4}\right\}$$
$$= \Phi(2)$$
$$= .9772$$

(b)
$$P\{X > -1\} = P\left\{\frac{X - 3}{4} > \frac{-1 - 3}{4}\right\}$$
$$= P\{Z > -1\}$$

$$= P\{Z < 1\}$$

$$= .8413$$

(c) $\qquad\qquad P\{2 < X < 7\} = P\left\{\dfrac{2-3}{4} < \dfrac{X-3}{4} < \dfrac{7-3}{4}\right\}$

$$= \Phi(1) - \Phi(-1/4)$$

$$= \Phi(1) - (1 - \Phi(1/4))$$

$$= .8413 + .5987 - 1 = .4400 \quad\blacksquare$$

EXAMPLE 5.5b Suppose that a binary message — either "0" or "1" — must be transmitted by wire from location A to location B. However, the data sent over the wire are subject to a channel noise disturbance and so to reduce the possibility of error, the value 2 is sent over the wire when the message is "1" and the value -2 is sent when the message is "0." If $x, x = \pm 2$, is the value sent at location A then R, the value received at location B, is given by $R = x + N$, where N is the channel noise disturbance. When the message is received at location B, the receiver decodes it according to the following rule:

if $R \geq .5$, then "1" is concluded

if $R < .5$, then "0" is concluded

Because the channel noise is often normally distributed, we will determine the error probabilities when N is a standard normal random variable.

There are two types of errors that can occur: One is that the message "1" can be incorrectly concluded to be "0" and the other that "0" is incorrectly concluded to be "1." The first type of error will occur if the message is "1" and $2 + N < .5$, whereas the second will occur if the message is "0" and $-2 + N \geq .5$.

Hence,

$$P\{\text{error}|\text{message is "1"}\} = P\{N < -1.5\}$$
$$= 1 - \Phi(1.5) = .0668$$

and

$$P\{\text{error}|\text{message is "0"}\} = P\{N > 2.5\}$$
$$= 1 - \Phi(2.5) = .0062 \quad\blacksquare$$

EXAMPLE 5.5c The power W dissipated in a resistor is proportional to the square of the voltage V. That is,

$$W = rV^2$$

where r is a constant. If $r = 3$, and V can be assumed (to a very good approximation) to be a normal random variable with mean 6 and standard deviation 1, find

(a) $E[W]$;

(b) $P\{W > 120\}$.

SOLUTION

(a)
$$E[W] = E[3V^2]$$
$$= 3E[V^2]$$
$$= 3(\text{Var}[V] + E^2[V])$$
$$= 3(1 + 36) = 111$$

(b)
$$P\{W > 120\} = P\{3V^2 > 120\}$$
$$= P\{V > \sqrt{40}\}$$
$$= P\{V - 6 > \sqrt{40} - 6\}$$
$$= P\{Z > .3246\}$$
$$= 1 - \Phi(.3246)$$
$$= .3727 \quad \blacksquare$$

Let us now compute the moment generating function of a normal random variable. To start, we compute the moment generating function of a standard normal random variable Z.

$$
\begin{aligned}
E[e^{tZ}] &= \int_{-\infty}^{\infty} e^{tx} \frac{1}{\sqrt{2\pi}} e^{-x^2/2} \, dx \\
&= \frac{1}{\sqrt{2\pi}} \int_{-\infty}^{\infty} e^{-(x^2 - 2tx)/2} \, dx \\
&= e^{-t^2/2} \frac{1}{\sqrt{2\pi}} \int_{-\infty}^{\infty} e^{-(x-t)^2/2} \, dx \\
&= e^{-t^2/2} \frac{1}{\sqrt{2\pi}} \int_{-\infty}^{\infty} e^{-y^2/2} \, dy \\
&= e^{-t^2/2}
\end{aligned}
$$

Now, if Z is a standard normal, then $X = \mu + \sigma Z$ is normal with mean μ and variance σ^2. Using the preceding, its moment generating function is

$$
\begin{aligned}
E[e^{tX}] &= E[e^{t\mu + t\sigma Z}] \\
&= E[e^{t\mu} e^{t\sigma Z}] \\
&= e^{t\mu} E[e^{t\sigma Z}] \\
&= e^{t\mu} e^{-(\sigma t)^2/2} \\
&= e^{\mu t - \sigma^2 t^2/2}
\end{aligned}
$$

Another important result is that the sum of independent normal random variables is also a normal random variable. To see this, suppose that $X_i, i = 1, \ldots, n$, are independent, with X_i being normal with mean μ_i and variance σ_i^2. The moment generating function of $\sum_{i=1}^{n} X_i$ is as follows.

$$E\left[e^{t \sum_{i=1}^{n} X_i}\right] = E\left[e^{tX_1} e^{tX_2} \cdots e^{tX_n}\right]$$

$$= \prod_{i=1}^{n} E\left[e^{tX_i}\right] \qquad \text{by independence}$$

$$= \prod_{i=1}^{n} e^{\mu_i t + \sigma_i^2 t^2 / 2}$$

$$= e^{\mu t + \sigma^2 t^2 / 2}$$

where

$$\mu = \sum_{i=1}^{n} \mu_i, \qquad \sigma^2 = \sum_{i=1}^{n} \sigma_i^2$$

Therefore, $\sum_{i=1}^{n} X_i$ has the same moment generating function as a normal random variable having mean μ and variance σ^2. Hence, from the one-to-one correspondence between moment generating functions and distributions, we can conclude that $\sum_{i=1}^{n} X_i$ is normal with mean $\sum_{i=1}^{n} \mu_i$ and variance $\sum_{i=1}^{n} \sigma_i^2$.

EXAMPLE 5.5d Data from the National Oceanic and Atmospheric Administration indicate that the yearly precipitation in Los Angeles is a normal random variable with a mean of 12.08 inches and a standard deviation of 3.1 inches.

(a) Find the probability that the total precipitation during the next 2 years will exceed 25 inches.

(b) Find the probability that next year's precipitation will exceed that of the following year by more than 3 inches.

Assume that the precipitation totals for the next 2 years are independent.

SOLUTION Let X_1 and X_2 be the precipitation totals for the next 2 years.

(a) Since $X_1 + X_2$ is normal with mean 24.16 and variance $2(3.1)^2 = 19.22$, it follows that

$$P\{X_1 + X_2 > 25\} = P\left\{\frac{X_1 + X_2 - 24.16}{\sqrt{19.22}} > \frac{25 - 24.16}{\sqrt{19.22}}\right\}$$

$$= P\{Z > .1916\}$$

$$\approx .4240$$

(b) Since $-X_2$ is a normal random variable with mean -12.08 and variance $(-1)^2(3.1)^2$, it follows that $X_1 - X_2$ is normal with mean 0 and variance 19.22. Hence,

$$P\{X_1 > X_2 + 3\} = P\{X_1 - X_2 > 3\}$$
$$= P\left\{\frac{X_1 - X_2}{\sqrt{19.22}} > \frac{3}{\sqrt{19.22}}\right\}$$
$$= P\{Z > .6843\}$$
$$\approx .2469$$

Thus there is a 42.4 percent chance that the total precipitation in Los Angeles during the next 2 years will exceed 25 inches, and there is a 24.69 percent chance that next year's precipitation will exceed that of the following year by more than 3 inches. ∎

For $\alpha \in (0, 1)$, let z_α be such that

$$P\{Z > z_\alpha\} = 1 - \Phi(z_\alpha) = \alpha$$

That is, the probability that a standard normal random variable is greater than z_α is equal to α (see Figure 5.9).

The value of z_α can, for any α, be obtained from Table A1. For instance, since

$$1 - \Phi(1.645) = .05$$
$$1 - \Phi(1.96) = .025$$
$$1 - \Phi(2.33) = .01$$

it follows that

$$z_{.05} = 1.645, \qquad z_{.025} = 1.96, \qquad z_{.01} = 2.33$$

Program 5.5b on the text disk can also be used to obtain the value of z_α.

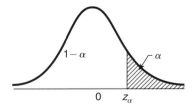

FIGURE 5.9 $P\{Z > z_\alpha\} = \alpha$.

Since

$$P\{Z < z_\alpha\} = 1 - \alpha$$

it follows that $100(1 - \alpha)$ percent of the time a standard normal random variable will be less than z_α. As a result, we call z_α the $100(1 - \alpha)$ *percentile* of the standard normal distribution.

5.6 EXPONENTIAL RANDOM VARIABLES

A continuous random variable whose probability density function is given, for some $\lambda > 0$, by

$$f(x) = \begin{cases} \lambda e^{-\lambda x} & \text{if } x \geq 0 \\ 0 & \text{if } x < 0 \end{cases}$$

is said to be an *exponential* random variable (or, more simply, is said to be exponentially distributed) with parameter λ. The cumulative distribution function $F(x)$ of an exponential random variable is given by

$$F(x) = P\{X \leq x\}$$

$$= \int_0^x \lambda e^{-\lambda y} \, dy$$

$$= 1 - e^{-\lambda x}, \qquad x \geq 0$$

The exponential distribution often arises, in practice, as being the distribution of the amount of time until some specific event occurs. For instance, the amount of time (starting from now) until an earthquake occurs, or until a new war breaks out, or until a telephone call you receive turns out to be a wrong number are all random variables that tend in practice to have exponential distributions (see Section 5.6.1 for an explanation).

The moment generating function of the exponential is given by

$$\phi(t) = E[e^{tX}]$$

$$= \int_0^\infty e^{tx} \lambda e^{-\lambda x} \, dx$$

$$= \lambda \int_0^\infty e^{-(\lambda - t)x} \, dx$$

$$= \frac{\lambda}{\lambda - t}, \qquad t < \lambda$$

Differentiation yields

$$\phi'(t) = \frac{\lambda}{(\lambda - t)^2}$$

$$\phi''(t) = \frac{2\lambda}{(\lambda - t)^3}$$

and so

$$E[X] = \phi'(0) = 1/\lambda$$

$$\text{Var}(X) = \phi''(0) - (E[X])^2$$

$$= 2/\lambda^2 - 1/\lambda^2$$

$$= 1/\lambda^2$$

Thus λ is the reciprocal of the mean, and the variance is equal to the square of the mean.

The key property of an exponential random variable is that it is memoryless, where we say that a nonnegative random variable X is *memoryless* if

$$P\{X > s + t | X > t\} = P\{X > s\} \qquad \text{for all } s, t \geq 0 \qquad (5.6.1)$$

To understand why Equation 5.6.1 is called the *memoryless property*, imagine that X represents the length of time that a certain item functions before failing. Now let us consider the probability that an item that is still functioning at age t will continue to function for at least an additional time s. Since this will be the case if the total functional lifetime of the item exceeds $t + s$ given that the item is still functioning at t, we see that

$$P\{\text{additional functional life of } t\text{-unit-old item exceeds } s\}$$
$$= P\{X > t + s | X > t\}$$

Thus, we see that Equation 5.6.1 states that the distribution of additional functional life of an item of age t is the same as that of a new item — in other words, when Equation 5.6.1 is satisfied, there is no need to remember the age of a functional item since as long as it is still functional it is "as good as new."

The condition in Equation 5.6.1 is equivalent to

$$\frac{P\{X > s + t, X > t\}}{P\{X > t\}} = P\{X > s\}$$

or

$$P\{X > s + t\} = P\{X > s\}P\{X > t\} \qquad (5.6.2)$$

When X is an exponential random variable, then

$$P\{X > x\} = e^{-\lambda x}, \qquad x > 0$$

and so Equation 5.6.2 is satisfied (since $e^{-\lambda(s+t)} = e^{-\lambda s}e^{-\lambda t}$). Hence, *exponentially distributed random variables are memoryless* (and in fact it can be shown that they are the only random variables that are memoryless).

EXAMPLE 5.6a Suppose that a number of miles that a car can run before its battery wears out is exponentially distributed with an average value of 10,000 miles. If a person desires to take a 5,000-mile trip, what is the probability that she will be able to complete her trip without having to replace her car battery? What can be said when the distribution is not exponential?

SOLUTION It follows, by the memoryless property of the exponential distribution, that the remaining lifetime (in thousands of miles) of the battery is exponential with parameter $\lambda = 1/10$. Hence the desired probability is

$$P\{\text{remaining lifetime} > 5\} = 1 - F(5)$$
$$= e^{-5\lambda}$$
$$= e^{-1/2} \approx .604$$

However, if the lifetime distribution F is not exponential, then the relevant probability is

$$P\{\text{lifetime} > t + 5 | \text{lifetime} > t\} = \frac{1 - F(t+5)}{1 - F(t)}$$

where t is the number of miles that the battery had been in use prior to the start of the trip. Therefore, if the distribution is not exponential, additional information is needed (namely, t) before the desired probability can be calculated. ■

For another illustration of the memoryless property, consider the following example.

EXAMPLE 5.6b A crew of workers has 3 interchangeable machines, of which 2 must be working for the crew to do its job. When in use, each machine will function for an exponentially distributed time having parameter λ before breaking down. The workers decide initially to use machines A and B and keep machine C in reserve to replace whichever of A or B breaks down first. They will then be able to continue working until one of the remaining machines breaks down. When the crew is forced to stop working because only one of the machines has not yet broken down, what is the probability that the still operable machine is machine C?

SOLUTION This can be easily answered, without any need for computations, by invoking the memoryless property of the exponential distribution. The argument is as follows: Consider the moment at which machine C is first put in use. At that time either A or B would have just broken down and the other one — call it machine 0 — will still be functioning. Now even though 0 would have already been functioning for some time, by the memoryless property of the exponential distribution, it follows that its remaining lifetime has the same distribution as that of a machine that is just being put into use. Thus, the remaining lifetimes of machine 0 and machine C have the same distribution and so, by symmetry, the probability that 0 will fail before C is $\frac{1}{2}$. ■

The following proposition presents another property of the exponential distribution.

PROPOSITION 5.6.1 If $X_1, X_2, \ldots, X_n$ are independent exponential random variables having respective parameters $\lambda_1, \lambda_2, \ldots, \lambda_n$, then $\min(X_1, X_2, \ldots, X_n)$ is exponential with parameter $\sum_{t=1}^{n} \lambda_i$.

Proof

Since the smallest value of a set of numbers is greater than x if and only if all values are greater than x, we have

$$
\begin{aligned}
P\{\min(X_1, X_2, \ldots, X_n) > x\} &= P\{X_1 > x, X_2 > x, \ldots, X_n > x\} \\
&= \prod_{i=1}^{n} P\{X_i > x\} \quad \text{by independence} \\
&= \prod_{i=1}^{n} e^{-\lambda_i x} \\
&= e^{-\sum_{i=1}^{n} \lambda_i x} \quad ■
\end{aligned}
$$

EXAMPLE 5.6c A series system is one that needs all of its components to function in order for the system itself to be functional. For an n-component series system in which the component lifetimes are independent exponential random variables with respective parameters $\lambda_1, \lambda_2, \ldots, \lambda_n$, what is the probability that the system survives for a time t?

SOLUTION Since the system life is equal to the minimal component life, it follows from Proposition 5.6.1 that

$$ P\{\text{system life exceeds } t\} = e^{-\sum_i \lambda_i t} \quad ■ $$

Another useful property of exponential random variables is that cX is exponential with parameter λ/c when X is exponential with parameter λ, and $c > 0$. This follows since

$$
\begin{aligned}
P\{cX \le x\} &= P\{X \le x/c\} \\
&= 1 - e^{-\lambda x/c}
\end{aligned}
$$

The parameter λ is called the *rate* of the exponential distribution.

*5.6.1 THE POISSON PROCESS

Suppose that "events" are occurring at random time points, and let $N(t)$ denote the number of events that occurs in the time interval $[0, t]$. These events are said to constitute a *Poisson process having rate* λ, $\lambda > 0$, if

(a) $N(0) = 0$

(b) The number of events that occur in disjoint time intervals are independent.

(c) The distribution of the number of events that occur in a given interval depends only on the length of the interval and not on its location.

(d) $\displaystyle\lim_{h \to 0} \frac{P\{N(h) = 1\}}{h} = \lambda$

(e) $\displaystyle\lim_{h \to 0} \frac{P\{N(h) \geq 2\}}{h} = 0$

Thus, Condition (a) states that the process begins at time 0. Condition (b), the *independent increment* assumption, states for instance that the number of events by time t [that is, $N(t)$] is independent of the number of events that occurs between t and $t + s$ [that is, $N(t + s) - N(t)$]. Condition (c), the *stationary increment* assumption, states that probability distribution of $N(t + s) - N(t)$ is the same for all values of t. Conditions (d) and (e) state that in a small interval of length h, the probability of one event occurring is approximately λh, whereas the probability of 2 or more is approximately 0.

We will now show that these assumptions imply that the number of events occurring in any interval of length t is a Poisson random variable with parameter λt. To be precise, let us call the interval $[0, t]$ and denote by $N(t)$ the number of events occurring in that interval. To obtain an expression for $P\{N(t) = k\}$, we start by breaking the interval $[0, t]$ into n nonoverlapping subintervals each of length t/n (Figure 5.10). Now there will be k events in $[0, t]$ if either

(i) $N(t)$ equals k and there is at most one event in each subinterval;

(ii) $N(t)$ equals k and at least one of the subintervals contains 2 or more events.

Since these two possibilities are clearly mutually exclusive, and since Condition (i) is equivalent to the statement that k of the n subintervals contain exactly 1 event and the other $n - k$ contain 0 events, we have that

$$P\{N(t) = k\} = P\{k \text{ of the } n \text{ subintervals contain exactly 1 event} \qquad (5.6.3)$$
$$\text{and the other } n - k \text{ contain 0 events}\} + P\{N(t) = k$$
$$\text{and at least 1 subinterval contains 2 or more events}\}$$

FIGURE 5.10

---

* Optional section.

Now it can be shown, using Condition (e), that

$$P\{N(t) = k \text{ and at least 1 subinterval contains 2 or more events}\}$$
$$\longrightarrow 0 \text{ as } n \to \infty \tag{5.6.4}$$

Also, it follows from Conditions (d) and (e) that

$$P\{\text{exactly 1 event in a subinterval}\} \approx \frac{\lambda t}{n}$$

$$P\{\text{0 events in a subinterval}\} \approx 1 - \frac{\lambda t}{n}$$

Hence, since the number of events that occur in different subintervals are independent [from Condition (b)], it follows that

$$P\{k \text{ of the subintervals contain exactly 1 event and the other } n - k \text{ contain 0 events}\}$$

$$\approx \binom{n}{k} \left(\frac{\lambda t}{n}\right)^k \left(1 - \frac{\lambda t}{n}\right)^{n-k} \tag{5.6.5}$$

with the approximation becoming exact as the number of subintervals, n, goes to ∞. However, the probability in Equation 5.6.5 is just the probability that a binomial random variable with parameters n and $p = \lambda t/n$ equals k. Hence, as n becomes larger and larger, this approaches the probability that a Poisson random variable with mean $n\lambda t/n = \lambda t$ equals k. Hence, from Equations 5.6.3, 5.6.4, and 5.6.5, we see upon letting n approach ∞ that

$$P\{N(t) = k\} = e^{-\lambda t}\frac{(\lambda t)^k}{k!}$$

We have shown:

PROPOSITION 5.6.2   For a Poisson process having rate λ

$$P\{N(t) = k\} = e^{-\lambda t}\frac{(\lambda t)^k}{k!}, \quad k = 0, 1, \dots$$

That is, the number of events in any interval of length t has a Poisson distribution with mean λt.

For a Poisson process, let X_1 denote the time of the first event. Further, for $n > 1$, let X_n denote the elapsed time between $(n - 1)$st and the nth events. The sequence $\{X_n, n = 1, 2, \dots\}$ is called the *sequence of interarrival times*. For instance, if $X_1 = 5$ and $X_2 = 10$, then the first event of the Poisson process would have occurred at time 5 and the second at time 15.

We now determine the distribution of the X_n. To do so, we first note that the event $\{X_1 > t\}$ takes place if and only if no events of the Poisson process occur in the interval $[0, t]$, and thus,

$$P\{X_1 > t\} = P\{N(t) = 0\} = e^{-\lambda t}$$

Hence, X_1 has an exponential distribution with mean $1/\lambda$. To obtain the distribution of X_2, note that

$$
\begin{aligned}
P\{X_2 > t | X_1 = s\} &= P\{0 \text{ events in } (s, s + t] | X_1 = s\} \\
&= P\{0 \text{ events in } (s, s + t]\} \\
&= e^{-\lambda t}
\end{aligned}
$$

where the last two equations followed from independent and stationary increments. There-fore, from the foregoing we see that X_2 is also an exponential random variable with mean $1/\lambda$, and furthermore, that X_2 is independent of X_1. Repeating the same argument yields:

PROPOSITION 5.6.3 $X_1, X_2, \ldots$ are independent exponential random variables, each with mean $1/\lambda$.

*5.6.2 THE PARETO DISTRIBUTION

If X is an exponential random variable with rate λ, then

$$Y = \alpha\, e^X$$

is said to be a *Pareto* random variable with parameters α and λ. The parameter $\lambda > 0$ is called the index parameter, and α is called the minimum parameter (because $P(Y \geq \alpha) = 1$). The distribution function of Y is derived as follows: For $y \geq \alpha$,

$$
\begin{aligned}
P\{Y > y\} &= P\{\alpha\, e^X > y\} \\
&= P\{e^X > y/\alpha\} \\
&= P\{X > \log(y/\alpha)\} \\
&= e^{-\lambda \log(y/\alpha)} \\
&= e^{-\log((y/\alpha)^\lambda)} \\
&= (\alpha/y)^\lambda
\end{aligned}
$$

Hence, the distribution function of Y is

$$F_Y(y) = 1 - P(Y > y) = 1 - \alpha^\lambda y^{-\lambda}, \quad y \geq \alpha$$

Differentiating the distribution function yields the density function of Y:

$$f_Y(y) = \lambda \alpha^\lambda y^{-(\lambda+1)}, \quad y \geq \alpha$$

---

* Optional section.

It can be shown (see Problem 5-49) that $E[Y] = \infty$ when $\lambda \leq 1$. When $\lambda > 1$, the mean is obtained as follows.

$$
\begin{aligned}
E[Y] &= \int_{\alpha}^{\infty} y \lambda \alpha^{\lambda} y^{-(\lambda+1)} \, dy \\
&= \lambda \alpha^{\lambda} \int_{\alpha}^{\infty} y^{-\lambda} \, dy \\
&= \alpha^{\lambda} \frac{\lambda}{1 - \lambda} y^{1-\lambda} |_{\alpha}^{\infty} \\
&= \alpha^{\lambda} \frac{\lambda}{\lambda - 1} \alpha^{1-\lambda} \\
&= \alpha \frac{\lambda}{\lambda - 1}
\end{aligned}
$$

An important feature of Pareto distributions is that for $y_0 > \alpha$ the conditional distribution of a Pareto random variable Y with parameters α and λ, given that it exceeds y_0, is the Pareto distribution with parameters y_0 and λ. To verify this, note for $y > y_0$ that

$$
P\{Y > y | Y > y_0\} = \frac{P\{Y > y, \, Y > y_0\}}{P\{Y > y_0\}} = \frac{P\{Y > y\}}{P\{Y > y_0\}} = \frac{\alpha^{\lambda} y^{-\lambda}}{\alpha^{\lambda} y_0^{-\lambda}} = y_0^{\lambda} y^{-\lambda}
$$

Thus, the conditional distribution is indeed Pareto with parameters y_0 and λ.

One of the earliest uses of the Pareto was as the distribution of the annual income of the members of a population. In fact, it has been widely supposed that incomes in many populations can be modeled as coming from a Pareto distribution with index parameter $\lambda = \log(5)/\log(4) \approx 1.161$. Under this supposition, it turns out that the total income of the top 20 percent of earners is 80 percent of the population's total income earnings, and that the top 20 percent of these high earners earn 80 percent of the total of all high earners income, and that the top 20 percent of these very high earners earn 80 percent of the total of all very high earners income, and so on.

To verify the preceding claim, let $y_{.8}$ be the 80 percentile of the Pareto distribution. Because $F_Y(y) = 1 - (\alpha/y)^{\lambda}$, we see that $.8 = F(y_{.8}) = 1 - (\alpha/y_{.8})^{\lambda}$, showing that

$$
(\alpha/y_{.8})^{\lambda} = .2 \quad \text{or} \quad (y_{.8}/\alpha)^{\lambda} = 5
$$

and thus

$$
y_{.8} = \alpha \, 5^{1/\lambda}
$$

Now suppose, from this point on, that $\lambda = \log(5)/\log(4)$, and note that $\log(4) = (1/\lambda)\log(5) = \log(5^{1/\lambda})$, showing that $4 = 5^{1/\lambda}$, or equivalently that $1/\lambda = \log_5(4)$. Hence,

$$
y_{.8} = \alpha \, 5^{\log_5(4)} = 4\alpha
$$

The average income of a randomly chosen individual in the top 20 percent is $E[Y | Y > y_{.8}]$, which is easily obtained by using that the conditional distribution of Y given that it exceeds $y_{.8}$ is Pareto with parameters $y_{.8}$ and λ. Using the previously derived formula for $E[Y]$, this yields that

$$E[Y|Y > y_{.8}] = y_{.8}\frac{\lambda}{\lambda - 1} = 4\alpha\,\frac{\lambda}{\lambda - 1}$$

To obtain $E[Y|Y < y_{.8}]$, the average income of a randomly chosen individual in the bottom 80 percent, we use the identity

$$E[Y] = E[Y|Y < y_{.8}](.8) + E[Y|Y > y_{.8}](.2)$$

Using the previously derived expressions for $E[Y]$ and $E[Y|Y > y_{.8}]$, the preceding equation yields that

$$\alpha\frac{\lambda}{\lambda - 1} = \frac{4}{5}E[Y|Y < y_{.8}] + \frac{4}{5}\alpha\,\frac{\lambda}{\lambda - 1}$$

showing that

$$E[Y|Y < y_{.8}] = \frac{\alpha}{4}\frac{\lambda}{\lambda - 1}$$

Thus,

$$E[Y|Y < y_{.8}] = \frac{1}{16}E[Y|Y > y_{.8}]$$

Hence, the average earnings of someone in the top 20 percent of income earned is 16 times that of someone in the lower 80 percent, thus showing that, although there are 4 times as many people in the lower earning group, the total income of the lower income group is only 20 percent of the total earnings of the population. (On average, for every 5 people in the population, 4 are in the lower 80 percent and 1 is in the upper 20 percent; the 4 in the lower earnings group earn on average a total of $4\frac{a}{4}\frac{\lambda}{\lambda-1} = a\frac{\lambda}{\lambda-1}$, whereas the one in the higher income group earns on average $4a\frac{\lambda}{\lambda-1}$. Thus, 4 out of every 5 dollars of the population's total income is earned by someone in the highest 20 percent.)

Because the conditional distribution of a high income earner (that is, one who earns more than $y_{.8}$) is Pareto with parameters $y_{.8}$ and λ, it also follows from the preceding that 80 percent of the total of the earnings of this group are earned by the top 20 percent of these high earners, and so on.

The Pareto distribution has been applied in a variety of areas. For instance, it has been used as the distribution of

(a) the file size of internet traffic (under the TCP protocol);

(b) the time to compete a job assigned to a supercomputer;

(c) the size of a meteorite;

(d) the yearly maximum one day rainfalls in different regions.

*5.7 THE GAMMA DISTRIBUTION

A random variable is said to have a gamma distribution with parameters $(\alpha, \lambda), \lambda > 0$, $\alpha > 0$, if its density function is given by

$$f(x) = \begin{cases} \dfrac{\lambda e^{-\lambda x}(\lambda x)^{\alpha-1}}{\Gamma(\alpha)} & x \geq 0 \\ 0 & x < 0 \end{cases}$$

where

$$\Gamma(\alpha) = \int_0^\infty \lambda e^{-\lambda x}(\lambda x)^{\alpha-1}\, dx$$

$$= \int_0^\infty e^{-y} y^{\alpha-1}\, dy \quad \text{(by letting } y = \lambda x)$$

The integration by parts formula $\int u\, dv = uv - \int v\, du$ yields, with $u = y^{\alpha-1}$, $dv = e^{-y} dy$, $v = -e^{-y}$, that for $\alpha > 1$,

$$\int_0^\infty e^{-y} y^{\alpha-1}\, dy = -e^{-y} y^{\alpha-1}\Big|_{y=0}^{y=\infty} + \int_0^\infty e^{-y}(\alpha-1) y^{\alpha-2}\, dy$$

$$= (\alpha-1)\int_0^\infty e^{-y} y^{\alpha-2}\, dy$$

or

$$\Gamma(\alpha) = (\alpha-1)\Gamma(\alpha-1) \tag{5.7.1}$$

When α is an integer — say, $\alpha = n$ — we can iterate the foregoing to obtain that

$$\Gamma(n) = (n-1)\Gamma(n-1)$$
$$= (n-1)(n-2)\Gamma(n-2) \qquad \text{by letting } \alpha = n-1 \text{ in Eq. 5.7.1}$$
$$= (n-1)(n-2)(n-3)\Gamma(n-3) \qquad \text{by letting } \alpha = n-2 \text{ in Eq. 5.7.1}$$
$$\vdots$$
$$= (n-1)!\,\Gamma(1)$$

Because

$$\Gamma(1) = \int_0^\infty e^{-y}\, dy = 1$$

we see that

$$\Gamma(n) = (n-1)!$$

The function $\Gamma(\alpha)$ is called the *gamma* function.

---

* Optional section.

It should be noted that when $\alpha = 1$, the gamma distribution reduces to the exponential with mean $1/\lambda$.

The moment generating function of a gamma random variable X with parameters (α, λ) is obtained as follows:

$$\phi(t) = E[e^{tX}]$$

$$= \frac{\lambda^\alpha}{\Gamma(\alpha)} \int_0^\infty e^{tx} e^{-\lambda x} x^{\alpha-1} \, dx$$

$$= \frac{\lambda^\alpha}{\Gamma(\alpha)} \int_0^\infty e^{-(\lambda-t)x} x^{\alpha-1} \, dx$$

$$= \left(\frac{\lambda}{\lambda - t}\right)^\alpha \frac{1}{\Gamma(\alpha)} \int_0^\infty e^{-y} y^{\alpha-1} \, dy \quad [\text{by } y = (\lambda - t)x]$$

$$= \left(\frac{\lambda}{\lambda - t}\right)^\alpha \tag{5.7.2}$$

where the final equality used that $e^{-y} y^{\alpha-1}/\Gamma(\alpha)$ is a density function, and thus integrates to 1.

Differentiation of Equation 5.7.2 yields

$$\phi'(t) = \frac{\alpha \lambda^\alpha}{(\lambda - t)^{\alpha+1}}$$

$$\phi''(t) = \frac{\alpha(\alpha + 1)\lambda^\alpha}{(\lambda - t)^{\alpha+2}}$$

Hence,

$$E[X] = \phi'(0) = \frac{\alpha}{\lambda} \tag{5.7.3}$$

$$\mathrm{Var}(X) = E[X^2] - (E[X])^2$$

$$= \phi''(0) - \left(\frac{\alpha}{\lambda}\right)^2$$

$$= \frac{\alpha(\alpha + 1)}{\lambda^2} - \frac{\alpha^2}{\lambda^2} = \frac{\alpha}{\lambda^2} \tag{5.7.4}$$

An important property of the gamma is that if X_1 and X_2 are independent gamma random variables having respective parameters (α_1, λ) and (α_2, λ), then $X_1 + X_2$ is a gamma random variable with parameters $(\alpha_1 + \alpha_2, \lambda)$. This result easily follows since

$$\phi_{X_1+X_2}(t) = E[e^{t(X_1+X_2)}] \tag{5.7.5}$$

$$= \phi_{X_1}(t)\phi_{X_2}(t)$$

$$= \left(\frac{\lambda}{\lambda - t}\right)^{\alpha_1} \left(\frac{\lambda}{\lambda - t}\right)^{\alpha_2} \quad \text{from Equation 5.7.2}$$

$$= \left(\frac{\lambda}{\lambda - t}\right)^{\alpha_1 + \alpha_2}$$

which is seen to be the moment generating function of a gamma $(\alpha_1 + \alpha_2, \lambda)$ random variable. Since a moment generating function uniquely characterizes a distribution, the result entails.

The foregoing result easily generalizes to yield the following proposition.

PROPOSITION 5.7.1   If $X_i, i = 1, \ldots, n$ are independent gamma random variables with respective parameters (α_i, λ), then $\sum_{i=1}^{n} X_i$ is gamma with parameters $\sum_{i=1}^{n} \alpha_i, \lambda$.

Since the gamma distribution with parameters $(1, \lambda)$ reduces to the exponential with the rate λ, we have thus shown the following useful result.

Corollary 5.7.2

If $X_1, \ldots, X_n$ are independent exponential random variables, each having rate λ, then $\sum_{i=1}^{n} X_i$ is a gamma random variable with parameters (n, λ).

EXAMPLE 5.7a   The lifetime of a battery is exponentially distributed with rate λ. If a stereo cassette requires one battery to operate, then the total playing time one can obtain from a total of n batteries is a gamma random variable with parameters (n, λ).   ■

Figure 5.11 presents a graph of the gamma $(\alpha, 1)$ density for a variety of values of α. It should be noted that as α becomes large, the density starts to resemble the normal density. This is theoretically explained by the central limit theorem, which will be presented in the next chapter.

5.8 DISTRIBUTIONS ARISING FROM THE NORMAL

5.8.1 THE CHI-SQUARE DISTRIBUTION

Definition

If $Z_1, Z_2, \ldots, Z_n$ are independent standard normal random variables, then X, defined by

$$X = Z_1^2 + Z_2^2 + \cdots + Z_n^2 \tag{5.8.1}$$

is said to have a *chi-square distribution with n degrees of freedom*. We will use the notation

$$X \sim \chi_n^2$$

to signify that X has a chi-square distribution with n degrees of freedom.

The chi-square distribution has the additive property that if X_1 and X_2 are independent chi-square random variables with n_1 and n_2 degrees of freedom, respectively, then $X_1 + X_2$ is chi-square with $n_1 + n_2$ degrees of freedom. This can be formally shown either by the use of moment generating functions or, most easily, by noting that $X_1 + X_2$ is the sum of squares of $n_1 + n_2$ independent standard normals and thus has a chi-square distribution with $n_1 + n_2$ degrees of freedom.

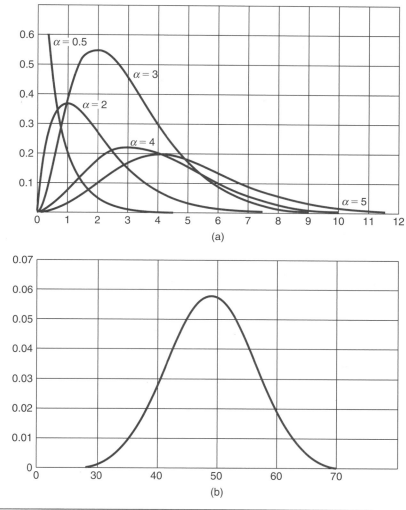

FIGURE 5.11   *Graphs of the gamma $(\alpha, 1)$ density for (a) $\alpha = .5, 2, 3, 4, 5$ and (b) $\alpha = 50$.*

If X is a chi-square random variable with n degrees of freedom, then for any $\alpha \in (0, 1)$, the quantity $\chi^2_{\alpha,n}$ is defined to be such that

$$P\{X \geq \chi^2_{\alpha,n}\} = \alpha$$

This is illustrated in Figure 5.12.

In Table A2 of the Appendix, we list $\chi^2_{\alpha,n}$ for a variety of values of α and n (including all those needed to solve problems and examples in this text). In addition, Programs 5.8.1a and 5.8.1b on the text disk can be used to obtain chi-square probabilities and the values of $\chi^2_{\alpha,n}$.

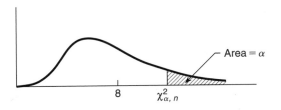

FIGURE 5.12   *The chi-square density function with 8 degrees of freedom.*

EXAMPLE 5.8a Determine $P\{\chi^2_{26} \leq 30\}$ when χ^2_{26} is a chi-square random variable with 26 degrees of freedom.

SOLUTION Using Program 5.8.1a gives the result

$$P\{\chi^2_{26} \leq 30\} = .7325 \quad \blacksquare$$

EXAMPLE 5.8b Find $\chi^2_{.05,15}$.

SOLUTION Use Program 5.8.1b to obtain:

$$\chi^2_{.05,15} = 24.996 \quad \blacksquare$$

EXAMPLE 5.8c Suppose that we are attempting to locate a target in three-dimensional space, and that the three coordinate errors (in meters) of the point chosen are independent normal random variables with mean 0 and standard deviation 2. Find the probability that the distance between the point chosen and the target exceeds 3 meters.

SOLUTION If D is the distance, then

$$D^2 = X_1^2 + X_2^2 + X_3^2$$

where X_i is the error in the ith coordinate. Since $Z_i = X_i/2, i = 1, 2, 3$, are all standard normal random variables, it follows that

$$P\{D^2 > 9\} = P\{Z_1^2 + Z_2^2 + Z_3^2 > 9/4\}$$

$$= P\{\chi^2_3 > 9/4\}$$

$$= .5222$$

where the final equality was obtained from Program 5.8.1a. $\blacksquare$

---

* Optional section.

*5.8.1.1 The Relation Between Chi-Square and Gamma Random Variables

Let us compute the moment generating function of a chi-square random variable with n degrees of freedom. To begin, we have, when $n = 1$, that

$$E[e^{tX}] = E[e^{tZ^2}] \text{ where } Z \sim \mathcal{N}(0, 1) \tag{5.8.2}$$

$$= \int_{-\infty}^{\infty} e^{tx^2} f_Z(x) \, dx$$

$$= \frac{1}{\sqrt{2\pi}} \int_{-\infty}^{\infty} e^{tx^2} e^{-x^2/2} \, dx$$

$$= \frac{1}{\sqrt{2\pi}} \int_{-\infty}^{\infty} e^{-x^2(1-2t)/2} \, dx$$

$$= \frac{1}{\sqrt{2\pi}} \int_{-\infty}^{\infty} e^{-x^2/2\bar{\sigma}^2} \, dx \quad \text{where } \bar{\sigma}^2 = (1 - 2t)^{-1}$$

$$= (1 - 2t)^{-1/2} \frac{1}{\sqrt{2\pi}\bar{\sigma}} \int_{-\infty}^{\infty} e^{-x^2/2\bar{\sigma}^2} \, dx$$

$$= (1 - 2t)^{-1/2}$$

where the last equality follows since the integral of the normal $(0, \bar{\sigma}^2)$ density equals 1. Hence, in the general case of n degrees of freedom

$$E[e^{tX}] = E\left[e^{t\sum_{i=1}^{n} Z_i^2}\right]$$

$$= E\left[\prod_{i=1}^{n} e^{tZ_i^2}\right]$$

$$= \prod_{i=1}^{n} E[e^{tZ_i^2}] \quad \text{by independence of the } Z_i$$

$$= \left(\frac{1/2}{1/2 - t}\right)^{n/2}$$

$$= (1 - 2t)^{-n/2} \quad \text{from Equation 5.8.2}$$

which we recognize as being the moment generating function of a gamma random variable with parameters $(n/2, 1/2)$. Hence, by the uniqueness of moment generating functions, it follows that these two distributions — chi-square with n degrees of freedom and gamma

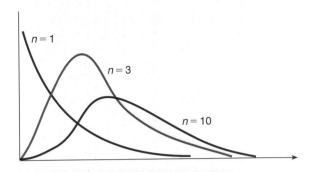

FIGURE 5.13   *The chi-square density function with n degrees of freedom.*

with parameters $n/2$ and $1/2$ — are identical, and thus we can conclude that the density of X is given by

$$f(x) = \frac{\frac{1}{2}e^{-x/2}\left(\frac{x}{2}\right)^{(n/2)-1}}{\Gamma\left(\frac{n}{2}\right)}, \quad x > 0$$

The chi-square density functions having 1, 3, and 10 degrees of freedom, respectively, are plotted in Figure 5.13.

Let us reconsider Example 5.8c, this time supposing that the target is located in the two-dimensional plane.

EXAMPLE 5.8d   When we attempt to locate a target in two-dimensional space, suppose that the coordinate errors are independent normal random variables with mean 0 and standard deviation 2. Find the probability that the distance between the point chosen and the target exceeds 3.

SOLUTION   If D is the distance and $X_i, i = 1, 2$, are the coordinate errors, then

$$D^2 = X_1^2 + X_2^2$$

Since $Z_i = X_i/2, i = 1, 2$, are standard normal random variables, we obtain

$$P\{D^2 > 9\} = P\{Z_1^2 + Z_2^2 > 9/4\} = P\{\chi_2^2 > 9/4\} = e^{-9/8} \approx .3247$$

where the preceding calculation used the fact that the chi-square distribution with 2 degrees of freedom is the same as the exponential distribution with parameter $1/2$. ∎

Since the chi-square distribution with n degrees of freedom is identical to the gamma distribution with parameters $\alpha = n/2$ and $\lambda = 1/2$, it follows from Equations 5.7.3 and 5.7.4 that the mean and variance of a random variable X having this distribution is

$$E[X] = n, \qquad \text{Var}(X) = 2n$$

5.8.2 The *t*-Distribution

If Z and χ_n^2 are independent random variables, with Z having a standard normal distribution and χ_n^2 having a chi-square distribution with n degrees of freedom, then the random variable T_n defined by

$$T_n = \frac{Z}{\sqrt{\chi_n^2/n}}$$

is said to have a *t-distribution with n degrees of freedom*. A graph of the density function of T_n is given in Figure 5.14 for $n = 1, 5$, and 10.

Like the standard normal density, the *t*-density is symmetric about zero. In addition, as n becomes larger, it becomes more and more like a standard normal density. To understand why, recall that χ_n^2 can be expressed as the sum of the squares of n standard normals, and so

$$\frac{\chi_n^2}{n} = \frac{Z_1^2 + \cdots + Z_n^2}{n}$$

where $Z_1, \ldots, Z_n$ are independent standard normal random variables. It now follows from the weak law of large numbers that, for large n, χ_n^2/n will, with probability close to 1, be approximately equal to $E[Z_i^2] = 1$. Hence, for n large, $T_n = Z/\sqrt{\chi_n^2/n}$ will have approximately the same distribution as Z.

Figure 5.15 shows a graph of the *t*-density function with 5 degrees of freedom compared with the standard normal density. Notice that the *t*-density has thicker "tails," indicating greater variability, than does the normal density.

The mean and variance of T_n can be shown to equal

$$E[T_n] = 0, \qquad n > 1$$
$$\mathrm{Var}(T_n) = \frac{n}{n-2}, \qquad n > 2$$

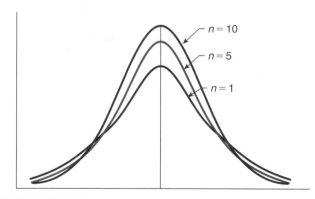

FIGURE 5.14 *Density function of T_n.*

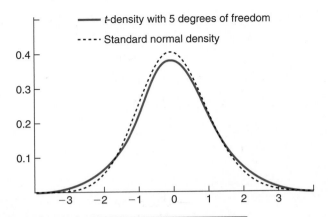

FIGURE 5.15   *Comparing standard normal density with the density of T_5.*

Thus the variance of T_n decreases to 1 — the variance of a standard normal random variable — as n increases to ∞. For $\alpha, 0 < \alpha < 1$, let $t_{\alpha,n}$ be such that

$$P\{T_n \geq t_{\alpha,n}\} = \alpha$$

It follows from the symmetry about zero of the t-density function that $-T_n$ has the same distribution as T_n, and so

$$
\begin{aligned}
\alpha &= P\{-T_n \geq t_{\alpha,n}\} \\
       &= P\{T_n \leq -t_{\alpha,n}\} \\
       &= 1 - P\{T_n > -t_{\alpha,n}\}
\end{aligned}
$$

Therefore,

$$P\{T_n \geq -t_{\alpha,n}\} = 1 - \alpha$$

leading to the conclusion that

$$-t_{\alpha,n} = t_{1-\alpha,n}$$

which is illustrated in Figure 5.16.

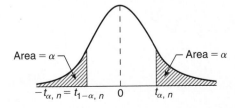

FIGURE 5.16 $t_{1-\alpha,n} = -t_{\alpha,n}.$

The values of $t_{\alpha,n}$ for a variety of values of n and α have been tabulated in Table A3 in the Appendix. In addition, Programs 5.8.2a and 5.8.2b on the text disk compute the t-distribution function and the values $t_{\alpha,n}$, respectively.

EXAMPLE 5.8e Find **(a)** $P\{T_{12} \leq 1.4\}$ and **(b)** $t_{.025,9}$.

SOLUTION Run Programs 5.8.2a and 5.8.2b to obtain the results.

   (a) .9066   **(b)** 2.2625 ■

5.8.3 THE *F*-DISTRIBUTION

If χ_n^2 and χ_m^2 are independent chi-square random variables with n and m degrees of freedom, respectively, then the random variable $F_{n,m}$ defined by

$$F_{n,m} = \frac{\chi_n^2/n}{\chi_m^2/m}$$

is said to have an *F-distribution with n and m degrees of freedom.*

For any $\alpha \in (0, 1)$, let $F_{\alpha,n,m}$ be such that

$$P\{F_{n,m} > F_{\alpha,n,m}\} = \alpha$$

This is illustrated in Figure 5.17.

The quantities $F_{\alpha,n,m}$ are tabulated in Table A4 of the Appendix for different values of n, m, and $\alpha \leq \frac{1}{2}$. If $F_{\alpha,n,m}$ is desired when $\alpha > \frac{1}{2}$, it can be obtained by using the following equalities:

$$\alpha = P\left\{\frac{\chi_n^2/n}{\chi_m^2/m} > F_{\alpha,n,m}\right\}$$

$$= P\left\{\frac{\chi_m^2/m}{\chi_n^2/n} < \frac{1}{F_{\alpha,n,m}}\right\}$$

$$= 1 - P\left\{\frac{\chi_m^2/m}{\chi_n^2/n} \geq \frac{1}{F_{\alpha,n,m}}\right\}$$

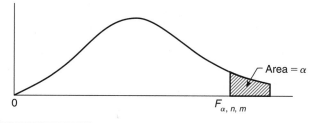

FIGURE 5.17 *Density function of $F_{n,m}$.*

or, equivalently,

$$P\left\{\frac{\chi_m^2/m}{\chi_n^2/n} \geq \frac{1}{F_{\alpha,n,m}}\right\} = 1 - \alpha \qquad (5.8.3)$$

But because $(\chi_m^2/m)/(\chi_n^2/n)$ has an F-distribution with degrees of freedom m and n, it follows that

$$1 - \alpha = P\left\{\frac{\chi_m^2/m}{\chi_n^2/n} \geq F_{1-\alpha,m,n}\right\}$$

implying, from Equation 5.8.3, that

$$\frac{1}{F_{\alpha,n,m}} = F_{1-\alpha,m,n}$$

Thus, for instance, $F_{.9,5,7} = 1/F_{.1,7,5} = 1/3.37 = .2967$ where the value of $F_{.1,7,5}$ was obtained from Table A4 of the Appendix.

Program 5.8.3 computes the distribution function of $F_{n,m}$.

EXAMPLE 5.8f   Determine $P\{F_{6,14} \leq 1.5\}$.

SOLUTION   Run Program 5.8.3 to obtain the solution .7518.   ■

*5.9 THE LOGISTICS DISTRIBUTION

A random variable X is said to have a *logistics* distribution with parameters μ and $v > 0$ if its distribution function is

$$F(x) = \frac{e^{(x-\mu)/v}}{1 + e^{(x-\mu)/v}}, \qquad -\infty < x < \infty$$

Differentiating $F(x) = 1 - 1/(1 + e^{(x-\mu)/v})$ yields the density function

$$f(x) = \frac{e^{(x-\mu)/v}}{v(1 + e^{(x-\mu)/v})^2}, \qquad -\infty < x < \infty$$

To obtain the mean of a logistics random variable,

$$E[X] = \int_{-\infty}^{\infty} x \frac{e^{(x-\mu)/v}}{v(1 + e^{(x-\mu)/v})^2} \, dx$$

---

* Optional section.

make the substitution $y = (x - \mu)/v$. This yields

$$E[X] = v \int_{-\infty}^{\infty} \frac{ye^y}{(1+e^y)^2} \, dy + \mu \int_{-\infty}^{\infty} \frac{e^y}{(1+e^y)^2} \, dy$$

$$= v \int_{-\infty}^{\infty} \frac{ye^y}{(1+e^y)^2} \, dy + \mu \qquad (5.9.1)$$

where the preceding equality used that $e^y/((1+e^y)^2)$ is the density function of a logistic random variable with parameters $\mu = 0$, $v = 1$ (such a random variable is called a *standard logistic*) and thus integrates to 1. Now,

$$\int_{-\infty}^{\infty} \frac{ye^y}{(1+e^y)^2} \, dy = \int_{-\infty}^{0} \frac{ye^y}{(1+e^y)^2} \, dy + \int_{0}^{\infty} \frac{ye^y}{(1+e^y)^2} \, dy$$

$$= -\int_{0}^{\infty} \frac{xe^{-x}}{(1+e^{-x})^2} \, dx + \int_{0}^{\infty} \frac{ye^y}{(1+e^y)^2} \, dy$$

$$= -\int_{0}^{\infty} \frac{xe^x}{(e^x+1)^2} \, dx + \int_{0}^{\infty} \frac{ye^y}{(1+e^y)^2} \, dy$$

$$= 0 \qquad (5.9.2)$$

where the second equality is obtained by making the substitution $x = -y$, and the third by multiplying the numerator and denominator by e^{2x}. From Equations 5.9.1 and 5.9.2 we obtain

$$E[X] = \mu$$

Thus μ is the mean of the logistic; v is called the dispersion parameter.

Problems

1. A satellite system consists of 4 components and can function adequately if at least 2 of the 4 components are in working condition. If each component is, independently, in working condition with probability .6, what is the probability that the system functions adequately?

2. A communications channel transmits the digits 0 and 1. However, due to static, the digit transmitted is incorrectly received with probability .2. Suppose that we want to transmit an important message consisting of one binary digit. To reduce the chance of error, we transmit 00000 instead of 0 and 11111 instead of 1. If the receiver of the message uses "majority" decoding, what is the probability

that the message will be incorrectly decoded? What independence assumptions are you making? (By majority decoding we mean that the message is decoded as "0" if there are at least three zeros in the message received and as "1" otherwise.)

3. If each voter is for Proposition A with probability .7, what is the probability that exactly 7 of 10 voters are for this proposition?

4. Suppose that a particular trait (such as eye color or left-handedness) of a person is classified on the basis of one pair of genes, and suppose that d represents a dominant gene and r a recessive gene. Thus, a person with dd genes is pure dominance, one with rr is pure recessive, and one with rd is hybrid. The pure dominance and the hybrid are alike in appearance. Children receive 1 gene from each parent. If, with respect to a particular trait, 2 hybrid parents have a total of 4 children, what is the probability that 3 of the 4 children have the outward appearance of the dominant gene?

5. At least one-half of an airplane's engines are required to function in order for it to operate. If each engine independently functions with probability p, for what values of p is a 4-engine plane more likely to operate than a 2-engine plane?

6. Let X be a binomial random variable with

$$E[X] = 7 \quad \text{and} \quad \text{Var}(X) = 2.1$$

Find

(a) $P\{X = 4\}$;
(b) $P\{X > 12\}$.

7. If X and Y are binomial random variables with respective parameters (n, p) and $(n, 1 - p)$, verify and explain the following identities:

(a) $P\{X \leq i\} = P\{Y \geq n - i\}$;
(a) $P\{X = k\} = P\{Y = n - k\}$.

8. If X is a binomial random variable with parameters n and p, where $0 < p < 1$, show that

(a) $P\{X = k + 1\} = \dfrac{p}{1 - p} \dfrac{n - k}{k + 1} P\{X = k\}, k = 0, 1, \ldots, n - 1.$
(b) As k goes from 0 to n, $P\{X = k\}$ first increases and then decreases, reaching its largest value when k is the largest integer less than or equal to $(n + 1)p$.

9. Derive the moment generating function of a binomial random variable and then use your result to verify the formulas for the mean and variance given in the text.

10. Compare the Poisson approximation with the correct binomial probability for the following cases:

  (a) $P\{X = 2\}$ when $n = 10, p = .1$;
  (b) $P\{X = 0\}$ when $n = 10, p = .1$;
  (c) $P\{X = 4\}$ when $n = 9, p = .2$.

11. If you buy a lottery ticket in 50 lotteries, in each of which your chance of winning a prize is $\frac{1}{100}$, what is the (approximate) probability that you will win a prize (a) at least once, (b) exactly once, and (c) at least twice?

12. The number of times that an individual contracts a cold in a given year is a Poisson random variable with parameter $\lambda = 3$. Suppose a new wonder drug (based on large quantities of vitamin C) has just been marketed that reduces the Poisson parameter to $\lambda = 2$ for 75 percent of the population. For the other 25 percent of the population, the drug has no appreciable effect on colds. If an individual tries the drug for a year and has 0 colds in that time, how likely is it that the drug is beneficial for him or her?

13. In the 1980s, an average of 121.95 workers died on the job each week. Give estimates of the following quantities:

  (a) the proportion of weeks having 130 deaths or more;
  (b) the proportion of weeks having 100 deaths or less.

  Explain your reasoning.

14. Approximately 80,000 marriages took place in the state of New York last year. Estimate the probability that for at least one of these couples

  (a) both partners were born on April 30;
  (b) both partners celebrated their birthday on the same day of the year.

  State your assumptions.

15. The game of frustration solitaire is played by turning the cards of a randomly shuffled deck of 52 playing cards over one at a time. Before you turn over the first card, say ace; before you turn over the second card, say two, before you turn over the third card, say three. Continue in this manner (saying ace again before turning over the fourteenth card, and so on). You lose if you ever turn over a card that matches what you have just said. Use the Poisson paradigm to approximate the probability of winning. (The actual probability is .01623.)

16. The probability of error in the transmission of a binary digit over a communication channel is $1/10^3$. Write an expression for the exact probability of more than 3 errors when transmitting a block of 10^3 bits. What is its approximate value? Assume independence.

17. If X is a Poisson random variable with mean λ, show that $P\{X = i\}$ first increases and then decreases as i increases, reaching its maximum value when i is the largest integer less than or equal to λ.

18. A contractor purchases a shipment of 100 transistors. It is his policy to test 10 of these transistors and to keep the shipment only if at least 9 of the 10 are in working condition. If the shipment contains 20 defective transistors, what is the probability it will be kept?

19. Let X denote a hypergeometric random variable with parameters n, m, and k. That is,

$$P\{X = i\} = \frac{\binom{n}{i}\binom{m}{k-i}}{\binom{n+m}{k}}, \qquad i = 0, 1, \ldots, \min(k, n)$$

(a) Derive a formula for $P\{X = i\}$ in terms of $P\{X = i - 1\}$.
(b) Use part (a) to compute $P\{X = i\}$ for $i = 0, 1, 2, 3, 4, 5$ when $n = m = 10$, $k = 5$, by starting with $P\{X = 0\}$.
(c) Based on the recursion in part (a), write a program to compute the hypergeometric distribution function.
(d) Use your program from part (c) to compute $P\{X \leq 10\}$ when $n = m = 30$, $k = 15$.

20. Independent trials, each of which is a success with probability p, are successively performed. Let X denote the first trial resulting in a success. That is, X will equal k if the first $k-1$ trials are all failures and the kth a success. X is called a *geometric random variable*. Compute

(a) $P\{X = k\}, k = 1, 2, \ldots$;
(b) $E[X]$.

Let Y denote the number of trials needed to obtain r successes. Y is called a *negative binomial random variable*. Compute

(c) $P\{Y = k\}, k = r, r + 1, \ldots$.

(*Hint*: In order for Y to equal k, how many successes must result in the first $k-1$ trials and what must be the outcome of trial k?)

(d) Show that

$$E[Y] = r/p$$

(*Hint*: Write $Y = Y_1 + \cdots + Y_r$ where Y_i is the number of trials needed to go from a total of $i - 1$ to a total of i successes.)

21. If U is uniformly distributed on $(0, 1)$, show that $a + (b - a)U$ is uniform on (a, b).

22. You arrive at a bus stop at 10 o'clock, knowing that the bus will arrive at some time uniformly distributed between 10 and 10:30. What is the probability that you will have to wait longer than 10 minutes? If at 10:15 the bus has not yet arrived, what is the probability that you will have to wait at least an additional 10 minutes?

23. If X is a normal random variable with parameters $\mu = 10, \sigma^2 = 36$, compute

    (a) $P\{X > 5\}$;
    (b) $P\{4 < X < 16\}$;
    (c) $P\{X < 8\}$;
    (d) $P\{X < 20\}$;
    (e) $P\{X > 16\}$.

24. The Scholastic Aptitude Test mathematics test scores across the population of high school seniors follow a normal distribution with mean 500 and standard deviation 100. If five seniors are randomly chosen, find the probability that (a) all scored below 600 and (b) exactly three of them scored above 640.

25. The annual rainfall (in inches) in a certain region is normally distributed with $\mu = 40, \sigma = 4$. What is the probability that in 2 of the next 4 years the rainfall will exceed 50 inches? Assume that the rainfalls in different years are independent.

26. The weekly demand for a product approximately has a normal distribution with mean 1,000 and standard deviation 200. The current on hand inventory is 2,200 and no deliveries will be occurring in the next two weeks. Assuming that the demands in different weeks are independent,

    (a) what is the probability that the demand in each of the next 2 weeks is less than 1,100?
    (b) what is the probability that the total of the demands in the next 2 weeks exceeds 2,200?

27. A certain type of lightbulb has an output that is normally distributed with mean 2,000 end foot candles and standard deviation 85 end foot candles. Determine a lower specification limit L so that only 5 percent of the lightbulbs produced will be below this limit. (That is, determine L so that $P\{X \geq L\} = .95$, where X is the output of a bulb.)

28. A manufacturer produces bolts that are specified to be between 1.19 and 1.21 inches in diameter. If its production process results in a bolt's diameter being normally distributed with mean 1.20 inches and standard deviation .005, what percentage of bolts will not meet specifications?

29. Let $I = \int_{-\infty}^{\infty} e^{-x^2/2}\, dx$.

(a) Show that for any μ and σ

$$\frac{1}{\sqrt{2\pi}\sigma} \int_{-\infty}^{\infty} e^{-(x-\mu)^2/2\sigma^2}\, dx = 1$$

is equivalent to $I = \sqrt{2\pi}$.

(b) Show that $I = \sqrt{2\pi}$ by writing

$$I^2 = \int_{-\infty}^{\infty} e^{-x^2/2}\, dx \int_{-\infty}^{\infty} e^{-y^2/2}\, dy = \int_{-\infty}^{\infty} \int_{-\infty}^{\infty} e^{-(x^2+y^2)/2}\, dx\, dy$$

and then evaluating the double integral by means of a change of variables to polar coordinates. (That is, let $x = r \cos\theta, y = r\sin\theta, dx\, dy = r\, dr\, d\theta$.)

30. A random variable X is said to have a lognormal distribution if $\log X$ is normally distributed. If X is lognormal with $E[\log X] = \mu$ and $\mathrm{Var}(\log X) = \sigma^2$, determine the distribution function of X. That is, what is $P\{X \le x\}$?

31. The salaries of pediatric physicians are approximately normally distributed. If 25 percent of these physicians earn below $180,000$ and 25 percent earn above $320,000$, what fraction earn

(a) below $250,000$;

(b) between $260,00$ and $300,000$?

32. The sample mean and sample standard deviation on your economics examination were 60 and 20, respectively; the sample mean and sample standard deviation on your statistics examination were 55 and 10, respectively. You scored 70 on the economics exam and 62 on the statistics exam. Assuming that the two histograms of test scores are approximately normal histograms,

(a) on which exam was your percentile score highest?

(b) approximate the percentage of the scores on the economics exam that were below your score.

(c) approximate the percentage of the scores on the statistics exam that were below your score.

33. Value at risk (VAR) has become a key concept in financial calculations. The VAR of an investment is defined as that value v such that there is only a 1 percent chance that the loss from the investment will exceed v.

(a) If the gain from an investment is a normal random variable with mean 10 and variance 49, determine the value at risk. (If X is the gain, then $-X$ is the loss.)

(b) Among a set of investments whose gains are all normally distributed show that the one having the smallest VAR is the one having the largest value of

$\mu - 2.33\sigma$, where μ and σ^2 are the mean and variance of the gain from the investment.

34. The annual rainfall in Cincinnati is normally distributed with mean 40.14 inches and standard deviation 8.7 inches.

    (a) What is the probability this year's rainfall will exceed 42 inches?

    (b) What is the probability that the sum of the next 2 years' rainfall will exceed 84 inches?

    (c) What is the probability that the sum of the next 3 years' rainfall will exceed 126 inches?

    (d) For parts (b) and (c), what independence assumptions are you making?

35. The height of adult women in the United States is normally distributed with mean 64.5 inches and standard deviation 2.4 inches. Find the probability that a randomly chosen woman is

    (a) less than 63 inches tall;

    (b) less than 70 inches tall;

    (c) between 63 and 70 inches tall.

    (d) Alice is 72 inches tall. What percentage of women is shorter than Alice?

    (e) Find the probability that the average of the heights of two randomly chosen women exceeds 66 inches.

    (f) Repeat part (e) for four randomly chosen women.

36. An IQ test produces scores that are normally distributed with mean value 100 and standard deviation 14.2. The top 1 percent of all scores are in what range?

37. The time (in hours) required to repair a machine is an exponentially distributed random variable with parameter $\lambda = 1$.

    (a) What is the probability that a repair time exceeds 2 hours?

    (b) What is the conditional probability that a repair takes at least 3 hours, given that its duration exceeds 2 hours?

38. The number of years a radio functions is exponentially distributed with parameter $\lambda = \frac{1}{8}$. If Jones buys a used radio, what is the probability that it will be working after an additional 10 years?

39. Jones figures that the total number of thousands of miles that a used auto can be driven before it would need to be junked is an exponential random variable with parameter $\frac{1}{20}$. Smith has a used car that he claims has been driven only 10,000 miles. If Jones purchases the car, what is the probability that she would get at least 20,000 additional miles out of it? Repeat under the assumption that the lifetime mileage of the car is not exponentially distributed but rather is (in thousands of miles) uniformly distributed over (0, 40).

*40. Let $X_1, X_2, \ldots, X_n$ denote the first n interarrival times of a Poisson process and set $S_n = \sum_{i=1}^{n} X_i$.

(a) What is the interpretation of S_n?

(b) Argue that the two events $\{S_n \leq t\}$ and $\{N(t) \geq n\}$ are identical.

(c) Use part (b) to show that

$$P\{S_n \leq t\} = 1 - \sum_{j=0}^{n-1} e^{-\lambda t}(\lambda t)^j/j!$$

(d) By differentiating the distribution function of S_n given in part (c), conclude that S_n is a gamma random variable with parameters n and λ. (This result also follows from Corollary 5.7.2.)

*41. Earthquakes occur in a given region in accordance with a Poisson process with rate 5 per year.

(a) What is the probability there will be at least two earthquakes in the first half of 2015?

(b) Assuming that the event in part (a) occurs, what is the probability that there will be no earthquakes during the first 9 months of 2016?

(c) Assuming that the event in part (a) occurs, what is the probability that there will be at least four earthquakes over the first 9 months of the year 2015?

*42. When shooting at a target in a two-dimensional plane, suppose that the horizontal miss distance is normally distributed with mean 0 and variance 4 and is independent of the vertical miss distance, which is also normally distributed with mean 0 and variance 4. Let D denote the distance between the point at which the shot lands and the target. Find $E[D]$.

43. If X is a chi-square random variable with 6 degrees of freedom, find

(a) $P\{X \leq 6\}$;

(b) $P\{3 \leq X \leq 9\}$.

44. If X and Y are independent chi-square random variables with 3 and 6 degrees of freedom, respectively, determine the probability that $X + Y$ will exceed 10.

45. Show that $\Gamma(1/2) = \sqrt{\pi}$ (*Hint*: Evaluate $\int_0^\infty e^{-x}x^{-1/2}\,dx$ by letting $x = y^2/2$, $dx = y\,dy$.)

46. If T has a t-distribution with 8 degrees of freedom, find **(a)** $P\{T \geq 1\}$, **(b)** $P\{T \leq 2\}$, and **(c)** $P\{-1 < T < 1\}$.

47. If T_n has a t-distribution with n degrees of freedom, show that T_n^2 has an F-distribution with 1 and n degrees of freedom.

---

* From optional sections.

48. Let Φ be the standard normal distribution function. If, for constants a and $b > 0$

$$P\{X \le x\} = \Phi\left(\frac{x - a}{b}\right)$$

characterize the distribution of X.

*49.** Suppose that Y has a Pareto distribution with minimal parameter α and index parameter λ.

   (a) Find $E[Y]$ when $\lambda > 1$, and show that $E[Y] = \infty$ when $\lambda \le 1$.
   (b) Find $\text{Var}(Y)$ when $\lambda > 2$.

*50.** Suppose that $Y = \alpha e^X$, where X is exponential with rate λ. Use the lack of memory property of the exponential to argue that the conditional distribution of Y given that $Y > y_0 > \alpha$ is Pareto with parameters y_0 and λ.

Chapter 6

DISTRIBUTIONS OF SAMPLING STATISTICS

6.1 INTRODUCTION

The science of statistics deals with drawing conclusions from observed data. For instance, a typical situation in a technological study arises when one is confronted with a large collection, or *population*, of items that have measurable values associated with them. By suitably *sampling* from this collection, and then analyzing the sampled items, one hopes to be able to draw some conclusions about the collection as a whole.

To use sample data to make inferences about an entire population, it is necessary to make some assumptions about the relationship between the two. One such assumption, which is often quite reasonable, is that there is an underlying (population) probability distribution such that the measurable values of the items in the population can be thought of as being independent random variables having this distribution. If the sample data are then chosen in a random fashion, then it is reasonable to suppose that they too are independent values from the distribution.

Definition

If $X_1, \ldots, X_n$ are independent random variables having a common distribution F, then we say that they constitute a *sample* (sometimes called a *random sample*) from the distribution F.

In most applications, the population distribution F will not be completely specified and one will attempt to use the data to make inferences about F. Sometimes it will be supposed that F is specified up to some unknown parameters (for instance, one might suppose that F was a normal distribution function having an unknown mean and variance, or that it is a Poisson distribution function whose mean is not given), and at other times it might be assumed that almost nothing is known about F (except maybe for assuming that it is a continuous, or a discrete, distribution). Problems in which the form of the underlying distribution is specified up to a set of unknown parameters are called *parametric* inference

207

problems, whereas those in which nothing is assumed about the form of F are called *nonparametric* inference problems.

EXAMPLE 6.1a   Suppose that a new process has just been installed to produce computer chips, and suppose that the successive chips produced by this new process will have useful lifetimes that are independent with a common unknown distribution F. Physical reasons sometimes suggest the parametric form of the distribution F; for instance, it may lead us to believe that F is a normal distribution, or that F is an exponential distribution. In such cases, we are confronted with a parametrical statistical problem in which we would want to use the observed data to estimate the parameters of F. For instance, if F were assumed to be a normal distribution, then we would want to estimate its mean and variance; if F were assumed to be exponential, we would want to estimate its mean. In other situations, there might not be any physical justification for supposing that F has any particular form; in this case the problem of making inferences about F would constitute a nonparametric inference problem.   ■

In this chapter, we will be concerned with the probability distributions of certain statistics that arise from a sample, where a *statistic* is a random variable whose value is determined by the sample data. Two important statistics that we will discuss are the sample mean and the sample variance. In Section 6.2, we consider the sample mean and derive its expectation and variance. We note that when the sample size is at least moderately large, the distribution of the sample mean is approximately normal. This follows from the central limit theorem, one of the most important theoretical results in probability, which is discussed in Section 6.3. In Section 6.4, we introduce the sample variance and determine its expected value. In Section 6.5, we suppose that the population distribution is normal and present the joint distribution of the sample mean and the sample variance. In Section 6.6, we suppose that we are sampling from a finite population of elements and explain what it means for the sample to be a "random sample." When the population size is large in relation to the sample size, we often treat it as if it were of infinite size; this is illustrated and its consequences are discussed.

6.2   THE SAMPLE MEAN

Consider a population of elements, each of which has a numerical value attached to it. For instance, the population might consist of the adults of a specified community and the value attached to each adult might be his or her annual income, or height, or age, and so on. We often suppose that the value associated with any member of the population can be regarded as being the value of a random variable having expectation μ and variance σ^2. The quantities μ and σ^2 are called the *population mean* and the *population variance*, respectively. Let $X_1, X_2, \ldots, X_n$ be a sample of values from this population. The sample mean is defined by

$$\overline{X} = \frac{X_1 + \cdots + X_n}{n}$$

Since the value of the sample mean $\overline{X}$ is determined by the values of the random variables in the sample, it follows that $\overline{X}$ is also a random variable. Its expected value and variance are obtained as follows:

$$E[\overline{X}] = E\left[\frac{X_1 + \cdots + X_n}{n}\right]$$

$$= \frac{1}{n}(E[X_1] + \cdots + E[X_n])$$

$$= \mu$$

and

$$\text{Var}(\overline{X}) = \text{Var}\left(\frac{X_1 + \cdots + X_n}{n}\right)$$

$$= \frac{1}{n^2}[\text{Var}(X_1) + \cdots + \text{Var}(X_n)] \quad \text{by independence}$$

$$= \frac{n\sigma^2}{n^2}$$

$$= \frac{\sigma^2}{n}$$

where μ and σ^2 are the population mean and variance, respectively. Hence, the expected value of the sample mean is the population mean μ whereas its variance is $1/n$ times the population variance. As a result, we can conclude that $\overline{X}$ is also centered about the population mean μ, but its spread becomes more and more reduced as the sample size increases. Figure 6.1 plots the probability density function of the sample mean from a standard normal population for a variety of sample sizes.

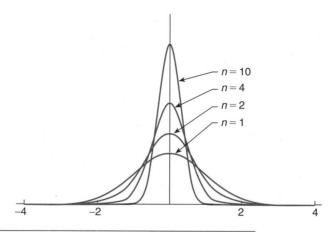

FIGURE 6.1 *Densities of sample means from a standard normal population.*

6.3   THE CENTRAL LIMIT THEOREM

In this section, we will consider one of the most remarkable results in probability — namely, the *central limit theorem*. Loosely speaking, this theorem asserts that the sum of a large number of independent random variables has a distribution that is approximately normal. Hence, it not only provides a simple method for computing approximate probabilities for sums of independent random variables, but it also helps explain the remarkable fact that the empirical frequencies of so many natural populations exhibit a bell-shaped (that is, a normal) curve.

In its simplest form, the central limit theorem is as follows:

Theorem 6.3.1 The Central Limit Theorem

Let $X_1, X_2, \ldots, X_n$ be a sequence of independent and identically distributed random variables each having mean μ and variance σ^2. Then for n large, the distribution of

$$X_1 + \cdots + X_n$$

is approximately normal with mean $n\mu$ and variance $n\sigma^2$.

It follows from the central limit theorem that

$$\frac{X_1 + \cdots + X_n - n\mu}{\sigma\sqrt{n}}$$

is approximately a standard normal random variable; thus, for n large,

$$P\left\{ \frac{X_1 + \cdots + X_n - n\mu}{\sigma\sqrt{n}} < x \right\} \approx P\{Z < x\}$$

where Z is a standard normal random variable.

EXAMPLE 6.3a An insurance company has 25,000 automobile policy holders. If the yearly claim of a policy holder is a random variable with mean 320 and standard deviation 540, approximate the probability that the total yearly claim exceeds 8.3 million.

SOLUTION Let X denote the total yearly claim. Number the policy holders, and let X_i denote the yearly claim of policy holder i. With $n = 25,000$, we have from the central limit theorem that $X = \sum_{i=1}^{n} X_i$ will have approximately a normal distribution with mean $320 \times 25,000 = 8 \times 10^6$ and standard deviation $540\sqrt{25,000} = 8.5381 \times 10^4$. Therefore,

$$P\{X > 8.3 \times 10^6\} = P\left\{ \frac{X - 8 \times 10^6}{8.5381 \times 10^4} > \frac{8.3 \times 10^6 - 8 \times 10^6}{8.5381 \times 10^4} \right\}$$

$$= P\left\{ \frac{X - 8 \times 10^6}{8.5381 \times 10^4} > \frac{.3 \times 10^6}{8.5381 \times 10^4} \right\}$$

$$\approx P\{Z > 3.51\} \qquad \text{where } Z \text{ is a standard normal}$$
$$\approx .00023$$

Thus, there are only 2.3 chances out of 10,000 that the total yearly claim will exceed 8.3 million. ∎

EXAMPLE 6.3b Civil engineers believe that W, the amount of weight (in units of 1,000 pounds) that a certain span of a bridge can withstand without structural damage resulting, is normally distributed with mean 400 and standard deviation 40. Suppose that the weight (again, in units of 1,000 pounds) of a car is a random variable with mean 3 and standard deviation .3. How many cars would have to be on the bridge span for the probability of structural damage to exceed .1?

SOLUTION Let P_n denote the probability of structural damage when there are n cars on the bridge. That is,

$$P_n = P\{X_1 + \cdots + X_n \geq W\}$$
$$= P\{X_1 + \cdots + X_n - W \geq 0\}$$

where X_i is the weight of the ith car, $i = 1, \ldots, n$. Now it follows from the central limit theorem that $\sum_{i=1}^{n} X_i$ is approximately normal with mean $3n$ and variance $.09n$. Hence, since W is independent of the $X_i, i = 1, \ldots, n$, and is also normal, it follows that $\sum_{i=1}^{n} X_i - W$ is approximately normal, with mean and variance given by

$$E\left[\sum_{1}^{n} X_i - W\right] = 3n - 400$$

$$\text{Var}\left(\sum_{1}^{n} X_i - W\right) = \text{Var}\left(\sum_{1}^{n} X_i\right) + \text{Var}(W) = .09n + 1,600$$

Thus,

$$P_n = P\left\{\frac{X_1 + \cdots + X_n - W - (3n - 400)}{\sqrt{.09n + 1,600}} \geq \frac{-(3n - 400)}{\sqrt{.09n + 1,600}}\right\}$$

$$\approx P\left\{Z \geq \frac{400 - 3n}{\sqrt{.09n + 1,600}}\right\}$$

where Z is a standard normal random variable. Now $P\{Z \geq 1.28\} \approx .1$, and so if the number of cars n is such that

$$\frac{400 - 3n}{\sqrt{.09n + 1,600}} \leq 1.28$$

or

$$n \geq 117$$

then there is at least 1 chance in 10 that structural damage will occur. ■

The central limit theorem is illustrated by Program 6.1 on the text disk. This program plots the probability mass function of the sum of n independent and identically distributed random variables that each take on one of the values 0, 1, 2, 3, 4. When using it, one enters the probabilities of these five values, and the desired value of n. Figures 6.2(a)–(f) give the resulting plot for a specified set of probabilities when $n = 1, 3, 5, 10, 25, 100$.

One of the most important applications of the central limit theorem is in regard to binomial random variables. Since such a random variable X having parameters (n, p) represents the number of successes in n independent trials when each trial is a success with probability p, we can express it as

$$X = X_1 + \cdots + X_n$$

where

$$X_i = \begin{cases} 1 & \text{if the } i\text{th trial is a success} \\ 0 & \text{otherwise} \end{cases}$$

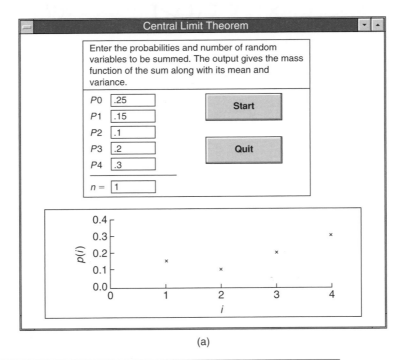

(a)

FIGURE 6.2   (a) $n = 1$, (b) $n = 3$, (c) $n = 5$, (d) $n = 10$, (e) $n = 25$, (f) $n = 100$.

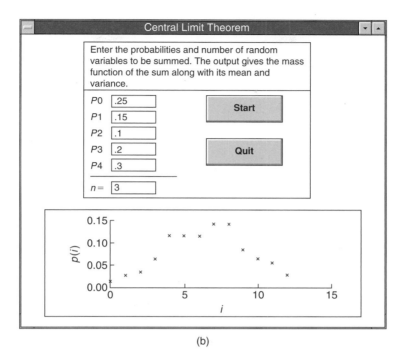

(b)

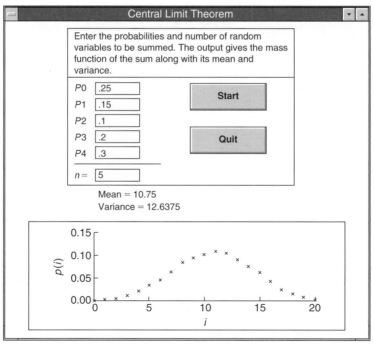

(c)

FIGURE 6.2 *(continued)*

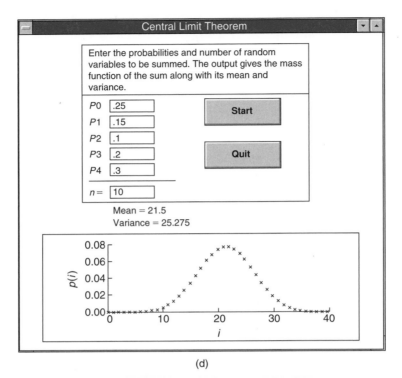

(d)

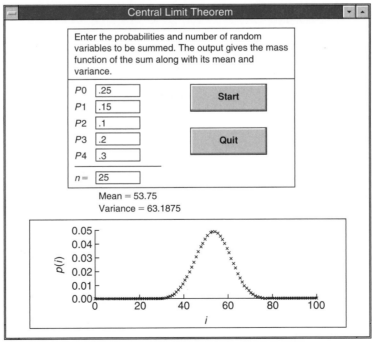

(e)

FIGURE 6.2   *(continued)*

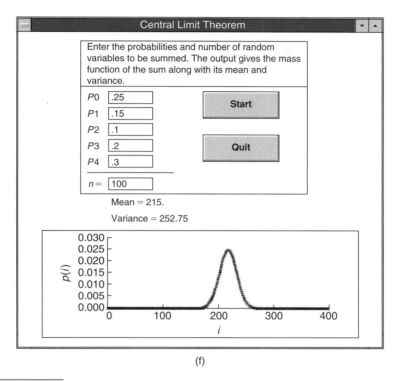

(f)

FIGURE 6.2   *(continued)*

Because

$$E[X_i] = p, \qquad \text{Var}(X_i) = p(1-p)$$

it follows from the central limit theorem that for n large

$$\frac{X - np}{\sqrt{np(1-p)}}$$

will approximately be a standard normal random variable [see Figure 6.3, which graphically illustrates how the probability mass function of a binomial (n, p) random variable becomes more and more "normal" as n becomes larger and larger].

EXAMPLE 6.3c   The ideal size of a first-year class at a particular college is 150 students. The college, knowing from past experience that, on the average, only 30 percent of those accepted for admission will actually attend, uses a policy of approving the applications of 450 students. Compute the probability that more than 150 first-year students attend this college.

SOLUTION   Let X denote the number of students that attend; then assuming that each accepted applicant will independently attend, it follows that X is a binomial random

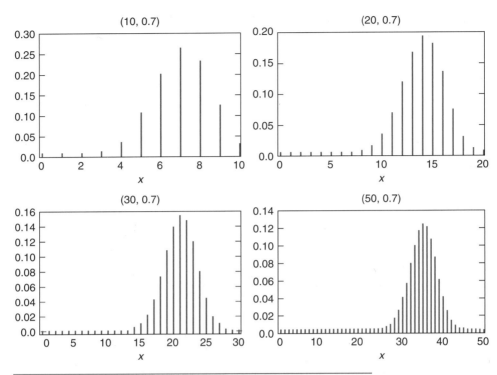

FIGURE 6.3   *Binomial probability mass functions converging to the normal density.*

variable with parameters $n = 450$ and $p = .3$. Since the binomial is a discrete and the normal a continuous distribution, it is best to compute $P\{X = i\}$ as $P\{i - .5 < X < i + .5\}$ when applying the normal approximation (this is called the continuity correction). This yields the approximation

$$P\{X > 150.5\} = P\left\{\frac{X - (450)(.3)}{\sqrt{450(.3)(.7)}} \geq \frac{150.5 - (450)(.3)}{\sqrt{450(.3)(.7)}}\right\}$$

$$\approx P\{Z > 1.59\} = .06$$

Hence, only 6 percent of the time do more than 150 of the first 450 accepted actually attend. ∎

It should be noted that we now have two possible approximations to binomial probabilities: The Poisson approximation, which yields a good approximation when n is large and p small, and the normal approximation, which can be shown to be quite good when $np(1 - p)$ is large. [The normal approximation will, in general, be quite good for values of n satisfying $np(1 - p) \geq 10$.]

6.3.1 APPROXIMATE DISTRIBUTION OF THE SAMPLE MEAN

Let $X_1, \ldots, X_n$ be a sample from a population having mean μ and variance σ^2. The central limit theorem can be used to approximate the distribution of the sample mean

$$\overline{X} = \sum_{i=1}^{n} X_i/n$$

Since a constant multiple of a normal random variable is also normal, it follows from the central limit theorem that $\overline{X}$ will be approximately normal when the sample size n is large. Since the sample mean has expected value μ and standard deviation $\sigma/\sqrt{n}$, it then follows that

$$\frac{\overline{X} - \mu}{\sigma/\sqrt{n}}$$

has approximately a standard normal distribution.

EXAMPLE 6.3d The weights of a population of workers have mean 167 and standard deviation 27.

(a) If a sample of 36 workers is chosen, approximate the probability that the sample mean of their weights lies between 163 and 170.

(b) Repeat part (a) when the sample is of size 144.

SOLUTION Let Z be a standard normal random variable.

(a) It follows from the central limit theorem that $\overline{X}$ is approximately normal with mean 167 and standard deviation $27/\sqrt{36} = 4.5$. Therefore, with Z being a standard normal random variable,

$$P\{163 < \overline{X} < 170\} = P\left\{\frac{163 - 167}{4.5} < \frac{\overline{X} - 167}{4.5} < \frac{170 - 167}{4.5}\right\}$$
$$\approx P\{-.8889 < Z < .8889\}$$
$$= P\{Z < .8889\} - P\{Z < -.8889\}$$
$$= 2P\{Z < .8889\} - 1$$
$$\approx .6259$$

(b) For a sample of size 144, the sample mean will be approximately normal with mean 167 and standard deviation $27/\sqrt{144} = 2.25$. Therefore,

$$P\{163 < \overline{X} < 170\} = P\left\{\frac{163 - 167}{2.25} < \frac{\overline{X} - 167}{2.25} < \frac{170 - 167}{2.25}\right\}$$
$$= P\left\{-1.7778 < \frac{\overline{X} - 167}{4.5} < 1.7778\right\}$$
$$\approx 2P\{Z < 1.7778\} - 1$$
$$\approx .9246$$

Thus increasing the sample size from 36 to 144 increases the probability from .6259 to .9246. ■

EXAMPLE 6.3e An astronomer wants to measure the distance from her observatory to a distant star. However, due to atmospheric disturbances, any measurement will not yield the exact distance d. As a result, the astronomer has decided to make a series of measurements and then use their average value as an estimate of the actual distance. If the astronomer believes that the values of the successive measurements are independent random variables with a mean of d light years and a standard deviation of 2 light years, how many measurements need she make to be at least 95 percent certain that her estimate is accurate to within $\pm .5$ light years?

SOLUTION If the astronomer makes n measurements, then $\overline{X}$, the sample mean of these measurements, will be approximately a normal random variable with mean d and standard deviation $2/\sqrt{n}$. Thus, the probability that it will lie between $d \pm .5$ is obtained as follows:

$$P\{-.5 < \overline{X} - d < .5\} = P\left\{\frac{-.5}{2/\sqrt{n}} < \frac{\overline{X} - d}{2/\sqrt{n}} < \frac{.5}{2/\sqrt{n}}\right\}$$

$$\approx P\{-\sqrt{n}/4 < Z < \sqrt{n}/4\}$$

$$= 2P\{Z < \sqrt{n}/4\} - 1$$

where Z is a standard normal random variable.

Thus, the astronomer should make n measurements, where n is such that

$$2P\{Z < \sqrt{n}/4\} - 1 \geq .95$$

or, equivalently,

$$P\{Z < \sqrt{n}/4\} \geq .975$$

Since $P\{Z < 1.96\} = .975$, it follows that n should be chosen so that

$$\sqrt{n}/4 \geq 1.96$$

That is, at least 62 observations are necessary. ■

6.3.2 How Large a Sample Is Needed?

The central limit theorem leaves open the question of how large the sample size n needs to be for the normal approximation to be valid, and indeed the answer depends on the population distribution of the sample data. For instance, if the underlying population distribution is normal, then the sample mean $\overline{X}$ will also be normal regardless of the sample size. A general rule of thumb is that one can be confident of the normal approximation whenever the sample size n is at least 30. That is, practically speaking, no matter how nonnormal the underlying population distribution is, the sample mean of a sample of size at least 30 will be approximately normal. In most cases, the normal approximation is

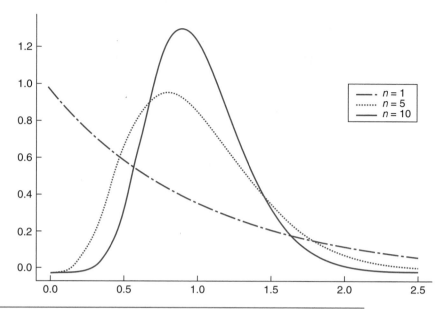

FIGURE 6.4 *Densities of the average of n exponential random variables having mean 1.*

valid for much smaller sample sizes. Indeed, a sample of size 5 will often suffice for the approximation to be valid. Figure 6.4 presents the distribution of the sample means from an exponential population distribution for samples of sizes $n = 1, 5, 10$.

6.4 THE SAMPLE VARIANCE

Let $X_1, \ldots, X_n$ be a random sample from a distribution with mean μ and variance σ^2. Let $\overline{X}$ be the sample mean, and recall the following definition from Section 2.3.2.

Definition

The statistic S^2, defined by

$$S^2 = \frac{\sum_{i=1}^{n} (X_i - \overline{X})^2}{n - 1}$$

is called the *sample variance*. $S = \sqrt{S^2}$ is called the *sample standard deviation*.

To compute $E[S^2]$, we use an identity that was proven in Section 2.3.2: For any numbers $x_1, \ldots, x_n$

$$\sum_{i=1}^{n} (x_i - \overline{x})^2 = \sum_{i=1}^{n} x_i^2 - n\overline{x}^2$$

where $\bar{x} = \sum_{i=1}^{n} x_i/n$. It follows from this identity that

$$(n-1)S^2 = \sum_{i=1}^{n} X_i^2 - n\bar{X}^2$$

Taking expectations of both sides of the preceding yields, upon using the fact that for any random variable W, $E[W^2] = \text{Var}(W) + (E[W])^2$,

$$(n-1)E[S^2] = E\left[\sum_{i=1}^{n} X_i^2\right] - nE[\bar{X}^2]$$

$$= nE[X_1^2] - nE[\bar{X}^2]$$

$$= n\text{Var}(X_1) + n(E[X_1])^2 - n\text{Var}(\bar{X}) - n(E[\bar{X}])^2$$

$$= n\sigma^2 + n\mu^2 - n(\sigma^2/n) - n\mu^2$$

$$= (n-1)\sigma^2$$

or

$$E[S^2] = \sigma^2$$

That is, the expected value of the sample variance S^2 is equal to the population variance σ^2.

6.5 SAMPLING DISTRIBUTIONS FROM A NORMAL POPULATION

Let $X_1, X_2, \ldots, X_n$ be a sample from a normal population having mean μ and variance σ^2. That is, they are independent and $X_i \sim \mathcal{N}(\mu, \sigma^2)$, $i = 1, \ldots, n$. Also let

$$\bar{X} = \sum_{i=1}^{n} X_i/n$$

and

$$S^2 = \frac{\sum_{i=1}^{n} (X_i - \bar{X})^2}{n-1}$$

denote the sample mean and sample variance, respectively. We would like to compute their distributions.

6.5.1 DISTRIBUTION OF THE SAMPLE MEAN

Since the sum of independent normal random variables is normally distributed, it follows that $\overline{X}$ is normal with mean

$$E[\overline{X}] = \sum_{i=1}^{n} \frac{E[X_i]}{n} = \mu$$

and variance

$$\text{Var}(\overline{X}) = \frac{1}{n^2} \sum_{i=1}^{n} \text{Var}(X_i) = \sigma^2/n$$

That is, $\overline{X}$, the average of the sample, is normal with a mean equal to the population mean but with a variance reduced by a factor of $1/n$. It follows from this that

$$\frac{\overline{X} - \mu}{\sigma/\sqrt{n}}$$

is a standard normal random variable.

6.5.2 JOINT DISTRIBUTION OF $\overline{X}$ AND S^2

In this section, we not only obtain the distribution of the sample variance S^2, but we also discover a fundamental fact about normal samples — namely, that $\overline{X}$ and S^2 are independent with $(n-1)S^2/\sigma^2$ having a chi-square distribution with $n-1$ degrees of freedom.

To start, for numbers $x_1, \ldots, x_n$, let $y_i = x_i - \mu, i = 1, \ldots, n$. Then as $\overline{y} = \overline{x} - \mu$, it follows from the identity

$$\sum_{i=1}^{n} (y_i - \overline{y})^2 = \sum_{i=1}^{n} y_i^2 - n\overline{y}^2$$

that

$$\sum_{i=1}^{n} (x_i - \overline{x})^2 = \sum_{i=1}^{n} (x_i - \mu)^2 - n(\overline{x} - \mu)^2$$

Now, if $X_1, \ldots, X_n$ is a sample from a normal population having mean μ and variance σ^2, then we obtain from the preceding identity that

$$\frac{\sum_{i=1}^{n} (X_i - \mu)^2}{\sigma^2} = \frac{\sum_{i=1}^{n} (X_i - \overline{X})^2}{\sigma^2} + \frac{n(\overline{X} - \mu)^2}{\sigma^2}$$

or, equivalently,

$$\sum_{i=1}^{n}\left(\frac{X_i - \mu}{\sigma}\right)^2 = \frac{\sum_{i=1}^{n}(X_i - \overline{X})^2}{\sigma^2} + \left[\frac{\sqrt{n}(\overline{X} - \mu)}{\sigma}\right]^2 \qquad (6.5.1)$$

Because $(X_i - \mu)/\sigma, i = 1, \ldots, n$ are independent standard normals, it follows that the left side of Equation 6.5.1 is a chi-square random variable with n degrees of freedom. Also, as shown in Section 6.5.1, $\sqrt{n}(\overline{X} - \mu)/\sigma$ is a standard normal random variable and so its square is a chi-square random variable with 1 degree of freedom. Thus Equation 6.5.1 equates a chi-square random variable having n degrees of freedom to the sum of two random variables, one of which is chi-square with 1 degree of freedom. But it has been established that the sum of two independent chi-square random variables is also chi-square with a degree of freedom equal to the sum of the two degrees of freedom. Thus, it would seem that there is a reasonable possibility that the two terms on the right side of Equation 6.5.1 are independent, with $\sum_{i=1}^{n}(X_i - \overline{X})^2/\sigma^2$ having a chi-square distribution with $n - 1$ degrees of freedom. Since this result can indeed be established, we have the following fundamental result.

Theorem 6.5.1

If $X_1, \ldots, X_n$ is a sample from a normal population having mean μ and variance σ^2, then $\overline{X}$ and S^2 are independent random variables, with $\overline{X}$ being normal with mean μ and variance σ^2/n and $(n - 1)S^2/\sigma^2$ being chi-square with $n - 1$ degrees of freedom.

Theorem 6.5.1 not only provides the distributions of $\overline{X}$ and S^2 for a normal population but also establishes the important fact that they are independent. In fact, it turns out that this independence of $\overline{X}$ and S^2 is a unique property of the normal distribution. Its importance will become evident in the following chapters.

EXAMPLE 6.5a   The time it takes a central processing unit to process a certain type of job is normally distributed with mean 20 seconds and standard deviation 3 seconds. If a sample of 15 such jobs is observed, what is the probability that the sample variance will exceed 12?

SOLUTION   Since the sample is of size $n = 15$ and $\sigma^2 = 9$, write

$$P\{S^2 > 12\} = P\left\{\frac{14S^2}{9} > \frac{14}{9} \cdot 12\right\}$$
$$= P\{\chi_{14}^2 > 18.67\}$$
$$= 1 - .8221 \qquad \text{from Program 5.8.1a}$$
$$= .1779 \quad \blacksquare$$

The following corollary of Theorem 6.5.1 will be quite useful in the following chapters.

Corollary 6.5.2

Let $X_1, \ldots, X_n$ be a sample from a normal population with mean μ. If $\overline{X}$ denotes the sample mean and S the sample standard deviation, then

$$\sqrt{n}\frac{(\overline{X} - \mu)}{S} \sim t_{n-1}$$

That is, $\sqrt{n}(\overline{X} - \mu)/S$ has a t-distribution with $n-1$ degrees of freedom.

Proof

Recall that a t-random variable with n degrees of freedom is defined as the distribution of

$$\frac{Z}{\sqrt{\chi_n^2/n}}$$

where Z is a standard normal random variable that is independent of χ_n^2, a chi-square random variable with n degrees of freedom. Because Theorem 6.5.1 gives that $\sqrt{n}(\overline{X}-\mu)/\sigma$ is a standard normal that is independent of $(n-1)S^2/\sigma^2$, which is chi-square with $n-1$ degrees of freedom, we can conclude that

$$\frac{\sqrt{n}(\overline{X} - \mu)/\sigma}{\sqrt{S^2/\sigma^2}} = \sqrt{n}\frac{(\overline{X} - \mu)}{S}$$

is a t-random variable with $n-1$ degrees of freedom. ∎

6.6 SAMPLING FROM A FINITE POPULATION

Consider a population of N elements, and suppose that p is the proportion of the population that has a certain characteristic of interest; that is, Np elements have this characteristic, and $N(1-p)$ do not. A sample of size n from this population is said to be a *random sample* if it is chosen in such a manner that each of the $\binom{N}{n}$ population subsets of size n is equally likely to be the sample. For instance, if the population consists of the three elements a, b, c, then a random sample of size 2 is one that is chosen so that each of the subsets $\{a, b\}, \{a, c\}$, and $\{b, c\}$ is equally likely to be the sample. A random subset can be chosen sequentially by letting its first element be equally likely to be any of the N elements of the population, then letting its second element be equally likely to be any of the remaining $N-1$ elements of the population, and so on.

Suppose now that a random sample of size n has been chosen from a population of size N. For $i = 1, \ldots, n$, let

$$X_i = \begin{cases} 1 & \text{if the } i\text{th member of the sample has the characteristic} \\ 0 & \text{otherwise} \end{cases}$$

Consider now the sum of the X_i; that is, consider

$$X = X_1 + X_2 + \cdots + X_n$$

Because the term X_i contributes 1 to the sum if the ith member of the sample has the characteristic and 0 otherwise, it follows that X is equal to the number of members of the sample that possess the characteristic. In addition, the sample mean

$$\overline{X} = X/n = \sum_{i=1}^{n} X_i/n$$

is equal to the proportion of the members of the sample that possess the characteristic.

Let us now consider the probabilities associated with the statistics X and $\overline{X}$. To begin, note that since each of the N members of the population is equally likely to be the ith member of the sample, it follows that

$$P\{X_i = 1\} = \frac{Np}{N} = p$$

Also,

$$P\{X_i = 0\} = 1 - P\{X_i = 1\} = 1 - p$$

That is, each X_i is equal to either 1 or 0 with respective probabilities p and $1 - p$.

It should be noted that the random variables $X_1, X_2, \ldots, X_n$ are not independent. For instance, since the second selection is equally likely to be any of the N members of the population, of which Np have the characteristic, it follows that the probability that the second selection has the characteristic is $Np/N = p$. That is, without any knowledge of the outcome of the first selection,

$$P\{X_2 = 1\} = p$$

However, the conditional probability that $X_2 = 1$, given that the first selection has the characteristic, is

$$P\{X_2 = 1 | X_1 = 1\} = \frac{Np - 1}{N - 1}$$

which is seen by noting that if the first selection has the characteristic, then the second selection is equally likely to be any of the remaining $N - 1$ elements, of which $Np - 1$ have the characteristic. Similarly, the probability that the second selection has the characteristic given that the first one does not is

$$P\{X_2 = 1 | X_1 = 0\} = \frac{Np}{N - 1}$$

Thus, knowing whether or not the first element of the random sample has the characteristic changes the probability for the next element. However, when the population size N

is large in relation to the sample size n, this change will be very slight. For instance, if $N = 1,000, p = .4$, then

$$P\{X_2 = 1|X_1 = 1\} = \frac{399}{999} = .3994$$

which is very close to the unconditional probability that $X_2 = 1$; namely,

$$P\{X_2 = 1\} = .4$$

Similarly, the probability that the second element of the sample has the characteristic given that the first does not is

$$P\{X_2 = 1|X_1 = 0\} = \frac{400}{999} = .4004$$

which is again very close to .4.

Indeed, it can be shown that when the population size N is large with respect to the sample size n, then $X_1, X_2, \ldots, X_n$ are approximately independent. Now if we think of each X_i as representing the result of a trial that is a success if X_i equals 1 and a failure otherwise, it follows that $X = \sum_{i=1}^{n} X_i$ can be thought of as representing the total number of successes in n trials. Hence, if the X_i were independent, then X would be a binomial random variable with parameters n and p. In other words, when the population size N is large in relation to the sample size n, then the distribution of the number of members of the sample that possess the characteristic is approximately that of a binomial random variable with parameters n and p.

REMARK

Of course, X is a hypergeometric random variable (Section 5.4); and so the preceding shows that a hypergeometric can be approximated by a binomial random variable when the number chosen is small in relation to the total number of elements.

> For the remainder of this text, we will suppose that the underlying population is large in relation to the sample size and we will take the distribution of X to be binomial.

By using the formulas given in Section 5.1 for the mean and standard deviation of a binomial random variable, we see that

$$E[X] = np \quad \text{and} \quad SD(X) = \sqrt{np(1-p)}$$

Moreover $\overline{X} = X/n$, the proportion of the sample that has the characteristic, has mean and variance given by

$$E[\overline{X}] = E[X]/n = p$$

and
$$\text{Var}(\overline{X}) = \text{Var}(X)/n^2 = p(1-p)/n$$

EXAMPLE 6.6a Suppose that 45 percent of the population favors a certain candidate in an upcoming election. If a random sample of size 200 is chosen, find

(a) the expected value and standard deviation of the number of members of the sample that favor the candidate;

(b) the probability that more than half the members of the sample favor the candidate.

SOLUTION

(a) The expected value and standard deviation of the proportion that favor the candidate are

$$E[X] = 200(.45) = 90, \quad SD(X) = \sqrt{200(.45)(1-.45)} = 7.0356$$

(b) Since X is binomial with parameters 200 and .45, the text disk gives the solution

$$P\{X \geq 101\} = .0681$$

If this program were not available, then the normal approximation to the binomial (Section 6.3) could be used:

$$P\{X \geq 101\} = P\{X \geq 100.5\} \quad \text{(the continuity correction)}$$
$$= P\left\{\frac{X-90}{7.0356} \geq \frac{100.5-90}{7.0356}\right\}$$
$$\approx P\{Z \geq 1.4924\}$$
$$\approx .0678$$

The solution obtained by the normal approximation is correct to 3 decimal places. ∎

Even when each element of the population has more than two possible values, it still remains true that if the population size is large in relation to the sample size, then the sample data can be regarded as being independent random variables from the population distribution.

EXAMPLE 6.6b According to the U.S. Department of Agriculture's *World Livestock Situation*, the country with the greatest per capita consumption of pork is Denmark. In 2013, the amount of pork consumed by a person residing in Denmark had a mean value of 147 pounds with a standard deviation of 62 pounds. If a random sample of 25 Danes is chosen, approximate the probability that the average amount of pork consumed by the members of this group in 2013 exceeded 150 pounds.

SOLUTION If we let X_i be the amount consumed by the ith member of the sample, $i = 1, \ldots, 25$, then the desired probability is

$$P\left\{\frac{X_1 + \cdots + X_{25}}{25} > 150\right\} = P\{\overline{X} > 150\}$$

where $\overline{X}$ is the sample mean of the 25 sample values. Since we can regard the X_i as being independent random variables with mean 147 and standard deviation 62, it follows from the central limit theorem that their sample mean will be approximately normal with mean 147 and standard deviation 62/5. Thus, with Z being a standard normal random variable, we have

$$P\{\overline{X} > 150\} = P\left\{\frac{\overline{X} - 147}{12.4} > \frac{150 - 147}{12.4}\right\}$$
$$\approx P\{Z > .242\}$$
$$\approx .404 \quad \blacksquare$$

Problems

1. Suppose that X_1, X_2, X_3 are independent with the common probability mass function

$$P\{X_i = 0\} = .2, \quad P\{X_i = 1\} = .3, \quad P\{X_i = 3\} = .5, \quad i = 1, 2, 3$$

   (a) Plot the probability mass function of $\overline{X}_2 = \dfrac{X_1 + X_2}{2}$.
   (b) Determine $E[\overline{X}_2]$ and $\text{Var}(\overline{X}_2)$.
   (c) Plot the probability mass function of $\overline{X}_3 = \dfrac{X_1 + X_2 + X_3}{3}$.
   (d) Determine $E[\overline{X}_3]$ and $\text{Var}(\overline{X}_3)$.

2. If 10 fair dice are rolled, approximate the probability that the sum of the values obtained (which ranges from 10 to 60) is between 30 and 40 inclusive.

3. Approximate the probability that the sum of 16 independent uniform $(0, 1)$ random variables exceeds 10.

4. A roulette wheel has 38 slots, numbered 0, 00, and 1 through 36. If you bet 1 on a specified number, you either win 35 if the roulette ball lands on that number or lose 1 if it does not. If you continually make such bets, approximate the

probability that

(a) you are winning after 34 bets;
(b) you are winning after 1,000 bets;
(c) you are winning after 100,000 bets.

Assume that each roll of the roulette ball is equally likely to land on any of the 38 numbers.

5. A highway department has enough salt to handle a total of 80 inches of snow-fall. Suppose the daily amount of snow has a mean of 1.5 inches and a standard deviation of .3 inches.

(a) Approximate the probability that the salt on hand will suffice for the next 50 days.
(b) What assumption did you make in solving part (a)?
(c) Do you think this assumption is justified? Explain briefly.

6. Fifty numbers are rounded off to the nearest integer and then summed. If the individual roundoff errors are uniformly distributed between $-.5$ and .5, what is the approximate probability that the resultant sum differs from the exact sum by more than 3?

7. A six-sided die, in which each side is equally likely to appear, is repeatedly rolled until the total of all rolls exceeds 400. Approximate the probability that this will require more than 140 rolls.

8. The amount of time that a certain type of battery functions is a random variable with mean 5 weeks and standard deviation 1.5 weeks. Upon failure, it is imme-diately replaced by a new battery. Approximate the probability that 13 or more batteries will be needed in a year.

9. The lifetime of a certain electrical part is a random variable with mean 100 hours and standard deviation 20 hours. If 16 such parts are tested, find the probability that the sample mean is

(a) less than 104;
(b) between 98 and 104 hours.

10. A tobacco company claims that the amount of nicotine in its cigarettes is a ran-dom variable with mean 2.2 mg and standard deviation .3 mg. However, the sample mean nicotine content of 100 randomly chosen cigarettes was 3.1 mg. What is the approximate probability that the sample mean would have been as high or higher than 3.1 if the company's claims were true?

11. The lifetime (in hours) of a type of electric bulb has expected value 500 and standard deviation 80. Approximate the probability that the sample mean of n such bulbs is greater than 525 when

(a) $n = 4$;
(b) $n = 16$;

(c) $n = 36$;

(d) $n = 64$.

12. An instructor knows from past experience that student exam scores have mean 77 and standard deviation 15. At present the instructor is teaching two separate classes — one of size 25 and the other of size 64.

(a) Approximate the probability that the average test score in the class of size 25 lies between 72 and 82.

(b) Repeat part (a) for a class of size 64.

(c) What is the approximate probability that the average test score in the class of size 25 is higher than that of the class of size 64?

(d) Suppose the average scores in the two classes are 76 and 83. Which class, the one of size 25 or the one of size 64, do you think was more likely to have averaged 83?

13. If X is binomial with parameters $n = 150$, $p = .6$, compute the exact value of $P\{X \leq 80\}$ and compare with its normal approximation both (a) making use of and (b) not making use of the continuity correction.

14. Each computer chip made in a certain plant will, independently, be defective with probability .25. If a sample of 1,000 chips is tested, what is the approximate probability that fewer than 200 chips will be defective?

15. A club basketball team will play a 60-game season. Thirty-two of these games are against class A teams and 28 are against class B teams. The outcomes of all the games are independent. The team will win each game against a class A opponent with probability .5, and it will win each game against a class B opponent with probability .7. Let X denote its total number of victories in the season.

(a) Is X a binomial random variable?

(b) Let X_A and X_B denote, respectively, the number of victories against class A and class B teams. What are the distributions of X_A and X_B?

(c) What is the relationship between X_A, X_B, and X?

(d) Approximate the probability that the team wins 40 or more games.

16. Argue, based on the central limit theorem, that a Poisson random variable having mean λ will approximately have a normal distribution with mean and variance both equal to λ when λ is large. If X is Poisson with mean 100, compute the exact probability that X is less than or equal to 116 and compare it with its normal approximation both when a continuity correction is utilized and when it is not. The convergence of the Poisson to the normal is indicated in Figure 6.5.

17. Use the text disk to compute $P\{X \leq 10\}$ when X is a binomial random variable with parameters $n = 100, p = .1$. Now compare this with its (a) Poisson and

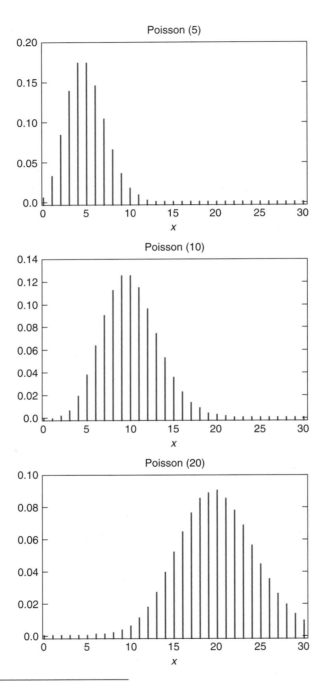

FIGURE 6.5   *Poisson probability mass functions.*

(b) normal approximation. In using the normal approximation, write the desired probability as $P\{X < 10.5\}$ so as to utilize the continuity correction.

18. The temperature at which a thermostat goes off is normally distributed with variance σ^2. If the thermostat is to be tested five times, find

   (a) $P\{S^2/\sigma^2 \le 1.8\}$
   (b) $P\{.85 \le S^2/\sigma^2 \le 1.15\}$

   where S^2 is the sample variance of the five data values.

19. In Problem 18, how large a sample would be necessary to ensure that the probability in part (a) is at least .95?

20. Consider two independent samples — the first of size 10 from a normal population having variance 4 and the second of size 5 from a normal population having variance 2. Compute the probability that the sample variance from the second sample exceeds the one from the first. (*Hint*: Relate it to the F-distribution.)

21. Twelve percent of the population is left-handed. Find the probability that there are between 10 and 14 left-handers in a random sample of 100 members of this population. That is, find $P\{10 \le X \le 14\}$, where X is the number of left-handers in the sample.

22. Fifty-two percent of the residents of a certain city are in favor of teaching evolution in high school. Find or approximate the probability that at least 50 percent of a random sample of size n is in favor of teaching evolution, when

   (a) $n = 10$;
   (b) $n = 100$;
   (c) $n = 1,000$;
   (d) $n = 10,000$.

23. The following table gives the percentages of individuals of a given city, categorized by gender, that follow certain negative health practices. Suppose a random sample of 300 men is chosen. Approximate the probability that

   (a) at least 150 of them rarely eat breakfast;
   (b) fewer than 100 of them smoke.

	Sleeps 6 Hours or Less per Night	Smoker	Rarely Eats Breakfast	Is 20 Percent or More Overweight
Men	22.7	28.4	45.4	29.6
Women	21.4	22.8	42.0	25.6

Source: U.S. National Center for Health Statistics, Health Promotion and Disease Prevention.

24. (Use the table from Problem 23.) Suppose a random sample of 300 women is chosen. Approximate the probability that

(a) at least 60 of them are overweight by 20 percent or more;

(b) fewer than 50 of them sleep 6 hours or less nightly.

25. (Use the table from Problem 23.) Suppose random samples of 300 women and
     of 300 men are chosen. Approximate the probability that more women than men
     rarely eat breakfast.

26. The following table uses data concerning the percentages of teenage male and
     female full-time workers whose annual salaries fall in different salary groupings.
     Suppose random samples of 1,000 men and 1,000 women were chosen. Use the
     table to approximate the probability that

(a) at least half of the women earned less than $20,000;

(b) more than half of the men earned $20,000 or more;

(c) more than half of the women and more than half of the men earned $20,000
     or more;

(d) 250 or fewer of the women earned at least $25,000;

(e) at least 200 of the men earned $50,000 or more;

(f) more women than men earned between $20,000 and $24,999.

Earnings Range	Percentage of Women	Percentage of Men
$4,999 or less	2.8	1.8
$5,000 to $9,999	10.4	4.7
$10,000 to $19,999	41.0	23.1
$20,000 to $24,999	16.5	13.4
$25,000 to $49,999	26.3	42.1
$50,000 and over	3.0	14.9

Source: U.S. Department of Commerce, Bureau of the Census.

27. In 1995 the percentage of the labor force that belonged to a union was 14.9. If
     five workers had been randomly chosen in that year, what is the probability that
     none of them would have belonged to a union? Compare your answer to what it
     would be for the year 1945, when an all-time high of 35.5 percent of the labor
     force belonged to a union.

28. The sample mean and sample standard deviation of all San Francisco student
     scores on the most recent Scholastic Aptitude Test examination in mathematics
     were 517 and 120. Approximate the probability that a random sample of 144
     students would have an average score exceeding

(a) 507;

(b) 517;

(c) 537;

(d) 550.

29. The average salary of newly graduated students with bachelor's degrees in chemical engineering is $53,600, with a standard deviation of $3,200. Approximate the probability that the average salary of a sample of 12 recently graduated chemical engineers exceeds $55,000.

30. A certain component is critical to the operation of an electrical system and must be replaced immediately upon failure. If the mean lifetime of this type of component is 100 hours and its standard deviation is 30 hours, how many of the components must be in stock so that the probability that the system is in continual operation for the next 2000 hours is at least .95?

PARAMETER ESTIMATION

7.1 INTRODUCTION

Let $X_1, \ldots, X_n$ be a random sample from a distribution F_θ that is specified up to a vector of unknown parameters θ. For instance, the sample could be from a Poisson distribution whose mean value is unknown; or it could be from a normal distribution having an unknown mean and variance. Whereas in probability theory it is usual to suppose that all of the parameters of a distribution are known, the opposite is true in statistics, where a central problem is to use the observed data to make inferences about the unknown parameters.

In Section 7.2, we present the *maximum likelihood* method for determining estimators of unknown parameters. The estimates so obtained are called *point estimates*, because they specify a single quantity as an estimate of θ. In Section 7.3, we consider the problem of obtaining *interval estimates*. In this case, rather than specifying a certain value as our estimate of θ, we specify an interval in which we estimate that θ lies. Additionally, we consider the question of how much *confidence* we can attach to such an interval estimate. We illustrate by showing how to obtain an interval estimate of the unknown mean of a normal distribution whose variance is specified. We then consider a variety of interval estimation problems. In Section 7.3.1, we present an interval estimate of the mean of a normal distribution whose variance is unknown. In Section 7.3.2, we obtain an interval estimate of the variance of a normal distribution. In Section 7.4, we determine an interval estimate for the difference of two normal means, both when their variances are assumed to be known and when they are assumed to be unknown (although in the latter case we suppose that the unknown variances are equal). In Sections 7.5 and the optional Section 7.6, we present interval estimates of the mean of a Bernoulli random variable and the mean of an exponential random variable.

In the optional Section 7.7, we return to the general problem of obtaining point estimates of unknown parameters and show how to evaluate an estimator by considering its mean square error. The bias of an estimator is discussed, and its relationship to the mean square error is explored.

In the optional Section 7.8, we consider the problem of determining an estimate of an unknown parameter when there is some prior information available. This is the *Bayesian* approach, which supposes that prior to observing the data, information about θ is always available to the decision maker, and that this information can be expressed in terms of a probability distribution on θ. In such a situation, we show how to compute the *Bayes estimator*, which is the estimator whose expected squared distance from θ is minimal.

7.2   MAXIMUM LIKELIHOOD ESTIMATORS

Any statistic used to estimate the value of an unknown parameter θ is called an *estimator* of θ. The observed value of the estimator is called the *estimate*. For instance, as we shall see, the usual estimator of the mean of a normal population, based on a sample $X_1, \ldots, X_n$ from that population, is the sample mean $\overline{X} = \sum_i X_i/n$. If a sample of size 3 yields the data $X_1 = 2, X_2 = 3, X_3 = 4$, then the estimate of the population mean, resulting from the estimator $\overline{X}$, is the value 3.

Suppose that the random variables $X_1, \ldots, X_n$, whose joint distribution is assumed given except for an unknown parameter θ, are to be observed. The problem of interest is to use the observed values to estimate θ. For example, the X_i's might be independent, exponential random variables each having the same unknown mean θ. In this case, the joint density function of the random variables would be given by

$$
\begin{aligned}
f(x_1, & x_2, \ldots, x_n) \\
&= f_{X_1}(x_1) f_{X_2}(x_2) \cdots f_{X_n}(x_n) \\
&= \frac{1}{\theta} e^{-x_1/\theta} \frac{1}{\theta} e^{-x_2/\theta} \cdots \frac{1}{\theta} e^{-x_n/\theta}, \qquad 0 < x_i < \infty, i = 1, \ldots, n \\
&= \frac{1}{\theta^n} \exp\left\{ -\sum_1^n x_i/\theta \right\}, \qquad 0 < x_i < \infty, i = 1, \ldots, n
\end{aligned}
$$

and the objective would be to estimate θ from the observed data $X_1, X_2, \ldots, X_n$.

A particular type of estimator, known as the *maximum likelihood* estimator, is widely used in statistics. It is obtained by reasoning as follows. Let $f(x_1, \ldots, x_n|\theta)$ denote the joint probability mass function of the random variables $X_1, X_2, \ldots, X_n$ when they are discrete, and let it be their joint probability density function when they are jointly continuous random variables. Because θ is assumed unknown, we also write f as a function of θ. Now since $f(x_1, \ldots, x_n|\theta)$ represents the likelihood that the values $x_1, x_2, \ldots, x_n$ will be observed when θ is the true value of the parameter, it would seem that a reasonable estimate of θ would be that value yielding the largest likelihood of the observed values. In other words, the maximum likelihood estimate $\hat{\theta}$ is defined to be that value of θ maximizing $f(x_1, \ldots, x_n|\theta)$ where $x_1, \ldots, x_n$ are the observed values. The function $f(x_1, \ldots, x_n|\theta)$ is often referred to as the *likelihood* function of θ.

In determining the maximizing value of θ, it is often useful to use the fact that $f(x_1, \ldots, x_n|\theta)$ and $\log[f(x_1, \ldots, x_n|\theta)]$ have their maximum at the same value of θ. Hence, we may also obtain $\hat{\theta}$ by maximizing $\log[f(x_1, \ldots, x_n|\theta)]$.

EXAMPLE 7.2a (Maximum Likelihood Estimator of a Bernoulli Parameter) Suppose that n independent trials, each of which is a success with probability p, are performed. What is the maximum likelihood estimator of p?

SOLUTION The data consist of the values of $X_1, \ldots, X_n$ where

$$X_i = \begin{cases} 1 & \text{if trial } i \text{ is a success} \\ 0 & \text{otherwise} \end{cases}$$

Now

$$P\{X_i = 1\} = p = 1 - P\{X_i = 0\}$$

which can be succinctly expressed as

$$P\{X_i = x\} = p^x(1-p)^{1-x}, \quad x = 0, 1$$

Hence, by the assumed independence of the trials, the likelihood (that is, the joint probability mass function) of the data is given by

$$\begin{aligned} f(x_1, \ldots, x_n|p) &= P\{X_1 = x_1, \ldots, X_n = x_n|p\} \\ &= p^{x_1}(1-p)^{1-x_1} \cdots p^{x_n}(1-p)^{1-x_n} \\ &= p^{\sum_1^n x_i}(1-p)^{n-\sum_1^n x_i}, \quad x_i = 0, 1, \quad i = 1, \ldots, n \end{aligned}$$

To determine the value of p that maximizes the likelihood, first take logs to obtain

$$\log f(x_1, \ldots, x_n|p) = \sum_1^n x_i \log p + \left(n - \sum_1^n x_i\right)\log(1-p)$$

Differentiation yields

$$\frac{d}{dp}\log f(x_1, \ldots, x_n|p) = \frac{\sum_1^n x_i}{p} - \frac{\left(n - \sum_1^n x_i\right)}{1-p}$$

Upon equating to zero and solving, we obtain that the maximum likelihood estimate $\hat{p}$ satisfies

$$\frac{\sum_{1}^{n} x_i}{\hat{p}} = \frac{n - \sum_{1}^{n} x_i}{1 - \hat{p}}$$

or

$$\hat{p} = \frac{\sum_{i=1}^{n} x_i}{n}$$

Hence, the maximum likelihood estimator of the unknown mean of a Bernoulli distribution is given by

$$d(X_1, \ldots, X_n) = \frac{\sum_{i=1}^{n} X_i}{n}$$

Since $\sum_{i=1}^{n} X_i$ is the number of successful trials, we see that the maximum likelihood estimator of p is equal to the proportion of the observed trials that result in successes. For an illustration, suppose that each RAM (random access memory) chip produced by a certain manufacturer is, independently, of acceptable quality with probability p. Then if out of a sample of 1,000 tested 921 are acceptable, it follows that the maximum likelihood estimate of p is .921. ■

EXAMPLE 7.2b   Two proofreaders were given the same manuscript to read. If proofreader 1 found n_1 errors, and proofreader 2 found n_2 errors, with $n_{1,2}$ of these errors being found by both proofreaders, estimate N, the total number of errors that are in the manuscript.

SOLUTION   Before we can estimate N we need to make some assumptions about the underlying probability model. So let us assume that the results of the proofreaders are independent, and that each error in the manuscript is independently found by proofreader i with probability p_i, $i = 1, 2$.

   To estimate N, we will start by deriving an estimator of p_1. To do so, note that each of the n_2 errors found by reader 2 will, independently, be found by proofreader 1 with probability p_i. Because proofreader 1 found $n_{1,2}$ of those n_2 errors, a reasonable estimate of p_1 is given by

$$\hat{p}_1 = \frac{n_{1,2}}{n_2}$$

However, because proofreader 1 found n_1 of the N errors in the manuscript, it is reasonable to suppose that p_1 is also approximately equal to $\frac{n_1}{N}$. Equating this to $\hat{p}_1$ gives that

$$\frac{n_{1,2}}{n_2} \approx \frac{n_1}{N}$$

or

$$N \approx \frac{n_1 n_2}{n_{1,2}}$$

Because the preceding estimate is symmetric in n_1 and n_2, it follows that it is the same no matter which proofreader is designated as proofreader 1.

An interesting application of the preceding occurred when two teams of researchers recently announced that they had decoded the human genetic code sequence. As part of their work both teams estimated that the human genome consisted of approximately 33,000 genes. Because both teams independently arrived at the same number, many scientists found this number believable. However, most scientists were quite surprised by this relatively small number of genes; by comparison it is only about twice as many as a fruit fly has. However, a closer inspection of the findings indicated that the two groups only agreed on the existence of about 17,000 genes. (That is, 17,000 genes were found by both teams.) Thus, based on our preceding estimator, we would estimate that the actual number of genes, rather than being 33,000, is

$$\frac{n_1 n_2}{n_{1,2}} = \frac{33,000 \times 33,000}{17,000} \approx 64,000$$

(Because there is some controversy about whether some of genes claimed to be found are actually genes, 64,000 should probably be taken as an upper bound on the actual number of genes.)

The estimation approach used when there are two proofreaders does not work when there are m proofreaders, when $m > 2$. Because, if for each i, we let $\hat{p}_i$ be the fraction of the errors found by at least one of the other proofreaders j, ($j \neq i$), that are also found by i, and then set that equal to $\frac{n_i}{N}$, then the estimate of N, namely $\frac{n_i}{\hat{p}_i}$, would differ for different values of i. Moreover, with this approach it is possible that we may have that $\hat{p}_i > \hat{p}_j$ even if proofreader i finds fewer errors than does proofreader j. For instance, for $m = 3$, suppose proofreaders 1 and 2 find exactly the same set of 10 errors whereas proofreader 3 finds 20 errors with only 1 of them in common with the set of errors found by the others. Then, because proofreader 1 (and 2) found 10 of the 29 errors found by at least one of the other proofreaders, $\hat{p}_i = 10/29$, $i = 1, 2$. On the other hand, because proofreader 3 only found 1 of the 10 errors found by the others, $\hat{p}_3 = 1/10$. Therefore, although proofreader 3 found twice the number of errors as did proofreader 1, the estimate of p_3 is less than that of p_1. To obtain more reasonable estimates, we could take the preceding values of $\hat{p}_i$, $i = 1, \ldots, m$, as preliminary estimates of the p_i. Now, let n_f be the number of errors that are found by at least one proofreader. Because n_f/N is the fraction of errors that are found by at least one proofreader, this should approximately equal $1 - \prod_{i=1}^{m}(1 - p_i)$, the probability that an error is found by at least one proofreader. Therefore, we have

$$\frac{n_f}{N} \approx 1 - \prod_{i=1}^{m}(1 - p_i)$$

suggesting that $N \approx \hat{N}$, where

$$\hat{N} = \frac{n_f}{1 - \prod_{i=1}^{m}(1 - \hat{p}_i)} \qquad (7.2.1)$$

With this estimate of N, we can then reset our estimates of the p_i by using

$$\hat{p}_i = \frac{n_i}{\hat{N}}, \qquad i = 1, \dots, m \qquad (7.2.2)$$

We can then reestimate N by using the new value (Equation 7.2.1). (The estimation need not stop here; each time we obtain a new estimate $\hat{N}$ of N we can use Equation 7.2.2 to obtain new estimates of the p_i, which can then be used to obtain a new estimate of N, and so on.) ■

EXAMPLE 7.2c (Maximum Likelihood Estimator of a Poisson Parameter) Suppose $X_1, \dots, X_n$ are independent Poisson random variables each having mean λ. Determine the maximum likelihood estimator of λ.

SOLUTION The likelihood function is given by

$$f(x_1, \dots, x_n|\lambda) = \frac{e^{-\lambda}\lambda^{x_1}}{x_1!} \cdots \frac{e^{-\lambda}\lambda^{x_n}}{x_n!}$$
$$= \frac{e^{-n\lambda}\lambda^{\Sigma_1^n x_i}}{x_1! \dots x_n!}$$

Thus,

$$\log f(x_1, \dots, x_n|\lambda) = -n\lambda + \sum_1^n x_i \log \lambda - \log c$$

where $c = \prod_{i=1}^{n} x_i!$ does not depend on λ. Differentiation yields

$$\frac{d}{d\lambda} \log f(x_1, \dots, x_n|\lambda) = -n + \frac{\sum_1^n x_i}{\lambda}$$

By equating to zero, we obtain that the maximum likelihood estimate $\hat{\lambda}$ equals

$$\hat{\lambda} = \frac{\sum\limits_{1}^{n} x_i}{n}$$

and so the maximum likelihood estimator is given by

$$d(X_1, \ldots, X_n) = \frac{\sum\limits_{i=1}^{n} X_i}{n}$$

For example, suppose that the number of people who enter a certain retail establishment in any day is a Poisson random variable having an unknown mean λ, which must be estimated. If after 20 days a total of 857 people have entered the establishment, then the maximum likelihood estimate of λ is $857/20 = 42.85$. That is, we estimate that on average, 42.85 customers will enter the establishment on a given day. ∎

EXAMPLE 7.2d The number of traffic accidents in Berkeley, California, in 10 randomly chosen nonrainy days in 1998 is as follows:

$$4, 0, 6, 5, 2, 1, 2, 0, 4, 3$$

Use these data to estimate the proportion of nonrainy days that had 2 or fewer accidents that year.

SOLUTION Since there are a large number of drivers, each of whom has a small probability of being involved in an accident in a given day, it seems reasonable to assume that the daily number of traffic accidents is a Poisson random variable. Since

$$\overline{X} = \frac{1}{10} \sum_{i=1}^{10} X_i = 2.7$$

it follows that the maximum likelihood estimate of the Poisson mean is 2.7. Since the long-run proportion of nonrainy days that have 2 or fewer accidents is equal to $P\{X \leq 2\}$, where X is the random number of accidents in a day, it follows that the desired estimate is

$$e^{-2.7}(1 + 2.7 + (2.7)^2/2) = .4936$$

That is, we estimate that a little less than half of the nonrainy days had 2 or fewer accidents. ∎

EXAMPLE 7.2e (Maximum Likelihood Estimator in a Normal Population) Suppose $X_1, \ldots, X_n$ are independent, normal random variables each with unknown mean μ and unknown standard deviation σ. The joint density is given by

$$f(x_1, \ldots, x_n | \mu, \sigma) = \prod_{i=1}^{n} \frac{1}{\sqrt{2\pi}\,\sigma} \exp\left[\frac{-(x_i - \mu)^2}{2\sigma^2}\right]$$

$$= \left(\frac{1}{2\pi}\right)^{n/2} \frac{1}{\sigma^n} \exp\left[\frac{-\sum_{1}^{n}(x_i - \mu)^2}{2\sigma^2}\right]$$

The logarithm of the likelihood is thus given by

$$\log f(x_1, \ldots, x_n | \mu, \sigma) = -\frac{n}{2} \log(2\pi) - n \log \sigma - \frac{\sum_{1}^{n}(x_i - \mu)^2}{2\sigma^2}$$

In order to find the value of μ and σ maximizing the foregoing, we compute

$$\frac{\partial}{\partial \mu} \log f(x_1, \ldots, x_n | \mu, \sigma) = \frac{\sum_{i=1}^{n}(x_i - \mu)}{\sigma^2}$$

$$\frac{\partial}{\partial \sigma} \log f(x_1, \ldots, x_n | \mu, \sigma) = -\frac{n}{\sigma} + \frac{\sum_{1}^{n}(x_i - \mu)^2}{\sigma^3}$$

Equating these equations to zero yields that

$$\hat{\mu} = \sum_{i=1}^{n} x_i / n$$

and

$$\hat{\sigma} = \left[\sum_{i=1}^{n}(x_i - \hat{\mu})^2 / n\right]^{1/2}$$

Hence, the maximum likelihood estimators of μ and σ are given, respectively, by

$$\overline{X} \quad \text{and} \quad \left[\sum_{i=1}^{n}(X_i - \overline{X})^2/n\right]^{1/2} \tag{7.2.3}$$

It should be noted that the maximum likelihood estimator of the standard deviation σ differs from the sample standard deviation

$$S = \left[\sum_{i=1}^{n}(X_i - \overline{X})^2/(n-1)\right]^{1/2}$$

in that the denominator in Equation 7.2.3 is $\sqrt{n}$ rather than $\sqrt{n-1}$. However, for n of reasonable size, these two estimators of σ will be approximately equal. ■

EXAMPLE 7.2f *Kolmogorov's law of fragmentation* states that the size of an individual particle in a large collection of particles resulting from the fragmentation of a mineral compound will have an approximate lognormal distribution, where a random variable X is said to have a *lognormal* distribution if $\log(X)$ has a normal distribution. The law, which was first noted empirically and then later given a theoretical basis by Kolmogorov, has been applied to a variety of engineering studies. For instance, it has been used in the analysis of the size of randomly chosen gold particles from a collection of gold sand. A less obvious application of the law has been to a study of the stress release in earthquake fault zones (see Lomnitz, C., "Global Tectonics and Earthquake Risk," *Developments in Geotectonics*, Elsevier, Amsterdam, 1979).

Suppose that a sample of 10 grains of metallic sand taken from a large sand pile have respective lengths (in millimeters):

$$2.2, \ 3.4, \ 1.6, \ 0.8, \ 2.7, \ 3.3, \ 1.6, \ 2.8, \ 2.5, \ 1.9$$

Estimate the percentage of sand grains in the entire pile whose length is between 2 and 3 mm.

SOLUTION Taking the natural logarithm of these 10 data values, the following transformed data set results

$$.7885, \ 1.2238, \ .4700, \ -.2231, \ .9933, \ 1.1939, \ .4700, \ 1.0296, \ .9163, \ .6419$$

Because the sample mean and sample standard deviation of these data are

$$\overline{x} = .7504, \quad s = .4351$$

it follows that the logarithm of the length of a randomly chosen grain has a normal distribution with mean approximately equal to .7504 and with standard deviation approximately equal to .4351. Hence, if X is the length of the grain, then

$$P\{2 < X < 3\} = P\{\log(2) < \log(X) < \log(3)\}$$

$$= P\left\{\frac{\log(2) - .7504}{.4351} < \frac{\log(X) - .7504}{.4351} < \frac{\log(3) - .7504}{.4351}\right\}$$

$$= P\left\{-.1316 < \frac{\log(X) - .7504}{.4351} < .8003\right\}$$

$$\approx \Phi(.8003) - \Phi(-.1316)$$

$$= .3405 \quad \blacksquare$$

The lognormal distribution is often assumed in situations where the random variable under interest can be regarded as the product of a large number of independent and identically distributed random variables. For instance, it is commonly used in finance as the distribution of the price of a security at some future time. To see why this might be reasonable, suppose that the current price of the security is s and that we are interested in $S(t)$, the price of the security after an additional time t. For a large value n, let $t_i = it/n$, and consider $S(t_1), \ldots, S(t_n)$, the prices of the security at the times $t_1, \ldots, t_n$. Now, a common assumption in finance is that the ratios $S(t_i)/S(t_{i-1})$ are approximately independent and identically distributed. Consequently, if we let $X_i = S(t_i)/S(t_{i-1})$, then writing

$$S(t) = S(t_n) \quad = \quad S(t_0) \cdot \frac{S(t_1)}{S(t_0)} \cdot \frac{S(t_2)}{S(t_1)} \cdots \frac{S(t_n)}{S(t_{n-1})}$$

$$= \quad s \prod_{i=1}^{n} X_i$$

we obtain, upon taking logarithms, that

$$\log(S(t)) = \log(s) + \sum_{i=1}^{n} \log(X_i)$$

Thus, by the central limit theorem $\log(S(t))$ will approximately have a normal distribution.

The lognormal distribution has also been shown to be a good fit for such random variables as length of patient stays in hospitals, and vehicle travel times.

In all of the foregoing examples, the maximum likelihood estimator of the population mean turned out to be the sample mean $\overline{X}$. To show that this is not always the situation, consider the following example.

EXAMPLE 7.2g (Estimating the Mean of a Uniform Distribution) Suppose $X_1, \ldots, X_n$ constitute a sample from a uniform distribution on $(0, \theta)$, where θ is unknown. Their joint density is thus

$$f(x_1, x_2, \ldots, x_n | \theta) = \begin{cases} \dfrac{1}{\theta^n} & 0 < x_i < \theta, \quad i = 1, \ldots, n \\ 0 & \text{otherwise} \end{cases}$$

This density is maximized by choosing θ as small as possible. Since θ must be at least as large as all of the observed values x_i, it follows that the smallest possible choice of θ is equal to $\max(x_1, x_2, \ldots, x_n)$. Hence, the maximum likelihood estimator of θ is

$$\hat{\theta} = \max(X_1, X_2, \ldots, X_n)$$

It easily follows from the foregoing that the maximum likelihood estimator of $\theta/2$, the mean of the distribution, is $\max(X_1, X_2, \ldots, X_n)/2$. ∎

*7.2.1 ESTIMATING LIFE DISTRIBUTIONS

Let X denote the age at death of a randomly chosen child born today. That is, $X = i$ if the newborn dies in its ith year, $i \geq 1$. To estimate the probability mass function of X, let λ_i denote the probability that a newborn who has survived his or her first $i - 1$ years dies in year i. That is,

$$\lambda_i = P\{X = i | X > i - 1\} = \frac{P\{X = i\}}{P\{X > i - 1\}}$$

Also, let

$$s_i = 1 - \lambda_i = \frac{P\{X > i\}}{P\{X > i - 1\}}$$

be the probability that a newborn who survives her first $i - 1$ years also survives year i. The quantity λ_i is called the *failure rate*, and s_i is called the *survival rate*, of an individual who is entering his or her ith year. Now,

$$s_1 s_2 \cdots s_i = P\{X > 1\} \frac{P\{X > 2\} P\{X > 3\}}{P\{X > 1\} P\{X > 2\}} \cdots \frac{P\{X > i\}}{P\{X > i - 1\}}$$
$$= P\{X > i\}$$

Therefore,

$$P\{X = n\} = P\{X > n - 1\} \lambda_n = s_1 \cdots s_{n-1}(1 - s_n)$$

Consequently, we can estimate the probability mass function of X by estimating the quantities s_i, $i = 1, \ldots, n$. The value s_i can be estimated by looking at all individuals in the

---

* Optional section.

population who reached age i 1 year ago, and then letting the estimate $\hat{s}_i$ be the fraction of them who are alive today. We would then use $\hat{s}_1\hat{s}_2\cdots\hat{s}_{n-1}(1-\hat{s}_n)$ as the estimate of $P\{X = n\}$. (Note that although we are using the most recent possible data to estimate the quantities s_i, our estimate of the probability mass function of the lifetime of a newborn assumes that the survival rate of the newborn when it reaches age i will be the same as last year's survival rate of someone of age i.)

The use of the survival rate to estimate a life distribution is also of importance in health studies with partial information. For instance, consider a study in which a new drug is given to a random sample of 12 lung cancer patients. Suppose that after some time we have the following data on the number of months of survival after starting the new drug:

$$4, 7^*, 9, 11^*, 12, 3, 14^*, 1, 8, 7, 5, 3^*$$

where x means that the patient died in month x after starting the drug treatment, and x^* means that the patient has taken the drug for x months and is still alive.

Let X equal the number of months of survival after beginning the drug treatment, and let

$$s_i = P\{X > i | X > i - 1\} = \frac{P\{X > i\}}{P\{X > i - 1\}}$$

To estimate s_i, the probability that a patient who has survived the first $i - 1$ months will also survive month i, we should take the fraction of those patients who began their ith month of drug taking and survived the month. For instance, because 11 of the 12 patients survived month 1, $\hat{s}_1 = 11/12$. Because all 11 patients who began month 2 survived, $\hat{s}_2 = 11/11$. Because 10 of the 11 patients who began month 3 survived, $\hat{s}_3 = 10/11$. Because 8 of the 9 patients who began their fourth month of taking the drug (the 9 being all but the ones labelled 1, 3, and 3^*) survived month 4, $\hat{s}_4 = 8/9$. Similar reasoning holds for the others, giving the following survival rate estimates:

$$\hat{s}_1 = 11/12$$
$$\hat{s}_2 = 11/11$$
$$\hat{s}_3 = 10/11$$
$$\hat{s}_4 = 8/9$$
$$\hat{s}_5 = 7/8$$
$$\hat{s}_6 = 7/7$$
$$\hat{s}_7 = 6/7$$
$$\hat{s}_8 = 4/5$$
$$\hat{s}_9 = 3/4$$
$$\hat{s}_{10} = 3/3$$
$$\hat{s}_{11} = 3/3$$

$$\hat{s}_{12} = 1/2$$
$$\hat{s}_{13} = 1/1$$
$$\hat{s}_{14} = 1/1$$

We can now use $\prod_{i=1}^{j} \hat{s}_i$ to estimate the probability that a drug taker survives at least j time periods, $j = 1, \ldots, 14$. For instance, our estimate of $P\{X > 6\}$ is 35/54.

7.3 INTERVAL ESTIMATES

Suppose that $X_1, \ldots, X_n$ is a sample from a normal population having unknown mean μ and known variance σ^2. It has been shown that $\overline{X} = \sum_{i=1}^{n} X_i/n$ is the maximum likelihood estimator for μ. However, we don't expect that the sample mean $\overline{X}$ will exactly equal μ, but rather that it will "be close." Hence, rather than a point estimate, it is sometimes more valuable to be able to specify an interval for which we have a certain degree of confidence that μ lies within. To obtain such an interval estimator, we make use of the probability distribution of the point estimator. Let us see how it works for the preceding situation.

In the foregoing, since the point estimator $\overline{X}$ is normal with mean μ and variance σ^2/n, it follows that

$$\frac{\overline{X} - \mu}{\sigma/\sqrt{n}} = \sqrt{n}\frac{(\overline{X} - \mu)}{\sigma}$$

has a standard normal distribution. Therefore,

$$P\left\{-1.96 < \sqrt{n}\frac{(\overline{X} - \mu)}{\sigma} < 1.96\right\} = .95$$

or, equivalently,

$$P\left\{-1.96\frac{\sigma}{\sqrt{n}} < \overline{X} - \mu < 1.96\frac{\sigma}{\sqrt{n}}\right\} = .95$$

Multiplying through by -1 yields the equivalent statement

$$P\left\{-1.96\frac{\sigma}{\sqrt{n}} < \mu - \overline{X} < 1.96\frac{\sigma}{\sqrt{n}}\right\} = .95$$

or, equivalently,

$$P\left\{\overline{X} - 1.96\frac{\sigma}{\sqrt{n}} < \mu < \overline{X} + 1.96\frac{\sigma}{\sqrt{n}}\right\} = .95$$

That is, 95 percent of the time the value of the sample average $\overline{X}$ will be such that the distance between it and the mean μ will be less than $1.96\,\sigma/\sqrt{n}$. If we now observe the sample and it turns out that $\overline{X} = \overline{x}$, then we say that "with 95 percent confidence"

$$\overline{x} - 1.96\frac{\sigma}{\sqrt{n}} < \mu < \overline{x} + 1.96\frac{\sigma}{\sqrt{n}} \tag{7.3.1}$$

That is, "with 95 percent confidence" we assert that the true mean lies within $1.96\,\sigma/\sqrt{n}$ of the observed sample mean. The interval

$$\left(\bar{x} - 1.96\frac{\sigma}{\sqrt{n}},\ \bar{x} + 1.96\frac{\sigma}{\sqrt{n}}\right)$$

is called a *95 percent confidence interval estimate* of μ.

EXAMPLE 7.3a Suppose that when a signal having value μ is transmitted from location A the value received at location B is normally distributed with mean μ and variance 4. That is, if μ is sent, then the value received is $\mu + N$ where N, representing noise, is normal with mean 0 and variance 4. To reduce error, suppose the same value is sent 9 times. If the successive values received are 5, 8.5, 12, 15, 7, 9, 7.5, 6.5, 10.5, let us construct a 95 percent confidence interval for μ.

Since

$$\bar{x} = \frac{81}{9} = 9$$

It follows, under the assumption that the values received are independent, that a 95 percent confidence interval for μ is

$$\left(9 - 1.96\frac{\sigma}{3},\ 9 + 1.96\frac{\sigma}{3}\right) = (7.69, 10.31)$$

Hence, we are "95 percent confident" that the true message value lies between 7.69 and 10.31. ■

The interval in Equation 7.3.1 is called a *two-sided confidence interval*. Sometimes, however, we are interested in determining a value so that we can assert with, say, 95 percent confidence, that μ is at least as large as that value.

To determine such a value, note that if Z is a standard normal random variable then

$$P\{Z < 1.645\} = .95$$

As a result,

$$P\left\{\sqrt{n}\frac{(\bar{X} - \mu)}{\sigma} < 1.645\right\} = .95$$

or

$$P\left\{\bar{X} - 1.645\frac{\sigma}{\sqrt{n}} < \mu\right\} = .95$$

Thus, a 95 percent *one-sided upper confidence interval* for μ is

$$\left(\bar{x} - 1.645\frac{\sigma}{\sqrt{n}}, \infty\right)$$

where $\bar{x}$ is the observed value of the sample mean.

A *one-sided lower confidence interval* is obtained similarly; when the observed value of the sample mean is $\bar{x}$, then the 95 percent one-sided lower confidence interval for μ is

$$\left(-\infty, \bar{x} + 1.645\frac{\sigma}{\sqrt{n}}\right)$$

EXAMPLE 7.3b Determine the upper and lower 95 percent confidence interval estimates of μ in Example 7.3a.

SOLUTION Since

$$1.645\frac{\sigma}{\sqrt{n}} = \frac{3.29}{3} = 1.097$$

the 95 percent upper confidence interval is

$$(9 - 1.097, \infty) = (7.903, \infty)$$

and the 95 percent lower confidence interval is

$$(-\infty, 9 + 1.097) = (-\infty, 10.097) \quad \blacksquare$$

We can also obtain confidence intervals of any specified level of confidence. To do so, recall that z_α is such that

$$P\{Z > z_\alpha\} = \alpha$$

when Z is a standard normal random variable. But this implies (see Figure 7.1) that for any α

$$P\{-z_{\alpha/2} < Z < z_{\alpha/2}\} = 1 - \alpha$$

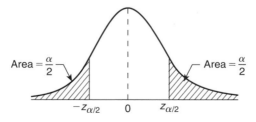

FIGURE 7.1 $P\{-z_{\alpha/2} < Z < z_{\alpha/2}\} = 1 - \alpha.$

As a result, we see that

$$P\left\{-z_{\alpha/2} < \sqrt{n}\frac{(\overline{X}-\mu)}{\sigma} < z_{\alpha/2}\right\} = 1-\alpha$$

or

$$P\left\{-z_{\alpha/2}\frac{\sigma}{\sqrt{n}} < \overline{X}-\mu < z_{\alpha/2}\frac{\sigma}{\sqrt{n}}\right\} = 1-\alpha$$

or

$$P\left\{-z_{\alpha/2}\frac{\sigma}{\sqrt{n}} < \mu-\overline{X} < z_{\alpha/2}\frac{\sigma}{\sqrt{n}}\right\} = 1-\alpha$$

That is,

$$P\left\{\overline{X}-z_{\alpha/2}\frac{\sigma}{\sqrt{n}} < \mu < \overline{X}+z_{\alpha/2}\frac{\sigma}{\sqrt{n}}\right\} = 1-\alpha$$

Hence, a $100(1-\alpha)$ percent two-sided confidence interval for μ is

$$\left(\overline{x}-z_{\alpha/2}\frac{\sigma}{\sqrt{n}}, \quad \overline{x}+z_{\alpha/2}\frac{\sigma}{\sqrt{n}}\right)$$

where $\overline{x}$ is the observed sample mean.

Similarly, knowing that $Z = \sqrt{n}\frac{(\overline{X}-\mu)}{\sigma}$ is a standard normal random variable, along with the identities

$$P\{Z > z_{\alpha}\} = \alpha$$

and

$$P\{Z < -z_{\alpha}\} = \alpha$$

results in one-sided confidence intervals of any desired level of confidence. Specifically, we obtain that

$$\left(\overline{x}-z_{\alpha}\frac{\sigma}{\sqrt{n}}, \quad \infty\right)$$

and

$$\left(-\infty, \quad \overline{x}+z_{\alpha}\frac{\sigma}{\sqrt{n}}\right)$$

are, respectively, $100(1-\alpha)$ percent one-sided upper and $100(1-\alpha)$ percent one-sided lower confidence intervals for μ.

EXAMPLE 7.3c Use the data of Example 7.3a to obtain a 99 percent confidence interval estimate of μ, along with 99 percent one-sided upper and lower intervals.

SOLUTION Since $z_{.005} = 2.58$, and

$$2.58\frac{\alpha}{\sqrt{n}} = \frac{5.16}{3} = 1.72$$

it follows that a 99 percent confidence interval for μ is

$$9 \pm 1.72$$

That is, the 99 percent confidence interval estimate is (7.28, 10.72).

Also, since $z_{.01} = 2.33$, a 99 percent upper confidence interval is

$$(9 - 2.33(2/3), \infty) = (7.447, \infty)$$

Similarly, a 99 percent lower confidence interval is

$$(-\infty, 9 + 2.33(2/3)) = (-\infty, 10.553) \quad \blacksquare$$

Sometimes we are interested in a two-sided confidence interval of a certain level, say $1 - \alpha$, and the problem is to choose the sample size n so that the interval is of a certain size. For instance, suppose that we want to compute an interval of length .1 that we can assert, with 99 percent confidence, contains μ. How large need n be? To solve this, note that as $z_{.005} = 2.58$ it follows that the 99 percent confidence interval for μ from a sample of size n is

$$\left(\bar{x} - 2.58\frac{\sigma}{\sqrt{n}}, \quad \bar{x} + 2.58\frac{\sigma}{\sqrt{n}} \right)$$

Hence, its length is

$$5.16\frac{\sigma}{\sqrt{n}}$$

Thus, to make the length of the interval equal to .1, we must choose

$$5.16\frac{\sigma}{\sqrt{n}} = .1$$

or

$$n = (51.6\,\sigma)^2$$

REMARK

The interpretation of "a $100(1 - \alpha)$ percent confidence interval" can be confusing. It should be noted that we are *not* asserting that the probability that $\mu \in (\bar{x} - 1.96\sigma/\sqrt{n}, \bar{x} + 1.96\sigma/\sqrt{n})$ is .95, for there are no random variables involved in this assertion. What we are asserting is that the technique utilized to obtain this interval is such that 95 percent of the time that it is employed it will result in an interval in which μ lies. In other words, before the data are observed we can assert that with probability .95 the interval that will be obtained will contain μ, whereas after the data are obtained we can only assert that the resultant interval indeed contains μ "with confidence .95."

EXAMPLE 7.3d From past experience it is known that the weights of salmon grown at a commercial hatchery are normal with a mean that varies from season to season but with a standard deviation that remains fixed at 0.3 pounds. If we want to be 95 percent certain that our estimate of the present season's mean weight of a salmon is correct to within ± 0.1 pounds, how large a sample is needed?

SOLUTION A 95 percent confidence interval estimate for the unknown mean μ, based on a sample of size n, is

$$\mu \in \left(\bar{x} - 1.96 \frac{\sigma}{\sqrt{n}}, \ \bar{x} + 1.96 \frac{\sigma}{\sqrt{n}} \right)$$

Because the estimate $\bar{x}$ is within $1.96(\sigma/\sqrt{n}) = .588/\sqrt{n}$ of any point in the interval, it follows that we can be 95 percent certain that $\bar{x}$ is within 0.1 of μ provided that

$$\frac{.588}{\sqrt{n}} \leq 0.1$$

That is, provided that

$$\sqrt{n} \geq 5.88$$

or

$$n \geq 34.57$$

That is, a sample size of 35 or larger will suffice. ■

7.3.1 Confidence Interval for a Normal Mean When the Variance Is Unknown

Suppose now that $X_1, \ldots, X_n$ is a sample from a normal distribution with unknown mean μ and unknown variance σ^2, and that we wish to construct a $100(1 - \alpha)$ percent confidence interval for μ. Since σ is unknown, we can no longer base our interval on the fact that $\sqrt{n}(\bar{X} - \mu)/\sigma$ is a standard normal random variable. However, by letting $S^2 = \sum_{i=1}^{n}(X_i - \bar{X})^2/(n - 1)$ denote the sample variance, then from Corollary 6.5.2 it follows that

$$\sqrt{n} \frac{(\bar{X} - \mu)}{S}$$

is a t-random variable with $n - 1$ degrees of freedom. Hence, from the symmetry of the t-density function (see Figure 7.2), we have that for any $\alpha \in (0, 1/2)$,

$$P \left\{ -t_{\alpha/2,n-1} < \sqrt{n} \frac{(\bar{X} - \mu)}{S} < t_{\alpha/2,n-1} \right\} = 1 - \alpha$$

or, equivalently,

$$P\{-t_{\alpha/2,n-1} \frac{S}{\sqrt{n}} < \bar{X} - \mu < t_{\alpha/2,n-1} \frac{S}{\sqrt{n}}\} = 1 - \alpha$$

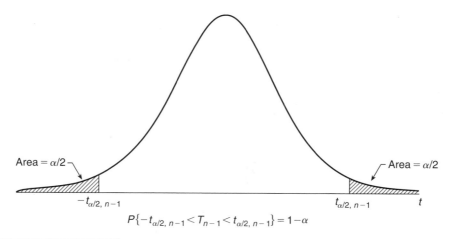

Area = $\alpha/2$

Area = $\alpha/2$

$-t_{\alpha/2,\,n-1}$

$t_{\alpha/2,\,n-1}$

t

$$P\{-t_{\alpha/2,\,n-1} < T_{n-1} < t_{\alpha/2,\,n-1}\} = 1 - \alpha$$

FIGURE 7.2   *t-density function.*

Multiplying all sides of the preceding by -1 and then adding $\bar{X}$ yields that

$$P\left\{\bar{X} - t_{\alpha/2,n-1}\frac{S}{\sqrt{n}} < \mu < \bar{X} + t_{\alpha/2,n-1}\frac{S}{\sqrt{n}}\right\} = 1 - \alpha$$

Thus, if it is observed that $\bar{X} = \bar{x}$ and $S = s$, then we can say that "with $100(1 - \alpha)$ percent confidence"

$$\mu \in \left(\bar{x} - t_{\alpha/2,n-1}\frac{s}{\sqrt{n}}, \ \bar{x} + t_{\alpha/2,n-1}\frac{s}{\sqrt{n}}\right)$$

EXAMPLE 7.3e   Let us again consider Example 7.3a but let us now suppose that when the value μ is transmitted at location A then the value received at location B is normal with mean μ and variance σ^2 but with σ^2 being unknown. If 9 successive values are, as in Example 7.3a, 5, 8.5, 12, 15, 7, 9, 7.5, 6.5, 10.5, compute a 95 percent confidence interval for μ.

SOLUTION   A simple calculation yields that

$$\bar{x} = 9$$

and

$$s^2 = \frac{\sum x_i^2 - 9(\bar{x})^2}{8} = 9.5$$

or

$$s = 3.082$$

Hence, as $t_{.025,8} = 2.306$, a 95 percent confidence interval for μ is

$$\left[9 - 2.306\frac{(3.082)}{3}, \ 9 + 2.306\frac{(3.082)}{3}\right] = (6.63, 11.37)$$

a larger interval than obtained in Example 7.3a. The reason why the interval just obtained is larger than the one in Example 7.3a is twofold. The primary reason is that we have a larger estimated variance than in Example 7.3a. That is, in Example 7.3a we assumed that σ^2 was known to equal 4, whereas in this example we assumed it to be unknown and our estimate of it turned out to be 9.5, which resulted in a larger confidence interval. In fact, the confidence interval would have been larger than in Example 7.3a even if our estimate of σ^2 was again 4 because by having to estimate the variance we need to utilize the t-distribution, which has a greater variance and thus a larger spread than the standard normal (which can be used when σ^2 is assumed known). For instance, if it had turned out that $\bar{x} = 9$ and $s^2 = 4$, then our confidence interval would have been

$$(9 - 2.306 \cdot \tfrac{2}{3}, 9 + 2.306 \cdot \tfrac{2}{3}) = (7.46, 10.54)$$

which is larger than that obtained in Example 7.3a.   ■

REMARKS

(a) The confidence interval for μ when σ is known is based on the fact that $\sqrt{n}(\bar{X} - \mu)/\sigma$ has a standard normal distribution. When σ is unknown, the foregoing approach is to estimate it by S and then use the fact that $\sqrt{n}(\bar{X} - \mu)/S$ has a t-distribution with $n - 1$ degrees of freedom.

(b) The length of a $100(1 - 2\alpha)$ percent confidence interval for μ is not always larger when the variance is unknown. For the length of such an interval is $2z_\alpha \sigma/\sqrt{n}$ when σ is known, whereas it is $2t_{\alpha, n-1} S/\sqrt{n}$ when σ is unknown; and it is certainly possible that the sample standard deviation S can turn out to be much smaller than σ. However, it can be shown that the mean length of the interval is longer when σ is unknown. That is, it can be shown that

$$t_{\alpha, n-1} E[S] \geq z_\alpha \sigma$$

Indeed, $E[S]$ is evaluated in Chapter 14 and it is shown, for instance, that

$$E[S] = \begin{cases} .94\,\sigma & \text{when } n = 5 \\ .97\,\sigma & \text{when } n = 9 \end{cases}$$

Since
$$z_{.025} = 1.96, \qquad t_{.025,4} = 2.78, \qquad t_{.025,8} = 2.31$$

the length of a 95 percent confidence interval from a sample of size 5 is $2 \times 1.96\sigma/\sqrt{5} = 1.75\sigma$ when σ is known, whereas its expected length is $2 \times 2.78 \times .94\sigma/\sqrt{5} = 2.34\sigma$ when σ is unknown — an increase of 33.7 percent. If the sample is of size 9, then the two values to compare are 1.31σ and 1.49σ — a gain of 13.7 percent.   ■

A one-sided upper confidence interval can be obtained by noting that

$$P\left\{\sqrt{n}\frac{(\overline{X} - \mu)}{S} < t_{\alpha,n-1}\right\} = 1 - \alpha$$

or

$$P\left\{\overline{X} - \mu < \frac{S}{\sqrt{n}}t_{\alpha,n-1}\right\} = 1 - \alpha$$

or

$$P\left\{\mu > \overline{X} - \frac{S}{\sqrt{n}}t_{\alpha,n-1}\right\} = 1 - \alpha$$

Hence, if it is observed that $\overline{X} = \overline{x}$, $S = s$, then we can assert "with $100(1 - \alpha)$ percent confidence" that

$$\mu \in \left(\overline{x} - \frac{s}{\sqrt{n}}t_{\alpha,n-1},\ \infty\right)$$

Similarly, a $100(1 - \alpha)$ lower confidence interval would be

$$\mu \in \left(-\infty,\ \overline{x} + \frac{s}{\sqrt{n}}t_{\alpha,n-1}\right)$$

Program 7.3.1 will compute both one- and two-sided confidence intervals for the mean of a normal distribution when the variance is unknown.

EXAMPLE 7.3f   Determine a 95 percent confidence interval for the average resting pulse of the members of a health club if a random selection of 15 members of the club yielded the data 54, 63, 58, 72, 49, 92, 70, 73, 69, 104, 48, 66, 80, 64, 77. Also determine a 95 percent lower confidence interval for this mean.

SOLUTION   We use Program 7.3.1 to obtain the solution (see Figure 7.3).   ∎

Our derivations of the $100(1 - \alpha)$ percent confidence intervals for the population mean μ have assumed that the population distribution is normal. However, even when this is not the case, if the sample size is reasonably large then the intervals obtained will still be approximate $100(1-\alpha)$ percent confidence intervals for μ. This is true because, by the central limit theorem, $\sqrt{n}(\overline{X} - \mu)/\sigma$ will have approximately a normal distribution, and $\sqrt{n}(\overline{X} - \mu)/S$ will have approximately a t-distribution.

EXAMPLE 7.3g   Simulation provides a powerful method for evaluating single and multi-dimensional integrals. For instance, let f be a function of an r-valued vector $(y_1, \ldots, y_r)$, and suppose that we want to estimate the quantity θ, defined by

$$\theta = \int_0^1 \int_0^1 \cdots \int_0^1 f(y_1, y_2, \ldots, y_r)\, dy_1\, dy_2, \ldots, dy_r$$

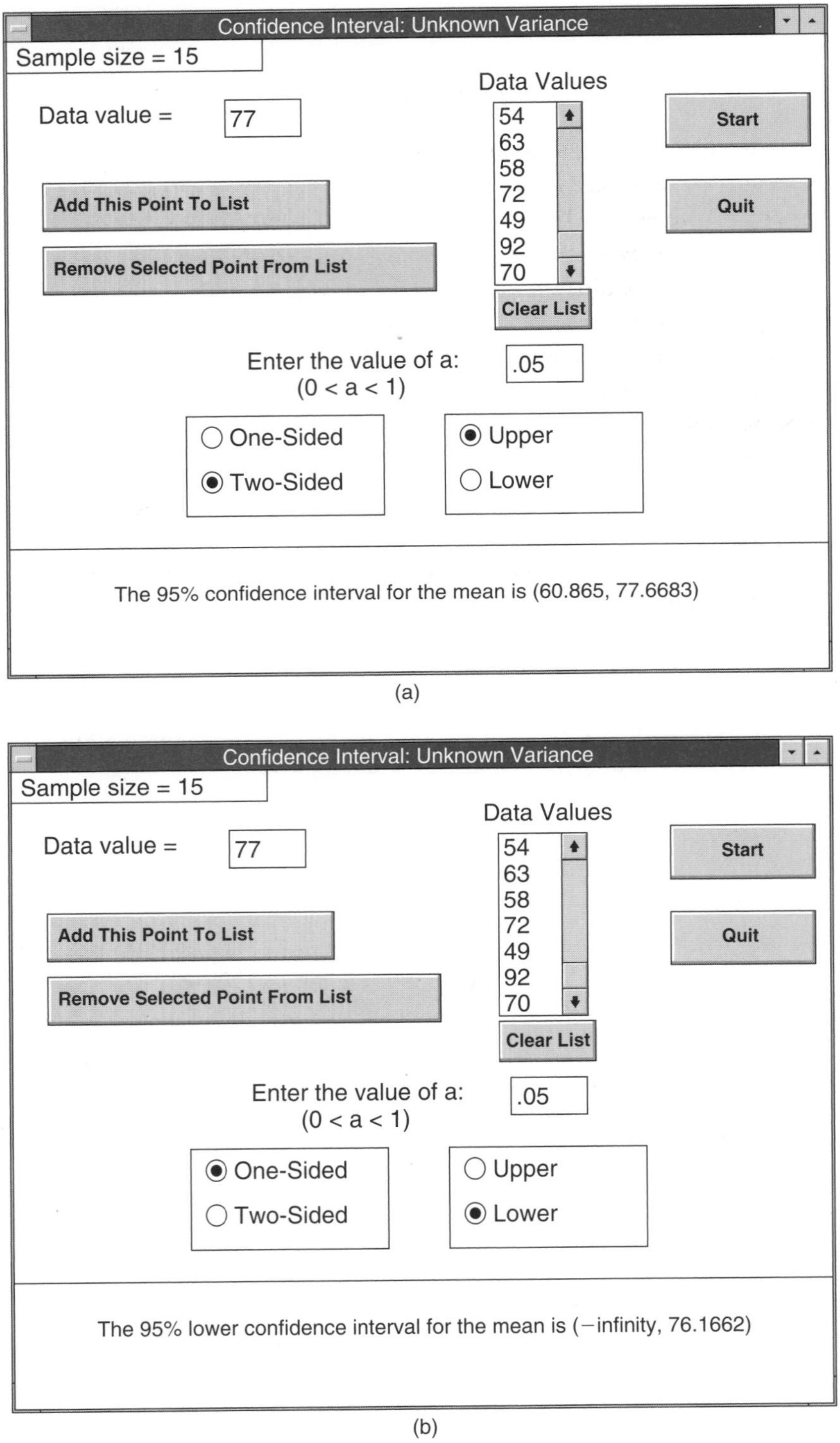

(a)

(b)

FIGURE 7.3 *(a) Two-sided and (b) lower 95 percent confidence intervals for Example 7.3f.*

To accomplish this, note that if $U_1, U_2, \ldots, U_r$ are independent uniform random variables on (0, 1), then

$$\theta = E[f(U_1, U_2, \ldots, U_r)]$$

Now, the values of independent uniform (0, 1) random variables can be approximated on a computer (by so-called *pseudo random numbers*); if we generate a vector of r of them, and evaluate f at this vector, then the value obtained, call it X_1, will be a random variable with mean θ. If we now repeat this process, then we obtain another value, call it X_2, which will have the same distribution as X_1. Continuing on, we can generate a sequence $X_1, X_2, \ldots, X_n$ of independent and identically distributed random variables with mean θ; we then use their observed values to estimate θ. This method of approximating integrals is called *Monte Carlo simulation*.

For instance, suppose we wanted to estimate the one-dimensional integral

$$\theta = \int_0^1 \sqrt{1 - y^2} \, dy = E[\sqrt{1 - U^2}]$$

where U is a uniform (0, 1) random variable. To do so, let $U_1, \ldots, U_{100}$ be independent uniform (0, 1) random variables, and set

$$X_i = \sqrt{1 - U_i^2}, \qquad i = 1, \ldots, 100$$

In this way, we have generated a sample of 100 random variables having mean θ. Suppose that the computer generated values of $U_1, \ldots, U_{100}$, resulting in $X_1, \ldots, X_{100}$ having sample mean .786 and sample standard deviation .03. Consequently, since $t_{.025,99} = 1.985$, it follows that a 95 percent confidence interval for θ would be given by

$$.786 \pm 1.985(.003)$$

As a result, we could assert, with 95 percent confidence, that θ (which can be shown to equal $\pi/4$) is between .780 and .792. ∎

7.3.2 PREDICTION INTERVALS

Suppose that $X_1, \ldots, X_n, X_{n+1}$ is a sample from a normal distribution with unknown mean μ and unknown variance σ^2. Suppose further that the values of $X_1, \ldots, X_n$ are to be observed and that we want to use them to predict the value of X_{n+1}. To begin, note that if the mean μ were known, then it would be the natural predictor for X_{n+1}. As it is not known, it seems natural to use its current estimator after observing $X_1, \ldots, X_n$, namely the average of these observed values, as the predicted value of X_{n+1}. That is, we should use the observed value of $\bar{X}_n = \sum_{i=1}^{n} X_i/n$, as the predicted value of X_{n+1}.

Suppose now that we want to determine an interval in which we predict, with a certain degree of confidence, that X_{n+1} will lie. To obtain such a prediction interval, note that as $\bar{X}_n$ is normal with mean μ and variance σ^2/n, and is independent of X_{n+1} which is normal with mean μ and variance σ^2, it follows that $X_{n+1} - \bar{X}_n$ is normal with mean 0 and variance $\sigma^2/n + \sigma^2$. Consequently,

$$\frac{X_{n+1} - \bar{X}_n}{\sigma\sqrt{1 + 1/n}} \qquad \text{is a standard normal random variable.}$$

Because this is independent of $S_n^2 = \sum_{i=1}^n (X_i - \bar{X}_n)^2/(n-1)$, it follows from the same argument used to establish Corollary 6.5.2, that replacing σ by its estimator S_n in the preceding expression will yield a t-random variable with $n-1$ degrees of freedom. That is,

$$\frac{X_{n+1} - \bar{X}_n}{S_n\sqrt{1 + 1/n}}$$

is a t-random variable with $n-1$ degrees of freedom. Hence, for any $\alpha \in (0, 1/2)$,

$$P\{-t_{\alpha/2,n-1} < \frac{X_{n+1} - \bar{X}_n}{S_n\sqrt{1 + 1/n}} < t_{\alpha/2,n-1}\} = 1 - \alpha$$

which is equivalent to

$$P\{\bar{X}_n - t_{\alpha/2,n-1}\, S_n\sqrt{1 + 1/n} < X_{n+1} < \bar{X}_n + t_{\alpha/2,n-1}\, S_n\sqrt{1 + 1/n}\}$$

Hence, if the observed values of $\bar{X}_n$ and S_n are, respectively, $\bar{x}_n$ and s_n, then we can predict, with $100(1-\alpha)$ percent confidence, that X_{n+1} will lie between $\bar{x}_n - t_{\alpha/2,n-1}\, s_n\sqrt{1 + 1/n}$ and $\bar{x}_n + t_{\alpha/2,n-1}\, s_n\sqrt{1 + 1/n}$. That is, with $100(1-\alpha)$ percent confidence, we can predict that

$$X_{n+1} \in \left(\bar{x}_n - t_{\alpha/2,n-1}\, s_n\sqrt{1 + 1/n},\ \ \bar{x}_n + t_{\alpha/2,n-1}\, s_n\sqrt{1 + 1/n} \right)$$

EXAMPLE 7.3h   The following are the number of steps walked in each of the last 7 days

$$6822 \quad 5333 \quad 7420 \quad 7432 \quad 6252 \quad 7005 \quad 6752$$

Assuming that the daily number of steps can be thought of as being independent realizations from a normal distribution, give a prediction interval that, with 95 percent confidence, will contain the number of steps that will be walked tomorrow.

SOLUTION   A simple calculation gives that the sample mean and sample variance of the 7 data values are

$$\bar{X}_7 = 6716.57 \qquad S_7 = 733.97$$

Because $t_{.025,6} = 2.447$, and $2.4447 \cdot 733.97\sqrt{1 + 1/7} = 1920.03$, we can predict, with 95 percent confidence, that tomorrow's number of steps will be between $6716.57 - 1920.03$ and $6716.57 + 1920.03$. That is, with 95 percent confidence, X_8 will lie in the interval $(4796.54,\ \ 8636.60)$.

7.3.3 CONFIDENCE INTERVALS FOR THE VARIANCE OF A NORMAL DISTRIBUTION

If $X_1, \ldots, X_n$ is a sample from a normal distribution having unknown parameters μ and σ^2, then we can construct a confidence interval for σ^2 by using the fact that

$$(n-1)\frac{S^2}{\sigma^2} \sim \chi^2_{n-1}$$

Hence,

$$P\left\{\chi^2_{1-\alpha/2,n-1} \leq (n-1)\frac{S^2}{\sigma^2} \leq \chi^2_{\alpha/2,n-1}\right\} = 1 - \alpha$$

or, equivalently,

$$P\left\{\frac{(n-1)S^2}{\chi^2_{\alpha/2,n-1}} \leq \sigma^2 \leq \frac{(n-1)S^2}{\chi^2_{1-\alpha/2,n-1}}\right\} = 1 - \alpha$$

Hence when $S^2 = s^2$, a $100(1-\alpha)$ percent confidence interval for σ^2 is

$$\left(\frac{(n-1)s^2}{\chi^2_{\alpha/2,n-1}}, \frac{(n-1)s^2}{\chi^2_{1-\alpha/2,n-1}}\right)$$

EXAMPLE 7.3i   A standardized procedure is expected to produce washers with very small deviation in their thicknesses. Suppose that 10 such washers were chosen and measured. If the thicknesses of these washers were, in inches,

.123	.133
.124	.125
.126	.128
.120	.124
.130	.126

what is a 90 percent confidence interval for the standard deviation of the thickness of a washer produced by this procedure?

SOLUTION   A computation gives that

$$S^2 = 1.366 \times 10^{-5}$$

Because $\chi^2_{.05,9} = 16.917$ and $\chi^2_{.95,9} = 3.334$, and because

$$\frac{9 \times 1.366 \times 10^{-5}}{16.917} = 7.267 \times 10^{-6}, \qquad \frac{9 \times 1.366 \times 10^{-5}}{3.334} = 36.875 \times 10^{-6}$$

it follows that, with confidence .90,

$$\sigma^2 \in (7.267 \times 10^{-6}, \quad 36.875 \times 10^{-6})$$

Taking square roots yields that, with confidence .90,

$$\sigma \in (2.696 \times 10^{-3}, \quad 6.072 \times 10^{-3}) \quad \blacksquare$$

One-sided confidence intervals for σ^2 are obtained by similar reasoning and are presented in Table 7.1, which sums up the results of this section.

7.4 ESTIMATING THE DIFFERENCE IN MEANS OF TWO NORMAL POPULATIONS

Let $X_1, X_2, \ldots, X_n$ be a sample of size n from a normal population having mean μ_1 and variance σ_1^2 and let $Y_1, \ldots, Y_m$ be a sample of size m from a different normal population having mean μ_2 and variance σ_2^2 and suppose that the two samples are independent of each other. We are interested in estimating $\mu_1 - \mu_2$.

Since $\overline{X} = \sum_{i=1}^{n} X_i/n$ and $\overline{Y} = \sum_{i=1}^{m} Y_i/m$ are the maximum likelihood estimators of μ_1 and μ_2 it seems intuitive (and can be proven) that $\overline{X} - \overline{Y}$ is the maximum likelihood estimator of $\mu_1 - \mu_2$.

To obtain a confidence interval estimator, we need the distribution of $\overline{X} - \overline{Y}$. Because

$$\overline{X} \sim \mathcal{N}(\mu_1, \sigma_1^2/n)$$
$$\overline{Y} \sim \mathcal{N}(\mu_2, \sigma_2^2/m)$$

it follows from the fact that the sum of independent normal random variables is also normal, that

$$\overline{X} - \overline{Y} \sim \mathcal{N}\left(\mu_1 - \mu_2, \frac{\sigma_1^2}{n} + \frac{\sigma_2^2}{m}\right)$$

TABLE 7.1 *100(1 − α) Percent Confidence Intervals*

$$X_1, \ldots, X_n \sim \mathcal{N}(\mu, \sigma^2)$$

$$\overline{X} = \sum_{i=1}^{n} X_i/n, \qquad S = \sqrt{\sum_{i=1}^{n}(X_i - \overline{X})^2/(n-1)}$$

Assumption	Parameter	Confidence Interval	Lower Interval	Upper Interval
σ^2 known	μ	$\overline{X} \pm z_{\alpha/2}\dfrac{\sigma}{\sqrt{n}}$	$\left(-\infty, \overline{X} + z_\alpha \dfrac{\sigma}{\sqrt{n}}\right)$	$\left(\overline{X} + z_\alpha \dfrac{\sigma}{\sqrt{n}}, \infty\right)$
σ^2 unknown	μ	$\overline{X} \pm t_{\alpha/2, n-1}\dfrac{S}{\sqrt{n}}$	$\left(-\infty, \overline{X} + t_{\alpha, n-1} \dfrac{S}{\sqrt{n}}\right)$	$\left(\overline{X} - t_{\alpha, n-1} \dfrac{S}{\sqrt{n}}, \infty\right)$
μ unknown	σ^2	$\left(\dfrac{(n-1)S^2}{\chi^2_{\alpha/2, n-1}}, \dfrac{(n-1)S^2}{\chi^2_{1-\alpha/2, n-1}}\right)$	$\left(0, \dfrac{(n-1)S^2}{\chi^2_{1-\alpha, n-1}}\right)$	$\left(\dfrac{(n-1)S^2}{\chi^2_{\alpha, n-1}}, \infty\right)$

Hence, assuming σ_1^2 and σ_2^2 are known, we have that

$$\frac{\overline{X} - \overline{Y} - (\mu_1 - \mu_2)}{\sqrt{\dfrac{\sigma_1^2}{n} + \dfrac{\sigma_2^2}{m}}} \sim \mathcal{N}(0, 1) \tag{7.4.1}$$

and so

$$P\left\{ -z_{\alpha/2} < \frac{\overline{X} - \overline{Y} - (\mu_1 - \mu_2)}{\sqrt{\dfrac{\sigma_1^2}{n} + \dfrac{\sigma_2^2}{m}}} < z_{\alpha/2} \right\} = 1 - \alpha$$

or, equivalently,

$$P\left\{ \overline{X} - \overline{Y} - z_{\alpha/2}\sqrt{\frac{\sigma_1^2}{n} + \frac{\sigma_2^2}{m}} < \mu_1 - \mu_2 < \overline{X} - \overline{Y} + z_{\alpha/2}\sqrt{\frac{\sigma_1^2}{n} + \frac{\sigma_2^2}{m}} \right\} = 1 - \alpha$$

Hence, if $\overline{X}$ and $\overline{Y}$ are observed to equal $\overline{x}$ and $\overline{y}$, respectively, then a $100(1-\alpha)$ two-sided confidence interval estimate for $\mu_1 - \mu_2$ is

$$\mu_1 - \mu_2 \in \left(\overline{x} - \overline{y} - z_{\alpha/2}\sqrt{\frac{\sigma_1^2}{n} + \frac{\sigma_2^2}{m}}, \ \overline{x} - \overline{y} + z_{\alpha/2}\sqrt{\frac{\sigma_1^2}{n} + \frac{\sigma_2^2}{m}} \right)$$

One-sided confidence intervals for $\mu_1 - \mu_2$ are obtained in a similar fashion, and we leave it for the reader to verify that a $100(1 - \alpha)$ percent one-sided interval is given by

$$\mu_1 - \mu_2 \in \left(-\infty, \ \overline{x} - \overline{y} + z_{\alpha}\sqrt{\sigma_1^2/n + \sigma_2^2/m} \right)$$

Program 7.4.1 will compute both one- and two-sided confidence intervals for $\mu_1 - \mu_2$.

EXAMPLE 7.4a Two different types of electrical cable insulation have recently been tested to determine the voltage level at which failures tend to occur. When specimens were subjected to an increasing voltage stress in a laboratory experiment, failures for the two types of cable insulation occurred at the following voltages:

Type A		Type B	
36	54	52	60
44	52	64	44
41	37	38	48
53	51	68	46
38	44	66	70
36	35	52	62
34	44		

Suppose that it is known that the amount of voltage that cables having type A insulation can withstand is normally distributed with unknown mean μ_A and known variance $\sigma_A^2 = 40$, whereas the corresponding distribution for type B insulation is normal with unknown mean μ_B and known variance $\sigma_B^2 = 100$. Determine a 95 percent confidence interval for $\mu_A - \mu_B$. Determine a value that we can assert, with 95 percent confidence, exceeds $\mu_A - \mu_B$.

SOLUTION   We run Program 7.4.1 to obtain the solution (see Figure 7.4).   ■

Let us suppose now that we again desire an interval estimator of $\mu_1 - \mu_2$ but that the population variances σ_1^2 and σ_2^2 are unknown. In this case, it is natural to try to replace σ_1^2 and σ_2^2 in Equation 7.4.1 by the sample variances

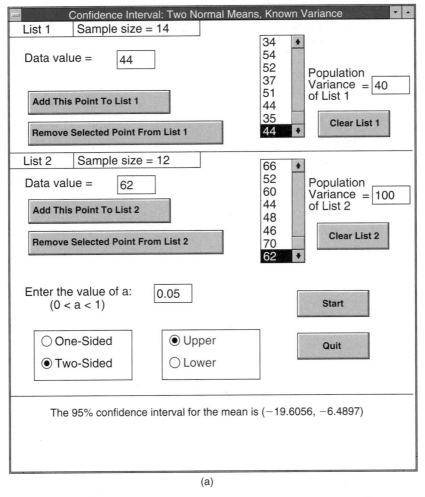

(a)

FIGURE 7.4   *(a) Two-sided and (b) lower 95 percent confidence intervals for Example 7.4a.*

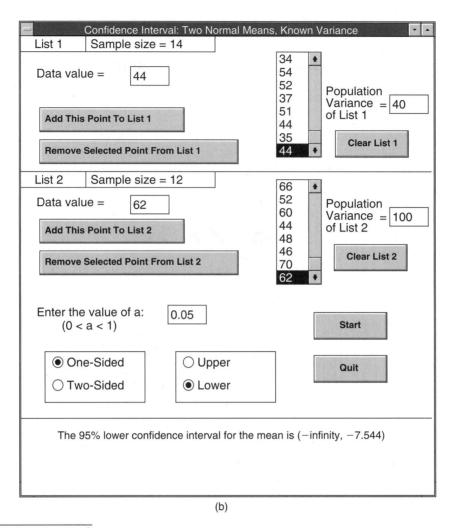

(b)

FIGURE 7.4   *(continued)*

$$S_1^2 = \sum_{i=1}^{n} \frac{(X_i - \overline{X})^2}{n-1}$$

$$S_2^2 = \sum_{i=1}^{m} \frac{(Y_i - \overline{Y})^2}{m-1}$$

That is, it is natural to base our interval estimate on something like

$$\frac{\overline{X} - \overline{Y} - (\mu_1 - \mu_2)}{\sqrt{S_1^2/n + S_2^2/m}}$$

However, to utilize the foregoing to obtain a confidence interval, we need its distribution and it must not depend on any of the unknown parameters σ_1^2 and σ_2^2. Unfortunately, this distribution is both complicated and does indeed depend on the unknown parameters σ_1^2 and σ_2^2. In fact, it is only in the special case when $\sigma_1^2 = \sigma_2^2$ that we will be able to obtain an interval estimator. So let us suppose that the population variances, though unknown, are equal and let σ^2 denote their common value. Now, from Theorem 6.5.1 it follows that

$$(n-1)\frac{S_1^2}{\sigma^2} \sim \chi_{n-1}^2$$

and

$$(m-1)\frac{S_2^2}{\sigma^2} \sim \chi_{m-1}^2$$

Also, because the samples are independent, it follows that these two chi-square random variables are independent. Hence, from the additive property of chi-square random variables, which states that the sum of independent chi-square random variables is also chi-square with a degree of freedom equal to the sum of their degrees of freedom, it follows that

$$(n-1)\frac{S_1^2}{\sigma^2} + (m-1)\frac{S_2^2}{\sigma^2} \sim \chi_{n+m-2}^2 \qquad (7.4.2)$$

Also, since

$$\overline{X} - \overline{Y} \sim \mathcal{N}\left(\mu_1 - \mu_2, \frac{\sigma^2}{n} + \frac{\sigma^2}{m}\right)$$

we see that

$$\frac{\overline{X} - \overline{Y} - (\mu_1 - \mu_2)}{\sqrt{\dfrac{\sigma^2}{n} + \dfrac{\sigma^2}{m}}} \sim \mathcal{N}(0,1) \qquad (7.4.3)$$

Now it follows from the fundamental result that in normal sampling $\overline{X}$ and S^2 are independent (Theorem 6.5.1), that $\overline{X}_1, S_1^2, \overline{X}_2, S_2^2$ are independent random variables. Hence, using the definition of a t-random variable (as the ratio of two independent random variables, the numerator being a standard normal and the denominator being the square root of a chi-square random variable divided by its degree of freedom parameter), it follows from Equations 7.4.2 and 7.4.3 that if we let

$$S_p^2 = \frac{(n-1)S_1^2 + (m-1)S_2^2}{n+m-2}$$

then

$$\frac{\overline{X} - \overline{Y} - (\mu_1 - \mu_2)}{\sqrt{\sigma^2(1/n + 1/m)}} \div \sqrt{S_p^2/\sigma^2} = \frac{\overline{X} - \overline{Y} - (\mu_1 - \mu_2)}{\sqrt{S_p^2(1/n + 1/m)}}$$

has a t-distribution with $n + m - 2$ degrees of freedom. Consequently,

$$P\left\{-t_{\alpha/2,n+m-2} \le \frac{\overline{X} - \overline{Y} - (\mu_1 - \mu_2)}{S_p\sqrt{1/n + 1/m}} \le t_{\alpha/2,n+m-2}\right\} = 1 - \alpha$$

Therefore, when the data result in the values $\overline{X} = \overline{x}, \overline{Y} = \overline{y}, S_p = s_p$, we obtain the following $100(1 - \alpha)$ percent confidence interval for $\mu_1 - \mu_2$:

$$\left(\overline{x} - \overline{y} - t_{\alpha/2,n+m-2}\, s_p\sqrt{1/n + 1/m}, \quad \overline{x} - \overline{y} + t_{\alpha/2,n+m-2}\, s_p\sqrt{1/n + 1/m}\right) \quad (7.4.4)$$

One-sided confidence intervals are similarly obtained.

Program 7.4.2 can be used to obtain both one- and two-sided confidence intervals for the difference in means in two normal populations having unknown but equal variances.

EXAMPLE 7.4b   There are two different techniques a given manufacturer can employ to produce batteries. A random selection of 12 batteries produced by technique I and of 14 produced by technique II resulted in the following capacities (in ampere hours):

Technique I		Technique II	
140	132	144	134
136	142	132	130
138	150	136	146
150	154	140	128
152	136	128	131
144	142	150	137
		130	135

Determine a 90 percent level two-sided confidence interval for the difference in means, assuming a common variance. Also determine a 95 percent upper confidence interval for $\mu_{\text{I}} - \mu_{\text{II}}$.

SOLUTION   We run Program 7.4.2 to obtain the solution (see Figure 7.5). ∎

REMARK

The confidence interval given by Equation 7.4.4 was obtained under the assumption that the population variances are equal; with σ^2 as their common value, it follows that

$$\frac{\overline{X} - \overline{Y} - (\mu_1 - \mu_2)}{\sqrt{\sigma^2/n + \sigma^2/m}} = \frac{\overline{X} - \overline{Y} - (\mu_1 - \mu_2)}{\sigma\sqrt{1/n + 1/m}}$$

has a standard normal distribution. However, since σ^2 is unknown this result cannot be immediately applied to obtain a confidence interval; σ^2 must first be estimated. To do so, note that both sample variances are estimators of σ^2; moreover, since S_1^2 has $n-1$ degrees of freedom and S_2^2 has $m-1$, the appropriate estimator is to take a weighted average of the two sample variances, with the weights proportional to these degrees of freedom. That is, the estimator of σ^2 is the *pooled estimator*

$$S_p^2 = \frac{(n-1)S_1^2 + (m-1)S_2^2}{n+m-2}$$

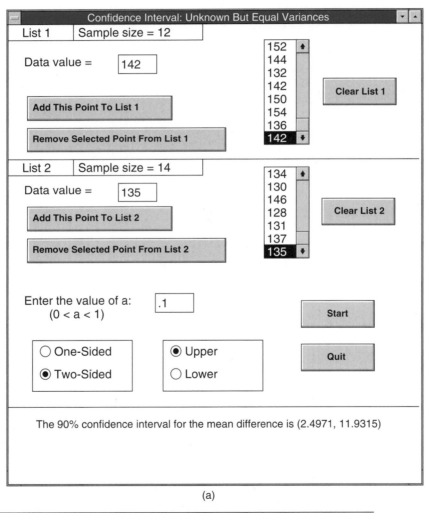

(a)

FIGURE 7.5   *(a) Two-sided and (b) upper 90 percent confidence intervals for Example 7.4b.*

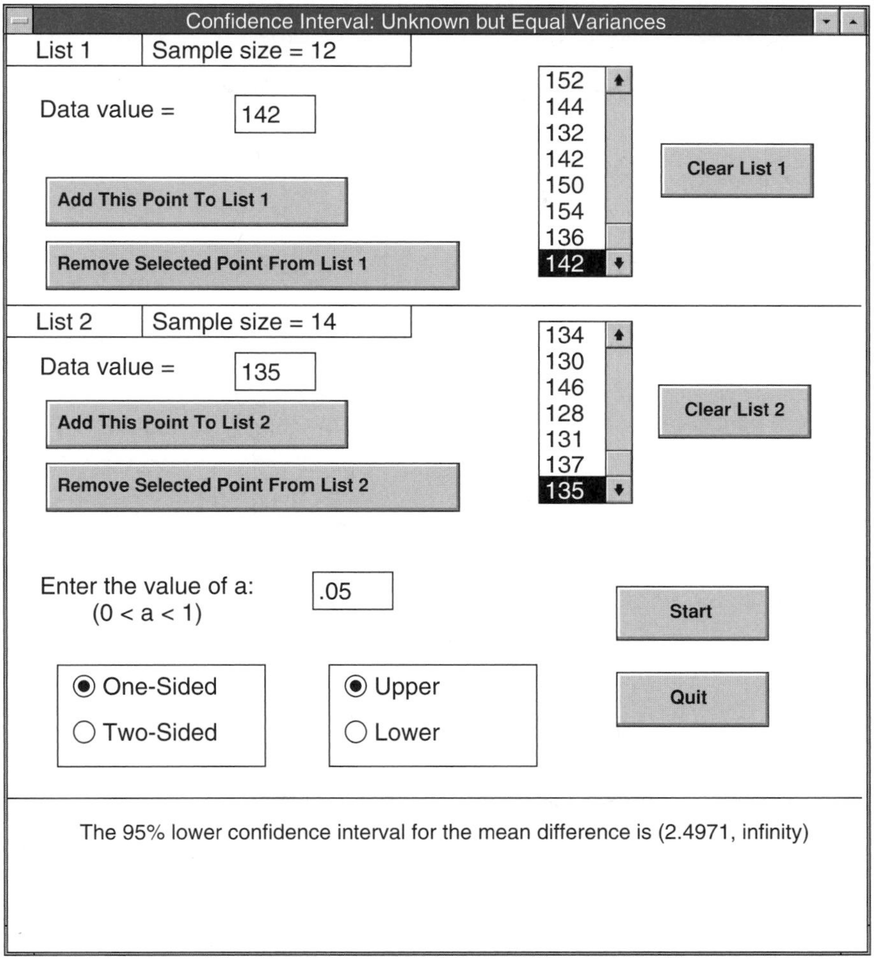

Confidence Interval: Unknown but Equal Variances		▾ ▴

List 1   Sample size = 12

Data value =   [142]

[Add This Point To List 1]

[Remove Selected Point From List 1]

```
152 ▲
144
132
142
150
154
136
142 ▼
```

[Clear List 1]

List 2   Sample size = 14

Data value =   [135]

[Add This Point To List 2]

[Remove Selected Point From List 2]

```
134 ▲
130
146
128
131
137
135 ▼
```

[Clear List 2]

Enter the value of a: [.05]
     (0 < a < 1)

[Start]

◉ One-Sided   ◉ Upper

○ Two-Sided   ○ Lower

[Quit]

The 95% lower confidence interval for the mean difference is (2.4971, infinity)

(b)

FIGURE 7.5   *(continued)*

and the confidence interval is then based on the statistic

$$\frac{\overline{X} - \overline{Y} - (\mu_1 - \mu_2)}{\sqrt{S_p^2}\sqrt{1/n + 1/m}}$$

which, by our previous analysis, has a t-distribution with $n + m - 2$ degrees of freedom. The results of this section are summarized in Table 7.2.

TABLE 7.2 *$100(1 - \sigma)$ Percent Confidence Intervals for $\mu_1 - \mu_2$*

$$X_1, \ldots, X_n \sim \mathcal{N}(\mu_1, \sigma_1^2)$$
$$Y_1, \ldots, Y_m \sim \mathcal{N}(\mu_2, \sigma_2^2)$$

$$\overline{X} = \sum_{i=1}^{n} X_i/n, \qquad S_1^2 = \sum_{i=1}^{n} (X_i - \overline{X})^2/(n-1)$$

$$\overline{Y} = \sum_{i=1}^{m} Y_i/n, \qquad S_2^2 = \sum_{i=1}^{m} (Y_i - \overline{Y})^2/(m-1)$$

Assumption	Confidence Interval
σ_1, σ_2 known	$\overline{X} - \overline{Y} \pm z_{\alpha/2}\sqrt{\sigma_1^2/n + \sigma_2^2/m}$
σ_1, σ_2 unknown but equal	$\overline{X} - \overline{Y} \pm t_{\alpha/2,\, n+m-2}\sqrt{\left(\dfrac{1}{n} + \dfrac{1}{m}\right)\dfrac{(n-1)S_1^2 + (m-1)S_2^2}{n+m-2}}$

Assumption	Lower Confidence Interval
σ_1, σ_2 known	$\left(-\infty,\ \overline{X} - \overline{Y} + z_\alpha\sqrt{\sigma_1^2/n + \sigma_2^2/m}\,\right)$
σ_1, σ_2 unknown but equal	$\left(-\infty,\ \overline{X} - \overline{Y} + t_{\alpha,\, n+m-2}\sqrt{\left(\dfrac{1}{n} + \dfrac{1}{m}\right)\dfrac{(n-1)S_1^2 + (m-1)S_2^2}{n+m-2}}\,\right)$

Note: Upper confidence intervals for $\mu_1 - \mu_2$ are obtained from lower confidence intervals for $\mu_2 - \mu_1$.

7.5   APPROXIMATE CONFIDENCE INTERVAL FOR THE MEAN OF A BERNOULLI RANDOM VARIABLE

Consider a population of items, each of which independently meets certain standards with some unknown probability p. If n of these items are tested to determine whether they meet the standards, how can we use the resulting data to obtain a confidence interval for p?

If we let X denote the number of the n items that meet the standards, then X is a binomial random variable with parameters n and p. Thus, when n is large, it follows by the normal approximation to the binomial that X is approximately normally distributed with mean np and variance $np(1 - p)$. Hence,

$$\frac{X - np}{\sqrt{np(1-p)}} \stackrel{\cdot}{\sim} \mathcal{N}(0, 1) \tag{7.5.1}$$

where $\overset{\cdot}{\sim}$ means "is approximately distributed as." Therefore, for any $\alpha \in (0, 1)$,

$$P\left\{-z_{\alpha/2} < \frac{X - np}{\sqrt{np(1-p)}} < z_{\alpha/2}\right\} \approx 1 - \alpha$$

and so if X is observed to equal x, then an approximate $100(1 - \alpha)$ percent confidence *region* for p is

$$\left\{p : -z_{\alpha/2} < \frac{x - np}{\sqrt{np(1-p)}} < z_{\alpha/2}\right\}$$

The foregoing region, however, is not an interval. To obtain a confidence *interval* for p, let $\hat{p} = X/n$ be the fraction of the items that meet the standards. From Example 7.2a, $\hat{p}$ is the maximum likelihood estimator of p, and so should be approximately equal to p. As a result, $\sqrt{n\hat{p}(1-\hat{p})}$ will be approximately equal to $\sqrt{np(1-p)}$ and so from Equation 7.5.1 we see that

$$\frac{X - np}{\sqrt{n\hat{p}(1-\hat{p})}} \overset{\cdot}{\sim} \mathcal{N}(0, 1)$$

Hence, for any $\alpha \in (0, 1)$ we have that

$$P\left\{-z_{\alpha/2} < \frac{X - np}{\sqrt{n\hat{p}(1-\hat{p})}} < z_{\alpha/2}\right\} \approx 1 - \alpha$$

or, equivalently,

$$P\{-z_{\alpha/2}\sqrt{n\hat{p}(1-\hat{p})} < np - X < z_{\alpha/2}\sqrt{n\hat{p}(1-\hat{p})}\} \approx 1 - \alpha$$

Dividing all sides of the preceding inequality by n, and using that $\hat{p} = X/n$, the preceding can be written as

$$P\{\hat{p} - z_{\alpha/2}\sqrt{\hat{p}(1-\hat{p})/n} < p < \hat{p} + z_{\alpha/2}\sqrt{\hat{p}(1-\hat{p})/n}\} \approx 1 - \alpha$$

which yields an approximate $100(1 - \alpha)$ percent confidence interval for p.

EXAMPLE 7.5a A sample of 100 transistors is randomly chosen from a large batch and tested to determine if they meet the current standards. If 80 of them meet the standards, then an approximate 95 percent confidence interval for p, the fraction of all the transistors that meet the standards, is given by

$$(.8 - 1.96\sqrt{.8(.2)/100}, \ .8 + 1.96\sqrt{.8(.2)/100}) = (.7216, .8784)$$

That is, with "95 percent confidence," between 72.16 and 87.84 percent of all transistors meet the standards. ■

EXAMPLE 7.5b In August 2013, the *New York Times* reported that a recent poll indicated that 52 percent of the population was in favor of the job performance of President Obama, with a margin of error of ±4 percent. What does this mean? Can we infer how many people were questioned?

SOLUTION It has become common practice for the news media to present 95 percent confidence intervals. Since $z_{.025} = 1.96$, a 95 percent confidence interval for p, the percentage of the population that is in favor of President Obama's job performance, is given by

$$\hat{p} \pm 1.96\sqrt{\hat{p}(1-\hat{p})/n} = .52 \pm 1.96\sqrt{.52(.48)/n}$$

where n is the size of the sample. Since the "margin of error" is ±4 percent, it follows that

$$1.96\sqrt{.52(.48)/n} = .04$$

or

$$n = \frac{(1.96)^2(.52)(.48)}{(.04)^2} = 599.29$$

That is, approximately 599 people were sampled, and 52 percent of them reported favorably on President Obama's job performance. ∎

We often want to specify an approximate $100(1-\alpha)$ percent confidence interval for p that is no greater than some given length, say b. The problem is to determine the appropriate sample size n to obtain such an interval. To do so, note that the length of the approximate $100(1-\alpha)$ percent confidence interval for p from a sample of size n is

$$2z_{\alpha/2}\sqrt{\hat{p}(1-\hat{p})/n}$$

which is approximately equal to $2z_{\alpha/2}\sqrt{p(1-p)/n}$. Unfortunately, p is not known in advance, and so we cannot just set $2z_{\alpha/2}\sqrt{p(1-p)/n}$ equal to b to determine the necessary sample size n. What we can do, however, is to first take a preliminary sample to obtain a rough estimate of p, and then use this estimate to determine n. That is, we use p^*, the proportion of the preliminary sample that meets the standards, as a preliminary estimate of p; we then determine the total sample size n by solving the equation

$$2z_{\alpha/2}\sqrt{p^*(1-p^*)/n} = b$$

Squaring both sides of the preceding yields that

$$(2z_{\alpha/2})^2 p^*(1-p^*)/n = b^2$$

or

$$n = \frac{(2z_{\alpha/2})^2 p^*(1 - p^*)}{b^2}$$

That is, if k items were initially sampled to obtain the preliminary estimate of p, then an additional $n - k$ (or 0 if $n \leq k$) items should be sampled.

EXAMPLE 7.5c A certain manufacturer produces computer chips; each chip is independently acceptable with some unknown probability p. To obtain an approximate 99 percent confidence interval for p, whose length is approximately .05, an initial sample of 30 chips has been taken. If 26 of these chips are of acceptable quality, then the preliminary estimate of p is 26/30. Using this value, a 99 percent confidence interval of length approximately .05 would require an approximate sample of size

$$n = \frac{4(z_{.005})^2}{(.05)^2} \frac{26}{30} \left(1 - \frac{26}{30}\right) = \frac{4(2.58)^2}{(.05)^2} \frac{26}{30} \frac{4}{30} = 1{,}231$$

Hence, we should now sample an additional 1,201 chips and if, for instance, 1,040 of them are acceptable, then the final 99 percent confidence interval for p is

$$\left(\frac{1{,}066}{1{,}231} - \sqrt{1{,}066\left(1 - \frac{1{,}066}{1{,}231}\right)} \frac{z_{.005}}{1{,}231}, \; \frac{1{,}066}{1{,}231} + \sqrt{1{,}066\left(1 - \frac{1{,}066}{1{,}231}\right)} \frac{z_{.005}}{1{,}231}\right)$$

or

$$p \in (.84091, .89101) \quad \blacksquare$$

REMARK

As shown, a $100(1 - \alpha)$ percent confidence interval for p will be of approximate length b when the sample size is

$$n = \frac{(2z_{\alpha/2})^2}{b^2} p(1 - p)$$

Now it is easily shown that the function $g(p) = p(1 - p)$ attains its maximum value of $\frac{1}{4}$, in the interval $0 \leq p \leq 1$, when $p = \frac{1}{2}$. Thus an upper bound on n is

$$n \leq \frac{(z_{\alpha/2})^2}{b^2}$$

and so by choosing a sample whose size is at least as large as $(z_{\alpha/2})^2/b^2$, one can be assured of obtaining a confidence interval of length no greater than b without need of any additional sampling. $\blacksquare$

One-sided approximate confidence intervals for p are also easily obtained; Table 7.3 gives the results.

TABLE 7.3 *Approximate $100(1 - \alpha)$ Percent Confidence Intervals for p*
X Is a Binomial (n, p) Random Variable
$$\hat{p} = X/n$$

Type of Interval	Confidence Interval
Two-sided	$\hat{p} \pm z_{\alpha/2}\sqrt{\hat{p}(1 - \hat{p})/n}$
One-sided lower	$\left(-\infty, \hat{p} + z_{\alpha}\sqrt{\hat{p}(1 - \hat{p})/n}\right)$
One-sided upper	$\left(\hat{p} - z_{\alpha}\sqrt{\hat{p}(1 - \hat{p})/n}, \infty\right)$

*7.6 CONFIDENCE INTERVAL OF THE MEAN OF THE EXPONENTIAL DISTRIBUTION

If $X_1, X_2, \ldots, X_n$ are independent exponential random variables each having mean θ, then it can be shown that the maximum likelihood estimator of θ is the sample mean $\sum_{i=1}^{n} X_i/n$. To obtain a confidence interval estimator of θ, recall from Section 5.7 that $\sum_{i=1}^{n} X_i$ has a gamma distribution with parameters n, $1/\theta$. This in turn implies (from the relationship between the gamma and chi-square distribution shown in Section 5.8.1.1) that

$$\frac{2}{\theta} \sum_{i=1}^{n} X_i \sim \chi_{2n}^2$$

Hence, for any $\alpha \in (0, 1)$

$$P\left\{\chi_{1-\alpha/2, 2n}^2 < \frac{2}{\theta} \sum_{i=1}^{n} X_i < \chi_{\alpha/2, 2n}^2\right\} = 1 - \alpha$$

or, equivalently,

$$P\left\{\frac{2\sum_{i=1}^{n} X_i}{\chi_{\alpha/2, 2n}^2} < \theta < \frac{2\sum_{i=1}^{n} X_i}{\chi_{1-\alpha/2, 2n}^2}\right\} = 1 - \alpha$$

Hence, a $100(1 - \alpha)$ percent confidence interval for θ is

$$\theta \in \left(\frac{2\sum_{i=1}^{n} X_i}{\chi_{\alpha/2, 2n}^2}, \frac{2\sum_{i=1}^{n} X_i}{\chi_{1-\alpha/2, 2n}^2}\right)$$

---

* Optional section.

EXAMPLE 7.6a The successive items produced by a certain manufacturer are assumed to have useful lives that (in hours) are independent with a common density function

$$f(x) = \frac{1}{\theta}e^{-x/\theta}, \quad 0 < x < \infty$$

If the sum of the lives of the first 10 items is equal to 1,740, what is a 95 percent confidence interval for the population mean θ?

SOLUTION From Program 5.8.1b (or Table A2), we see that

$$\chi^2_{.025,20} = 34.169, \qquad \chi^2_{.975,20} = 9.661$$

and so we can conclude, with 95 percent confidence, that

$$\theta \in \left(\frac{3480}{34.169}, \frac{3480}{9.661} \right)$$

or, equivalently,

$$\theta \in (101.847, 360.211) \quad \blacksquare$$

*7.7 EVALUATING A POINT ESTIMATOR

Let $\mathbf{X} = (X_1, \ldots, X_n)$ be a sample from a population whose distribution is specified up to an unknown parameter θ, and let $d = d(\mathbf{X})$ be an estimator of θ. How are we to determine its worth as an estimator of θ? One way is to consider the square of the difference between $d(\mathbf{X})$ and θ. However, since $(d(\mathbf{X}) - \theta)^2$ is a random variable, let us agree to consider $r(d, \theta)$, the *mean square error* of the estimator d, which is defined by

$$r(d, \theta) = E[(d(\mathbf{X}) - \theta)^2]$$

as an indication of the worth of d as an estimator of θ.

It would be nice if there were a single estimator d that minimized $r(d, \theta)$ for all possible values of θ. However, except in trivial situations, this will never be the case. For example, consider the estimator d^* defined by

$$d^*(X_1, \ldots, X_n) = 4$$

That is, no matter what the outcome of the sample data, the estimator d^* chooses 4 as its estimate of θ. While this seems like a silly estimator (since it makes no use of the data), it is, however, true that when θ actually equals 4, the mean square error of this estimator is 0.

---

* Optional section.

Thus, the mean square error of any estimator different from d^* must, in most situations, be larger than the mean square error of d^* when $\theta = 4$.

Although minimum mean square estimators rarely exist, it is sometimes possible to find an estimator having the smallest mean square error among all estimators that satisfy a certain property. One such property is that of unbiasedness.

Definition

Let $d = d(\mathbf{X})$ be an estimator of the parameter θ. Then

$$b_\theta(d) = E[d(\mathbf{X})] - \theta$$

is called the *bias* of d as an estimator of θ. If $b_\theta(d) = 0$ for all θ, then we say that d is an *unbiased* estimator of θ. In other words, an estimator is unbiased if its expected value always equals the value of the parameter it is attempting to estimate.

EXAMPLE 7.7a Let $X_1, X_2, \ldots, X_n$ be a random sample from a distribution having unknown mean θ. Then

$$d_1(X_1, X_2, \ldots, X_n) = X_1$$

and

$$d_2(X_1, X_2, \ldots, X_n) = \frac{X_1 + X_2 + \cdots + X_n}{n}$$

are both unbiased estimators of θ since

$$E[X_1] = E\left[\frac{X_1 + X_2 + \cdots + X_n}{n}\right] = \theta$$

More generally, $d_3(X_1, X_2, \ldots, X_n) = \sum_{i=1}^n \lambda_i X_i$ is an unbiased estimator of θ whenever $\sum_{i=1}^n \lambda_i = 1$. This follows since

$$E\left[\sum_{i=1}^n \lambda_i X_i\right] = \sum_{i=1}^n E[\lambda_i X_i]$$
$$= \sum_{i=1}^n \lambda_i E(X_i)$$
$$= \theta \sum_{i=1}^n \lambda_i$$
$$= \theta \quad \blacksquare$$

If $d(X_1, \ldots, X_n)$ is an unbiased estimator, then its mean square error is given by

$$r(d, \theta) = E[(d(\mathbf{X}) - \theta)^2]$$
$$= E[(d(\mathbf{X}) - E[d(\mathbf{X})])^2] \qquad \text{since } d \text{ is unbiased}$$
$$= \text{Var}(d(\mathbf{X}))$$

Thus the mean square error of an unbiased estimator is equal to its variance.

EXAMPLE 7.7b (Combining Independent Unbiased Estimators) Let d_1 and d_2 denote independent unbiased estimators of θ, having known variances σ_1^2 and σ_2^2. That is, for $i = 1, 2$,

$$E[d_i] = \theta, \qquad \text{Var}(d_i) = \sigma_i^2$$

Any estimator of the form

$$d = \lambda d_1 + (1 - \lambda) d_2$$

will also be unbiased. To determine the value of λ that results in d having the smallest possible mean square error, note that

$$r(d, \theta) = \text{Var}(d)$$
$$= \lambda^2 \, \text{Var}(d_1) + (1 - \lambda)^2 \, \text{Var}(d_2)$$
$$\qquad \text{by the independence of } d_1 \text{ and } d_2$$
$$= \lambda^2 \sigma_1^2 + (1 - \lambda)^2 \sigma_2^2$$

Differentiation yields that

$$\frac{d}{d\lambda} r(d, \theta) = 2\lambda \sigma_1^2 - 2(1 - \lambda)\sigma_2^2$$

To determine the value of λ that minimizes $r(d, \theta)$ — call it $\hat{\lambda}$ — set this equal to 0 and solve for λ to obtain

$$2\hat{\lambda}\sigma_1^2 = 2(1 - \hat{\lambda})\sigma_2^2$$

or

$$\hat{\lambda} = \frac{\sigma_2^2}{\sigma_1^2 + \sigma_2^2} = \frac{1/\sigma_1^2}{1/\sigma_1^2 + 1/\sigma_2^2}$$

In words, the optimal weight to give an estimator is inversely proportional to its variance (when all the estimators are unbiased and independent).

For an application of the foregoing, suppose that a conservation organization wants to determine the acidity content of a certain lake. To determine this quantity, they draw some water from the lake and then send samples of this water to n different laboratories. These laboratories will then, independently, test for acidity content by using their respective titration equipment, which is of differing precision. Specifically, suppose that d_i, the result of a titration test at laboratory i, is a random variable having mean θ, the true acidity of the sample water, and variance σ_i^2, $i = 1, \ldots, n$. If the quantities σ_i^2, $i = 1, \ldots, n$ are known to the conservation organization, then they should estimate the acidity of the sampled water from the lake by

$$d = \frac{\sum\limits_{i=1}^{n} d_i/\sigma_i^2}{\sum\limits_{i=1}^{n} 1/\sigma_i^2}$$

The mean square error of d is as follows:

$$r(d, \theta) = \text{Var}(d) \qquad \text{since } d \text{ is unbiased}$$

$$= \left(\sum_{i=1}^{n} 1/\sigma_i^2 \right)^{-2} \sum_{i=1}^{n} \left(\frac{1}{\sigma_i^2} \right)^2 \sigma_i^2$$

$$= \frac{1}{\sum\limits_{i=1}^{n} 1/\sigma_i^2} \qquad \blacksquare$$

A generalization of the result that the mean square error of an unbiased estimator is equal to its variance is that the mean square error of any estimator is equal to its variance plus the square of its bias. This follows since

$$r(d, \theta) = E[(d(\mathbf{X}) - \theta)^2]$$
$$= E[(d - E[d] + E[d] - \theta)^2]$$
$$= E[(d - E[d])^2 + (E[d] - \theta)^2 + 2(E[d] - \theta)(d - E[d])]$$
$$= E[(d - E[d])^2] + E[(E[d] - \theta)^2]$$
$$\quad + 2E[(E[d] - \theta)(d - E[d])]$$
$$= E[(d - E[d])^2] + (E[d] - \theta)^2 + 2(E[d] - \theta)E[d - E[d]]$$
$$\qquad \text{since } E[d] - \theta \text{ is constant}$$
$$= E[(d - E[d])^2] + (E[d] - \theta)^2$$

The last equality follows since

$$E[d - E[d]] = 0$$

Hence

$$r(d, \theta) = \text{Var}(d) + b_\theta^2(d)$$

EXAMPLE 7.7c Let $X_1, \ldots, X_n$ denote a sample from a uniform $(0, \theta)$ distribution, where θ is assumed unknown. Since

$$E[X_i] = \frac{\theta}{2}$$

a "natural" estimator to consider is the unbiased estimator

$$d_1 = d_1(\mathbf{X}) = \frac{2 \sum\limits_{i=1}^{n} X_i}{n}$$

Since $E[d_1] = \theta$, it follows that

$$r(d_1, \theta) = \text{Var}(d_1)$$

$$= \frac{4}{n} \text{Var}(X_i)$$

$$= \frac{4}{n} \frac{\theta^2}{12} \quad \text{since } \text{Var}(X_i) = \frac{\theta^2}{12}$$

$$= \frac{\theta^2}{3n}$$

A second possible estimator of θ is the maximum likelihood estimator, which, as shown in Example 7.2d, is given by

$$d_2 = d_2(\mathbf{X}) = \max_i X_i$$

To compute the mean square error of d_2 as an estimator of θ, we need to first compute its mean (so as to determine its bias) and variance. To do so, note that the distribution function of d_2 is as follows:

$$F_2(x) \equiv P\{d_2(\mathbf{X}) \leq x\}$$

$$= P\{\max_i X_i \leq x\}$$

$$= P\{X_i \leq x \text{ for all } i = 1, \ldots, n\}$$

$$= \prod_{i=1}^{n} P\{X_i \leq x\} \qquad \text{by independence}$$

$$= \left(\frac{x}{\theta}\right)^n \qquad x \leq \theta$$

Hence, upon differentiating, we obtain that the density function of d_2 is

$$f_2(x) = \frac{nx^{n-1}}{\theta^n}, x \leq \theta$$

Therefore,

$$E[d_2] = \int_0^\infty x f_2(x)\, dx = \frac{n}{\theta^n} \int_0^\theta x^n\, dx = \frac{n}{n+1}\theta \qquad (7.7.1)$$

Also

$$E[d_2^2] = \frac{n}{\theta^n} \int_0^\theta x^{n+1}\, dx = \frac{n}{n+2}\theta^2$$

and so

$$\mathrm{Var}(d_2) = \frac{n}{n+2}\theta^2 - \left(\frac{n}{n+1}\theta\right)^2 \qquad (7.7.2)$$

$$= n\theta^2 \left[\frac{1}{n+2} - \frac{n}{(n+1)^2}\right] = \frac{n\theta^2}{(n+2)(n+1)^2}$$

Hence

$$r(d_2, \theta) = (E(d_2) - \theta)^2 + \mathrm{Var}(d_2) \qquad (7.7.3)$$

$$= \frac{\theta^2}{(n+1)^2} + \frac{n\theta^2}{(n+2)(n+1)^2}$$

$$= \frac{\theta^2}{(n+1)^2}\left[1 + \frac{n}{n+2}\right]$$

$$= \frac{2\theta^2}{(n+1)(n+2)}$$

Since

$$\frac{2\theta^2}{(n+1)(n+2)} \leq \frac{\theta^2}{3n} \qquad n = 1, 2, \ldots$$

it follows that d_2 is a superior estimator of θ than is d_1.

Equation 7.7.1 suggests the use of even another estimator — namely, the unbiased estimator $(1 + 1/n)d_2(\mathbf{X}) = (1 + 1/n)\max_i X_i$. However, rather than considering this estimator directly, let us consider all estimators of the form

$$d_c(\mathbf{X}) = c \max_i X_i = c\, d_2(\mathbf{X})$$

where c is a given constant. The mean square error of this estimator is

$$
\begin{aligned}
r(d_c(\mathbf{X}), \theta) &= \operatorname{Var}(d_c(\mathbf{X})) + (E[d_c(\mathbf{X})] - \theta)^2 \\
&= c^2 \operatorname{Var}(d_2(\mathbf{X})) + (cE[d_2(\mathbf{X})] - \theta)^2 \\
&= \frac{c^2 n \theta^2}{(n+2)(n+1)^2} + \theta^2 \left(\frac{c n}{n+1} - 1 \right)^2
\end{aligned}
$$

by Equations 7.7.2 and 7.7.1   (7.7.4)

To determine the constant c resulting in minimal mean square error, we differentiate to obtain

$$
\frac{d}{dc} r(d_c(\mathbf{X}), \theta) = \frac{2c\, n \theta^2}{(n+2)(n+1)^2} + \frac{2\theta^2 n}{n+1} \left(\frac{c n}{n+1} - 1 \right)
$$

Equating this to 0 shows that the best constant c — call it c^* — is such that

$$
\frac{c^*}{n+2} + c^* n - (n+1) = 0
$$

or

$$
c^* = \frac{(n+1)(n+2)}{n^2 + 2n + 1} = \frac{n+2}{n+1}
$$

Substituting this value of c into Equation 7.7.4 yields that

$$
\begin{aligned}
r\left(\frac{n+2}{n+1} \max_i X_i, \theta \right) &= \frac{(n+2)n\theta^2}{(n+1)^4} + \theta^2 \left(\frac{n(n+2)}{(n+1)^2} - 1 \right)^2 \\
&= \frac{(n+2)n\theta^2}{(n+1)^4} + \frac{\theta^2}{(n+1)^4} \\
&= \frac{\theta^2}{(n+1)^2}
\end{aligned}
$$

A comparison with Equation 7.7.3 shows that the (biased) estimator $(n+2)/(n+1) \max_i X_i$ has about half the mean square error of the maximum likelihood estimator $\max_i X_i$. ∎

*7.8 THE BAYES ESTIMATOR

In certain situations it seems reasonable to regard an unknown parameter θ as being the value of a random variable from a given probability distribution. This usually arises when, prior to the observance of the outcomes of the data $X_1, \ldots, X_n$, we have some information about the value of θ and this information is expressible in terms of a probability distribution (called appropriately the *prior* distribution of θ). For instance, suppose that from

---

* Optional section.

past experience we know that θ is equally likely to be near any value in the interval $(0, 1)$. Hence, we could reasonably assume that θ is chosen from a uniform distribution on $(0, 1)$.

Suppose now that our prior feelings about θ are that it can be regarded as being the value of a continuous random variable having probability density function $p(\theta)$; and suppose that we are about to observe the value of a sample whose distribution depends on θ. Specifically, suppose that $f(x|\theta)$ represents the likelihood — that is, it is the probability mass function in the discrete case or the probability density function in the continuous case — that a data value is equal to x when θ is the value of the parameter. If the observed data values are $X_i = x_i, i = 1, \ldots, n$, then the updated, or conditional, probability density function of θ is as follows:

$$f(\theta|x_1, \ldots, x_n) = \frac{f(\theta, x_1, \ldots, x_n)}{f(x_1, \ldots, x_n)}$$

$$= \frac{p(\theta)f(x_1, \ldots, x_n|\theta)}{\int f(x_1, \ldots, x_n|\theta)p(\theta)\, d\theta}$$

The conditional density function $f(\theta|x_1, \ldots, x_n)$ is called the *posterior* density function. (Thus, before observing the data, one's feelings about θ are expressed in terms of the prior distribution, whereas once the data are observed, this prior distribution is updated to yield the posterior distribution.)

Now we have shown that whenever we are given the probability distribution of a random variable, the best estimate of the value of that random variable, in the sense of minimizing the expected squared error, is its mean. Therefore, it follows that the best estimate of θ, given the data values $X_i = x_i, i = 1, \ldots, n$, is the mean of the posterior distribution $f(\theta|x_1, \ldots, x_n)$. This estimator, called the *Bayes estimator*, is written as $E[\theta|X_1, \ldots, X_n]$. That is, if $X_i = x_i, i = 1, \ldots, n$, then the value of the Bayes estimator is

$$E[\theta|X_1 = x_1, \ldots, X_n = x_n] = \int \theta f(\theta|x_1, \ldots, x_n)\, d\theta$$

EXAMPLE 7.8a Suppose that $X_1, \ldots, X_n$ are independent Bernoulli random variables, each having probability mass function given by

$$f(x|\theta) = \theta^x(1 - \theta)^{1-x}, \qquad x = 0, 1$$

where θ is unknown. Further, suppose that θ is chosen from a uniform distribution on $(0, 1)$. Compute the Bayes estimator of θ.

SOLUTION We must compute $E[\theta|X_1, \ldots, X_n]$. Since the prior density of θ is the uniform density

$$p(\theta) = 1, \qquad 0 < \theta < 1$$

we have that the conditional density of θ given $X_1, \ldots, X_n$ is given by

$$
\begin{aligned}
f(\theta|x_1, \ldots, x_n) &= \frac{f(x_1, \ldots, x_n, \theta)}{f(x_1, \ldots, x_n)} \\
&= \frac{f(x_1, \ldots, x_n|\theta)p(\theta)}{\int_0^1 f(x_1, \ldots, x_n|\theta)p(\theta)\, d\theta} \\
&= \frac{\theta^{\sum_1^n x_i}(1-\theta)^{n-\sum_1^n x_i}}{\int_0^1 \theta^{\sum_1^n x_i}(1-\theta)^{n-\sum_1^n x_i}\, d\theta}
\end{aligned}
$$

Now it can be shown that for integral values m and r

$$
\int_0^1 \theta^m(1-\theta)^r\, d\theta = \frac{m!\, r!}{(m+r+1)!} \tag{7.8.1}
$$

Hence, upon letting $x = \sum_{i=1}^n x_i$

$$
f(\theta|x_1, \ldots, x_n) = \frac{(n+1)!\ \theta^x(1-\theta)^{n-x}}{x!\ (n-x)!} \tag{7.8.2}
$$

Therefore,

$$
\begin{aligned}
E[\theta|x_1, \ldots, x_n] &= \frac{(n+1)!}{x!(n-x)!} \int_0^1 \theta^{1+x}(1-\theta)^{n-x}\, d\theta \\
&= \frac{(n+1)!}{x!(n-x)!}\, \frac{(1+x)!(n-x)!}{(n+2)!} \qquad \text{from Equation 7.8.1} \\
&= \frac{x+1}{n+2}
\end{aligned}
$$

Thus, the Bayes estimator is given by

$$
E[\theta|X_1, \ldots, X_n] = \frac{\sum\limits_{i=1}^n X_i + 1}{n+2}
$$

As an illustration, if 10 independent trials, each of which results in a success with probability θ, result in 6 successes, then assuming a uniform $(0, 1)$ prior distribution on θ, the Bayes estimator of θ is 7/12 (as opposed, for instance, to the maximum likelihood estimator of 6/10). ∎

REMARK

The conditional distribution of θ given that $X_i = x_i, i = 1, \ldots, n$, whose density function is given by Equation 7.8.2, is called the beta distribution with parameters $\sum_{i=1}^{n} x_i + 1$, $n - \sum_{i=1}^{n} x_i + 1$. ∎

EXAMPLE 7.8b   Suppose $X_1, \ldots, X_n$ are independent normal random variables, each having unknown mean θ and known variance σ_0^2. If θ is itself selected from a normal population having known mean μ and known variance σ^2, what is the Bayes estimator of θ?

SOLUTION   In order to determine $E[\theta|X_1, \ldots, X_n]$, the Bayes estimator, we need first determine the conditional density of θ given the values of $X_1, \ldots, X_n$. Now

$$f(\theta|x_1, \ldots, x_n) = \frac{f(x_1, \ldots, x_n|\theta)p(\theta)}{f(x_1, \ldots, x_n)}$$

where

$$f(x_1, \ldots, x_n|\theta) = \frac{1}{(2\pi)^{n/2}\sigma_0^n} \exp\left\{ -\sum_{i=1}^{n}(x_i - \theta)^2/2\sigma_0^2 \right\}$$

$$p(\theta) = \frac{1}{\sqrt{2\pi}\sigma} \exp\{-(\theta - \mu)^2/2\sigma^2\}$$

and

$$f(x_1, \ldots, x_n) = \int_{-\infty}^{\infty} f(x_1, \ldots, x_n|\theta)p(\theta)\, d\theta$$

With the help of a little algebra, it can now be shown that this conditional density is a *normal* density with mean

$$E[\theta|X_1, \ldots, X_n] = \frac{n\sigma^2}{n\sigma^2 + \sigma_0^2}\overline{X} + \frac{\sigma_0^2}{n\sigma^2 + \sigma_0^2}\mu \qquad (7.8.3)$$

$$= \frac{\frac{n}{\sigma_0^2}}{\frac{n}{\sigma_0^2} + \frac{1}{\sigma^2}}\overline{X} + \frac{\frac{1}{\sigma^2}}{\frac{n}{\sigma_0^2} + \frac{1}{\sigma^2}}\mu$$

and variance

$$\mathrm{Var}(\theta|X_1, \ldots, X_n) = \frac{\sigma_0^2\sigma^2}{n\sigma^2 + \sigma_0^2}$$

Writing the Bayes estimator as we did in Equation 7.8.3 is informative, for it shows that it is a weighted average of $\overline{X}$, the sample mean, and μ, the *a priori* mean. In fact, the weights given to these two quantities are in proportion to the inverses of σ_0^2/n (the conditional variance of the sample mean $\overline{X}$ given θ) and σ^2 (the variance of the prior distribution). ∎

REMARK: ON CHOOSING A NORMAL PRIOR

As illustrated by Example 7.8b, it is computationally very convenient to choose a normal prior for the unknown mean θ of a normal distribution — for then the Bayes estimator is simply given by Equation 7.8.3. This raises the question of how one should go about determining whether there is a normal prior that reasonably represents one's prior feelings about the unknown mean.

To begin, it seems reasonable to determine the value — call it μ — that you *a priori* feel is most likely to be near θ. That is, we start with the mode (which equals the mean when the distribution is normal) of the prior distribution. We should then try to ascertain whether or not we believe that the prior distribution is symmetric about μ. That is, for each $a > 0$ do we believe that it is just as likely that θ will lie between $\mu - a$ and μ as it is that it will be between μ and $\mu + a$? If the answer is positive, then we accept, as a working hypothesis, that our prior feelings about θ can be expressed in terms of a prior distribution that is normal with mean μ. To determine σ, the standard deviation of the normal prior, think of an interval centered about μ that you *a priori* feel is 90 percent certain to contain θ. For instance, suppose you feel 90 percent (no more and no less) certain that θ will lie between $\mu - a$ and $\mu + a$. Then, since a normal random variable θ with mean μ and variance σ^2 is such that

$$P\left\{-1.645 < \frac{\theta - \mu}{\sigma} < 1.645\right\} = .90$$

or

$$P\{\mu - 1.645\sigma < \theta < \mu + 1.645\sigma\} = .90$$

it seems reasonable to take

$$1.645\sigma = a \quad \text{or} \quad \sigma = \frac{a}{1.645}$$

Thus, if your prior feelings can indeed be reasonably described by a normal distribution, then that distribution would have mean μ and standard deviation $\sigma = a/1.645$. As a test of whether this distribution indeed fits your prior feelings you might ask yourself such questions as whether you are 95 percent certain that θ will fall between $\mu - 1.96\sigma$ and $\mu + 1.96\sigma$, or whether you are 99 percent certain that θ will fall between $\mu - 2.58\sigma$ and $\mu + 2.58\sigma$, where these intervals are determined by the equalities

$$P\left\{-1.96 < \frac{\theta - \mu}{\sigma} < 1.96\right\} = .95$$

$$P\left\{-2.58 < \frac{\theta - \mu}{\sigma} < 2.58\right\} = .99$$

which hold when θ is normal with mean μ and variance σ^2.

EXAMPLE 7.8c Consider the likelihood function $f(x_1, \ldots, x_n|\theta)$ and suppose that θ is uniformly distributed over some interval (a, b). The posterior density of θ given $X_1, \ldots, X_n$ equals

$$f(\theta|x_1, \ldots, x_n) = \frac{f(x_1, \ldots, x_n|\theta)p(\theta)}{\int_a^b f(x_1, \ldots, x_n|\theta)p(\theta)\, d\theta}$$

$$= \frac{f(x_1, \ldots, x_n|\theta)}{\int_a^b f(x_1, \ldots, x_n|\theta)\, d\theta} \qquad a < \theta < b$$

Now the *mode* of a density $f(\theta)$ was defined to be that value of θ that maximizes $f(\theta)$. By the foregoing, it follows that the mode of the density $f(\theta|x_1, \ldots, x_n)$ is that value of θ maximizing $f(x_1, \ldots, x_n|\theta)$; that is, it is just the maximum likelihood estimate of θ [when it is constrained to be in (a, b)]. In other words, the maximum likelihood estimate equals the mode of the posterior distribution when a uniform prior distribution is assumed. ∎

If, rather than a point estimate, we desire an interval in which θ lies with a specified probability — say $1 - \alpha$ — we can accomplish this by choosing values a and b such that

$$\int_a^b f(\theta|x_1, \ldots, x_n)\, d\theta = 1 - \alpha$$

EXAMPLE 7.8d Suppose that if a signal of value s is sent from location A, then the signal value received at location B is normally distributed with mean s and variance 60. Suppose also that the value of a signal sent at location A is, *a priori*, known to be normally distributed with mean 50 and variance 100. If the value received at location B is equal to 40, determine an interval that will contain the actual value sent with probability .90.

SOLUTION It follows from Example 7.8b that the conditional distribution of S, the signal value sent, given that 40 is the value received, is normal with mean and variance given by

$$E[S|\text{data}] = \frac{1/60}{1/60 + 1/100}40 + \frac{1/100}{1/60 + 1/100}50 = 43.75$$

$$\text{Var}(S|\text{data}) = \frac{1}{1/60 + 1/100} = 37.5$$

Hence, given that the value received is 40, $(S - 43.75)/\sqrt{37.5}$ has a standard normal distribution and so

$$P\left\{-1.645 < \frac{S - 43.75}{\sqrt{37.5}} < 1.645|\text{data}\right\} = .90$$

or

$$P\{43.75 - 1.645\sqrt{37.5} < S < 43.75 + 1.645\sqrt{37.5}|\text{data}\} = .95$$

That is, with *probability* .90, the true signal sent lies within the interval (33.68, 53.82). ∎

Problems

1. Let $X_1, \ldots, X_n$ be a sample from the distribution whose density function is

$$f(x) = \begin{cases} e^{-(x-\theta)} & x \geq \theta \\ 0 & \text{otherwise} \end{cases}$$

   Determine the maximum likelihood estimator of θ.

2. Determine the maximum likelihood estimator of θ when $X_1, \ldots, X_n$ is a sample with density function

$$f(x) = \tfrac{1}{2} e^{-|x-\theta|}, \qquad -\infty < x < \infty$$

3. Let $X_1, \ldots, X_n$ be a sample from a normal μ, σ^2 population. Determine the maximum likelihood estimator of σ^2 when μ is known. What is the expected value of this estimator?

4. Determine the maximum likelihood estimates of a and λ when $X_1, \ldots, X_n$ is a sample from the Pareto density function

$$f(x) = \begin{cases} \lambda a^\lambda x^{-(\lambda+1)}, & \text{if } x \geq a \\ 0, & \text{if } x < a \end{cases}$$

5. Suppose that $X_1, \ldots, X_n$ are normal with mean μ_1; $Y_1, \ldots, Y_n$ are normal with mean μ_2; and $W_1, \ldots, W_n$ are normal with mean $\mu_1 + \mu_2$. Assuming that all $3n$ random variables are independent with a common variance, find the maximum likelihood estimators of μ_1 and μ_2.

6. River floods are often measured by their discharges (in units of feet cubed per second). The value v is said to be the value of a 100-year flood if

$$P\{D \geq v\} = .01$$

   where D is the discharge of the largest flood in a randomly chosen year. The following table gives the flood discharges of the largest floods of the Blackstone River in Woonsocket, Rhode Island, in each of the years from 1929 to 1965. Assuming that these discharges follow a lognormal distribution, estimate the value of a 100-year flood.

Annual Floods of the Blackstone River (1929–1965)

Year	Flood Discharge (ft^3/s)
1929	4,570
1930	1,970
1931	8,220
1932	4,530
1933	5,780
1934	6,560
1935	7,500
1936	15,000
1937	6,340
1938	15,100
1939	3,840
1940	5,860
1941	4,480
1942	5,330
1943	5,310
1944	3,830
1945	3,410
1946	3,830
1947	3,150
1948	5,810
1949	2,030
1950	3,620
1951	4,920
1952	4,090
1953	5,570
1954	9,400
1955	32,900
1956	8,710
1957	3,850
1958	4,970
1959	5,398
1960	4,780
1961	4,020
1962	5,790
1963	4,510
1964	5,520
1965	5,300

7. Recall that X is said to have a lognormal distribution with parameters μ and σ^2 if $\log(X)$ is normal with mean μ and variance σ^2. Suppose X is such a lognormal random variable.
   (a) Find $E[X]$.
   (b) Find $Var(X)$.

Hint: Make use of the formula for the moment generating function of a normal random variable.

(c) The following are, in minutes, travel times to work over a sequence of 10 days.

$$42, 28, 53, 57, 67, 39, 35, 50, 44, 39$$

Assuming an underlying lognormal distribution, use the data to estimate the mean travel time.

8. An electric scale gives a reading equal to the true weight plus a random error that is normally distributed with mean 0 and standard deviation $\sigma = .1$ mg. Suppose that the results of five successive weighings of the same object are as follows: 3.142, 3.163, 3.155, 3.150, 3.141.

   (a) Determine a 95 percent confidence interval estimate of the true weight.
   (b) Determine a 99 percent confidence interval estimate of the true weight.

9. The PCB concentration of a fish caught in Lake Michigan was measured by a technique that is known to result in an error of measurement that is normally distributed with a standard deviation of .08 ppm (parts per million). Suppose the results of 10 independent measurements of this fish are

$$11.2, 12.4, 10.8, 11.6, 12.5, 10.1, 11.0, 12.2, 12.4, 10.6$$

   (a) Give a 95 percent confidence interval for the PCB level of this fish.
   (b) Give a 95 percent lower confidence interval.
   (c) Give a 95 percent upper confidence interval.

10. The standard deviation of test scores on a certain achievement test is 11.3. If a random sample of 81 students had a sample mean score of 74.6, find a 90 percent confidence interval estimate for the average score of all students.

11. Let $X_1, \ldots, X_n, X_{n+1}$ be a sample from a normal population having an unknown mean μ and variance 1. Let $\bar{X}_n = \sum_{i=1}^{n} X_i/n$ be the average of the first n of them.

   (a) What is the distribution of $X_{n+1} - \bar{X}_n$?
   (b) If $\bar{X}_n = 4$, give an interval that, with 90 percent confidence, will contain the value of X_{n+1}.

12. If $X_1, \ldots, X_n$ is a sample from a normal population whose mean μ is unknown but whose variance σ^2 is known, show that $(-\infty, \bar{X} + z_\alpha \sigma/\sqrt{n})$ is a $100(1-\alpha)$ percent lower confidence interval for μ.

13. A sample of 20 cigarettes is tested to determine nicotine content and the average value observed was 1.2 mg. Compute a 99 percent two-sided confidence interval for the mean nicotine content of a cigarette if it is known that the standard deviation of a cigarette's nicotine content is $\sigma = .2$ mg.

14. In Problem 13, suppose that the population variance is not known in advance of the experiment. If the sample variance is .04, compute a 99 percent two-sided confidence interval for the mean nicotine content.

15. In Problem 14, compute a value c for which we can assert "with 99 percent confidence" that c is larger than the mean nicotine content of a cigarette.

16. Suppose that when sampling from a normal population having an unknown mean μ and unknown variance σ^2, we wish to determine a sample size n so as to guarantee that the resulting $100(1 - \alpha)$ percent confidence interval for μ will be of size no greater than A, for given values α and A. Explain how we can approximately do this by a double sampling scheme that first takes a subsample of size 30 and then chooses the total sample size by using the results of the first subsample.

17. The following data resulted from 24 independent measurements of the melting point of lead.

330°C	322°C	345°C
328.6°C	331°C	342°C
342.4°C	340.4°C	329.7°C
334°C	326.5°C	325.8°C
337.5°C	327.3°C	322.6°C
341°C	340°C	333°C
343.3°C	331°C	341°C
329.5°C	332.3°C	340°C

Assuming that the measurements can be regarded as constituting a normal sample whose mean is the true melting point of lead, determine a 95 percent two-sided confidence interval for this value. Also determine a 99 percent two-sided confidence interval.

18. The following are scores on IQ tests of a random sample of 18 students at a large eastern university.

130, 122, 119, 142, 136, 127, 120, 152, 141,
132, 127, 118, 150, 141, 133, 137, 129, 142

   (a) Construct a 95 percent confidence interval estimate of the average IQ score of all students at the university.
   (b) Construct a 95 percent lower confidence interval estimate.
   (c) Construct a 95 percent upper confidence interval estimate.

19. Suppose that a random sample of nine recently sold houses in a certain city resulted in a sample mean price of $222,000, with a sample standard deviation of $22,000. Give a 95 percent upper confidence interval for the mean price of all recently sold houses in this city.

20. A company self-insures its large fleet of cars against collisions. To determine its mean repair cost per collision, it has randomly chosen a sample of 16 accidents. If the average repair cost in these accidents is $2,200 with a sample standard deviation of $800, find a 90 percent confidence interval estimate of the mean cost per collision.

21. A standardized test is given annually to all sixth-grade students in the state of Washington. To determine the average score of students in her district, a school supervisor selects a random sample of 100 students. If the sample mean of these students' scores is 320 and the sample standard deviation is 16, give a 95 percent confidence interval estimate of the average score of students in that supervisor's district.

22. Each of 20 science students independently measured the melting point of lead. The sample mean and sample standard deviation of these measurements were (in degrees centigrade) 330.2 and 15.4, respectively. Construct (a) a 95 percent and (b) a 99 percent confidence interval estimate of the true melting point of lead.

23. A random sample of 300 CitiBank VISA cardholder accounts indicated a sample mean debt of $1,220 with a sample standard deviation of $840. Construct a 95 percent confidence interval estimate of the average debt of all cardholders.

24. In Problem 23, find the smallest value v that "with 90 percent confidence," exceeds the average debt per cardholder.

25. Verify the formula given in Table 7.1 for the $100(1 - \alpha)$ percent lower confidence interval for μ when σ is unknown.

26. The following are the daily number of steps taken by a certain individual in 20 weekdays.

2,100	1,984	2,072	1,898
1,950	1,992	2,096	2,103
2,043	2,218	2,244	2,206
2,210	2,152	1,962	2,007
2,018	2,106	1,938	1,956

Assuming that the daily number of steps is normally distributed, construct (a) a 95 percent and (b) a 99 percent two-sided confidence interval for the mean number of steps. (c) Determine the largest value v that, "with 95 percent confidence," will be less than the mean range.

27. Studies were conducted in Los Angeles to determine the carbon monoxide concentration near freeways. The basic technique used was to capture air samples in special bags and to then determine the carbon monoxide concentration by using a spectrophotometer. The measurements in ppm (parts per million) over a sampled period during the year were 102.2, 98.4, 104.1, 101, 102.2, 100.4, 98.6,

88.2, 78.8, 83, 84.7, 94.8, 105.1, 106.2, 111.2, 108.3, 105.2, 103.2, 99, 98.8. Compute a 95 percent two-sided confidence interval for the mean carbon monoxide concentration.

28. A set of 10 determinations, by a method devised by the chemist Karl Fischer, of the percentage of water in a methanol solution yielded the following data.

$$.50, .55, .53, .56, .54,$$
$$.57, .52, .60, .55, .58$$

Assuming normality, use these data to give a 95 percent confidence interval for the actual percentage.

29. Suppose that $U_1, U_2, \ldots$ is a sequence of independent uniform $(0,1)$ random variables, and define N by

$$N = \min\{n : U_1 + \cdots + U_n > 1\}$$

That is, N is the number of uniform $(0, 1)$ random variables that need to be summed to exceed 1. Use random numbers to determine the value of 36 random variables having the same distribution as N, and then use these data to obtain a 95 percent confidence interval estimate of $E[N]$. Based on this interval, guess the exact value of $E[N]$.

30. An important issue for a retailer is to decide when to reorder stock from a supplier. A common policy used to make the decision is of a type called s, S: The retailer orders at the end of a period if the on-hand stock is less than s, and orders enough to bring the stock up to S. The appropriate values of s and S depend on different cost parameters, such as inventory holding costs and the profit per item sold, as well as the distribution of the demand during a period. Consequently, it is important for the retailer to collect data relating to the parameters of the demand distribution. Suppose that the following data give the numbers of a certain type of item sold in each of 30 weeks.

$$14, 8, 12, 9, 5, 22, 15, 12, 16, 7, 10, 9, 15, 15, 12,$$
$$9, 11, 16, 8, 7, 15, 13, 9, 5, 18, 14, 10, 13, 7, 11$$

Assuming that the numbers sold each week are independent random variables from a common distribution, use the data to obtain a 95 percent confidence interval for the mean number sold in a week.

31. A random sample of 16 professors at a large private university yielded a sample mean annual salary of $90,450 with a sample standard deviation of $9,400. Determine a 95 percent confidence interval of the average salary of all professors at that university.

32. Let $X_1, \ldots, X_{n+1}$ be a sample from a population with mean μ and variance σ^2. As noted in the text, the natural predictor of X_{n+1} based on the data values $X_1, \ldots, X_n$ is $\bar{X}_n = \sum_{i=1}^{n} X_i/n$. Determine the mean square error of this predictor. That is, find $E[(X_{n+1} - \bar{X}_n)^2]$.

33. National Safety Council data show that the number of accidental deaths due to drowning in the United States in the years from 1990 to 1993 were (in units of one thousand) 5.2, 4.6, 4.3, 4.8. Use these data to give an interval that will, with 95 percent confidence, contain the number of such deaths in 1994.

34. The daily dissolved oxygen concentration for a water stream has been recorded over 30 days. If the sample average of the 30 values is 2.5 mg/liter and the sample standard deviation is 2.12 mg/liter, determine a value which, with 90 percent confidence, exceeds the mean daily concentration.

35. Verify the formulas given in Table 7.1 for the $100(1 - \alpha)$ percent lower and upper confidence intervals for σ^2.

36. The capacities (in ampere-hours) of 10 batteries were recorded as follows:

$$140, 136, 150, 144, 148, 152, 138, 141, 143, 151$$

(a) Estimate the population variance σ^2.
(b) Compute a 99 percent two-sided confidence interval for σ^2.
(c) Compute a value v that enables us to state, with 90 percent confidence, that σ^2 is less than v.

37. Find a 95 percent two-sided confidence interval for the variance of the diameter of a rivet based on the data given here.

6.68	6.66	6.62	6.72
6.76	6.67	6.70	6.72
6.78	6.66	6.76	6.72
6.76	6.70	6.76	6.76
6.74	6.74	6.81	6.66
6.64	6.79	6.72	6.82
6.81	6.77	6.60	6.72
6.74	6.70	6.64	6.78
6.70	6.70	6.75	6.79

Assume a normal population.

38. The following are independent samples from two normal populations, both of which have the same standard deviation σ.

$$16, 17, 19, 20, 18 \quad \text{and} \quad 3, 4, 8$$

Use them to estimate σ.

39. The amount of beryllium in a substance is often determined by the use of a photometric filtration method. If the weight of the beryllium is μ, then the value given by the photometric filtration method is normally distributed with mean μ and standard deviation σ. A total of eight independent measurements of 3.180 mg of beryllium gave the following results.

$$3.166, 3.192, 3.175, 3.180, 3.182, 3.171, 3.184, 3.177$$

   Use the preceding data to
   (a) estimate σ;
   (b) find a 90 percent confidence interval estimate of σ.

40. If $X_1, \ldots, X_n$ is a sample from a normal population, explain how to obtain a $100(1 - \alpha)$ percent confidence interval for the population variance σ^2 when the population mean μ is known. Explain in what sense knowledge of μ improves the interval estimator compared with when it is unknown.
   Repeat Problem 38 if it is known that the mean burning time is 53.6 seconds.

41. A civil engineer wishes to measure the compressive strength of two different types of concrete. A random sample of 10 specimens of the first type yielded the following data (in psi)

   Type 1:   3,250,   3,268,   4,302,   3,184,   3,266
               3,297,   3,332,   3,502,   3,064,   3,116

   whereas a sample of 10 specimens of the second yielded the data

   Type 2:   3,094,   3,106,   3,004,   3,066,   2,984,
               3,124,   3,316,   3,212,   3,380,   3,018

   If we assume that the samples are normal with a common variance, determine
   (a) a 95 percent two-sided confidence interval for $\mu_1 - \mu_2$, the difference in means;
   (b) a 95 percent one-sided upper confidence interval for $\mu_1 - \mu_2$;
   (c) a 95 percent one-sided lower confidence interval for $\mu_1 - \mu_2$.

42. Independent random samples are taken from the output of two machines on a production line. The weight of each item is of interest. From the first machine, a sample of size 36 is taken, with sample mean weight of 120 grams and a sample variance of 4. From the second machine, a sample of size 64 is taken, with a sample mean weight of 130 grams and a sample variance of 5. It is assumed that the weights of items from the first machine are normally distributed with mean μ_1 and variance σ^2 and that the weights of items from the second machine are normally distributed with mean μ_2 and variance σ^2 (that is, the variances are assumed to be equal). Find a 99 percent confidence interval for $\mu_1 - \mu_2$, the difference in population means.

43. Do Problem 42 when it is known in advance that the population variances are 4 and 5.

44. The following are the daily numbers of company website visits resulting from advertisements on two different types of media.

Type I		Type II	
481	572	526	537
506	561	511	582
527	501	556	605
661	487	542	558
501	524	491	578

Find a 99 percent confidence interval for the mean difference in daily visits assuming normality with unknown but equal variances.

45. If $X_1, \ldots, X_n$ is a sample from a normal population having known mean μ_1 and unknown variance σ_1^2, and $Y_1, \ldots, Y_m$ is an independent sample from a normal population having known mean μ_2 and unknown variance σ_2^2, determine a $100(1 - \alpha)$ percent confidence interval for σ_1^2/σ_2^2.

46. Two analysts took repeated readings on the hardness of city water. Assuming that the readings of analyst i constitute a sample from a normal population having variance $\sigma_i^2, i = 1, 2$, compute a 95 percent two-sided confidence interval for σ_1^2/σ_2^2 when the data are as follows:

Coded Measures of Hardness	
Analyst 1	Analyst 2
.46	.82
.62	.61
.37	.89
.40	.51
.44	.33
.58	.48
.48	.23
.53	.25
	.67
	.88

47. A problem of interest in baseball is whether a sacrifice bunt is a good strategy when there is a man on first base and no outs. Assuming that the bunter will be out but will be successful in advancing the man on base, we could compare the probability of scoring a run with a player on first base and no outs with the probability of scoring a run with a player on second base and one out.

The following data resulted from a study of randomly chosen major league baseball games played in 1959 and 1960.

(a) Give a 95 percent confidence interval estimate for the probability of scoring at least one run when there is a man on first and no outs.

(b) Give a 95 percent confidence interval estimate for the probability of scoring at least one run when there is a man on second and one out.

Base Occupied	Number of Outs	Number of Cases in Which 0 Runs Are Scored	Total Number of Cases
First	0	1,044	1,728
Second	1	401	657

48. A random sample of 1,200 engineers included 48 Hispanic Americans, 80 African Americans, and 204 females. Determine 90 percent confidence intervals for the proportion of all engineers who are

(a) female;

(b) Hispanic Americans or African Americans.

49. To estimate p, the proportion of all newborn babies that are male, the gender of 10,000 newborn babies was noted. If 5,106 of them were male, determine (a) a 90 percent and (b) a 99 percent confidence interval estimate of p.

50. An airline is interested in determining the proportion of its customers who are flying for reasons of business. If they want to be 90 percent certain that their estimate will be correct to within 2 percent, how large a random sample should they select?

51. A recent newspaper poll indicated that Candidate A is favored over Candidate B by a 53 to 47 percentage, with a margin of error of ±4 percent. The newspaper then stated that since the 6-point gap is larger than the margin of error, its readers can be certain that Candidate A is the current choice. Is this reasoning correct?

52. A market research firm is interested in determining the proportion of households that are watching a particular sporting event. To accomplish this task, they plan on using a telephone poll of randomly chosen households. How large a sample is needed if they want to be 90 percent certain that their estimate is correct to within ±.02?

53. In a recent study, 79 of 140 meteorites were observed to enter the atmosphere with a velocity of less than 25 miles per second. If we take $\hat{p} = 79/140$ as an estimate of the probability that an arbitrary meteorite that enters the atmosphere will have a speed less than 25 miles per second, what can we say, with 99 percent confidence, about the maximum error of our estimate?

54. A random sample of 100 items from a production line revealed 17 of them to be defective. Compute a 95 percent two-sided confidence interval for the probability that an item produced is defective. Determine also a 99 percent upper confidence interval for this value. What assumptions are you making?

55. Of 100 randomly detected cases of individuals having lung cancer, 67 died within 5 years of detection.

   (a) Estimate the probability that a person contracting lung cancer will die within 5 years.

   (b) How large an additional sample would be required to be 95 percent confident that the error in estimating the probability in part (a) is less than .02?

56. Derive $100(1 - \alpha)$ percent lower and upper confidence intervals for p, when the data consist of the values of n independent Bernoulli random variables with parameter p.

57. Suppose the lifetimes of batteries are exponentially distributed with mean θ. If the average of a sample of 10 batteries is 36 hours, determine a 95 percent two-sided confidence interval for θ.

58. Determine $100(1 - \alpha)$ percent one-sided upper and lower confidence intervals for θ in Problem 57.

59. Let $X_1, X_2, \ldots, X_n$ denote a sample from a population whose mean value θ is unknown. Use the results of Example 7.7b to argue that among all unbiased estimators of θ of the form $\sum_{i=1}^{n} \lambda_i X_i, \sum_{i=1}^{n} \lambda_i = 1$, the one with minimal mean square error has $\lambda_i \equiv 1/n, i = 1, \ldots, n$.

60. Consider two independent samples from normal populations having the same variance σ^2, of respective sizes n and m. That is, $X_1, \ldots, X_n$ and $Y_1, \ldots, Y_m$ are independent samples from normal populations each having variance σ^2. Let S_x^2 and S_y^2 denote the respective sample variances. Thus both S_x^2 and S_y^2 are unbiased estimators of σ^2. Show by using the results of Example 7.7b along with the fact that

$$\text{Var}(\chi_k^2) = 2k$$

where χ_k^2 is chi-square with k degrees of freedom, that the minimum mean square estimator of σ^2 of the form $\lambda S_x^2 + (1 - \lambda) S_y^2$ is

$$S_p^2 = \frac{(n-1)S_x^2 + (m-1)S_y^2}{n + m - 2}$$

This is called the *pooled estimator* of σ^2.

61. Consider two estimators d_1 and d_2 of a parameter θ. If $E[d_1] = \theta$, $\mathrm{Var}(d_1) = 6$ and $E[d_2] = \theta + 2$, $\mathrm{Var}(d_2) = 2$, which estimator should be preferred?

62. Suppose that the number of accidents occurring daily in a certain plant has a Poisson distribution with an unknown mean λ. Based on previous experience in similar industrial plants, suppose that a statistician's initial feelings about the possible value of λ can be expressed by an exponential distribution with parameter 1. That is, the prior density is

$$p(\lambda) = e^{-\lambda}, \qquad 0 < \lambda < \infty$$

Determine the Bayes estimate of λ if there are a total of 83 accidents over the next 10 days. What is the maximum likelihood estimate?

63. The functional lifetimes in hours of computer chips produced by a certain semiconductor firm are exponentially distributed with mean $1/\lambda$. Suppose that the prior distribution on λ is the gamma distribution with density function

$$g(x) = \frac{e^{-x}x^2}{2}, \qquad 0 < x < \infty$$

If the average life of the first 20 chips tested is 4.6 hours, compute the Bayes estimate of λ.

64. Each item produced will, independently, be defective with probability p. If the prior distribution on p is uniform on $(0, 1)$, compute the posterior probability that p is less than .2 given

(a) a total of 2 defectives out of a sample of size 10;
(b) a total of 1 defective out of a sample of size 10;
(c) a total of 10 defectives out of a sample of size 10.

65. The breaking strength of a certain type of cloth is to be measured for 10 specimens. The underlying distribution is normal with unknown mean θ but with a standard deviation equal to 3 psi. Suppose also that based on previous experience we feel that the unknown mean has a prior distribution that is normally distributed with mean 200 and standard deviation 2. If the average breaking strength of a sample of 20 specimens is 182 psi, determine a region that contains θ with probability .95.

Chapter 8

HYPOTHESIS TESTING

8.1 INTRODUCTION

As in the previous chapter, let us suppose that a random sample from a population distribution, specified except for a vector of unknown parameters, is to be observed. However, rather than wishing to explicitly estimate the unknown parameters, let us now suppose that we are primarily concerned with using the resulting sample to test some particular hypothesis concerning them. As an illustration, suppose that a construction firm has just purchased a large supply of cables that have been guaranteed to have an average breaking strength of at least 7,000 psi. To verify this claim, the firm has decided to take a random sample of 10 of these cables to determine their breaking strengths. They will then use the result of this experiment to ascertain whether or not they accept the cable manufacturer's hypothesis that the population mean is at least 7,000 pounds per square inch.

A statistical hypothesis is usually a statement about a set of parameters of a population distribution. It is called a hypothesis because it is not known whether or not it is true. A primary problem is to develop a procedure for determining whether or not the values of a random sample from this population are consistent with the hypothesis. For instance, consider a particular normally distributed population having an unknown mean value θ and known variance 1. The statement "θ is less than 1" is a statistical hypothesis that we could try to test by observing a random sample from this population. If the random sample is deemed to be consistent with the hypothesis under consideration, we say that the hypothesis has been "accepted"; otherwise we say that it has been "rejected."

Note that in accepting a given hypothesis we are not actually claiming that it is true but rather we are saying that the resulting data appear to be consistent with it. For instance, in the case of a normal $(\theta, 1)$ population, if a resulting sample of size 10 has an average value of 1.25, then although such a result cannot be regarded as being evidence in favor of the hypothesis "$\theta < 1$," it is not inconsistent with this hypothesis, which would thus be accepted. On the other hand, if the sample of size 10 has an average value of 3, then even though a sample value that large is possible when $\theta < 1$, it is so unlikely that it seems inconsistent with this hypothesis, which would thus be rejected.

8.2   SIGNIFICANCE LEVELS

Consider a population having distribution F_θ, where θ is unknown, and suppose we want to test a specific hypothesis about θ. We shall denote this hypothesis by H_0 and call it the *null hypothesis*. For example, if F_θ is a normal distribution function with mean θ and variance equal to 1, then two possible null hypotheses about θ are

(a) $H_0 : \theta = 1$

(b) $H_0 : \theta \le 1$

Thus the first of these hypotheses states that the population is normal with mean 1 and variance 1, whereas the second states that it is normal with variance 1 and a mean less than or equal to 1. Note that the null hypothesis in (a), when true, completely specifies the population distribution, whereas the null hypothesis in (b) does not. A hypothesis that, when true, completely specifies the population distribution is called a *simple* hypothesis; one that does not is called a *composite* hypothesis.

Suppose now that in order to test a specific null hypothesis H_0, a population sample of size n — say $X_1, \ldots, X_n$ — is to be observed. Based on these n values, we must decide whether or not to accept H_0. A test for H_0 can be specified by defining a region C in n-dimensional space with the proviso that the hypothesis is to be rejected if the random sample $X_1, \ldots, X_n$ turns out to lie in C and accepted otherwise. The region C is called the *critical region*. In other words, the statistical test determined by the critical region C is the one that

$$\text{accepts} \quad H_0 \quad \text{if} \quad (X_1, X_2, \ldots, X_n) \notin C$$

and

$$\text{rejects} \quad H_0 \quad \text{if} \quad (X_1, \ldots, X_n) \in C$$

For instance, a common test of the hypothesis that θ, the mean of a normal population with variance 1, is equal to 1 has a critical region given by

$$C = \{(X_1, \ldots, X_n) : |\bar{X} - 1| > 1.96/\sqrt{n}\} \tag{8.2.1}$$

Thus, this test calls for rejection of the null hypothesis that $\theta = 1$ when the sample average differs from 1 by more than 1.96 divided by the square root of the sample size.

It is important to note when developing a procedure for testing a given null hypothesis H_0 that, in any test, two different types of errors can result. The first of these, called a *type I error*, is said to result if the test incorrectly calls for rejecting H_0 when it is indeed correct. The second, called a *type II error*, results if the test calls for accepting H_0 when it is false.

Now, as was previously mentioned, the objective of a statistical test of H_0 is not to explicitly determine whether or not H_0 is true but rather to determine if its validity is consistent with the resultant data. Hence, with this objective it seems reasonable that H_0 should only be rejected if the resultant data are very unlikely when H_0 is true. The classical way of accomplishing this is to specify a value α and then require the test to have the property that whenever H_0 is true its probability of being rejected is never greater than α. The value α, called the *level of significance of the test*, is usually set in advance, with commonly chosen values being $\alpha = .1, .05, .005$. In other words, the classical approach to testing H_0 is to fix a significance level α and then require that the test have the property that the probability of a type I error occurring can never be greater than α.

Suppose now that we are interested in testing a certain hypothesis concerning θ, an unknown parameter of the population. Specifically, for a given set of parameter values w, suppose we are interested in testing

$$H_0 : \theta \in w$$

A common approach to developing a test of H_0, say at level of significance α, is to start by determining a point estimator of θ — say $d(\mathbf{X})$. The hypothesis is then rejected if $d(\mathbf{X})$ is "far away" from the region w. However, to determine how "far away" it need be to justify rejection of H_0, we need to determine the probability distribution of $d(\mathbf{X})$ when H_0 is true since this will usually enable us to determine the appropriate critical region so as to make the test have the required significance level α. For example, the test of the hypothesis that the mean of a normal $(\theta, 1)$ population is equal to 1, given by Equation 8.2.1, calls for rejection when the point estimate of θ — that is, the sample average — is farther than $1.96/\sqrt{n}$ away from 1. As we will see in the next section, the value $1.96/\sqrt{n}$ was chosen to meet a level of significance of $\alpha = .05$.

8.3   TESTS CONCERNING THE MEAN OF A NORMAL POPULATION

8.3.1 CASE OF KNOWN VARIANCE

Suppose that $X_1, \ldots, X_n$ is a sample of size n from a normal distribution having an unknown mean μ and a known variance σ^2 and suppose we are interested in testing the null hypothesis

$$H_0 : \mu = \mu_0$$

against the alternative hypothesis

$$H_1 : \mu \neq \mu_0$$

where μ_0 is some specified constant.

Since $\overline{X} = \sum_{i=1}^{n} X_i/n$ is a natural point estimator of μ, it seems reasonable to accept H_0 if $\overline{X}$ is not too far from μ_0. That is, the critical region of the test would be of the form

$$C = \{X_1, \ldots, X_n : |\overline{X} - \mu_0| > c\} \tag{8.3.1}$$

for some suitably chosen value c.

If we desire that the test has significance level α, then we must determine the critical value c in Equation 8.3.1 that will make the type I error equal to α. That is, c must be such that

$$P_{\mu_0}\{|\overline{X} - \mu_0| > c\} = \alpha \tag{8.3.2}$$

where we write P_{μ_0} to mean that the preceding probability is to be computed under the assumption that $\mu = \mu_0$. However, when $\mu = \mu_0$, $\overline{X}$ will be normally distributed with mean μ_0 and variance σ^2/n and so Z, defined by

$$Z \equiv \frac{\overline{X} - \mu_0}{\sigma/\sqrt{n}} = \frac{\sqrt{n}(\overline{X} - \mu_0)}{\sigma}$$

will have a standard normal distribution. Now Equation 8.3.2 is equivalent to

$$P\left\{|Z| > \frac{c\sqrt{n}}{\sigma}\right\} = \alpha$$

or, equivalently,

$$2P\left\{Z > \frac{c\sqrt{n}}{\sigma}\right\} = \alpha$$

where Z is a standard normal random variable. However, we know that

$$P\{Z > z_{\alpha/2}\} = \alpha/2$$

and so

$$\frac{c\sqrt{n}}{\sigma} = z_{\alpha/2}$$

or

$$c = \frac{z_{\alpha/2}\sigma}{\sqrt{n}}$$

Thus, the significance level α test is to reject H_0 if $|\overline{X} - \mu_0| > z_{\alpha/2}\sigma/\sqrt{n}$ and accept otherwise; or, equivalently, to

$$\begin{aligned} \text{reject} \quad & H_0 \quad \text{if} \quad \frac{\sqrt{n}}{\sigma}|\overline{X} - \mu_0| > z_{\alpha/2} \\[2mm] \text{accept} \quad & H_0 \quad \text{if} \quad \frac{\sqrt{n}}{\sigma}|\overline{X} - \mu_0| \leq z_{\alpha/2} \end{aligned} \tag{8.3.3}$$

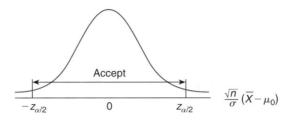

$$\frac{\sqrt{n}}{\sigma}(\overline{X}-\mu_0)$$

This can be pictorially represented as shown in Figure 8.1, where we have superimposed the standard normal density function [which is the density of the test statistic $\sqrt{n}(\overline{X} - \mu_0)/\sigma$ when H_0 is true].

EXAMPLE 8.3a It is known that if a signal of value μ is sent from location A, then the value received at location B is normally distributed with mean μ and standard deviation 2. That is, the random noise added to the signal is an $N(0, 4)$ random variable. There is reason for the people at location B to suspect that the signal value $\mu = 8$ will be sent today. Test this hypothesis if the same signal value is independently sent five times and the average value received at location B is $\overline{X} = 9.5$.

SOLUTION Suppose we are testing at the 5 percent level of significance. To begin, we compute the test statistic

$$\frac{\sqrt{n}}{\sigma}|\overline{X} - \mu_0| = \frac{\sqrt{5}}{2}(1.5) = 1.68$$

Since this value is less than $z_{.025} = 1.96$, the hypothesis is accepted. In other words, the data are not inconsistent with the null hypothesis in the sense that a sample average as far from the value 8 as observed would be expected, when the true mean is 8, over 5 percent of the time. Note, however, that if a less stringent significance level were chosen — say $\alpha = .1$ — then the null hypothesis would have been rejected. This follows since $z_{.05} = 1.645$, which is less than 1.68. Hence, if we would have chosen a test that had a 10 percent chance of rejecting H_0 when H_0 was true, then the null hypothesis would have been rejected.

The "correct" level of significance to use in a given situation depends on the individual circumstances involved in that situation. For instance, if rejecting a null hypothesis H_0 would result in large costs that would thus be lost if H_0 were indeed true, then we might elect to be quite conservative and so choose a significance level of .05 or .01. Also, if we initially feel strongly that H_0 was correct, then we would require very stringent data evidence to the contrary for us to reject H_0. (That is, we would set a very low significance level in this situation.) ■

The test given by Equation 8.3.3 can be described as follows: For any observed value of the test statistic $\sqrt{n}|\overline{X} - \mu_0|/\sigma$, call it v, the test calls for rejection of the null hypothesis if the probability that the test statistic would be as large as v when H_0 is true is less than or equal to the significance level α. From this, it follows that we can determine whether or not to accept the null hypothesis by computing, first, the value of the test statistic and, second, the probability that a standard normal would (in absolute value) exceed that quantity. This probability — called the p-value of the test — gives the critical significance level in the sense that H_0 will be accepted if the significance level α is less than the p-value and rejected if it is greater than or equal.

In practice, the significance level is often not set in advance but rather the data are looked at to determine the resultant p-value. Sometimes, this critical significance level is clearly much larger than any we would want to use, and so the null hypothesis can be readily accepted. At other times the p-value is so small that it is clear that the hypothesis should be rejected.

EXAMPLE 8.3b   In Example 8.3a, suppose that the average of the 5 values received is $\overline{X} = 8.5$. In this case,

$$\frac{\sqrt{n}}{\sigma}|\overline{X} - \mu_0| = \frac{\sqrt{5}}{4} = .559$$

Since

$$P\{|Z| > .559\} = 2P\{Z > .559\}$$
$$= 2 \times .288 = .576$$

it follows that the p-value is .576 and thus the null hypothesis H_0 that the signal sent has value 8 would be accepted at any significance level $\alpha < .576$. Since we would clearly never want to test a null hypothesis using a significance level as large as .576, H_0 would be accepted.

On the other hand, if the average of the data values were 11.5, then the p-value of the test that the mean is equal to 8 would be

$$P\{|Z| > 1.75\sqrt{5}\} = P\{|Z| > 3.913\}$$
$$\approx .00005$$

For such a small p-value, the hypothesis that the value 8 was sent is rejected.   ∎

We have not yet talked about the probability of a type II error — that is, the probability of accepting the null hypothesis when the true mean μ is unequal to μ_0. This probability will depend on the value of μ, and so let us define $\beta(\mu)$ by

$$
\begin{aligned}
\beta(\mu) &= P_\mu\{\text{acceptance of } H_0\} \\
&= P_\mu\left\{\left|\frac{\overline{X} - \mu_0}{\sigma/\sqrt{n}}\right| \leq z_{\alpha/2}\right\} \\
&= P_\mu\left\{-z_{\alpha/2} \leq \frac{\overline{X} - \mu_0}{\sigma/\sqrt{n}} \leq z_{\alpha/2}\right\}
\end{aligned}
$$

The function $\beta(\mu)$ is called the *operating characteristic* (or OC) *curve* and represents the probability that H_0 will be accepted when the true mean is μ.

To compute this probability, we use the fact that $\overline{X}$ is normal with mean μ and variance σ^2/n and so

$$
Z \equiv \frac{\overline{X} - \mu}{\sigma/\sqrt{n}} \sim \mathcal{N}(0, 1)
$$

Hence,

$$
\begin{aligned}
\beta(\mu) &= P_\mu\left\{-z_{\alpha/2} \leq \frac{\overline{X} - \mu_0}{\sigma/\sqrt{n}} \leq z_{\alpha/2}\right\} \\
&= P_\mu\left\{-z_{\alpha/2} - \frac{\mu}{\sigma/\sqrt{n}} \leq \frac{\overline{X} - \mu_0 - \mu}{\sigma/\sqrt{n}} \leq z_{\alpha/2} - \frac{\mu}{\sigma/\sqrt{n}}\right\} \\
&= P_\mu\left\{-z_{\alpha/2} - \frac{\mu}{\sigma/\sqrt{n}} \leq Z - \frac{\mu_0}{\sigma/\sqrt{n}} \leq z_{\alpha/2} - \frac{\mu}{\sigma/\sqrt{n}}\right\} \\
&= P\left\{\frac{\mu_0 - \mu}{\sigma/\sqrt{n}} - z_{\alpha/2} \leq Z \leq \frac{\mu_0 - \mu}{\sigma/\sqrt{n}} + z_{\alpha/2}\right\} \\
&= \Phi\left(\frac{\mu_0 - \mu}{\sigma/\sqrt{n}} + z_{\alpha/2}\right) - \Phi\left(\frac{\mu_0 - \mu}{\sigma/\sqrt{n}} - z_{\alpha/2}\right)
\end{aligned} \tag{8.3.4}
$$

where Φ is the standard normal distribution function.

For a fixed significance level α, the OC curve given by Equation 8.3.4 is symmetric about μ_0 and indeed will depend on μ only through $\sqrt{n}\,|\mu - \mu_0|/\sigma$. This curve with the abscissa changed from μ to $d = \sqrt{n}\,|\mu - \mu_0|/\sigma$ is presented in Figure 8.2 when $\alpha = .05$.

EXAMPLE 8.3c For the problem presented in Example 8.3a, let us determine the probability of accepting the null hypothesis that $\mu = 8$ when the actual value sent is 10. To do so, we compute

$$
\frac{\sqrt{n}}{\sigma}(\mu_0 - \mu) = -\frac{\sqrt{5}}{2} \times 2 = -\sqrt{5}
$$

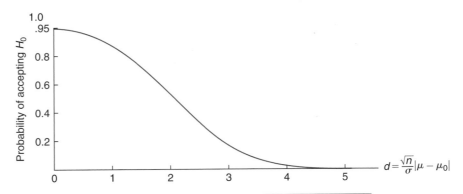

FIGURE 8.2   *The OC curve for the two-sided normal test for significance level $\alpha = .05$.*

As $z_{.025} = 1.96$, the desired probability is, from Equation 8.3.4,

$$\Phi(-\sqrt{5} + 1.96) - \Phi(-\sqrt{5} - 1.96)$$
$$= 1 - \Phi(\sqrt{5} - 1.96) - [1 - \Phi(\sqrt{5} + 1.96)]$$
$$= \Phi(4.196) - \Phi(.276)$$
$$= .392 \quad \blacksquare$$

REMARK

The function $1 - \beta(\mu)$ is called the *power-function* of the test. Thus, for a given value μ, the power of the test is equal to the probability of rejection when μ is the true value. $\blacksquare$

The operating characteristic function is useful in determining how large the random sample need be to meet certain specifications concerning type II errors. For instance, suppose that we desire to determine the sample size n necessary to ensure that the probability of accepting $H_0 : \mu = \mu_0$ when the true mean is actually μ_1 is approximately β. That is, we want n to be such that

$$\beta(\mu_1) \approx \beta$$

But from Equation 8.3.4, this is equivalent to

$$\Phi\left(\frac{\sqrt{n}(\mu_0 - \mu_1)}{\sigma} + z_{\alpha/2}\right) - \Phi\left(\frac{\sqrt{n}(\mu_0 - \mu_1)}{\sigma} - z_{\alpha/2}\right) \approx \beta \qquad (8.3.5)$$

Although the foregoing cannot be analytically solved for n, a solution can be obtained by using the standard normal distribution table. In addition, an approximation for n can be derived from Equation 8.3.5 as follows. To start, suppose that $\mu_1 > \mu_0$. Then, because this implies that

$$\frac{\sqrt{n}(\mu_0 - \mu_1)}{\sigma} - z_{\alpha/2} \leq -z_{\alpha/2}$$

it follows, since Φ is an increasing function, that

$$\Phi\left(\frac{\sqrt{n}(\mu_0 - \mu_1)}{\sigma} - z_{\alpha/2}\right) \leq \Phi(-z_{\alpha/2}) = P\{Z \leq -z_{\alpha/2}\} = P\{Z \geq z_{\alpha/2}\} = \alpha/2$$

Hence, we can take

$$\Phi\left(\frac{\sqrt{n}(\mu_0 - \mu_1)}{\sigma} - z_{\alpha/2}\right) \approx 0$$

and so from Equation 8.3.5

$$\beta \approx \Phi\left(\frac{\sqrt{n}(\mu_0 - \mu_1)}{\sigma} + z_{\alpha/2}\right) \tag{8.3.6}$$

or, since

$$\beta = P\{Z > z_\beta\} = P\{Z < -z_\beta\} = \Phi(-z_\beta)$$

we obtain from Equation 8.3.6 that

$$-z_\beta \approx (\mu_0 - \mu_1)\frac{\sqrt{n}}{\sigma} + z_{\alpha/2}$$

or

$$n \approx \frac{(z_{\alpha/2} + z_\beta)^2 \sigma^2}{(\mu_1 - \mu_0)^2} \tag{8.3.7}$$

In fact, the same approximation would result when $\mu_1 < \mu_0$ (the details are left as an exercise) and so Equation 8.3.7 is in all cases a reasonable approximation to the sample size necessary to ensure that the type II error at the value $\mu = \mu_1$ is approximately equal to β.

EXAMPLE 8.3d For the problem of Example 8.3a, how many signals need be sent so that the .05 level test of $H_0 : \mu = 8$ has at least a 75 percent probability of rejection when $\mu = 9.2$?

SOLUTION Since $z_{.025} = 1.96, z_{.25} = .67$, the approximation 8.3.7 yields

$$n \approx \frac{(1.96 + .67)^2}{(1.2)^2} 4 = 19.21$$

Hence a sample of size 20 is needed. From Equation 8.3.4, we see that with $n = 20$

$$\beta(9.2) = \Phi\left(-\frac{1.2\sqrt{20}}{2} + 1.96\right) - \Phi\left(-\frac{1.2\sqrt{20}}{2} - 1.96\right)$$

$$= \Phi(-.723) - \Phi(-4.643)$$

$$\approx 1 - \Phi(.723)$$

$$\approx .235$$

Therefore, if the message is sent 20 times, then there is a 76.5 percent chance that the null hypothesis $\mu = 8$ will be rejected when the true mean is 9.2. ∎

8.3.1.1   ONE-SIDED TESTS

In testing the null hypothesis that $\mu = \mu_0$, we have chosen a test that calls for rejection when $\overline{X}$ is far from μ_0. That is, a very small value of $\overline{X}$ or a very large value appears to make it unlikely that μ (which $\overline{X}$ is estimating) could equal μ_0. However, what happens when the only alternative to μ being equal to μ_0 is for μ to be greater than μ_0? That is, what happens when the alternative hypothesis to $H_0 : \mu = \mu_0$ is $H_1 : \mu > \mu_0$? Clearly, in this latter case we would not want to reject H_0 when $\overline{X}$ is small (since a small $\overline{X}$ is more likely when H_0 is true than when H_1 is true). Thus, in testing

$$H_0 : \mu = \mu_0 \qquad \text{versus} \qquad H_1 : \mu > \mu_0 \qquad\qquad (8.3.8)$$

we should reject H_0 when $\overline{X}$, the point estimate of μ_0, is much greater than μ_0. That is, the critical region should be of the following form:

$$C = \{(X_1, \ldots, X_n) : \overline{X} - \mu_0 > c\}$$

Since the probability of rejection should equal α when H_0 is true (that is, when $\mu = \mu_0$), we require that c be such that

$$P_{\mu_0}\{\overline{X} - \mu_0 > c\} = \alpha \qquad\qquad (8.3.9)$$

But since

$$Z = \frac{\overline{X} - \mu_0}{\sigma/\sqrt{n}} = \frac{\sqrt{n}(\overline{X} - \mu_0)}{\sigma}$$

has a standard normal distribution when H_0 is true, Equation 8.3.9 is equivalent to

$$P\left\{Z > \frac{c\sqrt{n}}{\sigma}\right\} = \alpha$$

when Z is a standard normal. But since

$$P\{Z > z_\alpha\} = \alpha$$

we see that

$$c = \frac{z_\alpha \sigma}{\sqrt{n}}$$

Hence, the test of the hypothesis 8.3.8 is to reject H_0 if $\overline{X} - \mu_0 > z_\alpha \sigma / \sqrt{n}$, and accept otherwise; or, equivalently, to

$$
\begin{aligned}
\text{accept} \quad & H_0 \quad \text{if} \quad \frac{\sqrt{n}}{\sigma}(\overline{X} - \mu_0) \leq z_\alpha \\
\text{reject} \quad & H_0 \quad \text{if} \quad \frac{\sqrt{n}}{\sigma}(\overline{X} - \mu_0) > z_\alpha
\end{aligned}
\tag{8.3.10}
$$

This is called a *one-sided* critical region (since it calls for rejection only when $\overline{X}$ is large). Correspondingly, the hypothesis testing problem

$$
\begin{aligned}
H_0 &: \mu = \mu_0 \\
H_1 &: \mu > \mu_0
\end{aligned}
$$

is called a one-sided testing problem (in contrast to the *two-sided* problem that results when the alternative hypothesis is $H_1 : \mu \neq \mu_0$).

To compute the *p*-value in the one-sided test, Equation 8.3.10, we first use the data to determine the value of the statistic $\sqrt{n}(\overline{X} - \mu_0)/\sigma$. The *p*-value is then equal to the probability that a standard normal would be at least as large as this value.

EXAMPLE 8.3e Suppose in Example 8.3a that we know in advance that the signal value is at least as large as 8. What can be concluded in this case?

SOLUTION To see if the data are consistent with the hypothesis that the mean is 8, we test

$$
H_0 : \mu = 8
$$

against the one-sided alternative

$$
H_1 : \mu > 8
$$

The value of the test statistic is $\sqrt{n}(\overline{X} - \mu_0)/\sigma = \sqrt{5}(9.5 - 8)/2 = 1.68$, and the *p*-value is the probability that a standard normal would exceed 1.68, namely,

$$
p\text{-value} = 1 - \Phi(1.68) = .0465
$$

the test would call for rejection at all significance levels greater than or equal to .0465, it would, for instance, reject the null hypothesis at the $\alpha = .05$ level of significance. ∎

The operating characteristic function of the one-sided test, Equation 8.3.10,

$$
\beta(\mu) = P_\mu\{\text{accepting } H_0\}
$$

can be obtained as follows:

$$\beta(\mu) = P_\mu\left\{\overline{X} \le \mu_0 + z_\alpha \frac{\sigma}{\sqrt{n}}\right\}$$

$$= P\left\{\frac{\sqrt{n}(\overline{X} - \mu_0)}{\sigma} \le \frac{\sqrt{n}(\mu_0 - \mu)}{\sigma} + z_\alpha\right\}$$

$$= P\left\{Z \le \frac{\sqrt{n}(\mu_0 - \mu)}{\sigma} + z_\alpha\right\}, \quad Z \sim \mathcal{N}(0, 1)$$

where the last equation follows since $\sqrt{n}(\overline{X} - \mu)/\sigma$ has a standard normal distribution. Hence we can write

$$\beta(\mu) = \Phi\left(\frac{\sqrt{n}(\mu_0 - \mu)}{\sigma} + z_\alpha\right)$$

Since Φ, being a distribution function, is increasing in its argument, it follows that $\beta(\mu)$ decreases in μ, which is intuitively pleasing since it certainly seems reasonable that the larger the true mean μ, the less likely it should be to conclude that $\mu \le \mu_0$. Also since $\Phi(z_\alpha) = 1 - \alpha$, it follows that

$$\beta(\mu_0) = 1 - \alpha$$

The test given by Equation 8.3.10, which was designed to test $H_0 : \mu = \mu_0$ versus $H_1 : \mu > \mu_0$, can also be used to test, at level of significance α, the one-sided hypothesis

$$H_0 : \mu \le \mu_0$$

versus

$$H_1 : \mu > \mu_0$$

To verify that it remains a level α test, we need to show that the probability of rejection is never greater than α when H_0 is true. That is, we must verify that

$$1 - \beta(\mu) \le \alpha \qquad \text{for all } \mu \le \mu_0$$

or

$$\beta(\mu) \ge 1 - \alpha \qquad \text{for all } \mu \le \mu_0$$

But it has previously been shown that for the test given by Equation 8.3.10, $\beta(\mu)$ decreases in μ and $\beta(\mu_0) = 1 - \alpha$. This gives that

$$\beta(\mu) \ge \beta(\mu_0) = 1 - \alpha \qquad \text{for all } \mu \le \mu_0$$

which shows that the test given by Equation 8.3.10 remains a level α test for $H_0 : \mu \le \mu_0$ against the alternative hypothesis $H_1 : \mu > \mu_0$.

REMARK

We can also test the one-sided hypothesis

$$H_0 : \mu = \mu_0 \quad (\text{or } \mu \geq \mu_0) \qquad \text{versus} \quad H_1 : \mu < \mu_0$$

at significance level α by

$$\text{accepting} \quad H_0 \quad \text{if} \quad \frac{\sqrt{n}}{\sigma}(\overline{X} - \mu_0) \geq -z_\alpha$$

$$\text{rejecting} \quad H_0 \quad \text{otherwise}$$

This test can alternatively be performed by first computing the value of the test statistic $\sqrt{n}(\overline{X} - \mu_0)/\sigma$. The p-value would then equal the probability that a standard normal would be less than this value, and the hypothesis would be rejected at any significance level greater than or equal to this p-value.

EXAMPLE 8.3f All cigarettes presently on the market have an average nicotine content of at least 1.6 mg per cigarette. A firm that produces cigarettes claims that it has discovered a new way to cure tobacco leaves that will result in the average nicotine content of a cigarette being less than 1.6 mg. To test this claim, a sample of 20 of the firm's cigarettes were analyzed. If it is known that the standard deviation of a cigarette's nicotine content is .8 mg, what conclusions can be drawn, at the 5 percent level of significance, if the average nicotine content of the 20 cigarettes is 1.54?

Note: The above raises the question of how we would know in advance that the standard deviation is .8. One possibility is that the variation in a cigarette's nicotine content is due to variability in the amount of tobacco in each cigarette and not on the method of curing that is used. Hence, the standard deviation can be known from previous experience.

SOLUTION We must first decide on the appropriate null hypothesis. As was previously noted, our approach to testing is not symmetric with respect to the null and the alternative hypotheses since we consider only tests having the property that their probability of rejecting the null hypothesis when it is true will never exceed the significance level α. Thus, whereas rejection of the null hypothesis is a strong statement about the data not being consistent with this hypothesis, an analogous statement cannot be made when the null hypothesis is accepted. Hence, since in the preceding example we would like to endorse the producer's claims only when there is substantial evidence for it, we should take this claim as the alternative hypothesis. That is, we should test

$$H_0 : \mu \geq 1.6 \qquad \text{versus} \qquad H_1 : \mu < 1.6$$

Now, the value of the test statistic is

$$\sqrt{n}(\overline{X} - \mu_0)/\sigma = \sqrt{20}(1.54 - 1.6)/.8 = -.336$$

and so the p-value is given by

$$p\text{-value} = P\{Z < -.336\}, \qquad Z \sim N(0,1)$$
$$= .368$$

Since this value is greater than .05, the foregoing data do not enable us to reject, at the .05 percent level of significance, the hypothesis that the mean nicotine content exceeds 1.6 mg. In other words, the evidence, although supporting the cigarette producer's claim, is not strong enough to prove that claim. ■

REMARKS

(a) There is a direct analogy between confidence interval estimation and hypothesis testing. For instance, for a normal population having mean μ and known variance σ^2, we have shown in Section 7.3 that a $100(1-\alpha)$ percent confidence interval for μ is given by

$$\mu \in \left(\bar{x} - z_{\alpha/2} \frac{\sigma}{\sqrt{n}}, \ \bar{x} + z_{\alpha/2} \frac{\sigma}{\sqrt{n}} \right)$$

where $\bar{x}$ is the observed sample mean. More formally, the preceding confidence interval statement is equivalent to

$$P\left\{ \mu \in \left(\overline{X} - z_{\alpha/2} \frac{\sigma}{\sqrt{n}}, \ \overline{X} + z_{\alpha/2} \frac{\sigma}{\sqrt{n}} \right) \right\} = 1 - \alpha$$

Hence, if $\mu = \mu_0$, then the probability that μ_0 will fall in the interval

$$\left(\overline{X} - z_{\alpha/2} \frac{\sigma}{\sqrt{n}}, \ \overline{X} + z_{\alpha/2} \frac{\sigma}{\sqrt{n}} \right)$$

is $1 - \alpha$, implying that a significance level α test of $H_0 : \mu = \mu_0$ versus $H_1 : \mu \neq \mu_0$ is to reject H_0 when

$$\mu_0 \notin \left(\overline{X} - z_{\alpha/2} \frac{\sigma}{\sqrt{n}}, \ \overline{X} + z_{\alpha/2} \frac{\sigma}{\sqrt{n}} \right)$$

Similarly, since a $100(1-\alpha)$ percent one-sided confidence interval for μ is given by

$$\mu \in \left(\overline{X} - z_{\alpha} \frac{\sigma}{\sqrt{n}}, \ \infty \right)$$

it follows that an α-level significance test of $H_0 : \mu \leq \mu_0$ versus $H_1 : \mu > \mu_0$ is to reject H_0 when $\mu_0 \notin (\overline{X} - z_{\alpha}\sigma/\sqrt{n}, \infty)$ — that is, when $\mu_0 < \overline{X} - z_{\alpha}\sigma/\sqrt{n}$.

TABLE 8.1 $X_1, \ldots, X_n$ Is a Sample from a $\mathcal{N}(\mu, \sigma^2)$ Population
σ^2 Is Known, $\overline{X} = \sum\limits_{i=1}^{n} X_i/n$

H_0	H_1	Test Statistic TS	Significance Level α Test	p-Value if $TS = t$
$\mu = \mu_0$	$\mu \neq \mu_0$	$\sqrt{n}(\overline{X} - \mu_0)/\sigma$	Reject if $\|TS\| > z_{\alpha/2}$	$2P\{Z \geq \|t\|\}$
$\mu \leq \mu_0$	$\mu > \mu_0$	$\sqrt{n}(\overline{X} - \mu_0)/\sigma$	Reject if $TS > z_\alpha$	$P\{Z \geq t\}$
$\mu \geq \mu_0$	$\mu < \mu_0$	$\sqrt{n}(\overline{X} - \mu_0)/\sigma$	Reject if $TS < -z_\alpha$	$P\{Z \leq t\}$

Z is a standard normal random variable.

(b) A Remark on Robustness   A test that performs well even when the underlying assumptions on which it is based are violated is said to be *robust*. For instance, the tests of Sections 8.3.1 and 8.3.1.1 were derived under the assumption that the underlying population distribution is normal with known variance σ^2. However, in deriving these tests, this assumption was used only to conclude that $\overline{X}$ also has a normal distribution. But, by the central limit theorem, it follows that for a reasonably large sample size, $\overline{X}$ will approximately have a normal distribution no matter what the underlying distribution. Thus we can conclude that these tests will be relatively robust for any population distribution with variance σ^2.

   Table 8.1 summarizes the tests of this subsection.

8.3.2 CASE OF UNKNOWN VARIANCE: THE *t*-TEST

Up to now we have supposed that the only unknown parameter of the normal population distribution is its mean. However, the more common situation is one where the mean μ and variance σ^2 are both unknown. Let us suppose this to be the case and again consider a test of the hypothesis that the mean is equal to some specified value μ_0. That is, consider a test of

$$H_0 : \mu = \mu_0$$

versus the alternative

$$H_1 : \mu \neq \mu_0$$

It should be noted that the null hypothesis is not a simple hypothesis since it does not specify the value of σ^2.

   As before, it seems reasonable to reject H_0 when the sample mean $\overline{X}$ is far from μ_0. However, how far away it need be to justify rejection will depend on the variance σ^2. Recall that when the value of σ^2 was known, the test called for rejecting H_0 when $|\overline{X} - \mu_0|$ exceeded $z_{\alpha/2}\sigma/\sqrt{n}$ or, equivalently, when

$$\left| \frac{\sqrt{n}(\overline{X} - \mu_0)}{\sigma} \right| > z_{\alpha/2}$$

Now when σ^2 is no longer known, it seems reasonable to estimate it by

$$S^2 = \frac{\sum_{i=1}^{n}(X_i - \overline{X})^2}{n - 1}$$

and then to reject H_0 when

$$\left| \frac{\overline{X} - \mu_0}{S/\sqrt{n}} \right|$$

is large.

To determine how large a value of the statistic

$$\left| \frac{\sqrt{n}(\overline{X} - \mu_0)}{S} \right|$$

to require for rejection, in order that the resulting test have significance level α, we must determine the probability distribution of this statistic when H_0 is true. However, as shown in Section 6.5, the statistic T, defined by

$$T = \frac{\sqrt{n}(\overline{X} - \mu_0)}{S}$$

has, when $\mu = \mu_0$, a t-distribution with $n - 1$ degrees of freedom. Hence,

$$P_{\mu_0}\left\{ -t_{\alpha/2, n-1} \leq \frac{\sqrt{n}(\overline{X} - \mu_0)}{S} \leq t_{\alpha/2, n-1} \right\} = 1 - \alpha \qquad (8.3.11)$$

where $t_{\alpha/2, n-1}$ is the 100 $\alpha/2$ upper percentile value of the t-distribution with $n - 1$ degrees of freedom. (That is, $P\{T_{n-1} \geq t_{\alpha/2, n-1}\} = P\{T_{n-1} \leq -t_{\alpha/2, n-1}\} = \alpha/2$ when T_{n-1} has a t-distribution with $n - 1$ degrees of freedom.) From Equation 8.3.11 we see that the appropriate significance level α test of

$$H_0 : \mu = \mu_0 \qquad \text{versus} \qquad H_1 : \mu \neq \mu_0$$

is, when σ^2 is unknown, to

$$\text{accept} \quad H_0 \quad \text{if} \quad \left| \frac{\sqrt{n}(\overline{X} - \mu_0)}{S} \right| \leq t_{\alpha/2, n-1}$$

$$(8.3.12)$$

$$\text{reject} \quad H_0 \quad \text{if} \quad \left| \frac{\sqrt{n}(\overline{X} - \mu_0)}{S} \right| > t_{\alpha/2, n-1}$$

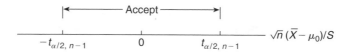

FIGURE 8.3   *The two-sided t-test.*

The test defined by Equation 8.3.12 is called a *two-sided t-test.* It is pictorially illustrated in Figure 8.3.

If we let t denote the observed value of the test statistic $T = \sqrt{n}(\overline{X} - \mu_0)/S$, then the p-value of the test is the probability that $|T|$ would exceed $|t|$ when H_0 is true. That is, the p-value is the probability that the absolute value of a t-random variable with $n - 1$ degrees of freedom would exceed $|t|$. The test then calls for rejection at all significance levels higher than the p-value and acceptance at all lower significance levels.

Program 8.3.2 computes the value of the test statistic and the corresponding p-value. It can be applied both for one- and two-sided tests. (The one-sided material will be presented shortly.)

EXAMPLE 8.3g   Among a clinic's patients having blood cholesterol levels ranging in the medium to high range (at least 220 milliliters per deciliter of serum), volunteers were recruited to test a new drug designed to reduce blood cholesterol. A group of 50 volunteers was given the drug for 1 month and the changes in their blood cholesterol levels were noted. If the average change was a reduction of 14.8 with a sample standard deviation of 6.4, what conclusions can be drawn?

SOLUTION   Let us start by testing the hypothesis that the change could be due solely to chance — that is, that the 50 changes constitute a normal sample with mean 0. Because the value of the t-statistic used to test the hypothesis that a normal mean is equal to 0 is

$$T = \sqrt{n}\,\overline{X}/S = \sqrt{50}\ 14.8/6.4 = 16.352$$

is clear that we should reject the hypothesis that the changes were solely due to chance. Unfortunately, however, we are not justified at this point in concluding that the changes were due to the specific drug used and not to some other possibility. For instance, it is well known that any medication received by a patient (whether or not this medication is directly relevant to the patient's suffering) often leads to an improvement in the patient's condition — the so-called placebo effect. Also, another possibility that may need to be taken into account would be the weather conditions during the month of testing, for it is certainly conceivable that this affects blood cholesterol level. Indeed, it must be concluded that the foregoing was a very poorly designed experiment, for in order to test whether a specific treatment has an effect on a disease that may be affected by many things, we should try to design the experiment so as to neutralize all other possible causes. The accepted approach for accomplishing this is to divide the volunteers at random into two groups — one group to receive the drug and the other to receive a placebo (that is, a tablet that looks and tastes like the actual drug but has no physiological effect). The volunteers should not be told whether they are in the actual or control group, and indeed it is best if even the clinicians do not have this information (the so-called double-blind test) so as not to allow their own biases to play a role. Since the two groups are chosen at random from among the volunteers, we can now hope that on average all factors affecting the two groups will be the same except that one received the actual drug and the other a placebo. Hence, any difference in performance between the groups can be attributed to the drug. ∎

EXAMPLE 8.3h A public health official claims that the mean home water use is 350 gallons a day. To verify this claim, a study of 20 randomly selected homes was instigated with the result that the average daily water uses of these 20 homes were as follows:

340	344	362	375
356	386	354	364
332	402	340	355
362	322	372	324
318	360	338	370

Do the data contradict the official's claim?

SOLUTION To determine if the data contradict the official's claim, we need to test

$$H_0 : \mu = 350 \qquad \text{versus} \qquad H_1 : \mu \neq 350$$

This can be accomplished by running Program 8.3.2 or, if it is incovenient to utilize, by noting first that the sample mean and sample standard deviation of the preceding data set are

$$\overline{X} = 353.8, \qquad S = 21.8478$$

Thus, the value of the test statistic is

$$T = \frac{\sqrt{20}(3.8)}{21.8478} = .7778$$

Because this is less than $t_{.05,19} = 1.730$, the null hypothesis is accepted at the 10 percent level of significance. Indeed, the p-value of the test data is

$$p\text{-value} = P\{|T_{19}| > .7778\} = 2P\{T_{19} > .7778\} = .4462$$

indicating that the null hypothesis would be accepted at any reasonable significance level, and thus that the data are not inconsistent with the claim of the health official. ∎

We can use a one-sided t-test to test the hypothesis

$$H_0 : \mu = \mu_0 \qquad (\text{or } H_0 : \mu \leq \mu_0)$$

against the one-sided alternative

$$H_1 : \mu > \mu_0$$

The significance level α test is to

$$\text{accept} \quad H_0 \quad \text{if} \quad \frac{\sqrt{n}(\overline{X} - \mu_0)}{S} \leq t_{\alpha, n-1}$$

$$\text{reject} \quad H_0 \quad \text{if} \quad \frac{\sqrt{n}(\overline{X} - \mu_0)}{S} > t_{\alpha, n-1}$$

(8.3.13)

If $\sqrt{n}(\overline{X} - \mu_0)/S = v$, then the p-value of the test is the probability that a t-random variable with $n - 1$ degrees of freedom would be at least as large as v.

The significance level α test of

$$H_0 : \mu = \mu_0 \qquad (\text{or } H_0 : \mu \geq \mu_0)$$

versus the alternative

$$H_1 : \mu < \mu_0$$

is to

$$\text{accept} \quad H_0 \quad \text{if} \quad \frac{\sqrt{n}(\overline{X} - \mu_0)}{S} \geq -t_{\alpha, n-1}$$

$$\text{reject} \quad H_0 \quad \text{if} \quad \frac{\sqrt{n}(\overline{X} - \mu_0)}{S} < -t_{\alpha, n-1}$$

The p-value of this test is the probability that a t-random variable with $n - 1$ degrees of freedom would be less than or equal to the observed value of $\sqrt{n}(\overline{X} - \mu_0)/S$.

EXAMPLE 8.3i The manufacturer of a new fiberglass tire claims that its average life will be at least 40,000 miles. To verify this claim a sample of 12 tires is tested, with their lifetimes (in 1,000s of miles) being as follows:

Tire	1	2	3	4	5	6	7	8	9	10	11	12
Life	36.1	40.2	33.8	38.5	42	35.8	37	41	36.8	37.2	33	36

Test the manufacturer's claim at the 5 percent level of significance.

SOLUTION To determine whether the foregoing data are consistent with the hypothesis that the mean life is at least 40,000 miles, we will test

$$H_0 : \mu \geq 40,000 \qquad \text{versus} \qquad H_1 : \mu < 40,000$$

A computation gives that

$$\overline{X} = 37.2833, \qquad S = 2.7319$$

and so the value of the test statistic is

$$T = \frac{\sqrt{12}(37.2833 - 40)}{2.7319} = -3.4448$$

Since this is less than $-t_{.05,11} = -1.796$, the null hypothesis is rejected at the 5 percent level of significance. Indeed, the p-value of the test data is

$$p\text{-value} = P\{T_{11} < -3.4448\} = P\{T_{11} > 3.4448\} = .0028$$

indicating that the manufacturer's claim would be rejected at any significance level greater than .003. ∎

The preceding could also have been obtained by using Program 8.3.2, as illustrated in Figure 8.4.

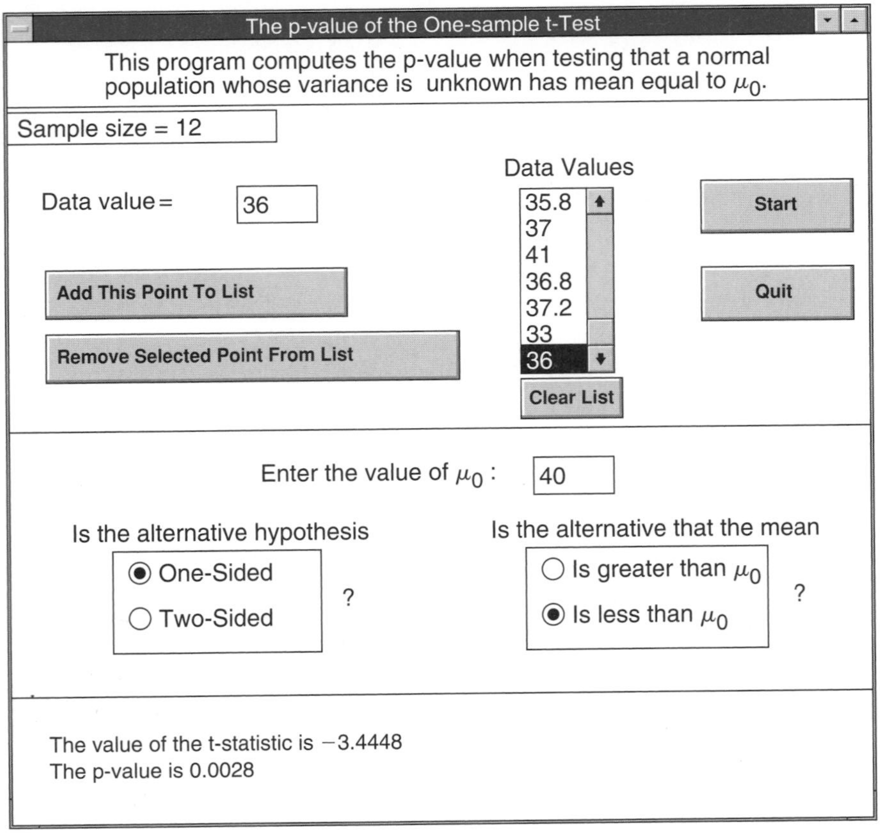

FIGURE 8.4

EXAMPLE 8.3j In a single-server queueing system in which customers arrive according to a Poisson process, the long-run average queueing delay per customer depends on the service distribution through its mean and variance. Indeed, if μ is the mean service time, and σ^2 is the variance of a service time, then the average amount of time that a customer spends waiting in queue is given by

$$\frac{\lambda(\mu^2 + \sigma^2)}{2(1 - \lambda\mu)}$$

provided that $\lambda\mu < 1$, where λ is the arrival rate. (The average delay is infinite if $\lambda\mu \geq 1$.) As can be seen by this formula, the average delay is quite large when μ is only slightly smaller than $1/\lambda$, where, since λ is the arrival *rate*, $1/\lambda$ is the average time between arrivals.

Suppose that the owner of a service station will hire a second server if it can be shown that the average service time exceeds 8 minutes. The following data give the service times (in minutes) of 28 customers of this queueing system. Do they indicate that the mean service time is greater than 8 minutes?

$$8.6, 9.4, 5.0, 4.4, 3.7, 11.4, 10.0, 7.6, 14.4, 12.2, 11.0, 14.4, 9.3, 10.5,$$
$$10.3, 7.7, 8.3, 6.4, 9.2, 5.7, 7.9, 9.4, 9.0, 13.3, 11.6, 10.0, 9.5, 6.6$$

SOLUTION Let us use the preceding data to test the null hypothesis that the mean service time is less than or equal to 8 minutes. A small p-value will then be strong evidence that the mean service time is greater than 8 minutes. Running Program 8.3.2 on these data shows that the value of the test statistic is 2.257, with a resulting p-value of .016. Such a small p-value is certainly strong evidence that the mean service time exceeds 8 minutes. ∎

Table 8.2 summarizes the tests of this subsection.

TABLE 8.2 $X_1, \ldots, X_n$ *Is a Sample from a* $\mathcal{N}(\mu, \sigma^2)$ *Population*

σ^2 *Is Unknown,* $\overline{X} = \sum\limits_{i=1}^{n} X_i/n$ $S^2 = \sum\limits_{i=1}^{n} (X_i - \overline{X})^2/(n-1)$

H_0	H_1	Test Statistic TS	Significance Level α Test	p-Value if $TS = t$
$\mu = \mu_0$	$\mu \neq \mu_0$	$\sqrt{n}(\overline{X} - \mu_0)/S$	Reject if $\|TS\| > t_{\alpha/2, n-1}$	$2P\{T_{n-1} \geq \|t\|\}$
$\mu \leq \mu_0$	$\mu > \mu_0$	$\sqrt{n}(\overline{X} - \mu_0)/S$	Reject if $TS > t_{\alpha, n-1}$	$P\{T_{n-1} \geq t\}$
$\mu \geq \mu_0$	$\mu < \mu_0$	$\sqrt{n}(\overline{X} - \mu_0)/S$	Reject if $TS < -t_{\alpha, n-1}$	$P\{T_{n-1} \leq t\}$

T_{n-1} *is a t-random variable with* $n-1$ *degrees of freedom:* $P\{T_{n-1} > t_{\alpha, n-1}\} = \alpha$.

8.4 TESTING THE EQUALITY OF MEANS OF TWO NORMAL POPULATIONS

A common situation faced by a practicing engineer is one in which she must decide whether two different approaches lead to the same solution. Often such a situation can be modeled as a test of the hypothesis that two normal populations have the same mean value.

8.4.1 CASE OF KNOWN VARIANCES

Suppose that $X_1, \ldots, X_n$ and $Y_1, \ldots, Y_m$ are independent samples from normal populations having unknown means μ_x and μ_y but known variances σ_x^2 and σ_y^2. Let us consider the problem of testing the hypothesis

$$H_0 : \mu_x = \mu_y$$

versus the alternative

$$H_1 : \mu_x \neq \mu_y$$

Since $\overline{X}$ is an estimate of μ_x and $\overline{Y}$ of μ_y, it follows that $\overline{X} - \overline{Y}$ can be used to estimate $\mu_x - \mu_y$. Hence, because the null hypothesis can be written as $H_0 : \mu_x - \mu_y = 0$, it seems reasonable to reject it when $\overline{X} - \overline{Y}$ is far from zero. That is, the form of the test should be to

$$
\begin{aligned}
\text{reject} \quad H_0 \quad &\text{if} \quad |\overline{X} - \overline{Y}| > c \\
\text{accept} \quad H_0 \quad &\text{if} \quad |\overline{X} - \overline{Y}| \leq c
\end{aligned}
\tag{8.4.1}
$$

for some suitably chosen value c.

To determine that value of c that would result in the test in Equations 8.4.1 having a significance level α, we need determine the distribution of $\overline{X} - \overline{Y}$ when H_0 is true. However, as was shown in Section 7.3.2,

$$\overline{X} - \overline{Y} \sim \mathcal{N}\left(\mu_x - \mu_y, \frac{\sigma_x^2}{n} + \frac{\sigma_y^2}{m}\right)$$

which implies that

$$\frac{\overline{X} - \overline{Y} - (\mu_x - \mu_y)}{\sqrt{\dfrac{\sigma_x^2}{n} + \dfrac{\sigma_y^2}{m}}} \sim \mathcal{N}(0, 1) \tag{8.4.2}$$

Hence, when H_0 is true (and so $\mu_x - \mu_y = 0$), it follows that

$$(\overline{X} - \overline{Y}) \Bigg/ \sqrt{\frac{\sigma_x^2}{n} + \frac{\sigma_y^2}{m}}$$

has a standard normal distribution, and thus

$$P_{H_0}\left\{-z_{\alpha/2} \leq \frac{\overline{X} - \overline{Y}}{\sqrt{\dfrac{\sigma_x^2}{n} + \dfrac{\sigma_y^2}{m}}} \leq z_{\alpha/2}\right\} = 1 - \alpha \qquad (8.4.3)$$

From Equation 8.4.3, we obtain that the significance level α test of $H_0 : \mu_x = \mu_y$ versus $H_1 : \mu_x \neq \mu_y$ is

$$\text{accept} \quad H_0 \quad \text{if} \quad \frac{|\overline{X} - \overline{Y}|}{\sqrt{\sigma_x^2/n + \sigma_y^2/m}} \leq z_{\alpha/2}$$

$$\text{reject} \quad H_0 \quad \text{if} \quad \frac{|\overline{X} - \overline{Y}|}{\sqrt{\sigma_x^2/n + \sigma_y^2/m}} \geq z_{\alpha/2}$$

Program 8.4.1 will compute the value of the test statistic $(\overline{X} - \overline{Y})/\sqrt{\sigma_x^2/n + \sigma_y^2/m}$.

EXAMPLE 8.4a Two new methods for producing a tire have been proposed. To ascertain which is superior, a tire manufacturer produces a sample of 10 tires using the first method and a sample of 8 using the second. The first set is to be road tested at location A and the second at location B. It is known from past experience that the lifetime of a tire that is road tested at one of these locations is normally distributed with a mean life due to the tire but with a variance due (for the most part) to the location. Specifically, it is known that the lifetimes of tires tested at location A are normal with standard deviation equal to 4,000 kilometers, whereas those tested at location B are normal with $\sigma = 6,000$ kilometers. If the manufacturer is interested in testing the hypothesis that there is no appreciable difference in the mean life of tires produced by either method, what conclusion should be drawn at the 5 percent level of significance if the resulting data are as given in Table 8.3?

TABLE 8.3   *Tire Lives in Units of 100 Kilometers*

Tires Tested at A	Tires Tested at B
61.1	62.2
58.2	56.6
62.3	66.4
64	56.2
59.7	57.4
66.2	58.4
57.8	57.6
61.4	65.4
62.2	
63.6	

SOLUTION   A simple computation (or the use of Program 8.4.1) shows that the value of the test statistic is .066. For such a small value of the test statistic (which has a standard normal distribution when H_0 is true), it is clear that the null hypothesis is accepted. ∎

It follows from Equation 8.4.1 that a test of the hypothesis $H_0 : \mu_x = \mu_y$ (or $H_0 : \mu_x \le \mu_y$) against the one-sided alternative $H_1 : \mu_x > \mu_y$ would be to

$$\text{accept} \quad H_0 \quad \text{if} \quad \overline{X} - \overline{Y} \le z_\alpha \sqrt{\frac{\sigma_x^2}{n} + \frac{\sigma_y^2}{m}}$$

$$\text{reject} \quad H_0 \quad \text{if} \quad \overline{X} - \overline{Y} > z_\alpha \sqrt{\frac{\sigma_x^2}{n} + \frac{\sigma_y^2}{m}}$$

8.4.2 CASE OF UNKNOWN VARIANCES

Suppose again that $X_1, \ldots, X_n$ and $Y_1, \ldots, Y_m$ are independent samples from normal populations having respective parameters (μ_x, σ_x^2) and (μ_y, σ_y^2), but now suppose that all four parameters are unknown. We will once again consider a test of

$$H_0 : \mu_x = \mu_y \qquad \text{versus} \qquad H_1 : \mu_x \ne \mu_y$$

To determine a significance level α test of H_0 we will need to make the additional assumption that the unknown variances σ_x^2 and σ_y^2 are equal. Let σ^2 denote their value — that is,

$$\sigma^2 = \sigma_x^2 = \sigma_y^2$$

As before, we would like to reject H_0 when $\overline{X} - \overline{Y}$ is "far" from zero. To determine how far from zero it needs to be, let

$$S_x^2 = \frac{\displaystyle\sum_{i=1}^{n}(X_i - \overline{X})^2}{n - 1}$$

$$S_y^2 = \frac{\displaystyle\sum_{i=1}^{m}(Y_i - \overline{Y})^2}{m - 1}$$

denote the sample variances of the two samples. Then, as was shown in Section 7.3.2,

$$\frac{\overline{X} - \overline{Y} - (\mu_x - \mu_y)}{\sqrt{S_p^2(1/n + 1/m)}} \sim t_{n+m-2}$$

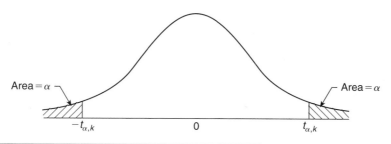

FIGURE 8.5 *Density of a t-random variable with k degrees of freedom.*

where S_p^2, the *pooled* estimator of the common variance σ^2, is given by

$$S_p^2 = \frac{(n-1)S_x^2 + (m-1)S_y^2}{n+m-2}$$

Hence, when H_0 is true, and so $\mu_x - \mu_y = 0$, the statistic

$$T \equiv \frac{\overline{X} - \overline{Y}}{\sqrt{S_p^2(1/n + 1/m)}}$$

has a *t*-distribution with $n + m - 2$ degrees of freedom. From this, it follows that we can test the hypothesis that $\mu_x = \mu_y$ as follows:

$$\text{accept} \quad H_0 \quad \text{if} \quad |T| \le t_{\alpha/2,\, n+m-2}$$
$$\text{reject} \quad H_0 \quad \text{if} \quad |T| > t_{\alpha/2,\, n+m-2}$$

where $t_{\alpha/2,\, n+m-2}$ is the 100 $\alpha/2$ percentile point of a *t*-random variable with $n + m - 2$ degrees of freedom (see Figure 8.5).

Alternatively, the test can be run by determining the *p*-value. If T is observed to equal v, then the resulting *p*-value of the test of H_0 against H_1 is given by

$$p\text{-value} = P\{|T_{n+m-2}| \ge |v|\}$$
$$= 2P\{T_{n+m-2} \ge |v|\}$$

where T_{n+m-2} is a *t*-random variable having $n + m - 2$ degrees of freedom.

If we are interested in testing the one-sided hypothesis

$$H_0 : \mu_x \le \mu_y \quad \text{versus} \quad H_1 : \mu_x > \mu_y$$

then H_0 will be rejected at large values of T. Thus the significance level α test is to

$$\text{reject} \quad H_0 \quad \text{if} \quad T \ge t_{\alpha,\, n+m-2}$$
$$\text{not reject} \quad H_0 \quad \text{otherwise}$$

If the value of the test statistic T is v, then the p-value is given by

$$p\text{-value} = P\{T_{n+m-2} \geq v\}$$

Program 8.4.2 computes both the value of the test statistic and the corresponding p-value.

EXAMPLE 8.4b   Twenty-two volunteers at a cold research institute caught a cold after having been exposed to various cold viruses. A random selection of 10 of these volunteers was given tablets containing 1 gram of vitamin C. These tablets were taken four times a day. The control group consisting of the other 12 volunteers was given placebo tablets that looked and tasted exactly the same as the vitamin C tablets. This was continued for each volunteer until a doctor, who did not know if the volunteer was receiving the vitamin C or the placebo tablets, decided that the volunteer was no longer suffering from the cold. The length of time the cold lasted was then recorded.

At the end of this experiment, the following data resulted.

Treated with Vitamin C	Treated with Placebo
5.5	6.5
6.0	6.0
7.0	8.5
6.0	7.0
7.5	6.5
6.0	8.0
7.5	7.5
5.5	6.5
7.0	7.5
6.5	6.0
	8.5
	7.0

Do the data listed prove that taking 4 grams daily of vitamin C reduces the mean length of time a cold lasts? At what level of significance?

SOLUTION   To prove the above hypothesis, we would need to reject the null hypothesis in a test of

$$H_0 : \mu_p \leq \mu_c \qquad \text{versus} \qquad H_1 : \mu_p > \mu_c$$

where μ_c is the mean time a cold lasts when the vitamin C tablets are taken and μ_p is the mean time when the placebo is taken. Assuming that the variance of the length of the cold is the same for the vitamin C patients and the placebo patients, we test the above by running Program 8.4.2. This yields the information shown in Figure 8.6. Thus H_0 would be rejected at the 5 percent level of significance.

```
┌──────────────────────────────────────────────────────────────┐
│              The p-value of the Two-sample t-Test      ▼  ▲    │
│  List 1     Sample size = 10                                   │
│                                         ┌──────────┐           │
│   Data value =    ┌─────────┐           │ 6      ▲ │           │
│                   │  6.5    │           │ 7.5      │           │
│                   └─────────┘           │ 6        │           │
│                                         │ 7.5      │           │
│   ┌───────────────────────────────┐     │ 5.5      │           │
│   │  Add This Point To List 1     │     │ 7        │  ┌──────────┐ │
│   └───────────────────────────────┘     │ 6.5    ▼ │  │ Clear List 1 │
│                                         └──────────┘  └──────────┘ │
│   ┌───────────────────────────────────┐                       │
│   │ Remove Selected Point From List 1 │                       │
│   └───────────────────────────────────┘                       │
│                                                                │
│  List 2     Sample size = 12            ┌──────────┐           │
│   Data value =    ┌─────────┐           │ 8      ▲ │           │
│                   │  7      │           │ 7.5      │           │
│                   └─────────┘           │ 6.5      │           │
│   ┌───────────────────────────────┐     │ 7.5      │           │
│   │  Add This Point To List 2     │     │ 6        │           │
│   └───────────────────────────────┘     │ 8.5      │  ┌──────────┐ │
│   ┌───────────────────────────────────┐ │ 7      ▼ │  │ Clear List 2 │
│   │ Remove Selected Point From List 2 │ └──────────┘  └──────────┘ │
│   └───────────────────────────────────┘                       │
│                                                                │
│  Is the alternative  ┌─────────────────┐                       │
│  hypothesis          │ ● One-Sided     │  ?   ┌──────────┐     │
│                      │ ○ Two-Sided     │      │  Start   │     │
│                      └─────────────────┘      └──────────┘     │
│  Is the alternative  ┌─────────────────┐                       │
│  that the mean       │ ○ Is greater than│  the mean ┌────────┐ │
│  of sample 1         │ ● Is less than  │  of sample 2?│ Quit │ │
│                      └─────────────────┘      └──────────┘     │
│                                                                │
│   The value of the t-statistic is −1.898695                    │
│   The p-value is 0.03607                                       │
└──────────────────────────────────────────────────────────────┘
```

FIGURE 8.6

Of course, if it were not convenient to run Program 8.4.2 then we could have performed the test by first computing the values of the statistics $\overline{X}, \overline{Y}, S_x^2, S_y^2$, and S_p^2, where the X sample corresponds to those receiving vitamin C and the Y sample to those receiving a placebo. These computations would give the values

$$\overline{X} = 6.450, \qquad \overline{Y} = 7.125$$
$$S_x^2 = .581, \qquad S_y^2 = .778$$

Therefore,

$$S_p^2 = \frac{9}{20}S_x^2 + \frac{11}{20}S_y^2 = .689$$

and the value of the test statistic is

$$TS = \frac{-.675}{\sqrt{.689(1/10 + 1/12)}} = -1.90$$

Since $t_{.05,20} = 1.725$, the null hypothesis is rejected at the 5 percent level of significance. That is, at the 5 percent level of significance the evidence is significant in establishing that vitamin C reduces the mean time that a cold persists.   ∎

EXAMPLE 8.4c   Reconsider Example 8.4a, but now suppose that the population variances are unknown but equal.

SOLUTION   Using Program 8.4.2 yields that the value of the test statistic is 1.028, and the resulting p-value is

$$p\text{-value} = P\{T_{16} > 1.028\} = .3192$$

Thus, the null hypothesis is accepted at any significance level less than .3192.   ∎

8.4.3 Case of Unknown and Unequal Variances

Let us now suppose that the population variances σ_x^2 and σ_y^2 are not only unknown but also cannot be considered to be equal. In this situation, since S_x^2 is the natural estimator of σ_x^2 and S_y^2 of σ_y^2, it would seem reasonable to base our test of

$$H_0 : \mu_x = \mu_y \qquad \text{versus} \qquad H_1 : \mu_x \neq \mu_y$$

on the test statistic

$$\frac{\overline{X} - \overline{Y}}{\sqrt{\dfrac{S_x^2}{n} + \dfrac{S_y^2}{m}}} \tag{8.4.4}$$

However, the foregoing has a complicated distribution, which, even when H_0 is true, depends on the unknown parameters, and thus cannot be generally employed. The one situation in which we can utilize the statistic of Equation 8.4.4 is when n and m are both large. In such a case, it can be shown that when H_0 is true Equation 8.4.4 will have *approximately* a standard normal distribution. Hence, when n and m are large an *approximate* level α test of $H_0 : \mu_x = \mu_y$ versus $H_1 : \mu_x \neq \mu_y$ is to

$$\text{accept} \quad H_0 \quad \text{if} \quad -z_{\alpha/2} \leq \frac{\overline{X} - \overline{Y}}{\sqrt{\dfrac{S_x^2}{n} + \dfrac{S_y^2}{m}}} \leq z_{\alpha/2}$$

$$\text{reject} \quad \text{otherwise}$$

The problem of determining an exact level α test of the hypothesis that the means of two normal populations, having unknown and not necessarily equal variances, are equal is known as the Behrens-Fisher problem. There is no completely satisfactory solution known.

Table 8.4 presents the two-sided tests of this section.

TABLE 8.4 $X_1, \ldots, X_n$ Is a Sample from a $\mathcal{N}(\mu_1, \sigma_1^2)$ Population; $Y_1, \ldots, Y_m$ Is a Sample from a $\mathcal{N}(\mu_2, \sigma_2^2)$ Population

The Two Population Samples Are Independent
to Test
$H_0 : \mu_1 = \mu_2$ versus $H_0 : \mu_1 \neq \mu_2$

Assumption	Test Statistic TS	Significance Level α Test	p-Value if $TS = t$
σ_1, σ_2 known	$\dfrac{\overline{X} - \overline{Y}}{\sqrt{\sigma_1^2/n + \sigma_2^2/m}}$	Reject if $|TS| > z_{\alpha/2}$	$2P\{Z \geq |t|\}$
$\sigma_1 = \sigma_2$	$\dfrac{\overline{X} - \overline{Y}}{\sqrt{\frac{(n-1)S_1^2 + (m-1)S_2^2}{n+m-2}}\sqrt{1/n + 1/m}}$	Reject if $|TS| > t_{\alpha/2, n+m-2}$	$2P\{T_{n+m-2} \geq |t|\}$
n, m large	$\dfrac{\overline{X} - \overline{Y}}{\sqrt{S_1^2/n + S_2^2/m}}$	Reject if $|TS| > z_{\alpha/2}$	$2P\{Z \geq |t|\}$

8.4.4 THE PAIRED t-TEST

Suppose we are interested in determining whether the installation of a certain antipollution device will affect a car's mileage. To test this, a collection of n cars that do not have this device are gathered. Each car's mileage per gallon is then determined both before and after the device is installed. How can we test the hypothesis that the antipollution control has no effect on gas consumption?

The data can be described by the n pairs $(X_i, Y_i), i = 1, \ldots, n$, where X_i is the gas consumption of the ith car before installation of the pollution control device, and Y_i of the same car after installation. Because each of the n cars will be inherently different, we cannot treat $X_1, \ldots, X_n$ and $Y_1, \ldots, Y_n$ as being independent samples. For example, if we know that X_1 is large (say, 40 miles per gallon), we would certainly expect that Y_1 would also probably be large. Thus, we cannot employ the earlier methods presented in this section.

One way in which we can test the hypothesis that the antipollution device does not affect gas mileage is to let the data consist of each car's difference in gas mileage. That is, let $W_i = X_i - Y_i, i = 1, \ldots, n$. Now, if there is no effect from the device, it should follow that the W_i would have mean 0. Hence, we can test the hypothesis of no effect by testing

$$H_0 : \mu_w = 0 \qquad \text{versus} \qquad H_1 : \mu_w \neq 0$$

where $W_1, \ldots, W_n$ are assumed to be a sample from a normal population having unknown mean μ_w and unknown variance σ_w^2. But the t-test described in Section 8.3.2 shows that this can be tested by

$$\text{accepting} \quad H_0 \quad \text{if} \quad -t_{\alpha/2,\, n-1} < \sqrt{n}\frac{\overline{W}}{S_w} < t_{\alpha/2,n-1}$$

$$\text{rejecting} \quad H_0 \quad \text{otherwise}$$

EXAMPLE 8.4d An industrial safety program was recently instituted in the computer chip industry. The average weekly loss (averaged over 1 month) in labor-hours due to accidents in 10 similar plants both before and after the program are as follows:

Plant	Before	After	$A - B$
1	30.5	23	−7.5
2	18.5	21	2.5
3	24.5	22	−2.5
4	32	28.5	−3.5
5	16	14.5	−1.5
6	15	15.5	.5
7	23.5	24.5	1
8	25.5	21	−4.5
9	28	23.5	−4.5
10	18	16.5	−1.5

Determine, at the 5 percent level of significance, whether the safety program has been proven to be effective.

SOLUTION To determine this, we will test

$$H_0 : \mu_A - \mu_B \geq 0 \qquad \text{versus} \qquad H_1 : \mu_A - \mu_B < 0$$

because this will enable us to see whether the null hypothesis that the safety program has not had a beneficial effect is a reasonable possibility. To test this, we run Program 8.3.2, which gives the value of the test statistic as −2.266, with

$$p\text{-value} = P\{T_q \leq -2.266\} = .025$$

Since the p-value is less than .05, the hypothesis that the safety program has not been effective is rejected and so we can conclude that its effectiveness has been established (at least for any significance level greater than .025). ∎

Note that the paired-sample t-test can be used even though the samples are not independent and the population variances are unequal.

8.5 HYPOTHESIS TESTS CONCERNING THE VARIANCE OF A NORMAL POPULATION

Let $X_1, \ldots, X_n$ denote a sample from a normal population having unknown mean μ and unknown variance σ^2, and suppose we desire to test the hypothesis

$$H_0 : \sigma^2 = \sigma_0^2$$

versus the alternative

$$H_1 : \sigma^2 \neq \sigma_0^2$$

for some specified value σ_0^2.

To obtain a test, recall (as was shown in Section 6.5) that $(n-1)S^2/\sigma^2$ has a chi-square distribution with $n-1$ degrees of freedom. Hence, when H_0 is true

$$\frac{(n-1)S^2}{\sigma_0^2} \sim \chi_{n-1}^2$$

Because $P\{\chi_{n-1}^2 < \chi_{\alpha/2,n-1}^2\} = 1 - \alpha/2$ and $P\{\chi_{n-1}^2 < \chi_{1-\alpha/2,n-1}^2\} = \alpha/2$, it follows that

$$P_{H_0} \left\{ \chi_{1-\alpha/2,n-1}^2 \leq \frac{(n-1)S^2}{\sigma_0^2} \leq \chi_{\alpha/2,n-1}^2 \right\} = 1 - \alpha$$

Therefore, a significance level α test is to

$$\text{accept} \quad H_0 \quad \text{if} \quad \chi_{1-\alpha/2,n-1}^2 \leq \frac{(n-1)S^2}{\sigma_0^2} \leq \chi_{\alpha/2,n-1}^2$$

$$\text{reject} \quad H_0 \quad \text{otherwise}$$

The preceding test can be implemented by first computing the value of the test statistic $(n-1)S^2/\sigma_0^2$ — call it c. Then compute the probability that a chi-square random variable with $n-1$ degrees of freedom would be (a) less than and (b) greater than c. If either of these probabilities is less than $\alpha/2$, then the hypothesis is rejected. In other words, the p-value of the test data is

$$p\text{-value} = 2\min(P\{\chi_{n-1}^2 < c\}, 1 - P\{\chi_{n-1}^2 < c\})$$

The quantity $P\{\chi_{n-1}^2 < c\}$ can be obtained from Program 5.8.1.A. The p-value for a one-sided test is similarly obtained.

EXAMPLE 8.5a A machine that automatically controls the amount of ribbon on a tape has recently been installed. This machine will be judged to be effective if the standard deviation σ of the amount of ribbon on a tape is less than .15 cm. If a sample of 20 tapes yields a sample variance of $S^2 = .025$ cm^2, are we justified in concluding that the machine is ineffective?

SOLUTION We will test the hypothesis that the machine is effective, since a rejection of this hypothesis will then enable us to conclude that it is ineffective. Since we are thus interested in testing

$$H_0 : \sigma^2 \leq .0225 \quad \text{versus} \quad H_1 : \sigma^2 > .0225$$

it follows that we would want to reject H_0 when S^2 is large. Hence, the p-value of the preceding test data is the probability that a chi-square random variable with 19 degrees of freedom would exceed the observed value of $19S^2/.0225 = 19 \times .025/.0225 = 21.111$. That is,

$$p\text{-value} = P\{\chi^2_{19} > 21.111\}$$
$$= 1 - .6693 = .3307 \quad \text{from Program 5.8.1.A}$$

Therefore, we must conclude that the observed value of $S^2 = .025$ is not large enough to reasonably preclude the possibility that $\sigma^2 \leq .0225$, and so the null hypothesis is accepted. ∎

8.5.1 TESTING FOR THE EQUALITY OF VARIANCES OF TWO NORMAL POPULATIONS

Let $X_1, \ldots, X_n$ and $Y_1, \ldots, Y_m$ denote independent samples from two normal populations having respective (unknown) parameters μ_x, σ_x^2 and μ_y, σ_y^2 and consider a test of

$$H_0 : \sigma_x^2 = \sigma_y^2 \quad \text{versus} \quad H_1 : \sigma_x^2 \neq \sigma_y^2$$

If we let

$$S_x^2 = \frac{\sum\limits_{i=1}^{n} (X_i - \overline{X})^2}{n-1}$$

$$S_y^2 = \frac{\sum\limits_{i=1}^{m} (Y_i - \overline{Y})^2}{m-1}$$

denote the sample variances, then as shown in Section 6.5, $(n-1)S_x^2/\sigma_x^2$ and $(m-1)S_y^2/\sigma_y^2$ are independent chi-square random variables with $n-1$ and $m-1$ degrees of freedom, respectively. Therefore, $(S_x^2/\sigma_x^2)/(S_y^2/\sigma_y^2)$ has an F-distribution with parameters $n-1$ and $m-1$. Hence, when H_0 is true

$$S_x^2/S_y^2 \sim F_{n-1,m-1}$$

and so

$$P_{H_0}\{F_{1-\alpha/2,n-1,m-1} \leq S_x^2/S_y^2 \leq F_{\alpha/2,n-1,m-1}\} = 1 - \alpha$$

Thus, a significance level α test of H_0 against H_1 is to

$$\begin{array}{lll} \text{accept} & H_0 & \text{if} \quad F_{1-\alpha/2,n-1,m-1} < S_x^2/S_y^2 < F_{\alpha/2,n-1,m-1} \\ \text{reject} & H_0 & \text{otherwise} \end{array}$$

   The preceding test can be effected by first determining the value of the test statistic S_x^2/S_y^2, say its value is v, and then computing $P\{F_{n-1,m-1} \leq v\}$ where $F_{n-1,m-1}$ is an F-random variable with parameters $n - 1$, $m - 1$. If this probability is either less than $\alpha/2$ (which occurs when S_x^2 is significantly less than S_y^2) or greater than $1 - \alpha/2$ (which occurs when S_x^2 is significantly greater than S_y^2), then the hypothesis is rejected. In other words, the p-value of the test data is

$$p\text{-value} = 2 \min(P\{F_{n-1,m-1} < v\}, 1 - P\{F_{n-1,m-1} < v\})$$

The test now calls for rejection whenever the significance level α is at least as large as the p-value.

EXAMPLE 8.5b   There are two different choices of a catalyst to stimulate a certain chemical process. To test whether the variance of the yield is the same no matter which catalyst is used, a sample of 10 batches is produced using the first catalyst, and 12 using the second. If the resulting data are $S_1^2 = .14$ and $S_2^2 = .28$, can we reject, at the 5 percent level, the hypothesis of equal variance?

SOLUTION   Program 5.8.3, which computes the F cumulative distribution function, yields that

$$P\{F_{9,11} \leq .5\} = .1539$$

Hence,

$$\begin{aligned} p\text{-value} &= 2 \min\{.1539, .8461\} \\ &= .3074 \end{aligned}$$

and so the hypothesis of equal variance cannot be rejected.   ■

8.6 HYPOTHESIS TESTS IN BERNOULLI POPULATIONS

The binomial distribution is frequently encountered in engineering problems. For a typical example, consider a production process that manufactures items that can be classified in one of two ways — either as acceptable or as defective. An assumption often made is that each item produced will, independently, be defective with probability p, and so the number of defects in a sample of n items will thus have a binomial distribution with parameters (n, p). We will now consider a test of

$$H_0 : p \leq p_0 \quad \text{versus} \quad H_1 : p > p_0$$

where p_0 is some specified value.

If we let X denote the number of defects in the sample of size n, then it is clear that we wish to reject H_0 when X is large. To see how large it needs to be to justify rejection at the α level of significance, note that

$$P\{X \geq k\} = \sum_{i=k}^{n} P\{X = i\} = \sum_{i=k}^{n} \binom{n}{i} p^i (1 - p)^{n-i}$$

Now it is certainly intuitive (and can be proven) that $P\{X \geq k\}$ is an increasing function of p — that is, the probability that the sample will contain at least k errors increases in the defect probability p. Using this, we see that when H_0 is true (and so $p \leq p_0$),

$$P\{X \geq k\} \leq \sum_{i=k}^{n} \binom{n}{i} p_0^i (1 - p_0)^{n-i}$$

Hence, a significance level α test of $H_0 : p \leq p_0$ versus $H_1 : p > p_0$ is to reject H_0 when

$$X \geq k^*$$

where k^* is the smallest value of k for which $\sum_{i=k}^{n} \binom{n}{i} p_0^i (1 - p_0)^{n-i} \leq \alpha$. That is,

$$k^* = \min \left\{ k : \sum_{i=k}^{n} \binom{n}{i} p_0^i (1 - p_0)^{n-i} \leq \alpha \right\}$$

This test can best be performed by first determining the value of the test statistic — say, $X = x$ — and then computing the p-value given by

$$p\text{-value} = P\{B(n, p_0) \geq x\}$$
$$= \sum_{i=x}^{n} \binom{n}{i} p_0^i (1 - p_0)^{n-i}$$

EXAMPLE 8.6a   A computer chip manufacturer claims that no more than 2 percent of the chips it sends out are defective. An electronics company, impressed with this claim, has purchased a large quantity of such chips. To determine if the manufacturer's claim can be taken literally, the company has decided to test a sample of 300 of these chips. If 10 of these 300 chips are found to be defective, should the manufacturer's claim be rejected?

SOLUTION   Let us test the claim at the 5 percent level of significance. To see if rejection is called for, we need to compute the probability that the sample of size 300 would have resulted in 10 or more defectives when p is equal to .02. (That is, we compute the p-value.) If this probability is less than or equal to .05, then the manufacturer's claim should be rejected. Now

$$P_{.02}\{X \geq 10\} = 1 - P_{.02}\{X < 10\}$$

$$= 1 - \sum_{i=0}^{9} \binom{300}{i} (.02)^i (.98)^{300-i}$$

$$= .0818 \quad \text{from Program 3.1}$$

and so the manufacturer's claim cannot be rejected at the 5 percent level of significance. ∎

EXAMPLE 8.6b In an attempt to show that proofreader A is superior to proofreader B, both proofreaders were given the same manuscript to read. If proofreader A found 28 errors, and proofreader B found 18, with 10 of these errors being found by both, can we conclude that A is the superior proofreader?

SOLUTION To begin note that A found 18 errors that B missed, and that B found 8 that A missed. Hence, a total of 26 errors were found by just a single proofreader. Now, if A and B were equally competent then they would be equally likely to be the sole finder of an error found by just one of them. Consequently, if A and B were equally competent then each of the 26 singly found errors would have been found by A with probability 1/2. Hence, to establish that A is the superior proofreader the result of 18 successes in 26 trials must be strong enough to reject the null hypothesis when testing

$$H_0 : p \leq 1/2 \quad \text{versus} \quad H_1 : p > 1/2$$

where p is a Bernoulli probability that a trial is a success. Because the resultant p-value for the data cited is

$$p\text{-value} = P\{\text{Bin}(26, .5) \geq 18\} = .0378$$

the null hypothesis would be rejected at the 5 percent level of significance, thus enabling one to conclude (at that level of significance) that A is the superior proofreader. ∎

When the sample size n is large, we can derive an *approximate* significance level α test of $H_0 : p \leq p_0$ versus $H_1 : p > p_0$ by using the normal approximation to the binomial. It works as follows: Because when n is large X will have approximately a normal distribution with mean and variance

$$E[X] = np, \qquad \text{Var}(X) = np(1 - p)$$

it follows that

$$\frac{X - np}{\sqrt{np(1 - p)}}$$

will have approximately a standard normal distribution. Therefore, an approximate significance level α test would be to reject H_0 if

$$\frac{X - np_0}{\sqrt{np_0(1 - p_0)}} \geq z_\alpha$$

Equivalently, one can use the normal approximation to approximate the p-value.

EXAMPLE 8.6c   In Example 8.6a, $np_0 = 300(.02) = 6$, and $\sqrt{np_0(1 - p_0)} = \sqrt{5.88}$. Consequently, the p-value that results from the data $X = 10$ is

$$\begin{aligned}
p\text{-value} &= P_{.02}\{X \geq 10\} \\
&= P_{.02}\{X \geq 9.5\} \\
&= P_{.02}\left\{\frac{X - 6}{\sqrt{5.88}} \geq \frac{9.5 - 6}{\sqrt{5.88}}\right\} \\
&\approx P\{Z \geq 1.443\} \\
&= .0745
\end{aligned}$$

Thus, whereas the exact p-value is .0818, the normal approximation gives the value .0745.   ∎

Suppose now that we want to test the null hypothesis that p is equal to some specified value; that is, we want to test

$$H_0 : p = p_0 \quad \text{versus} \quad H_1 : p \neq p_0$$

If X, a binomial random variable with parameters n and p, is observed to equal x, then a significance level α test would reject H_0 if the value x was either significantly larger or significantly smaller than what would be expected when p is equal to p_0. More precisely, the test would reject H_0 if either

$$P\{\text{Bin}(n, p_0) \geq x\} \leq \alpha/2 \quad \text{or} \quad P\{\text{Bin}(n, p_0) \leq x\} \leq \alpha/2$$

In other words, the p-value when $X = x$ is

$$p\text{-value} = 2 \min(P\{\text{Bin}(n, p_0) \geq x\}, P\{\text{Bin}(n, p_0) \leq x\})$$

EXAMPLE 8.6d   Historical data indicate that 4 percent of the components produced at a certain manufacturing facility are defective. A particularly acrimonious labor dispute has recently been concluded, and management is curious about whether it will result in any change in this figure of 4 percent. If a random sample of 500 items indicated 16 defectives (3.2 percent), is this significant evidence, at the 5 percent level of significance, to conclude that a change has occurred?

SOLUTION To be able to conclude that a change has occurred, the data need to be strong enough to reject the null hypothesis when we are testing

$$H_0 : p = .04 \quad \text{versus} \quad H_1 : p \neq .04$$

where p is the probability that an item is defective. The p-value of the observed data of 16 defectives in 500 items is

$$p\text{-value} = 2\min\{P\{X \leq 16\}, P\{X \geq 16\}\}$$

where X is a binomial $(500, .04)$ random variable. Since $500 \times .04 = 20$, we see that

$$p\text{-value} = 2P\{X \leq 16\}$$

Since X has mean 20 and standard deviation $\sqrt{20(.96)} = 4.38$, it is clear that twice the probability that X will be less than or equal to 16 — a value less than one standard deviation lower than the mean — is not going to be small enough to justify rejection. Indeed, it can be shown that

$$p\text{-value} = 2P\{X \leq 16\} = .432$$

and so there is not sufficient evidence to reject the hypothesis that the probability of a defective item has remained unchanged. ∎

8.6.1 TESTING THE EQUALITY OF PARAMETERS IN TWO BERNOULLI POPULATIONS

Suppose there are two distinct methods for producing a certain type of chip; and suppose that chips produced by the first method will, independently, be defective with probability p_1, with the corresponding probability being p_2 for those produced by the second method. To test the hypothesis that $p_1 = p_2$, a sample of n_1 chips is produced using method 1 and n_2 using method 2.

Let X_1 denote the number of defective chips obtained from the first sample and X_2 for the second. Thus, X_1 and X_2 are independent binomial random variables with respective parameters (n_1, p_1) and (n_2, p_2). Suppose that $X_1 + X_2 = k$ and so there have been a total of k defectives. Now, if H_0 is true, then each of the $n_1 + n_2$ chips produced will have the same probability of being defective, and so the determination of the k defectives will have the same distribution as a random selection of a sample of size k from a population of $n_1 + n_2$ items of which n_1 are white and n_2 are black. In other words, given a total of k defectives, the conditional distribution of the number of defective chips obtained from method 1 will, when H_0 is true, have the following hypergeometric distribution[*]:

$$P_{H_0}\{X_1 = i | X_1 + X_2 = k\} = \frac{\binom{n_1}{i}\binom{n_2}{k-i}}{\binom{n_1 + n_2}{k}}, \quad i = 0, 1, \dots, k \qquad (8.6.1)$$

Now, in testing

$$H_0 : p_1 = p_2 \quad \text{versus} \quad H_1 : p_1 \neq p_2$$

it seems reasonable to reject the null hypothesis when the proportion of defective chips produced by method 1 is much different from the proportion of defectives obtained under method 2. Therefore, if there is a total of k defectives, then we would expect, when H_0 is true, that X_1/n_1 (the proportion of defective chips produced by method 1) would be close to $(k - X_1)/n_2$ (the proportion of defective chips produced by method 2). Because X_1/n_1 and $(k - X_1)/n_2$ will be farthest apart when X_1 is either very small or very large, it thus seems that a reasonable significance level α test of Equation 8.6.1 is as follows. If $X_1 + X_2 = k$, then one should

$$\begin{aligned} \text{reject} \quad &H_0 \quad \text{if either} \quad P\{X \leq x_1\} \leq \alpha/2 \quad \text{or} \quad P\{X \geq x_1\} \leq \alpha/2 \\ \text{accept} \quad &H_0 \quad \text{otherwise} \end{aligned}$$

where X is a hypergeometric random variable with probability mass function

$$P\{X = i\} = \frac{\binom{n_1}{i}\binom{n_2}{k-i}}{\binom{n_1 + n_2}{k}} \quad i = 0, 1, \dots, k \qquad (8.6.2)$$

In other words, this test will call for rejection if the significance level is at least as large as the p-value given by

$$p\text{-value} = 2 \min(P\{X \leq x_1\}, P\{X \geq x_1\}) \qquad (8.6.3)$$

This is called the *Fisher-Irwin test*.

---

[*] See Example 5.3b for a formal verification of Equation 8.6.1.

COMPUTATIONS FOR THE FISHER-IRWIN TEST

To utilize the Fisher-Irwin test, we need to be able to compute the hypergeometric distribution function. To do so, note that with X having mass function Equation 8.6.2,

$$\frac{P\{X = i + 1\}}{P\{X = i\}} = \frac{\binom{n_1}{i+1}\binom{n_2}{k-i-1}}{\binom{n_1}{i}\binom{n_2}{k-i}} \tag{8.6.4}$$

$$= \frac{(n_1 - i)(k - i)}{(i + 1)(n_2 - k + i + 1)} \tag{8.6.5}$$

where the verification of the final equality is left as an exercise.

Program 8.6.1 uses the preceding identity to compute the p-value of the data for the Fisher-Irwin test of the equality of two Bernoulli probabilities. The program will work best if the Bernoulli outcome that is called unsuccessful (or defective) is the one whose probability is less than .5. For instance, if over half the items produced are defective, then rather than testing that the defect probability is the same in both samples, one should test that the probability of producing an acceptable item is the same in both samples.

EXAMPLE 8.6e Suppose that method 1 resulted in 20 unacceptable transistors out of 100 produced, whereas method 2 resulted in 12 unacceptable transistors out of 100 produced. Can we conclude from this, at the 10 percent level of significance, that the two methods are equivalent?

SOLUTION Upon running Program 8.6.1, we obtain that

$$p\text{-value} = .1763$$

Hence, the hypothesis that the two methods are equivalent cannot be rejected. ∎

The ideal way to test the hypothesis that the results of two different treatments are identical is to randomly divide a group of people into a set that will receive the first treatment and one that will receive the second. However, such randomization is not always possible. For instance, if we want to study whether drinking alcohol increases the risk of prostate cancer, we cannot instruct a randomly chosen sample to drink alcohol. An alternative way to study the hypothesis is to use an *observational* study that begins by randomly choosing a set of drinkers and one of nondrinkers. These sets are followed for a period of time and the resulting data are then used to test the hypothesis that members of the two groups have the same risk for prostate cancer.

Our next sample illustrates another way of performing an observational study.

EXAMPLE 8.6f In 1970, the researchers Herbst, Ulfelder, and Poskanzer (H-U-P) suspected that vaginal cancer in young women, a rather rare disease, might be caused by one's mother having taken the drug diethylstilbestrol (usually referred to as DES) while pregnant. To study this possibility, the researchers could have performed an observational study by searching for a (treatment) group of women whose mothers took DES when pregnant and a (control) group of women whose mothers did not. They could then observe these groups for a period of time and use the resulting data to test the hypothesis that the probabilities of contracting vaginal cancer are the same for both groups. However, because vaginal cancer is so rare (in both groups) such a study would require a large number of individuals in both groups and would probably have to continue for many years to obtain significant results. Consequently, H-U-P decided on a different type of observational study. They uncovered 8 women between the ages of 15 and 22 who had vaginal cancer. Each of these women (called cases) was then matched with 4 others, called *referents or controls*. Each of the referents of a case was free of the cancer and was born within 5 days in the same hospital and in the same type of room (either private or public) as the case. Arguing that if DES had no effect on vaginal cancer then the probability, call it p_c, that the mother of a case took DES would be the same as the probability, call it p_r, that the mother of a referent took DES, the researchers H-U-P decided to test

$$H_0 : p_c = p_r \quad \text{against} \quad H_1 : p_c \neq p_r$$

Discovering that 7 of the 8 cases had mothers who took DES while pregnant, while none of the 32 referents had mothers who took the drug, the researchers (see Herbst, A., Ulfelder, H., and Poskanzer, D., "Adenocarcinoma of the Vagina: Association of Maternal Stilbestrol Therapy with Tumor Appearance in Young Women," *New England Journal of Medicine*, **284**, 878–881, 1971) concluded that there was a strong association between DES and vaginal cancer. (The *p*-value for these data is approximately 0.) ■

When n_1 and n_2 are large, an approximate level α test of $H_0 : p_1 = p_2$, based on the normal approximation to the binomial, is outlined in Problem 63.

8.7 TESTS CONCERNING THE MEAN OF A POISSON DISTRIBUTION

Let X denote a Poisson random variable having mean λ and consider a test of

$$H_0 : \lambda = \lambda_0 \quad \text{versus} \quad H_1 : \lambda \neq \lambda_0$$

If the observed value of X is $X = x$, then a level α test would reject H_0 if either

$$P_{\lambda_0}\{X \geq x\} \leq \alpha/2 \quad \text{or} \quad P_{\lambda_0}\{X \leq x\} \leq \alpha/2 \tag{8.7.1}$$

where P_{λ_0} means that the probability is computed under the assumption that the Poisson mean is λ_0. It follows from Equation 8.7.1 that the p-value is given by

$$p\text{-value} = 2\min(P_{\lambda_0}\{X \geq x\}, P_{\lambda_0}\{X \leq x\})$$

The calculation of the preceding probabilities that a Poisson random variable with mean λ_0 is greater (less) than or equal to x can be obtained by using Program 5.2.

EXAMPLE 8.7a Management's claim that the mean number of defective computer chips produced daily is not greater than 25 is in dispute. Test this hypothesis, at the 5 percent level of significance, if a sample of 5 days revealed 28, 34, 32, 38, and 22 defective chips.

SOLUTION Because each individual computer chip has a very small chance of being defective, it is probably reasonable to suppose that the daily number of defective chips is approximately a Poisson random variable, with mean, say, λ. To see whether or not the manufacturer's claim is credible, we shall test the hypothesis

$$H_0 : \lambda \leq 25 \quad \text{versus} \quad H_1 : \lambda > 25$$

Now, under H_0, the total number of defective chips produced over a 5-day period is Poisson distributed (since the sum of independent Poisson random variables is Poisson) with a mean no greater than 125. Since this number is equal to 154, it follows that the p-value of the data is given by

$$
\begin{aligned}
p\text{-value} &= P_{125}\{X \geq 154\} \\
&= 1 - P_{125}\{X \leq 153\} \\
&= .0066 \quad \text{from Program 5.2}
\end{aligned}
$$

Therefore, the manufacturer's claim is rejected at the 5 percent (as it would be even at the 1 percent) level of significance. ∎

REMARK

If Program 5.2 is not available, one can use the fact that a Poisson random variable with mean λ is, for large λ, approximately normally distributed with a mean and variance equal to λ.

8.7.1 TESTING THE RELATIONSHIP BETWEEN TWO POISSON PARAMETERS

Let X_1 and X_2 be independent Poisson random variables with respective means λ_1 and λ_2, and consider a test of

$$H_0 : \lambda_2 = c\lambda_1 \quad \text{versus} \quad H_1 : \lambda_2 \neq c\lambda_1$$

for a given constant c. Our test of this is a conditional test (similar in spirit to the Fisher-Irwin test of Section 8.6.1), which is based on the fact that the conditional distribution of X_1 given the sum of X_1 and X_2 is binomial. More specifically, we have the following proposition.

PROPOSITION 8.7.1

$$P\{X_1 = k | X_1 + X_2 = n\} = \binom{n}{k} [\lambda_1/(\lambda_1 + \lambda_2)]^k [\lambda_2/(\lambda_1 + \lambda_2)]^{n-k}$$

Proof

$$P\{X_1 = k | X_1 + X_2 = n\}$$

$$= \frac{P\{X_1 = k, X_1 + X_2 = n\}}{P\{X_1 + X_2 = n\}}$$

$$= \frac{P\{X_1 = k, X_2 = n - k\}}{P\{X_1 + X_2 = n\}}$$

$$= \frac{P\{X_1 = k\} P\{X_2 = n - k\}}{P\{X_1 + X_2 = n\}} \quad \text{by independence}$$

$$= \frac{\exp\{-\lambda_1\} \lambda_1^k/k! \exp\{-\lambda_2\} \lambda_2^{n-k}/(n-k)!}{\exp\{-(\lambda_1 + \lambda_2)\}(\lambda_1 + \lambda_2)^n/n!}$$

$$= \frac{n!}{(n-k)!k!} [\lambda_1/(\lambda_1 + \lambda_2)]^k [\lambda_2/(\lambda_1 + \lambda_2)]^{n-k}$$

where the next to last equality follows because the sum of independent Poisson random variables is also Poisson. ∎

It follows from Proposition 8.7.1 that, if H_0 is true, then the conditional distribution of X_1 given that $X_1 + X_2 = n$ is the binomial distribution with parameters n and $p = 1/(1 + c)$. From this we can conclude that if $X_1 + X_2 = n$, then H_0 should be rejected if the observed value of X_1, call it x_1, is such that either

$$P\{\text{Bin}(n, 1/(1 + c)) \geq x_1\} \leq \alpha/2$$

or

$$P\{\text{Bin}(n, 1/(1 + c)) \leq x_1\} \leq \alpha/2$$

EXAMPLE 8.7b An industrial concern runs two large plants. If the number of accidents during the past 8 weeks at plant 1 were 16, 18, 9, 22, 17, 19, 24, 8 while the number of accidents during the last 6 weeks at plant 2 were 22, 18, 26, 30, 25, 28, can we conclude, at the 5 percent level of significance, that the safety conditions differ from plant to plant?

SOLUTION Since there is a small probability of an industrial accident in any given minute, it would seem that the weekly number of such accidents should have approximately a Poisson distribution. If we let X_1 denote the total number of accidents during an 8-week period at plant 1, and let X_2 be the number during a 6-week period at plant 2, then if the safety conditions did not differ at the two plants we would have that

$$\lambda_2 = \tfrac{3}{4}\lambda_1$$

where $\lambda_i \equiv E[X_i], i = 1, 2$. Hence, as $X_1 = 133, X_2 = 149$ it follows that the p-value of the test of

$$H_0 : \lambda_2 = \tfrac{3}{4}\lambda_1 \quad \text{versus} \quad H_1 : \lambda_2 \neq \tfrac{3}{4}\lambda_1$$

is given by

$$p\text{-value} = 2\min\left(P\{\text{Bin}(282, \tfrac{4}{7}) \geq 133\}, P\{\text{Bin}(282, \tfrac{4}{7}) \leq 133\}\right)$$
$$= 9.408 \times 10^{-4}$$

Thus, the hypothesis that the safety conditions at the two plants are equivalent is rejected. ∎

Problems

1. Consider a trial in which a jury must decide between the hypothesis that the defendant is guilty and the hypothesis that he or she is innocent.

   (a) In the framework of hypothesis testing and the U.S. legal system, which of the hypotheses should be the null hypothesis?

   (b) What do you think would be an appropriate significance level in this situation?

2. A colony of laboratory mice consists of several thousand mice. The average weight of all the mice is 32 grams with a standard deviation of 4 grams. A laboratory assistant was asked by a scientist to select 25 mice for an experiment. However, before performing the experiment the scientist decided to weigh the mice as an indicator of whether the assistant's selection constituted a random sample or whether it was made with some unconscious bias (perhaps the mice selected were the ones that were slowest in avoiding the assistant, which might indicate some inferiority about this group). If the sample mean of the

25 mice was 30.4, would this be significant evidence, at the 5 percent level of significance, against the hypothesis that the selection constituted a random sample?

3. A population distribution is known to have standard deviation 20. Determine the p-value of a test of the hypothesis that the population mean is equal to 50, if the average of a sample of 64 observations is
(a) 52.5; (b) 55.0; (c) 57.5.

4. In a certain chemical process, it is very important that a particular solution that is to be used as a reactant have a pH of exactly 8.20. A method for determining pH that is available for solutions of this type is known to give measurements that are normally distributed with a mean equal to the actual pH and with a standard deviation of .02. Suppose 10 independent measurements yielded the following pH values:

8.18	8.17
8.16	8.15
8.17	8.21
8.22	8.16
8.19	8.18

(a) What conclusion can be drawn at the $\alpha = .10$ level of significance?
(b) What about at the $\alpha = .05$ level of significance?

5. The mean breaking strength of a certain type of fiber is required to be at least 200 psi. Past experience indicates that the standard deviation of breaking strength is 5 psi. If a sample of 8 pieces of fiber yielded breakage at the following pressures,

210	198
195	202
197.4	196
199	195.5

would you conclude, at the 5 percent level of significance, that the fiber is unacceptable? What about at the 10 percent level of significance?

6. It is known that the average height of a man residing in the United States is 5 feet 10 inches and the standard deviation is 3 inches. To test the hypothesis that men in your city are "average," a sample of 20 men have been chosen. The heights of the men in the sample follow:

Man	Height in	Inches	Man
1	72	70.4	11
2	68.1	76	12
3	69.2	72.5	13
4	72.8	74	14
5	71.2	71.8	15
6	72.2	69.6	16
7	70.8	75.6	17
8	74	70.6	18
9	66	76.2	19
10	70.3	77	20

What do you conclude? Explain what assumptions you are making.

7. Suppose in Problem 4 that we wished to design a test so that if the pH were really equal to 8.20, then this conclusion will be reached with probability equal to .95. On the other hand, if the pH differs from 8.20 by .03 (in either direction), we want the probability of picking up such a difference to exceed .95.

   (a) What test procedure should be used?
   (b) What is the required sample size?
   (c) If $\bar{x} = 8.31$, what is your conclusion?
   (d) If the actual pH is 8.32, what is the probability of concluding that the pH is not 8.20, using the foregoing procedure?

8. Verify that the approximation in Equation 8.3.7 remains valid even when $\mu_1 < \mu_0$.

9. A British pharmaceutical company, Glaxo Holdings, has recently developed a new drug for migraine headaches. Among the claims Glaxo made for its drug, called somatriptan, was that the mean time it takes for it to enter the bloodstream is less than 10 minutes. To convince the Food and Drug Administration of the validity of this claim, Glaxo conducted an experiment on a randomly chosen set of migraine sufferers. To prove its claim, what should they have taken as the null and what as the alternative hypothesis?

10. The weights of salmon grown at a commercial hatchery are normally distributed with a standard deviation of 1.2 pounds. The hatchery claims that the mean weight of this year's crop is at least 7.6 pounds. Suppose a random sample of 16 fish yielded an average weight of 7.2 pounds. Is this strong enough evidence to reject the hatchery's claims at the

   (a) 5 percent level of significance;
   (b) 1 percent level of significance?
   (c) What is the p-value?

11. Consider a test of $H_0 : \mu \leq 100$ versus $H_1 : \mu > 100$. Suppose that a sample of size 20 has a sample mean of $\overline{X} = 105$. Determine the p-value of this outcome if the population standard deviation is known to equal
(a) 5; (b) 10; (c) 15.

12. An advertisement for a new toothpaste claims that it reduces cavities of children in their cavity-prone years. Cavities per year for this age group are normal with mean 3 and standard deviation 1. A study of 2,500 children who used this toothpaste found an average of 2.95 cavities per child. Assume that the standard deviation of the number of cavities of a child using this new toothpaste remains equal to 1.

(a) Are these data strong enough, at the 5 percent level of significance, to establish the claim of the toothpaste advertisement?

(b) Do the data convince you to switch to this new toothpaste?

13. There is some variability in the amount of phenobarbital in each capsule sold by a manufacturer. However, the manufacturer claims that the mean value is 20.0 mg. To test this, a sample of 25 pills yielded a sample mean of 19.7 with a sample standard deviation of 1.3. What inference would you draw from these data? In particular, are the data strong enough evidence to discredit the claim of the manufacturer? Use the 5 percent level of significance.

14. Twenty years ago, entering male high school students of Central High could do an average of 24 pushups in 60 seconds. To see whether this remains true today, a random sample of 36 freshmen was chosen. If their average was 22.5 with a sample standard deviation of 3.1, can we conclude that the mean is no longer equal to 24? Use the 5 percent level of significance.

15. The mean response time of a species of pigs to a stimulus is .8 seconds. Twenty-eight pigs were given 2 oz of alcohol and then tested. If their average response time was 1.0 seconds with a standard deviation of .3 seconds, can we conclude that alcohol affects the mean response time? Use the 5 percent level of significance.

16. Suppose that team A and team B are to play a National Football League game and that team A is favored by f points. Let $S(A)$ and $S(B)$ denote the scores of teams A and B, and let $X = S(A) - S(B) - f$. That is, X is the amount by which team A beats the point spread. It has been claimed that the distribution of X is normal with mean 0 and standard deviation 14. Use data from randomly chosen football games to test this hypothesis.

17. A medical scientist believes that the average basal temperature of (outwardly) healthy individuals has increased over time and is now greater than 98.6 degrees Fahrenheit (37 degrees Celsius). To prove this, she has randomly selected 100 healthy individuals. If their mean temperature is 98.74 with a sample standard deviation of 1.1 degrees, does this prove her claim at the 5 percent level? What about at the 1 percent level?

18. Use the results of a Sunday's worth of NFL professional football games to test the hypothesis that the average number of points scored by winning teams is less than or equal to 28. Use the 5 percent level of significance.

19. Use the results of a Sunday's worth of major league baseball scores to test the hypothesis that the average number of runs scored by winning teams is at least 5.6. Use the 5 percent level of significance.

20. A car is advertised as having a gas mileage rating of at least 30 miles/gallon in highway driving. If the miles per gallon obtained in 10 independent experiments are 26, 24, 20, 25, 27, 25, 28, 30, 26, 33, should you believe the advertisement? What assumptions are you making?

21. A producer specifies that the mean lifetime of a certain type of battery is at least 240 hours. A sample of 18 such batteries yielded the following data.

237 242 232
242 248 230
244 243 254
262 234 220
225 236 232
218 228 240

Assuming that the life of the batteries is approximately normally distributed, do the data indicate that the specifications are not being met?

22. Use the data of Example 2.3i of Chapter 2 to test the null hypothesis that the average noise level directly outside of Grand Central Station is less than or equal to 80 decibels.

23. An oil company claims that the sulfur content of its diesel fuel is at most .15 percent. To check this claim, the sulfur contents of 40 randomly chosen samples were determined; the resulting sample mean and sample standard deviation were .162 and .040. Using the 5 percent level of significance, can we conclude that the company's claims are invalid?

24. A company supplies plastic sheets for industrial use. A new type of plastic has been produced and the company would like to claim that the average stress resistance of this new product is at least 30.0, where stress resistance is measured in pounds per square inch (psi) necessary to crack the sheet. The following random sample was drawn off the production line. Based on this sample, would the claim clearly be unjustified?

30.1 32.7 22.5 27.5
27.7 29.8 28.9 31.4
31.2 24.3 26.4 22.8
29.1 33.4 32.5 21.7

Assume normality and use the 5 percent level of significance.

25. It is claimed that a certain type of bipolar transistor has a mean value of current gain that is at least 210. A sample of these transistors is tested. If the sample mean value of current gain is 200 with a sample standard deviation of 35, would the claim be rejected at the 5 percent level of significance if

(a) the sample size is 25;

(b) the sample size is 64?

26. A manufacturer of capacitors claims that the breakdown voltage of these capacitors has a mean value of at least 100 V. A test of 12 of these capacitors yielded the following breakdown voltages:

$$96, 98, 105, 92, 111, 114, 99, 103, 95, 101, 106, 97$$

Do these results prove the manufacturer's claim? Do they disprove them?

27. A sample of 10 fish were caught at lake A and their PCB concentrations were measured using a certain technique. The resulting data in parts per million were

$$\text{Lake A: } 11.5, 10.8, 11.6, 9.4, 12.4, 11.4, 12.2, 11, 10.6, 10.8$$

In addition, a sample of 8 fish were caught at lake B and their levels of PCB were measured by a different technique than that used at lake A. The resultant data were

$$\text{Lake B: } 11.8, 12.6, 12.2, 12.5, 11.7, 12.1, 10.4, 12.6$$

If it is known that the measuring technique used at lake A has a variance of .09 whereas the one used at lake B has a variance of .16, could you reject (at the 5 percent level of significance) a claim that the two lakes are equally contaminated?

28. A method for measuring the pH level of a solution yields a measurement value that is normally distributed with a mean equal to the actual pH of the solution and with a standard deviation equal to .05. An environmental pollution scientist claims that two different solutions come from the same source. If this were so, then the pH level of the solutions would be equal. To test the plausibility of this claim, 10 independent measurements were made of the pH level for both solutions, with the following data resulting.

Measurements of Solution A	Measurements of Solution B
6.24	6.27
6.31	6.25
6.28	6.33
6.30	6.27
6.25	6.24
6.26	6.31
6.24	6.28
6.29	6.29
6.22	6.34
6.28	6.27

(a) Do the data disprove the scientist's claim? Use the 5 percent level of significance.

(b) What is the p-value?

29. The following are the values of independent samples from two different populations.

Sample 1	122, 114, 130, 165, 144, 133, 139, 142, 150
Sample 2	108, 125, 122, 140, 132, 120, 137, 128, 138

Let μ_1 and μ_2 be the respective means of the two populations. Find the p-value of the test of the null hypothesis

$$H_0 : \mu_1 \leq \mu_2$$

versus the alternative

$$H_1 : \mu_1 > \mu_2$$

when the population standard deviations are $\sigma_1 = 10$ and
(a) $\sigma_2 = 5$; **(b)** $\sigma_2 = 10$; **(c)** $\sigma_2 = 20$.

30. The data below give the lifetimes in hundreds of hours of samples of two types of electronic tubes. Past lifetime data of such tubes have shown that they can often be modeled as arising from a lognormal distribution. That is, the logarithms of the data are normally distributed. Assuming that variance of the logarithms is equal

for the two populations, test, at the 5 percent level of significance, the hypothesis
that the two population distributions are identical.

Type 1	32, 84, 37, 42, 78, 62, 59, 74
Type 2	39, 111, 55, 106, 90, 87, 85

31. The viscosity of two different brands of car oil is measured and the following data
resulted:

Brand 1	10.62, 10.58, 10.33, 10.72, 10.44, 10.74
Brand 2	10.50, 10.52, 10.58, 10.62, 10.55, 10.51, 10.53

Test the hypothesis that the mean viscosity of the two brands is equal, assuming
that the populations have normal distributions with equal variances.

32. It is argued that the resistance of wire A is greater than the resistance of wire B.
You make tests on each wire with the following results.

Wire A	Wire B
.140 ohm	.135 ohm
.138	.140
.143	.136
.142	.142
.144	.138
.137	.140

What conclusion can you draw at the 10 percent significance level? Explain what
assumptions you are making.

In Problems 33 through 40, assume that the population distributions are normal and have equal variances.

33. Twenty-five men between the ages of 25 and 30, who were participating in a well-
known heart study carried out in Framingham, Massachusetts, were randomly
selected. Of these, 11 were smokers and 14 were not. The following data refer to
readings of their systolic blood pressure.

Smokers	Nonsmokers
124	130
134	122
136	128
125	129
133	118
127	122
135	116
131	127
133	135
125	120
118	122
	120
	115
	123

Use these data to test the hypothesis that the mean blood pressures of smokers and nonsmokers are the same.

34. In a 1943 experiment (Whitlock and Bliss, "A Bioassay Technique for Anti-helminthics," *Journal of Parasitology*, **29**, pp. 48–58) 10 albino rats were used to study the effectiveness of carbon tetrachloride as a treatment for worms. Each rat received an injection of worm larvae. After 8 days, the rats were randomly divided into two groups of 5 each; each rat in the first group received a dose of .032 cc of carbon tetrachloride, whereas the dosage for each rat in the second group was .063 cc. Two days later the rats were killed, and the number of adult worms in each rat was determined. The numbers detected in the group receiving the .032 dosage were

$$421, 462, 400, 378, 413$$

whereas they were

$$207, 17, 412, 74, 116$$

for those receiving the .063 dosage. Do the data prove that the larger dosage is more effective than the smaller?

35. A professor claims that the average starting salary of industrial engineering graduating seniors is greater than that of civil engineering graduates. To study this claim, samples of 16 industrial engineers and 16 civil engineers, all of whom graduated in 2006, were chosen and sample members were queried about their starting salaries. If the industrial engineers had a sample mean salary of $72,700 and a sample standard deviation of $2,400, and the civil engineers had a sample mean

salary of $71,400 and a sample standard deviation of $2,200, has the professor's claim been verified? Find the appropriate p-value.

36. In a certain experimental laboratory, a method A for producing gasoline from crude oil is being investigated. Before completing experimentation, a new method B is proposed. All other things being equal, it was decided to abandon A in favor of B only if the average yield of the latter was clearly greater. The yield of both processes is assumed to be normally distributed. However, there has been insufficient time to ascertain their true standard deviations, although there appears to be no reason why they cannot be assumed equal. Cost considerations impose size limits on the size of samples that can be obtained. If a 1 percent significance level is all that is allowed, what would be your recommendation based on the following random samples? The numbers represent percent yield of crude oil.

A	23.2, 26.6, 24.4, 23.5, 22.6, 25.7, 25.5
B	25.7, 27.7, 26.2, 27.9, 25.0, 21.4, 26.1

37. A study was instituted to learn how the diets of women changed during the winter and the summer. A random group of 12 women were observed during the month of July and the percentage of each woman's calories that came from fat was determined. Similar observations were made on a different randomly selected group of size 12 during the month of January. The results were as follows:

July	32.2, 27.4, 28.6, 32.4, 40.5, 26.2, 29.4, 25.8, 36.6, 30.3, 28.5, 32.0
January	30.5, 28.4, 40.2, 37.6, 36.5, 38.8, 34.7, 29.5, 29.7, 37.2, 41.5, 37.0

Test the hypothesis that the mean fat percentage intake is the same for both months. Use the **(a)** 5 percent level of significance and **(b)** 1 percent level of significance.

38. To learn about the feeding habits of bats, 22 bats were tagged and tracked by radio. Of these 22 bats, 12 were female and 10 were male. The distances flown (in meters) between feedings were noted for each of the 22 bats, and the following summary statistics were obtained.

Female Bats	Male Bats
$n = 12$	$m = 10$
$\overline{X} = 180$	$\overline{Y} = 136$
$S_x = 92$	$S_y = 86$

Test the hypothesis that the mean distance flown between feedings is the same for the populations of both male and of female bats. Use the 5 percent level of significance.

39. The following data summary was obtained from a comparison of the lead content of human hair removed from adult individuals that had died between 1880 and 1920 with the lead content of present-day adults. The data are in units of micrograms, equal to one-millionth of a gram.

	1880–1920	Today
Sample size:	30	100
Sample mean:	48.5	26.6
Sample standard deviation:	14.5	12.3

(a) Do the above data establish, at the 1 percent level of significance, that the mean lead content of human hair is less today than it was in the years between 1880 and 1920? Clearly state what the null and alternative hypotheses are.

(b) What is the p-value for the hypothesis test in part (a)?

40. Sample weights (in pounds) of newborn babies born in two adjacent counties in western Pennsylvania yielded the following data.

$$n = 53, \qquad m = 44$$
$$\overline{X} = 6.8, \qquad \overline{Y} = 7.2$$
$$S^2 = 5.2, \qquad S^2 = 4.9$$

Consider a test of the hypothesis that the mean weight of newborns is the same in both counties. What is the resulting p-value?

41. To verify the hypothesis that blood lead levels tend to be higher for children whose parents work in a factory that uses lead in the manufacturing process, researchers examined lead levels in the blood of 33 children whose parents worked in a battery manufacturing factory (Morton, D., Saah, A., Silberg, S., Owens, W., Roberts, M., and Saah, M., "Lead Absorption in Children of Employees in a Lead-Related Industry," *American Journal of Epidemiology*, **115**, 549–555, 1982). Each of these children was then *matched* by another child who was of similar age, lived in a similar neighborhood, had a similar exposure to traffic, but whose parent did not work with lead. The blood levels of the 33 cases (sample 1) as well as those of the 33 controls (sample 2) were then used to test the hypothesis that the average blood levels of these groups are the same. If the resulting sample means and sample standard deviations were

$$\bar{x}_1 = .015, \quad s_1 = .004, \quad \bar{x}_2 = .006, \quad s_2 = .006$$

find the resulting p-value. Assume a common variance.

42. Ten pregnant women were given an injection of pitocin to induce labor. Their systolic blood pressures immediately before and after the injection were:

Patient	Before	After	Patient	Before	After
1	134	140	6	140	138
2	122	130	7	118	124
3	132	135	8	127	126
4	130	126	9	125	132
5	128	134	10	142	144

Do the data indicate that injection of this drug changes blood pressure?

43. A question of medical importance is whether jogging leads to a reduction in one's pulse rate. To test this hypothesis, 8 nonjogging volunteers agreed to begin a 1-month jogging program. After the month their pulse rates were determined and compared with their earlier values. If the data are as follows, can we conclude that jogging has had an effect on the pulse rates?

Subject	1	2	3	4	5	6	7	8
Pulse Rate Before	74	86	98	102	78	84	79	70
Pulse Rate After	70	85	90	110	71	80	69	74

44. If $X_1, \ldots, X_n$ is a sample from a normal population having unknown parameters μ and σ^2, devise a significance level α test of

$$H_0 = \sigma^2 \leq \sigma_0^2$$

versus the alternative

$$H_1 = \sigma^2 > \sigma_0^2$$

for a given positive value σ_0^2.

45. In Problem 44, explain how the test would be modified if the population mean μ were known in advance.

46. A gun-like apparatus has recently been designed to replace needles in administering vaccines. The apparatus can be set to inject different amounts of the serum, but because of random fluctuations the actual amount injected is normally distributed with a mean equal to the setting and with an unknown variance σ^2. It has been decided that the apparatus would be too dangerous to use if σ exceeds .10. If a random sample of 50 injections resulted in a sample standard deviation of .08, should use of the new apparatus be discontinued? Suppose the level of significance is $\alpha = .10$. Comment on the appropriate choice of a significance level for this problem, as well as the appropriate choice of the null hypothesis.

47. A pharmaceutical house produces a certain drug item whose weight has a standard deviation of .5 milligrams. The company's research team has proposed a new method of producing the drug. However, this entails some costs and will be adopted only if there is strong evidence that the standard deviation of the weight of the items will drop to below .4 milligrams. If a sample of 10 items is produced and has the following weights, should the new method be adopted?

5.728	5.731
5.722	5.719
5.727	5.724
5.718	5.726
5.723	5.722

48. The production of large electrical transformers and capacitators requires the use of polychlorinated biphenyls (PCBs), which are extremely hazardous when released into the environment. Two methods have been suggested to monitor the levels of PCB in fish near a large plant. It is believed that each method will result in a normal random variable that depends on the method. Test the hypothesis at the $\alpha = .10$ level of significance that both methods have the same variance, if a given fish is checked 8 times by each method with the following data (in parts per million) recorded.

Method 1	6.2, 5.8, 5.7, 6.3, 5.9, 6.1, 6.2, 5.7
Method 2	6.3, 5.7, 5.9, 6.4, 5.8, 6.2, 6.3, 5.5

49. In Problem 31, test the hypothesis that the populations have the same variances.

50. If $X_1, \ldots, X_n$ is a sample from a normal population with variance σ_x^2, and $Y_1, \ldots, Y_n$ is an independent sample from normal population with variance σ_y^2, develop a significance level α test of

$$H_0 : \sigma_x^2 < \sigma_y^2 \qquad \text{versus} \qquad H_1 : \sigma_x^2 > \sigma_y^2$$

51. The amount of surface wax on each side of waxed paper bags is believed to be normally distributed. However, there is reason to believe that there is greater variation in the amount on the inner side of the paper than on the outside. A sample of 75 observations of the amount of wax on each side of these bags is obtained and the following data recorded.

Wax in Pounds per Unit Area of Sample	
Outside Surface	Inside Surface
$\bar{x} = .948$	$\bar{y} = .652$
$\sum x_i^2 = 91$	$\sum y_i^2 = 82$

Conduct a test to determine whether or not the variability of the amount of wax on the inner surface is greater than the variability of the amount on the outer surface ($\alpha = .05$).

52. In a famous experiment to determine the efficacy of aspirin in preventing heart attacks, 22,000 healthy middle-aged men were randomly divided into two equal groups, one of which was given a daily dose of aspirin and the other a placebo that looked and tasted identical to the aspirin. The experiment was halted at a time when 104 men in the aspirin group and 189 in the control group had had heart attacks. Use these data to test the hypothesis that the taking of aspirin does not change the probability of having a heart attack.

53. In the study of Problem 52, it also resulted that 119 from the aspirin group and 98 from the control group suffered strokes. Are these numbers significant to show that taking aspirin changes the probability of having a stroke?

54. A standard drug is known to be effective in 72 percent of the cases in which it is used to treat a certain infection. A new drug has been developed and testing has found it to be effective in 42 cases out of 50. Is this strong enough evidence to prove that the new drug is more effective than the old one? Find the relevant p-value.

55. Three independent news services are running a poll to determine if over half the population supports an initiative concerning limitations on driving automobiles in the downtown area. Each wants to see if the evidence indicates that over half the population is in favor. As a result, all three services will be testing

$$H_0 : p \le .5 \quad \text{versus} \quad H_1 : p > .5$$

where p is the proportion of the population in favor of the initiative.

(a) Suppose the first news organization samples 100 people, of which 56 are in favor of the initiative. Is this strong enough evidence, at the 5 percent level of significance, to reject the null hypothesis and so establish that over half the population favors the initiative?

(b) Suppose the second news organization samples 120 people, of which 68 are in favor of the initiative. Is this strong enough evidence, at the 5 percent level of significance, to reject the null hypothesis?

(c) Suppose the third news organization samples 110 people, of which 62 are in favor of the initiative. Is this strong enough evidence, at the 5 percent level of significance, to reject the null hypothesis?

(d) Suppose the news organizations combine their samples, to come up with a sample of 330 people, of which 186 support the initiative. Is this strong enough evidence, at the 5 percent level of significance, to reject the null hypothesis?

56. It has been a long held belief that the proportion of California births of African America mothers that result in twins is about 1.32 percent. (The twinning rate appears to be influenced by the ethnicity of the mother; claims are that it is 1.05 for Caucasian Americans, and 0.72 percent for Asian Americans.) A public health scientist believes that this number is no longer correct and has decided to test the null hypothesis that the proportion is 1.32 percent by gathering data on the next 1, 000 recorded birthing events, where twin birds are regarded as a single birthing event, in California.

   (a) What is the minimal number of twin births that would have to be observed in order to reject the null hypothesis at the 5 percent level of significance?

   (b) What is the probability the null hypothesis will be rejected if the actual twinning rate is 1.80?

57. An ambulance service claims that at least 45 percent of its calls involve life-threatening emergencies. To check this claim, a random sample of 200 calls was selected from the service's files. If 70 of these calls involved life-threatening emergencies, is the service's claim believable at the

   (a) 5 percent level of significance;

   (b) 1 percent level of significance?

58. A standard drug is known to be effective in 75 percent of the cases in which it is used to treat a certain infection. A new drug has been developed and has been found to be effective in 42 cases out of 50. Based on this, would you accept, at the 5 percent level of significance, the hypothesis that the two drugs are of equal effectiveness? What is the p-value?

59. Do Problem 58 by using a test based on the normal approximation to the binomial.

60. In a study of the effect of two chemotherapy treatments on the survival of patients with multiple myeloma, each of 156 patients was equally likely to be given either one of the two treatments. As reported by Lipsitz, Dear, Laird, and Molenberghs in a 1998 paper in *Biometrics*, the result of this was that 39 of the 72 patients given the first treatment and 44 of the 84 patients given the second treatment survived for over 5 years.

   (a) Use these data to test the null hypothesis that the two treatments are equally effective.

   (b) Is the fact that 72 of the patients received one of the treatments while 84 received the other consistent with the claim that the determination of the treatment to be given to each patient was made in a totally random fashion?

61. Let X_1 denote a binomial random variable with parameters (n_1, p_1) and X_2 an independent binomial random variable with parameters (n_2, p_2). Develop a test, using the same approach as in the Fisher-Irwin test, of

$$H_0 : p_1 \leq p_2$$

versus the alternative

$$H_1 : p_1 > p_2$$

62. Verify that Equation 8.6.5 follows from Equation 8.6.4.

63. Let X_1 and X_2 be binomial random variables with respective parameters n_1, p_1 and n_2, p_2. Show that when n_1 and n_2 are large, an approximate level α test of $H_0 : p_1 = p_2$ versus $H_1 : p_1 \neq p_2$ is as follows:

$$\text{reject}\quad H_0 \quad \text{if} \quad \frac{|X_1/n_1 - X_2/n_2|}{\sqrt{\dfrac{X_1 + X_2}{n_1 + n_2}\left(1 - \dfrac{X_1 + X_2}{n_1 + n_2}\right)\left(\dfrac{1}{n_1} + \dfrac{1}{n_2}\right)}} > z_{\alpha/2}$$

Hint: **(a)** Argue first that when n_1 and n_2 are large

$$\frac{\dfrac{X_1}{n_1} - \dfrac{X_2}{n_2} - (p_1 - p_2)}{\sqrt{\dfrac{p_1(1 - p_1)}{n_1} + \dfrac{p_2(1 - p_2)}{n_2}}} \mathrel{\dot\sim} \mathcal{N}(0, 1)$$

where $\dot\sim$ means "approximately has the distribution."

(b) Now argue that when H_0 is true and so $p_1 = p_2$, their common value can be best estimated by $(X_1 + X_2)/(n_1 + n_2)$.

64. Use the approximate test given in Problem 63 on the data of Problem 60.

65. Patients suffering from cancer must often decide whether to have their tumors treated with surgery or with radiation. A factor in their decision is the 5-year survival rates for these treatments. Surprisingly, it has been found that patients' decisions often seem to be affected by whether they are told the 5-year survival rates or the 5-year death rates (even though the information content is identical). For instance, in an experiment a group of 200 male prostate cancer patients were randomly divided into two groups of size 100 each. Each member of the first group was told that the 5-year survival rate for those electing surgery was 77 percent, whereas each member of the second group was told that the 5-year death rate for those electing surgery was 23 percent. Both groups were given the same information about radiation therapy. If it resulted that 24 members of the first group and 12 of the second group elected to have surgery, what conclusions would you draw?

66. The following data refer to Larry Bird's results when shooting a pair of free throws in basketball. During two consecutive seasons in the National Basketball Association, Bird shot a pair of free throws on 338 occasions. On 251 occasions he made both shots; on 34 occasions he made the first shot but missed the second one; on 48 occasions he missed the first shot but made the second one; on 5 occasions he missed both shots.

    (a) Use these data to test the hypothesis that Bird's probability of making the first shot is equal to his probability of making the second shot.
    (b) Use these data to test the hypothesis that Bird's probability of making the second shot is the same regardless of whether he made or missed the first one.

67. In the 1970s, the U.S. Veterans Administration (Murphy, 1977) conducted an experiment comparing coronary artery bypass surgery with medical drug therapy as treatments for coronary artery disease. The experiment involved 596 patients, of whom 286 were randomly assigned to receive surgery, with the remaining 310 assigned to drug therapy. A total of 252 of those receiving surgery, and a total of 270 of those receiving drug therapy were still alive 3 years after treatment. Use these data to test the hypothesis that the survival probabilities are equal.

68. Test the hypothesis, at the 5 percent level of significance, that the yearly number of earthquakes felt on a certain island has mean 52 if the readings for the past 8 years are 46, 62, 60, 58, 47, 50, 59, 49. Assume an underlying Poisson distribution and give an explanation to justify this assumption.

69. In 1995, the Fermi Laboratory announced the discovery of the top quark, the last of six quarks predicted by the "standard model of physics." The evidence for its existence was statistical in nature and involved signals created when antiprotons and protons were forced to collide. In a *Physical Review Letters* paper documenting the evidence, Abe, Akimoto, and Akopian (known in physics circle as the three A's) based their conclusion on a theoretical analysis that indicated that the number of decay events in a certain time interval would have a Poisson distribution with a mean equal to 6.7 if a top quark did not exist and with a larger mean if it did exist. In a careful analysis of the data the three A's showed that the actual count was 27. Is this strong enough evidence to prove the hypothesis that the mean of the Poisson distribution was greater than 6.7?

70. For the following data, sample 1 is from a Poisson distribution with mean λ_1 and sample 2 is from a Poisson distribution with mean λ_2. Test the hypothesis that $\lambda_1 = \lambda_2$.

Sample 1	24, 32, 29, 33, 40, 28, 34, 36
Sample 2	42, 36, 41

71. A scientist looking into the effect of smoking on heart disease has chosen a large random sample of smokers and of nonsmokers. She plans to study these two groups for 5 years to see if the number of heart attacks among the members of the smokers' group is significantly greater than the number among the nonsmokers. Such a result, the scientist feels, should be strong evidence of an association between smoking and heart attacks. Given that

   (a) older people are at greater risk of heart disease than are younger people; and
   (b) as a group, smokers tend to be somewhat older than nonsmokers,

would the scientist be justified in her conclusion? Explain how the experimental design can be improved so that meaningful conclusions can be drawn.

72. A researcher wants to analyze the average yearly increase in a stock over a 20-year period. To do so, she plans to randomly choose 100 stocks from the listing of current stocks, discarding any that were not in existence 20 years ago. She will then compare the current price of each stock with its price 20 years ago to determine its percentage increase. Do you think this is a valid method to study the average increase in the price of a stock?

Chapter 9

REGRESSION

9.1 INTRODUCTION

Many engineering and scientific problems are concerned with determining a relationship between a set of variables. For instance, in a chemical process, we might be interested in the relationship between the output of the process, the temperature at which it occurs, and the amount of catalyst employed. Knowledge of such a relationship would enable us to predict the output for various values of temperature and amount of catalyst.

In many situations, there is a single *response* variable Y, also called the *dependent* variable, which depends on the value of a set of *input*, also called *independent*, variables $x_1, \ldots, x_r$. The simplest type of relationship between the dependent variable Y and the input variables $x_1, \ldots, x_r$ is a linear relationship. That is, for some constants $\beta_0, \beta_1, \ldots, \beta_r$ the equation

$$Y = \beta_0 + \beta_1 x_1 + \cdots + \beta_r x_r \tag{9.1.1}$$

would hold. If this was the relationship between Y and the $x_i, i = 1, \ldots, r$, then it would be possible (once the β_i were learned) to exactly predict the response for any set of input values. However, in practice, such precision is almost never attainable, and the most that one can expect is that Equation 9.1.1 would be valid *subject to random error*. By this we mean that the explicit relationship is

$$Y = \beta_0 + \beta_1 x_1 + \cdots + \beta_r x_r + e \tag{9.1.2}$$

where e, representing the random error, is assumed to be a random variable having mean 0. Indeed, another way of expressing Equation 9.1.2 is as follows:

$$E[Y|\mathbf{x}] = \beta_0 + \beta_1 x_1 + \cdots + \beta_r x_r$$

where $\mathbf{x} = (x_1, \ldots, x_r)$ is the set of independent variables, and $E[Y|\mathbf{x}]$ is the expected response given the inputs $\mathbf{x}$.

Equation 9.1.2 is called a *linear regression equation*. We say that it describes the regression of Y on the set of independent variables $x_1, \ldots, x_r$. The quantities $\beta_0, \beta_1, \ldots, \beta_r$ are called the *regression coefficients*, and must usually be estimated from a set of data. A regression equation containing a single independent variable — that is, one in which $r = 1$ — is called a *simple regression equation*, whereas one containing many independent variables is called a *multiple regression equation*.

Thus, a simple linear regression model supposes a linear relationship between the mean response and the value of a single independent variable. It can be expressed as

$$Y = \alpha + \beta x + e$$

where x is the value of the independent variable, also called the input level, Y is the response, and e, representing the random error, is a random variable having mean 0.

EXAMPLE 9.1a   Consider the following 10 data pairs (x_i, y_i), $i = 1, \ldots, 10$, relating y, the percent yield of a laboratory experiment, to x, the temperature at which the experiment was run.

i	x_i	y_i	i	x_i	y_i
1	100	45	6	150	68
2	110	52	7	160	75
3	120	54	8	170	76
4	130	63	9	180	92
5	140	62	10	190	88

A plot of y_i versus x_i — called a *scatter diagram* — is given in Figure 9.1. As this scatter diagram appears to reflect, subject to random error, a linear relation between y and x, it seems that a simple linear regression model would be appropriate.   ■

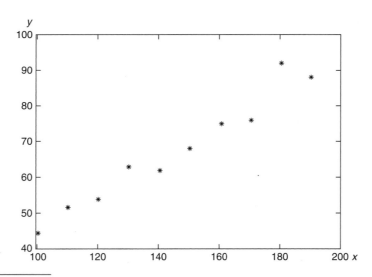

FIGURE 9.1   *Scatter plot.*

9.2 LEAST SQUARES ESTIMATORS OF THE REGRESSION PARAMETERS

Suppose that the responses Y_i corresponding to the input values $x_i, i = 1, \ldots, n$ are to be observed and used to estimate α and β in a simple linear regression model. To determine estimators of α and β we reason as follows: If A is the estimator of α and B of β, then the estimator of the response corresponding to the input variable x_i would be $A + Bx_i$. Since the actual response is Y_i, the squared difference is $(Y_i - A - Bx_i)^2$, and so if A and B are the estimators of α and β, then the sum of the squared differences between the estimated responses and the actual response values — call it SS — is given by

$$SS = \sum_{i=1}^{n}(Y_i - A - Bx_i)^2$$

The method of least squares chooses as estimators of α and β the values of A and B that minimize SS. To determine these estimators, we differentiate SS first with respect to A and then to B as follows:

$$\frac{\partial SS}{\partial A} = -2\sum_{i=1}^{n}(Y_i - A - Bx_i)$$

$$\frac{\partial SS}{\partial B} = -2\sum_{i=1}^{n}x_i(Y_i - A - Bx_i)$$

Setting these partial derivatives equal to zero yields the following equations for the minimizing values A and B:

$$\sum_{i=1}^{n}Y_i = nA + B\sum_{i=1}^{n}x_i \tag{9.2.1}$$

$$\sum_{i=1}^{n}x_iY_i = A\sum_{i=1}^{n}x_i + B\sum_{i=1}^{n}x_i^2$$

The Equations 9.2.1 are known as the *normal equations*. If we let

$$\overline{Y} = \sum_{i}Y_i/n, \qquad \overline{x} = \sum_{i}x_i/n$$

then we can write the first normal equation as

$$A = \overline{Y} - B\overline{x} \tag{9.2.2}$$

Substituting this value of A into the second normal equation yields

$$\sum_i x_i Y_i = (\overline{Y} - B\overline{x})n\overline{x} + B\sum_i x_i^2$$

or

$$B\left(\sum_i x_i^2 - n\overline{x}^2\right) = \sum_i x_i Y_i - n\overline{x}\,\overline{Y}$$

or

$$B = \frac{\sum\limits_i x_i Y_i - n\overline{x}\,\overline{Y}}{\sum\limits_i x_i^2 - n\overline{x}^2}$$

Hence, using Equation 9.2.2 and the fact that $n\overline{Y} = \sum_{i=1}^n Y_i$, we have proven the following proposition.

PROPOSITION 9.2.1   The least squares estimators of β and α corresponding to the data set $x_i, Y_i, i = 1, \ldots, n$ are, respectively,

$$B = \frac{\sum\limits_{i=1}^n x_i Y_i - \overline{x}\sum\limits_{i=1}^n Y_i}{\sum\limits_{i=1}^n x_i^2 - n\overline{x}^2}$$

$$A = \overline{Y} - B\overline{x}$$

The straight line $A + Bx$ is called the estimated regression line.

Program 9.2 computes the least squares estimators A and B. It also gives the user the option of computing some other statistics whose values will be needed in the following sections.

EXAMPLE 9.2a   The raw material used in the production of a certain synthetic fiber is stored in a location without a humidity control. Measurements of the relative humidity in the storage location and the moisture content of a sample of the raw material were taken over 15 days with the following data (in percentages) resulting.

Relative humidity	46	53	29	61	36	39	47	49	52	38	55	32	57	54	44
Moisture content	12	15	7	17	10	11	11	12	14	9	16	8	18	14	12

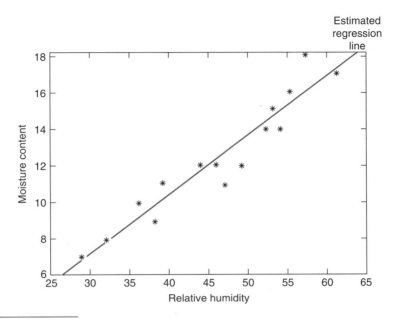

FIGURE 9.2   *Example 9.2a.*

These data are plotted in Figure 9.2. To compute the least squares estimator and the estimated regression line, we run Program 9.2; results are shown in Figure 9.3. ■

9.3   DISTRIBUTION OF THE ESTIMATORS

To specify the distribution of the estimators A and B, it is necessary to make additional assumptions about the random errors aside from just assuming that their mean is 0. The usual approach is to assume that the random errors are independent normal random variables having mean 0 and variance σ^2. That is, we suppose that if Y_i is the response corresponding to the input value x_i, then $Y_1, \ldots, Y_n$ are independent and

$$Y_i \sim \mathcal{N}(\alpha + \beta x_i, \sigma^2)$$

Note that the foregoing supposes that the variance of the random error does not depend on the input value but rather is a constant. This value σ^2 is not assumed to be known but rather must be estimated from the data.

Since the least squares estimator B of β can be expressed as

$$B = \frac{\sum_i (x_i - \bar{x}) Y_i}{\sum_i x_i^2 - n\bar{x}^2} \tag{9.3.1}$$

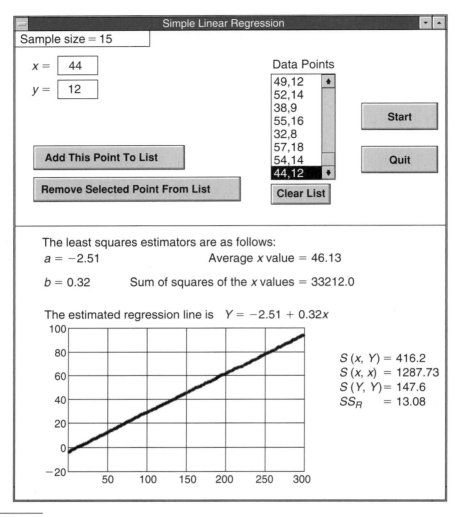

FIGURE 9.3

we see that it is a linear combination of the independent normal random variables Y_i, $i = 1, \ldots, n$ and so is itself normally distributed. Using Equation 9.3.1, the mean and variance of B are computed as follows:

$$E[B] = \frac{\sum_i (x_i - \bar{x})E[Y_i]}{\sum_i x_i^2 - n\bar{x}^2}$$

$$= \frac{\sum_i (x_i - \bar{x})(\alpha + \beta x_i)}{\sum_i x_i^2 - n\bar{x}^2}$$

$$= \frac{\alpha \sum_i (x_i - \bar{x}) + \beta \sum_i x_i(x_i - \bar{x})}{\sum_i x_i^2 - n\bar{x}^2}$$

$$= \beta \frac{\left[\sum_i x_i^2 - \bar{x} \sum_i x_i\right]}{\sum_i x_i^2 - n\bar{x}^2} \qquad \text{since } \sum_i (x_i - \bar{x}) = 0$$

$$= \beta$$

Thus $E[B] = \beta$ and so B is an unbiased estimator of β. We will now compute the variance of B.

$$\text{Var}(B) = \frac{\text{Var}\left(\sum_{i=1}^{n} (x_i - \bar{x}) Y_i\right)}{\left(\sum_{i=1}^{n} x_i^2 - n\bar{x}^2\right)^2}$$

$$= \frac{\sum_{i=1}^{n} (x_i - \bar{x})^2 \, \text{Var}(Y_i)}{\left(\sum_{i=1}^{n} x_i^2 - n\bar{x}^2\right)^2} \qquad \text{by independence}$$

$$= \frac{\sigma^2 \sum_{i=1}^{n} (x_i - \bar{x})^2}{\left(\sum_{i=1}^{n} x_i^2 - n\bar{x}^2\right)^2}$$

$$= \frac{\sigma^2}{\sum_{i=1}^{n} x_i^2 - n\bar{x}^2} \qquad (9.3.2)$$

where the final equality results from the use of the identity

$$\sum_{i=1}^{n} (x_i - \bar{x})^2 = \sum_{i=1}^{n} x_i^2 - n\bar{x}^2$$

Using Equation 9.3.1 along with the relationship

$$A = \sum_{i=1}^{n} \frac{Y_i}{n} - B\bar{x}$$

shows that A can also be expressed as a linear combination of the independent normal random variables $Y_i, i = 1, \ldots, n$ and is thus also normally distributed. Its mean is obtained from

$$E[A] = \sum_{i=1}^{n} \frac{E[Y_i]}{n} - \bar{x}E[B]$$

$$= \sum_{i=1}^{n} \frac{(\alpha + \beta x_i)}{n} - \bar{x}\beta$$

$$= \alpha + \beta\bar{x} - \bar{x}\beta$$

$$= \alpha$$

Thus A is also an unbiased estimator. The variance of A is computed by first expressing A as a linear combination of the Y_i. The result (whose details are left as an exercise) is that

$$\text{Var}(A) = \frac{\sigma^2 \sum_{i=1}^{n} x_i^2}{n \left(\sum_{i=1}^{n} x_i^2 - n\bar{x}^2 \right)} \qquad (9.3.3)$$

The quantities $Y_i - A - Bx_i, i = 1, \ldots, n$, which represent the differences between the actual responses (that is, the Y_i) and their least squares estimators (that is, $A + Bx_i$) are called the *residuals*. The sum of squares of the residuals

$$SS_R = \sum_{i=1}^{n}(Y_i - A - Bx_i)^2$$

can be utilized to estimate the unknown error variance σ^2. Indeed, it can be shown that

$$\frac{SS_R}{\sigma^2} \sim \chi_{n-2}^2$$

That is, SS_R/σ^2 has a chi-square distribution with $n-2$ degrees of freedom, which implies that

$$E\left[\frac{SS_R}{\sigma^2} \right] = n - 2$$

or

$$E\left[\frac{SS_R}{n-2}\right] = \sigma^2$$

Thus $SS_R/(n-2)$ is an unbiased estimator of σ^2. In addition, it can be shown that SS_R is independent of the pair A and B.

REMARKS

A plausibility argument as to why SS_R/σ^2 might have a chi-square distribution with $n-2$ degrees of freedom and be independent of A and B runs as follows. Because the Y_i are independent normal random variables, it follows that $(Y_i - E[Y_i])/\sqrt{\mathrm{Var}(Y_i)}, i = 1, \ldots, n$ are independent standard normals and so

$$\sum_{i=1}^{n} \frac{(Y_i - E[Y_i])^2}{\mathrm{Var}(Y_i)} = \sum_{i=1}^{n} \frac{(Y_i - \alpha - \beta x_i)^2}{\sigma^2} \sim \chi_n^2$$

Now if we substitute the estimators A and B for α and β, then 2 degrees of freedom are lost, and so it is not an altogether surprising result that SS_R/σ^2 has a chi-square distribution with $n-2$ degrees of freedom.

The fact that SS_R is independent of A and B is quite similar to the fundamental result that in normal sampling $\overline{X}$ and S^2 are independent. Indeed this latter result states that if $Y_1, \ldots, Y_n$ is a normal sample with population mean μ and variance σ_2, then if in the sum of squares $\sum_{i=1}^{n}(Y_i - \mu)^2/\sigma^2$, which has a chi-square distribution with n degrees of freedom, one substitutes the estimator $\overline{Y}$ for μ to obtain the new sum of squares $\sum_i (Y_i - \overline{Y})^2/\sigma^2$, then this quantity [equal to $(n-1)S^2/\sigma^2$] will be independent of $\overline{Y}$ and will have a chi-square distribution with $n-1$ degrees of freedom. Since SS_R/σ^2 is obtained by substituting the estimators A and B for α and β in the sum of squares $\sum_{i=1}^{n}(Y_i - \alpha - \beta x_i)^2/\sigma^2$, it is not unreasonable to expect that this quantity might be independent of A and B.

When the Y_i are normal random variables, the least squares estimators are also the maximum likelihood estimators. To verify this remark, note that the joint density of $Y_1, \ldots, Y_n$ is given by

$$f_{Y_1,\ldots,Y_n}(y_1,\ldots,y_n) = \prod_{i=1}^{n} f_{Y_i}(y_i)$$

$$= \prod_{i=1}^{n} \frac{1}{\sqrt{2\pi}\sigma} e^{-(y_i - \alpha - \beta x_i)^2/2\sigma^2}$$

$$= \frac{1}{(2\pi)^{n/2}\sigma^n} e^{-\sum_{i=1}^{n}(y_i - \alpha - \beta x_i)^2/2\sigma^2}$$

Consequently, the maximum likelihood estimators of α and β are precisely the values of α and β that minimize $\sum_{i=1}^{n}(y_i - \alpha - \beta x_i)^2$. That is, they are the least squares estimators.

Notation

If we let

$$S_{xY} = \sum_{i=1}^{n}(x_i - \overline{x})(Y_i - \overline{Y}) = \sum_{i=1}^{n}x_i Y_i - n\overline{x}\,\overline{Y}$$

$$S_{xx} = \sum_{i=1}^{n}(x_i - \overline{x})^2 = \sum_{i=1}^{n}x_i^2 - n\overline{x}^2$$

$$S_{YY} = \sum_{i=1}^{n}(Y_i - \overline{Y})^2 = \sum_{i=1}^{n}Y_i^2 - n\overline{Y}^2$$

then the least squares estimators can be expressed as

$$B = \frac{S_{xY}}{S_{xx}}$$

$$A = \overline{Y} - B\overline{x}$$

The following computational identity for SS_R, the sum of squares of the residuals, can be established.

Computational Identity for SS_R

$$SS_R = \frac{S_{xx}S_{YY} - S_{xY}^2}{S_{xx}} \tag{9.3.4}$$

The following proposition sums up the results of this section.

PROPOSITION 9.3.1 Suppose that the responses $Y_i, i = 1, \ldots, n$ are independent normal random variables with means $\alpha + \beta x_i$ and common variance σ^2. The least squares estimators of β and α

$$B = \frac{S_{xY}}{S_{xx}}, \qquad A = \overline{Y} - B\overline{x}$$

are distributed as follows:

$$A \sim \mathcal{N}\left(\alpha, \ \frac{\sigma^2 \sum_i x_i^2}{n S_{xx}}\right)$$

$$B \sim \mathcal{N}(\beta, \ \sigma^2/S_{xx})$$

In addition, if we let

$$SS_R = \sum_i (Y_i - A - Bx_i)^2$$

denote the sum of squares of the residuals, then

$$\frac{SS_R}{\sigma^2} \sim \chi^2_{n-2}$$

and SS_R is independent of the least squares estimators A and B. Also, SS_R can be computed from

$$SS_R = \frac{S_{xx}S_{YY} - (S_{xY})^2}{S_{xx}}$$

Program 9.2 will compute the least squares estimators A and B as well as $\bar{x}, \sum_i x_i^2,$ S_{xx}, S_{xY}, S_{YY}, and SS_R.

EXAMPLE 9.3a　The following data relate x, the moisture of a wet mix of a certain product, to Y, the density of the finished product.

x_i	y_i
5	7.4
6	9.3
7	10.6
10	15.4
12	18.1
15	22.2
18	24.1
20	24.8

Fit a linear curve to these data. Also determine SS_R.

SOLUTION　A plot of the data and the estimated regression line is shown in Figure 9.4. To solve the foregoing, run Program 9.2; results are shown in Figure 9.5.　∎

9.4　STATISTICAL INFERENCES ABOUT THE REGRESSION PARAMETERS

Using Proposition 9.3.1, it is a simple matter to devise hypothesis tests and confidence intervals for the regression parameters.

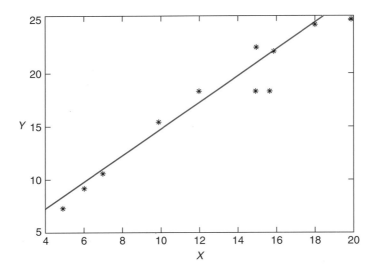

9.4.1 INFERENCES CONCERNING β

An important hypothesis to consider regarding the simple linear regression model

$$Y = \alpha + \beta x + e$$

is the hypothesis that $\beta = 0$. Its importance derives from the fact that it is equivalent to stating that the mean response does not depend on the input, or, equivalently, that there is no regression on the input variable. To test

$$H_0 : \beta = 0 \qquad \text{versus} \qquad H_1 : \beta \neq 0$$

note that, from Proposition 9.3.1,

$$\frac{B - \beta}{\sqrt{\sigma^2 / S_{xx}}} = \sqrt{S_{xx}} \frac{(B - \beta)}{\sigma} \sim \mathcal{N}(0, 1) \tag{9.4.1}$$

and is independent of

$$\frac{SS_R}{\sigma^2} \sim \chi^2_{n-2}$$

Hence, from the definition of a t-random variable it follows that

$$\frac{\sqrt{S_{xx}}(B - \beta)/\sigma}{\sqrt{\dfrac{SS_R}{\sigma^2(n-2)}}} = \sqrt{\frac{(n-2)S_{xx}}{SS_R}}(B - \beta) \sim t_{n-2} \tag{9.4.2}$$

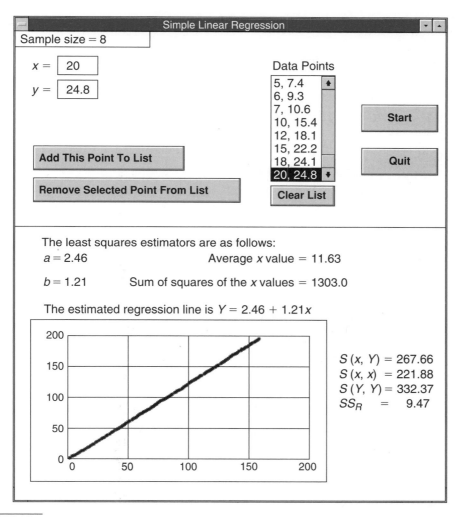

Simple Linear Regression

Sample size = 8

x = 20

y = 24.8

Data Points

5, 7.4
6, 9.3
7, 10.6
10, 15.4
12, 18.1
15, 22.2
18, 24.1
20, 24.8

Start

Quit

Add This Point To List

Remove Selected Point From List

Clear List

The least squares estimators are as follows:

$a = 2.46$ Average x value = 11.63

$b = 1.21$ Sum of squares of the x values = 1303.0

The estimated regression line is $Y = 2.46 + 1.21x$

$S(x, Y) = 267.66$
$S(x, x) = 221.88$
$S(Y, Y) = 332.37$
$SS_R = 9.47$

FIGURE 9.5

That is, $\sqrt{(n-2)S_{xx}/SS_R}\,(B - \beta)$ has a t-distribution with $n - 2$ degrees of freedom. Therefore, if H_0 is true (and so $\beta = 0$), then

$$\sqrt{\frac{(n-2)S_{xx}}{SS_R}}\,B \sim t_{n-2}$$

which gives rise to the following test of H_0.

Hypothesis Test of H_0: $\beta = 0$

A significance level γ test of H_0 is to

$$\begin{array}{ll} \text{reject} & H_0 \quad \text{if} \quad \sqrt{\dfrac{(n-2)S_{xx}}{SS_R}}|B| > t_{\gamma/2,n-2} \\ \text{accept} & H_0 \quad \text{otherwise} \end{array}$$

This test can be performed by first computing the value of the test statistic $\sqrt{(n-2)S_{xx}/SS_R}\,|B|$ — call its value v — and then rejecting H_0 if the desired significance level is at least as large as

$$p\text{-value} = P\{|T_{n-2}| > v\}$$
$$= 2P\{T_{n-2} > v\}$$

where T_{n-2} is a t-random variable with $n - 2$ degrees of freedom. This latter probability can be obtained by using Program 5.8.2a.

EXAMPLE 9.4a An individual claims that the fuel consumption of his automobile does not depend on how fast the car is driven. To test the plausibility of this hypothesis, the car was tested at various speeds between 45 and 70 miles per hour. The miles per gallon attained at each of these speeds was determined, with the following data resulting:

Speed	Miles per Gallon
45	24.2
50	25.0
55	23.3
60	22.0
65	21.5
70	20.6
75	19.8

Do these data refute the claim that the mileage per gallon of gas is unaffected by the speed at which the car is being driven?

SOLUTION Suppose that a simple linear regression model

$$Y = \alpha + \beta x + e$$

relates Y, the miles per gallon of the car, to x, the speed at which it is being driven. Now, the claim being made is that the regression coefficient β is equal to 0. To see if the data are strong enough to refute this claim, we need to see if it leads to a rejection of the null hypothesis when testing

$$H_0 : \beta = 0 \qquad \text{versus} \qquad H_1 : \beta \neq 0$$

To compute the value of the test statistic, we first compute the values of S_{xx}, S_{YY}, and S_{xY}. A hand calculation yields that

$$S_{xx} = 700, \qquad S_{YY} = 21.757, \qquad S_{xY} = -119$$

Using Equation 9.3.4 gives

$$SS_R = [S_{xx}S_{YY} - S_{xY}^2]/S_{xx}$$
$$= [700(21.757) - (119)^2]/700 = 1.527$$

Because

$$B = S_{xY}/S_{xx} = -119/700 = -.17$$

the value of the test statistic is

$$TS = \sqrt{5(700)/1.527}(.17) = 8.139$$

Since, from Table A2 of the Appendix, $t_{.005,5} = 4.032$, it follows that the hypothesis $\beta = 0$ is rejected at the 1 percent level of significance. Thus, the claim that the mileage does not depend on the speed at which the car is driven is rejected; there is strong evidence that increased speeds lead to decreased mileages. ∎

A confidence interval estimator for β is easily obtained from Equation 9.4.2. Indeed, it follows from Equation 9.4.2 that for any $a, 0 < a < 1$,

$$P\left\{-t_{a/2,n-2} < \sqrt{\frac{(n-2)S_{xx}}{SS_R}}(B-\beta) < t_{a/2,n-2}\right\} = 1-a$$

or, equivalently,

$$P\left\{B - \sqrt{\frac{SS_R}{(n-2)S_{xx}}}t_{a/2,n-2} < \beta < B + \sqrt{\frac{SS_R}{(n-2)S_{xx}}}t_{a/2,n-2}\right\} = 1-a$$

which yields the following.

Confidence Interval for β

A $100(1-a)$ percent confidence interval estimator of β is

$$\left(B - \sqrt{\frac{SS_R}{(n-2)S_{xx}}}t_{a/2,n-2}, B + \sqrt{\frac{SS_R}{(n-2)S_{xx}}}t_{a/2,n-2}\right)$$

REMARK

The result that

$$\frac{B - \beta}{\sqrt{\sigma^2/S_{xx}}} \sim \mathcal{N}(0, 1)$$

cannot be immediately applied to make inferences about β since it involves the unknown parameter σ^2. Instead, what we do is use the preceding statistic with σ^2 replaced by its estimator $SS_R/(n - 2)$, which has the effect of changing the distribution of the statistic from the standard normal to the t-distribution with $n - 2$ degrees of freedom.

EXAMPLE 9.4b   Derive a 95 percent confidence interval estimate of β in Example 9.4a.

SOLUTION   Since $t_{.025,5} = 2.571$, it follows from the computations of this example that the 95 percent confidence interval is

$$-.170 \pm 2.571 \sqrt{\frac{1.527}{3,500}} = -.170 \pm .054$$

That is, we can be 95 percent confident that β lies between $-.224$ and $-.116$.   ■

9.4.1.1   REGRESSION TO THE MEAN

The term *regression* was originally employed by Francis Galton while describing the laws of inheritance. Galton believed that these laws caused population extremes to "regress toward the mean." By this he meant that children of individuals having extreme values of a certain characteristic would tend to have less extreme values of this characteristic than their parent.

   If we assume a linear regression relationship between the characteristic of the offspring (Y) and that of the parent (x), then a regression to the mean will occur when the regression parameter β is between 0 and 1. That is, if

$$E[Y] = \alpha + \beta x$$

and $0 < \beta < 1$, then $E[Y]$ will be smaller than x when x is large and greater than x when x is small. That this statement is true can be easily checked either algebraically or by plotting the two straight lines

$$y = \alpha + \beta x$$

and

$$y = x$$

A plot indicates that, when $0 < \beta < 1$, the line $y = \alpha + \beta x$ is above the line $y = x$ for small values of x and is below it for large values of x.

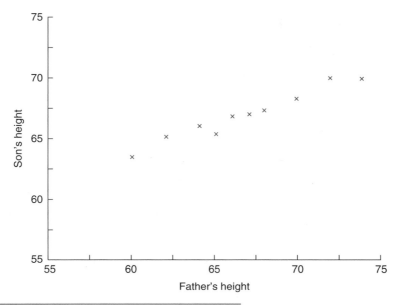

FIGURE 9.6 *Scatter diagram of son's height versus father's height.*

EXAMPLE 9.4c To illustrate Galton's thesis of regression to the mean, the British statistician Karl Pearson plotted the heights of 10 randomly chosen sons versus that of their fathers. The resulting data (in inches) were as follows.

Fathers' height	60	62	64	65	66	67	68	70	72	74
Sons' height	63.6	65.2	66	65.5	66.9	67.1	67.4	68.3	70.1	70

A scatter diagram representing these data is presented in Figure 9.6.

Note that whereas the data appear to indicate that taller fathers tend to have taller sons, it also appears to indicate that the sons of fathers who are either extremely short or extremely tall tend to be more "average" than their fathers — that is, there is a "regression toward the mean."

We will determine whether the preceding data are strong enough to prove that there is a regression toward the mean by taking this statement as the alternative hypothesis. That is, we will use the above data to test

$$H_0 : \beta \geq 1 \qquad \text{versus} \qquad H_1 : \beta < 1$$

which is equivalent to a test of

$$H_0 : \beta = 1 \qquad \text{versus} \qquad H_1 : \beta < 1$$

It now follows from Equation 9.4.2 that when $\beta = 1$, the test statistic

$$TS = \sqrt{8S_{xx}/SS_R}(B - 1)$$

has a t-distribution with 8 degrees of freedom. The significance level α test will reject H_0 when the value of TS is sufficiently small (since this will occur when B, the estimator of β, is sufficiently smaller than 1). Specifically, the test is to

$$\text{reject} \quad H_0 \quad \text{if} \quad \sqrt{8S_{xx}/SS_R}(B - 1) < -t_{\alpha,8}$$

Program 9.2 gives that

$$\sqrt{8S_{xx}/SS_R}(B - 1) = 30.2794(.4646 - 1) = -16.21$$

Since $t_{.01,8} = 2.896$, we see that

$$TS < -t_{.01,8}$$

and so the null hypothesis that $\beta \geq 1$ is rejected at the 1 percent level of significance. In fact, the p-value is

$$p\text{-value} = P\{T_8 \leq -16.213\} \approx 0$$

and so the null hypothesis that $\beta \geq 1$ is rejected at almost any significance level, thus establishing a regression toward the mean (see Figure 9.7).

A modern biological explanation for the regression to the mean phenomenon would roughly go along the lines of noting that as an offspring obtains a random selection of

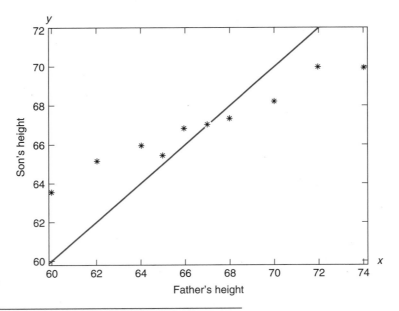

FIGURE 9.7   *Example 9.4c for x small, y > x. For x large, y < x.*

one-half of its parents' genes, it follows that the offspring of, say, a very tall parent would, by chance, tend to have fewer "tall" genes than its parent.

While the most important applications of the regression to the mean phenomenon concern the relationship between the biological characteristics of an offspring and that of its parents, this phenomenon also arises in situations where we have two sets of data referring to the same variables. ■

EXAMPLE 9.4d The data of Table 9.1 relate the number of motor vehicle deaths occurring in 12 counties in the northwestern United States in the years 1988 and 1989.

A glance at Figure 9.8 indicates that in 1989 there was, for the most part, a reduction in the number of deaths in those counties that had a large number of motor deaths in 1988. Similarly, there appears to have been an increase in those counties that had a low value in 1988. Thus, we would expect that a regression to the mean is in effect. In fact, running Program 9.2 yields that the estimated regression equation is

$$y = 74.589 + .276x$$

showing that the estimated value of β indeed appears to be less than 1.

One must be careful when considering the reason behind the regression to the mean phenomenon in the preceding data. For instance, it might be natural to suppose that those counties that had a large number of deaths caused by motor vehicles in 1988 would have made a large effort — perhaps by improving the safety of their roads or by making people more aware of the potential dangers of unsafe driving — to reduce this number. In addition, we might suppose that those counties that had the fewest number of deaths in 1988 might have "rested on their laurels" and not made much of an effort to further improve their numbers — and as a result had an increase in the number of casualties the following year.

TABLE 9.1   *Motor Vehicle Deaths, Northwestern United States, 1988 and 1989*

County	Deaths in 1988	Deaths in 1989
1	121	104
2	96	91
3	85	101
4	113	110
5	102	117
6	118	108
7	90	96
8	84	102
9	107	114
10	112	96
11	95	88
12	101	106

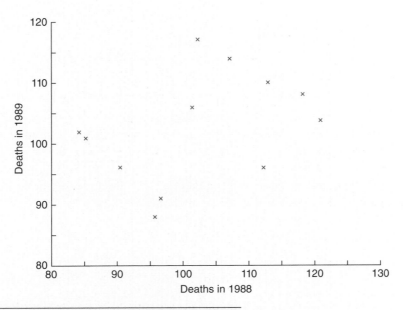

FIGURE 9.8   *Scatter diagram of 1989 deaths versus 1988 deaths.*

While this supposition might be correct, it is important to realize that a regression to the mean would probably have occurred even if none of the counties had done anything out of the ordinary. Indeed, it could very well be the case that those counties having large numbers of casualties in 1988 were just very unlucky in that year and thus a decrease in the next year was just a return to a more normal result for them. (For an analogy, if 9 heads result when 10 fair coins are flipped then it is quite likely that another flip of these 10 coins will result in fewer than 9 heads.) Similarly, those counties having few deaths in 1988 might have been "lucky" that year and a more normal result in 1989 would thus lead to an increase.

The mistaken belief that regression to the mean is due to some outside influence when it is in reality just due to "chance" occurs frequently enough that it is often referred to as the *regression fallacy*. ■

9.4.2 INFERENCES CONCERNING α

The determination of confidence intervals and hypothesis tests for α is accomplished in exactly the same manner as was done for β. Specifically, Proposition 9.3.1 can be used to show that

$$\sqrt{\frac{n(n-2)S_{xx}}{SS_R \sum_i x_i^2}}\,(A-\alpha) \tag{9.4.3}$$

which leads to the following confidence interval estimator of α.

Confidence Interval Estimator of α

The $100(1 - a)$ percent confidence interval for α is the interval

$$A \pm t_{\alpha/2, n-2} \sqrt{\frac{SS_R \sum_i x_i^2}{n(n-2)S_{xx}}}$$

Hypothesis tests concerning α are easily obtained from Equation 9.4.3, and their development is left as an exercise.

9.4.3 Inferences Concerning the Mean Response $\alpha + \beta x_0$

It is often of interest to use the data pairs (x_i, Y_i), $i = 1, \ldots, n$, to estimate $\alpha + \beta x_0$, the mean response for a given input level x_0. If it is a point estimator that is desired, then the natural estimator is $A + B x_0$, which is an unbiased estimator since

$$E[A + B x_0] = E[A] + x_0 E[B] = \alpha + \beta x_0$$

However, if we desire a confidence interval, or are interested in testing some hypothesis about this mean response, then it is necessary to first determine the probability distribution of the estimator $A + B x_0$. We now do so.

Using the expression for B given by Equation 9.3.1 yields that

$$B = c \sum_{i=1}^{n} (x_i - \bar{x}) Y_i$$

where

$$c = \frac{1}{\sum_{i=1}^{n} x_i^2 - n\bar{x}^2} = \frac{1}{S_{xx}}$$

Since

$$A = \bar{Y} - B\bar{x}$$

we see that

$$A + B x_0 = \frac{\sum_{i=1}^{n} Y_i}{n} - B(\bar{x} - x_0)$$

$$= \sum_{i=1}^{n} Y_i \left[\frac{1}{n} - c(x_i - \bar{x})(\bar{x} - x_0) \right]$$

Since the Y_i are independent normal random variables, the foregoing equation shows that $A + B x_0$ can be expressed as a linear combination of independent normal random variables and is thus itself normally distributed. Because we already know its mean, we need only compute its variance, which is accomplished as follows:

$$\text{Var}(A + Bx_0) = \sum_{i=1}^{n} \left[\frac{1}{n} - c(x_i - \bar{x})(\bar{x} - x_0) \right]^2 \text{Var}(Y_i)$$

$$= \sigma^2 \sum_{i=1}^{n} \left[\frac{1}{n^2} + c^2(\bar{x} - x_0)^2(x_i - \bar{x})^2 - 2c(x_i - \bar{x})\frac{(\bar{x} - x_0)}{n} \right]$$

$$= \sigma^2 \left[\frac{1}{n} + c^2(\bar{x} - x_0)^2 \sum_{i=1}^{n}(x_i - \bar{x})^2 - 2c(\bar{x} - x_0) \sum_{i=1}^{n}\frac{(x_i - \bar{x})}{n} \right]$$

$$= \sigma^2 \left[\frac{1}{n} + \frac{(\bar{x} - x_0)^2}{S_{xx}} \right]$$

where the last equality followed from

$$\sum_{i=1}^{n}(x_i - \bar{x})^2 = \sum_{i=1}^{n} x_i^2 - n\bar{x}^2 = 1/c = S_{xx}, \qquad \sum_{i=1}^{n}(x_i - \bar{x}) = 0$$

Hence, we have shown that

$$A + Bx_0 \sim \mathcal{N}\left(\alpha + \beta x_0, \sigma^2 \left[\frac{1}{n} + \frac{(x_0 - \bar{x})^2}{S_{xx}} \right] \right) \tag{9.4.4}$$

In addition, because $A + Bx_0$ is independent of

$$SS_R/\sigma^2 \sim \chi_{n-2}^2$$

it follows that

$$\frac{A + Bx_0 - (\alpha + \beta x_0)}{\sqrt{\frac{1}{n} + \frac{(x_0 - \bar{x})^2}{S_{xx}}} \sqrt{\frac{SS_R}{n - 2}}} \sim t_{n-2} \tag{9.4.5}$$

Equation 9.4.5 can now be used to obtain the following confidence interval estimator of $\alpha + \beta x_0$.

Confidence Interval Estimator of $\alpha + \beta x_0$

With $100(1 - a)$ percent confidence, $\alpha + \beta x_0$ will lie within

$$A + Bx_0 \pm \sqrt{\frac{1}{n} + \frac{(x_0 - \bar{x})^2}{S_{xx}}} \sqrt{\frac{SS_R}{n - 2}} t_{a/2, n-2}$$

EXAMPLE 9.4e Using the data of Example 9.4c, determine a 95 percent confidence interval for the average height of all males whose fathers are 68 inches tall.

SOLUTION Since the observed values are

$$n = 10, \qquad x_0 = 68, \qquad \bar{x} = 66.8, \qquad S_{xx} = 171.6, \qquad SS_R = 1.49721$$

we see that

$$\sqrt{\frac{1}{n} + \frac{(x_0 - \bar{x})^2}{S_{xx}}} \sqrt{\frac{SS_R}{n-2}} = .1424276$$

Also, because

$$t_{.025,8} = 2.306, \qquad A + Bx_0 = 67.56751$$

we obtain the following 95 percent confidence interval:

$$\alpha + \beta x_0 \in (67.239, 67.896) \quad \blacksquare$$

9.4.4 PREDICTION INTERVAL OF A FUTURE RESPONSE

It is often the case that it is more important to estimate the actual value of a future response rather than its mean value. For instance, if an experiment is to be performed at temperature level x_0, then we would probably be more interested in predicting $Y(x_0)$, the yield from this experiment, than we would be in estimating the expected yield — $E[Y(x_0)] = \alpha + \beta x_0$. (On the other hand, if a series of experiments were to be performed at input level x_0, then we would probably want to estimate $\alpha + \beta x_0$, the mean yield.)

Suppose first that we are interested in a single value (as opposed to an interval) to use as a predictor of $Y(x_0)$, the response at level x_0. Now, it is clear that the best predictor of $Y(x_0)$ is its mean value $\alpha + \beta x_0$. [Actually, this is not so immediately obvious since one could argue that the best predictor of a random variable is (1) its mean — which minimizes the expected square of the difference between the predictor and the actual value; or (2) its median — which minimizes the expected absolute difference between the predictor and the actual value; or (3) its mode — which is the most likely value to occur. However, as the mean, median, and mode of a normal random variable are all equal — and the response is, by assumption, normally distributed — there is no doubt in this situation.] Since α and β are not known, it seems reasonable to use their estimators A and B and thus use $A + Bx_0$ as the predictor of a new response at input level x_0.

Let us now suppose that rather than being concerned with determining a single value to predict a response, we are interested in finding a prediction interval that, with a given degree of confidence, will contain the response. To obtain such an interval, let Y denote the future response whose input level is x_0 and consider the probability distribution of the response minus its predicted value — that is, the distribution of $Y - A - B x_0$. Now,

$$Y \sim \mathcal{N}(\alpha + \beta x_0, \sigma^2)$$

and, as was shown in Section 9.4.3,

$$A + B x_0 \sim \mathcal{N}\left(\alpha + \beta x_0, \sigma^2 \left[\frac{1}{n} + \frac{(x_0 - \bar{x})^2}{S_{xx}}\right]\right)$$

Hence, because Y is independent of the earlier data values $Y_1, Y_2, \ldots, Y_n$ that were used to determine A and B, it follows that Y is independent of $A + B x_0$ and so

$$Y - A - B x_0 \sim \mathcal{N}\left(0, \sigma^2 \left[1 + \frac{1}{n} + \frac{(x_0 - \bar{x})^2}{S_{xx}}\right]\right)$$

or, equivalently,

$$\frac{Y - A - B x_0}{\sigma \sqrt{\dfrac{n+1}{n} + \dfrac{(x_0 - \bar{x})^2}{S_{xx}}}} \sim \mathcal{N}(0, 1) \tag{9.4.6}$$

Now, using once again the result that SS_R is independent of A and B (and also of Y) and

$$\frac{SS_R}{\sigma^2} \sim \chi_{n-2}^2$$

we obtain, by the usual argument, upon replacing σ^2 in Equation 9.4.6 by its estimator $SS_R/(n-2)$ that

$$\frac{Y - A - B x_0}{\sqrt{\dfrac{n+1}{n} + \dfrac{(x_0 - \bar{x})^2}{S_{xx}}} \sqrt{\dfrac{SS_R}{n-2}}} \sim t_{n-2}$$

and so, for any value $a, 0 < a < 1$,

$$P\left\{ -t_{a/2, n-2} < \frac{Y - A - B x_0}{\sqrt{\dfrac{n+1}{n} + \dfrac{(x_0 - \bar{x})^2}{S_{xx}}} \sqrt{\dfrac{SS_R}{n-2}}} < t_{a/2, n-2} \right\} = 1 - a$$

That is, we have just established the following.

Prediction Interval for a Response at the Input Level x_0

Based on the response values Y_i corresponding to the input values $x_i, i = 1, 2, \ldots, n$: With $100(1-a)$ percent confidence, the response Y at the input level x_0 will be contained in the interval

$$A + B x_0 \pm t_{a/2, n-2} \sqrt{\left[\frac{n+1}{n} + \frac{(x_0 - \bar{x})^2}{S_{xx}}\right] \frac{SS_R}{n-2}}$$

EXAMPLE 9.4f In Example 9.4c, suppose we want an interval that we can "be 95 percent certain" will contain the height of a given male whose father is 68 inches tall. A simple computation now yields the prediction interval

$$Y(68) \in 67.568 \pm 1.050$$

or, with 95 percent confidence, the person's height will be between 66.518 and 68.618. ∎

REMARKS

(a) There is often some confusion about the difference between a confidence and a prediction interval. A confidence interval is an interval that does contain, with a given degree of confidence, a fixed parameter of interest. A prediction interval, on the other hand, is an interval that will contain, again with a given degree of confidence, a random variable of interest.

(b) One should not make predictions about responses at input levels that are far from those used to obtain the estimated regression line. For instance, the data of Example 9.4c should not be used to predict the height of a male whose father is 42 inches tall.

9.4.5 SUMMARY OF DISTRIBUTIONAL RESULTS

We now summarize the distributional results of this section.

$$\text{Model: } Y = \alpha + \beta x + e, \quad e \sim \mathcal{N}(0, \sigma^2)$$
$$\text{Data: } (x_i, Y_i), \quad i = 1, 2, \ldots, n$$

Inferences About	Use the Distributional Result
β	$\sqrt{\dfrac{(n-2)S_{xx}}{SS_r}}(B - \beta) \sim t_{n-2}$
α	$\sqrt{\dfrac{n(n-2)S_{xx}}{\sum_i x_i^2 SS_R}}(A - \alpha) \sim t_{n-2}$

Inferences About	Use the Distributional Result
$\alpha + \beta x_0$	$\dfrac{A + B x_0 - \alpha - \beta x_0}{\sqrt{\left(\dfrac{1}{n} + \dfrac{(x_0 - \bar{x})^2}{S_{xx}} \right) \left(\dfrac{SS_R}{n-2} \right)}} \sim t_{n-2}$
$Y(x_0)$	$\dfrac{Y(x_0) - A - B x_0}{\sqrt{\left(1 + \dfrac{1}{n} + \dfrac{(x_0 - \bar{x})^2}{S_{xx}} \right) \left(\dfrac{SS_R}{n-2} \right)}} \sim t_{n-2}$

9.5 THE COEFFICIENT OF DETERMINATION AND THE SAMPLE CORRELATION COEFFICIENT

Suppose we wanted to measure the amount of variation in the set of response values $Y_1, \ldots, Y_n$ corresponding to the set of input values $x_1, \ldots, x_n$. A standard measure in statistics of the amount of variation in a set of values $Y_1, \ldots, Y_n$ is given by the quantity

$$S_{YY} = \sum_{i=1}^{n} (Y_i - \overline{Y})^2$$

For instance, if all the Y_i are equal — and thus are all equal to $\overline{Y}$ — then S_{YY} would equal 0.

The variation in the values of the Y_i arises from two factors. First, because the input values x_i are different, the response variables Y_i all have different mean values, which will result in some variation in their values. Second, the variation also arises from the fact that even when the differences in the input values are taken into account, each of the response variables Y_i has variance σ^2 and thus will not exactly equal the predicted value at its input x_i.

Let us consider now the question as to how much of the variation in the values of the response variables is due to the different input values, and how much is due to the inherent variance of the responses even when the input values are taken into account. To answer this question, note that the quantity

$$SS_R = \sum_{i=1}^{n} (Y_i - A - B x_i)^2$$

measures the remaining amount of variation in the response values after the different input values have been taken into account.

Thus,

$$S_{YY} - SS_R$$

represents the amount of variation in the response variables that is *explained* by the different input values, and so the quantity R^2 defined by

$$R^2 = \frac{S_{YY} - SS_R}{S_{YY}}$$

$$= 1 - \frac{SS_R}{S_{YY}}$$

represents the proportion of the variation in the response variables that is explained by the different input values. R^2 is called the *coefficient of determination.*

The coefficient of determination R^2 will have a value between 0 and 1. A value of R^2 near 1 indicates that most of the variation of the response data is explained by the different input values, whereas a value of R^2 near 0 indicates that little of the variation is explained by the different input values.

EXAMPLE 9.5a   In Example 9.4c, which relates the height of a son to that of his father, the output from Program 9.2 yielded that

$$S_{YY} = 38.521, \qquad SS_R = 1.497$$

Thus,

$$R^2 = 1 - \frac{1.497}{38.531} = .961$$

In other words, 96 percent of the variation of the heights of the 10 individuals is explained by the heights of their fathers. The remaining (unexplained) 4 percent of the variation is due to the variance of a son's height even when the father's height is taken into account. (That is, it is due to σ^2, the variance of the error random variable.) ∎

The value of R^2 is often used as an indicator of how well the regression model fits the data, with a value near 1 indicating a good fit, and one near 0 indicating a poor fit. In other words, if the regression model is able to explain most of the variation in the response data, then it is considered to fit the data well.

Recall that in Section 2.6 we defined the sample correlation coefficient r of the set of data pairs $(x_i, Y_i), i = 1, \ldots, n$, by

$$r = \frac{\sum_{i=1}^{n}(x_i - \bar{x})(Y_i - \overline{Y})}{\sqrt{\sum_{i=1}^{n}(x_i - \bar{x})^2 \sum_{i=1}^{n}(Y_i - \overline{Y})^2}}$$

It was noted that r provided a measure of the degree to which high values of x are paired with high values of Y and low values of x with low values of Y. A value of r

near $+1$ indicated that large x values were strongly associated with large Y values and small x values were strongly associated with small Y values, whereas a value near -1 indicated that large x values were strongly associated with small Y values and small x values with large Y values.

In the notation of this chapter,

$$r = \frac{S_{xY}}{\sqrt{S_{xx}S_{YY}}}$$

Upon using identity (9.3.4):

$$SS_R = \frac{S_{xx}S_{YY} - S_{xY}^2}{S_{xx}}$$

we see that

$$
\begin{aligned}
r^2 &= \frac{S_{xY}^2}{S_{xx}S_{YY}} \\
&= \frac{S_{xx}S_{YY} - SS_R S_{xx}}{S_{xx}S_{YY}} \\
&= 1 - \frac{SS_R}{S_{YY}} \\
&= R^2
\end{aligned}
$$

That is,

$$|r| = \sqrt{R^2}$$

and so, except for its sign indicating whether it is positive or negative, the sample correlation coefficient is equal to the square root of the coefficient of determination. The sign of r is the same as that of B.

The above gives additional meaning to the sample correlation coefficient. For instance, if a data set has its sample correlation coefficient r equal to .9, then this implies that a simple linear regression model for these data explains 81 percent (since $R^2 = .9^2 = .81$) of the variation in the response values. That is, 81 percent of the variation in the response values is explained by the different input values.

9.6 ANALYSIS OF RESIDUALS: ASSESSING THE MODEL

The initial step for ascertaining whether or not the simple linear regression model

$$Y = \alpha + \beta x + e, \qquad e \sim \mathcal{N}(0, \sigma^2)$$

is appropriate in a given situation is to investigate the scatter diagram. Indeed, this is often sufficient to convince one that the regression model is or is not correct. When the scatter diagram does not by itself rule out the preceding model, then the least squares estimators A and B should be computed and the residual $Y_i - (A + Bx_i), i = 1, \ldots, n$ analyzed. The analysis begins by normalizing, or standardizing, the residuals by dividing them by $\sqrt{SS_R/(n-2)}$, the estimate of the standard deviation of the Y_i. The resulting quantities

$$\frac{Y_i - (A + Bx_i)}{\sqrt{SS_R/(n-2)}}, \qquad i = 1, \ldots, n$$

are called the *standardized residuals*.

When the simple linear regression model is correct, the standardized residuals are approximately independent standard normal random variables, and thus should be randomly distributed about 0 with about 95 percent of their values being between -2 and $+2$ (since $P\{-1.96 < Z < 1.96\} = .95$). In addition, a plot of the standardized residuals should not indicate any distinct pattern. Indeed, any indication of a distinct pattern should make one suspicious about the validity of the assumed simple linear regression model.

Figure 9.9 presents three different scatter diagrams and their associated standardized residuals. The first of these, as indicated both by its scatter diagram and the random nature of its standardized residuals, appears to fit the straight-line model quite well. The second residual plot shows a discernible pattern, in that the residuals appear to be first

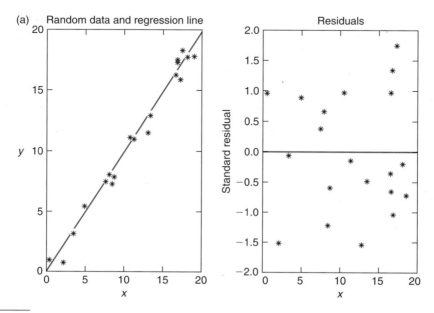

FIGURE 9.9

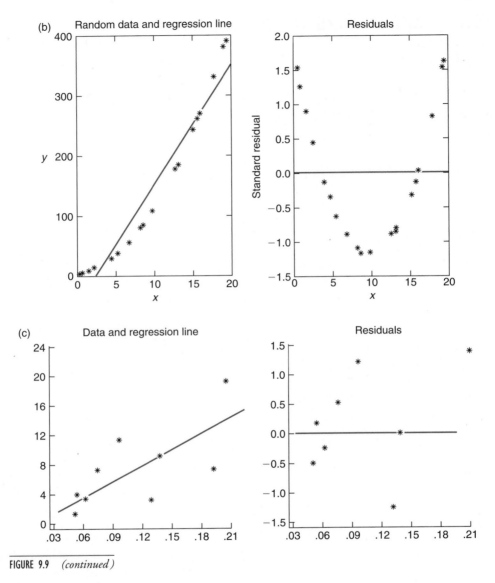

FIGURE 9.9   *(continued)*

decreasing and then increasing as the input level increases. This often means that higher-order (than just linear) terms are needed to describe the relationship between the input and response. Indeed, this is also indicated by the scatter diagram in this case. The third standardized residual plot also shows a pattern, in that the absolute value of the residuals, and thus their squares, appear to be increasing, as the input level increases. This often indicates that the variance of the response is not constant but, rather, increases with the input level.

9.7 TRANSFORMING TO LINEARITY

In many situations, it is clear that the mean response is not a linear function of the input level. In such cases, if the form of the relationship can be determined it is sometimes possible, by a change of variables, to transform it into a linear form. For instance, in certain applications it is known that $W(t)$, the amplitude of a signal a time t after its origination, is approximately related to t by the functional form

$$W(t) \approx ce^{-dt}$$

On taking logarithms, this can be expressed as

$$\log W(t) \approx \log c - dt$$

If we now let

$$Y = \log W(t)$$
$$\alpha = \log c$$
$$\beta = -d$$

then the foregoing can be modeled as a regression of the form

$$Y = \alpha + \beta t + e$$

The regression parameters α and β would then be estimated by the usual least squares approach and the original functional relationships can be predicted from

$$W(t) \approx e^{A+Bt}$$

EXAMPLE 9.7a The following table gives the percentages of a chemical that were used up when an experiment was run at various temperatures (in degrees Celsius). Use it to estimate the percentage of the chemical that would be used up if the experiment were to be run at 350 degrees.

Temperature	Percentage
5°	.061
10°	.113
20°	.192
30°	.259
40°	.339
50°	.401
60°	.461
80°	.551

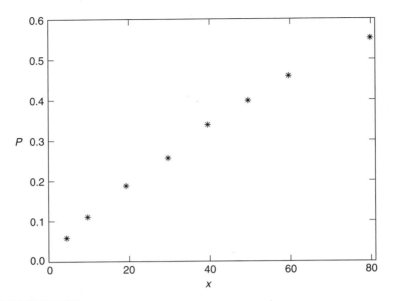

SOLUTION   Let $P(x)$ be the percentage of the chemical that is used up when the experiment is run at $10x$ degrees. Even though a plot of $P(x)$ looks roughly linear (see Figure 9.10), we can improve upon the fit by considering a nonlinear relationship between x and $P(x)$. Specifically, let us consider a relationship of the form

$$1 - P(x) \approx c(1 - d)^x$$

That is, let us suppose that the percentage of the chemical that survives an experiment run at temperature x approximately decreases at an exponential rate when x increases. Taking logs, the preceding can be written as

$$\log(1 - P(x)) \approx \log(c) + x\log(1 - d)$$

Thus, setting

$$Y = -\log(1 - P)$$
$$\alpha = -\log c$$
$$\beta = -\log(1 - d)$$

we obtain the usual regression equation

$$Y = \alpha + \beta x + e$$

TABLE 9.2

Temperature	$-\log(1 - P)$
5°	.063
10°	.120
20°	.213
30°	.300
40°	.414
50°	.512
60°	.618
80°	.801

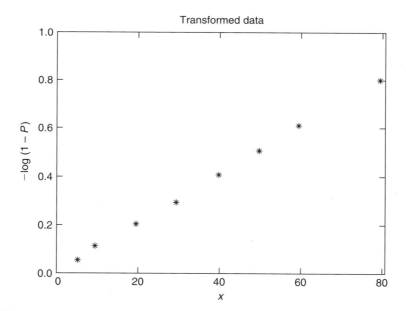

FIGURE 9.11

To see whether the data support this model, we can plot $-\log(1 - P)$ versus x. The transformed data are presented in Table 9.2 and the graph in Figure 9.11.

Running Program 9.2 yields that the least square estimates of α and β are

$$A = .0154$$

$$B = .0099$$

TABLE 9.3

x	P	$\hat{P}$	$P - \hat{P}$
5	.061	.063	−.002
10	.113	.109	.040
20	.192	.193	−.001
30	.259	.269	−.010
40	.339	.339	.000
50	.401	.401	.000
60	.461	.458	.003
80	.551	.556	−.005

Transforming this back into the original variable gives that the estimates of c and d are

$$\hat{c} = e^{-A} = .9847$$
$$1 - \hat{d} = e^{-B} = .9901$$

and so the estimated functional relationship is

$$\hat{P} = 1 - .9847(.9901)^x$$

The residuals $P - \hat{P}$ are presented in Table 9.3.   ∎

9.8 WEIGHTED LEAST SQUARES

In the regression model

$$Y = \alpha + \beta x + e$$

it often turns out that the variance of a response is not constant but rather depends on its input level. If these variances are known — at least up to a proportionality constant — then the regression parameters α and β should be estimated by minimizing a weighted sum of squares. Specifically, if

$$\text{Var}(Y_i) = \frac{\sigma^2}{w_i}$$

then the estimators A and B should be chosen to minimize

$$\sum_i \frac{[Y_i - (A + Bx_i)]^2}{\text{Var}(Y_i)} = \frac{1}{\sigma^2} \sum_i w_i (Y_i - A - Bx_i)^2$$

On taking partial derivatives with respect to A and B and setting them equal to 0, we obtain the following equations for the minimizing A and B.

$$\sum_i w_i Y_i = A \sum_i w_i + B \sum_i w_i x_i \qquad (9.8.1)$$

$$\sum_i w_i x_i Y_i = A \sum_i w_i x_i + B \sum_i w_i x_i^2$$

These equations are easily solved to yield the least squares estimators.

EXAMPLE 9.8a To develop a feel as to why the estimators should be obtained by minimizing the weighted sum of squares rather than the ordinary sum of squares, consider the following situation. Suppose that $X_1, \ldots, X_n$ are independent normal random variables each having mean μ and variance σ^2. Suppose further that the X_i are not directly observable but rather only Y_1 and Y_2, defined by

$$Y_1 = X_1 + \cdots + X_k, \qquad Y_2 = X_{k+1} + \cdots + X_n, \qquad k < n$$

are directly observable. Based on Y_1 and Y_2, how should we estimate μ?

Whereas the best estimator of μ is clearly $\overline{X} = \sum_{i=1}^{n} X_i/n = (Y_1 + Y_2)/n$, let us see what the ordinary least squares estimator would be. Since

$$E[Y_1] = k\mu, \qquad E[Y_2] = (n-k)\mu$$

the least squares estimator of μ would be that value of μ that minimizes

$$(Y_1 - k\mu)^2 + (Y_2 - [n-k]\mu)^2$$

On differentiating and setting equal to zero, we see that the least squares estimator of μ — call it $\hat{\mu}$ — is such that

$$-2k(Y_1 - k\hat{\mu}) - 2(n-k)[Y_2 - (n-k)\hat{\mu}] = 0$$

or

$$[k^2 + (n-k)^2]\hat{\mu} = kY_1 + (n-k)Y_2$$

or

$$\hat{\mu} = \frac{kY_1 + (n-k)Y_2}{k^2 + (n-k)^2}$$

Thus we see that while the ordinary least squares estimator is an unbiased estimator of μ — since

$$E[\hat{\mu}] = \frac{kE[Y_1] + (n-k)E[Y_2]}{k^2 + (n-k)^2} = \frac{k^2\mu + (n-k)^2\mu}{k^2 + (n-k)^2} = \mu,$$

it is not the best estimator $\overline{X}$.

Now let us determine the estimator produced by minimizing the weighted sum of squares. That is, let us determine the value of μ — call it μ_w — that minimizes

$$\frac{(Y_1 - k\mu)^2}{\text{Var}(Y_1)} + \frac{[Y_2 - (n-k)\mu]^2}{\text{Var}(Y_2)}$$

Since

$$\text{Var}(Y_1) = k\sigma^2, \qquad \text{Var}(Y_2) = (n-k)\sigma^2$$

this is equivalent to choosing μ to minimize

$$\frac{(Y_1 - k\mu)^2}{k} + \frac{[Y_2 - (n-k)\mu]^2}{n-k}$$

Upon differentiating and then equating to 0, we see that μ_w, the minimizing value, satisfies

$$\frac{-2k(Y_1 - k\mu_w)}{k} - \frac{2(n-k)[Y_2 - (n-k)\mu_w]}{n-k} = 0$$

or

$$Y_1 + Y_2 = n\mu_w$$

or

$$\mu_w = \frac{Y_1 + Y_2}{n}$$

That is, the weighted least squares estimator is indeed the preferred estimator $(Y_1 + Y_2)/n = \overline{X}$. ∎

REMARKS

(a) Assuming normally distributed data, the weighted least squares estimators are precisely the maximum likelihood estimators. This follows because the joint density of the data $Y_1, \ldots, Y_n$ is

$$f_{Y_1,\ldots,Y_n}(y_1, \ldots, y_n) = \prod_{i=1}^{n} \frac{1}{\sqrt{2\pi}\,(\sigma/\sqrt{w_i})} e^{-(y_i - \alpha - \beta x_i)^2/(2\sigma^2/w_i)}$$

$$= \frac{\sqrt{w_1 \cdots w_n}}{(2\pi)^{n/2}\sigma^n} e^{-\sum_{i=1}^{n} w_i(y_i - \alpha - \beta x_i)^2/2\sigma^2}$$

Consequently, the maximum likelihood estimators of α and β are precisely the values of α and β that minimize the weighted sum of squares $\sum_{i=1}^{n} w_i(y_i - \alpha - \beta x_i)^2$.

(b) The weighted sum of squares can also be seen as the relevant quantity to be minimized by multiplying the regression equation

$$Y = \alpha + \beta x + e$$

by $\sqrt{w}$. This results in the equation

$$Y\sqrt{w} = \alpha\sqrt{w} + \beta x\sqrt{w} + e\sqrt{w}$$

Now, in this latter equation the error term $e\sqrt{w}$ has mean 0 and constant variance. Hence, the natural least squares estimators of α and β would be the values of A and B that minimize

$$\sum_i (Y_i\sqrt{w_i} - A\sqrt{w_i} - Bx_i\sqrt{w_i})^2 = \sum_i w_i(Y_i - A - Bx_i)^2$$

(c) The weighted least squares approach puts the greatest emphasis on those data pairs having the greatest weights (and thus the smallest variance in their error term). ■

At this point it might appear that the weighted least squares approach is not particularly useful since it requires a knowledge, up to a constant, of the variance of a response at an arbitrary input level. However, by analyzing the model that generates the data, it is often possible to determine these values. This will be indicated by the following two examples.

EXAMPLE 9.8b The following data represent travel times in a downtown area of a certain city. The independent, or input, variable is the distance to be traveled.

Distance (miles)	.5	1	1.5	2	3	4	5	6	8	10
Travel time (minutes)	15.0	15.1	16.5	19.9	27.7	29.7	26.7	35.9	42	49.4

Assuming a linear relationship of the form

$$Y = \alpha + \beta x + e$$

between Y, the travel time, and x, the distance, how should we estimate α and β? To utilize the weighted least squares approach we need to know, up to a multiplicative constant, the variance of Y as a function of x. We will now present an argument that $\mathrm{Var}(Y)$ should be proportional to x.

SOLUTION Let d denote the length of a city block. Thus a trip of distance x will consist of x/d blocks. If we let $Y_i, i = 1, \ldots, x/d$, denote the time it takes to traverse block i, then the total travel time can be expressed as

$$Y = Y_1 + Y_2 + \cdots + Y_{x/d}$$

Now in many applications it is probably reasonable to suppose that the Y_i are independent random variables with a common variance, and thus,

$$\mathrm{Var}(Y) = \mathrm{Var}(Y_1) + \cdots + \mathrm{Var}(Y_{x/d})$$

$$= (x/d)\mathrm{Var}(Y_1) \qquad \text{since } \mathrm{Var}(Y_i) = \mathrm{Var}(Y_1)$$

$$= x\sigma^2, \qquad \text{where } \sigma^2 = \mathrm{Var}(Y_1)/d$$

Thus, it would seem that the estimators A and B should be chosen so as to minimize

$$\sum_i \frac{(Y_i - A - Bx_i)^2}{x_i}$$

Using the preceding data with the weights $w_i = 1/x_i$, the least squares Equations 9.8.1 are

$$104.22 = 5.34A + 10B$$

$$277.9 = 10A + 41B$$

which yield the solution

$$A = 12.561, \qquad B = 3.714$$

A graph of the estimated regression line $12.561 + 3.714x$ along with the data points is presented in Figure 9.12. As a qualitative check of our solution, note that the regression line fits the data pairs best when the input levels are small, which is as it should be since the weights are inversely proportional to the inputs. ∎

EXAMPLE 9.8c Consider the relationship between Y, the number of accidents on a heavily traveled highway, and x, the number of cars traveling on the highway. After a little thought it would probably seem to most that the linear model

$$Y = \alpha + \beta x + e$$

would be appropriate. However, as there does not appear to be any *a priori* reason why $\mathrm{Var}(Y)$ should not depend on the input level x, it is not clear that we would be justified in using the ordinary least squares approach to estimate α and β. Indeed, we will now argue that a weighted least squares approach with weights $1/x$ should be employed — that is, we should choose A and B to minimize

$$\sum_i \frac{(Y_i - A - Bx_i)^2}{x_i}$$

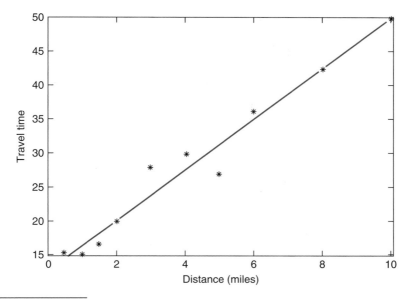

FIGURE 9.12   *Example 9.8b.*

The rationale behind this claim is that it seems reasonable to suppose that Y has approximately a Poisson distribution. This is so since we can imagine that each of the x cars will have a small probability of causing an accident and so, for large x, the number of accidents should be approximately a Poisson random variable. Since the variance of a Poisson random variable is equal to its mean, we see that

$$\text{Var}(Y) \simeq E[Y] \qquad \text{since } Y \text{ is approximately Poisson}$$
$$= \alpha + \beta x$$
$$\simeq \beta x \qquad \text{for large } x \quad \blacksquare$$

REMARKS

(a) Another technique that is often employed when the variance of the response depends on the input level is to attempt to stabilize the variance by an appropriate transformation. For example, if Y is a Poisson random variable with mean λ, then it can be shown [see Remark (b)] that $\sqrt{Y}$ has approximate variance .25 no matter what the value of λ. Based on this fact, one might try to model $E[\sqrt{Y}]$ as a linear function of the input. That is, one might consider the model

$$\sqrt{Y} = \alpha + \beta x + e$$

(b) Proof that $\text{Var}(\sqrt{Y}) \approx .25$ when Y is Poisson with mean λ. Consider the Taylor series expansion of $g(y) = \sqrt{y}$ about the value λ. By ignoring all terms beyond the second derivative term, we obtain that

$$g(y) \approx g(\lambda) + g'(\lambda)(y - \lambda) + \frac{g''(\lambda)(y - \lambda)^2}{2} \qquad (9.8.2)$$

Since

$$g'(\lambda) = \tfrac{1}{2}\lambda^{-1/2}, \qquad g''(\lambda) = -\tfrac{1}{4}\lambda^{-3/2}$$

we obtain, on evaluating Equation 9.8.2 at $y = Y$, that

$$\sqrt{Y} \approx \sqrt{\lambda} + \tfrac{1}{2}\lambda^{-1/2}(Y - \lambda) - \tfrac{1}{8}\lambda^{-3/2}(Y - \lambda)^2$$

Taking expectations, and using the results that

$$E[Y - \lambda] = 0, \quad E[(Y - \lambda)^2] = \text{Var}(Y) = \lambda$$

yields that

$$E[\sqrt{Y}] \approx \sqrt{\lambda} - \frac{1}{8\sqrt{\lambda}}$$

Hence

$$(E[\sqrt{Y}])^2 \approx \lambda + \frac{1}{64\lambda} - \frac{1}{4}$$
$$\approx \lambda - \frac{1}{4}$$

and so

$$\text{Var}(\sqrt{Y}) = E[Y] - (E[\sqrt{Y}])^2$$
$$\approx \lambda - \left(\lambda - \frac{1}{4}\right)$$
$$= \frac{1}{4}$$

9.9 POLYNOMIAL REGRESSION

In situations where the functional relationship between the response Y and the independent variable x cannot be adequately approximated by a linear relationship, it is sometimes possible to obtain a reasonable fit by considering a polynomial relationship. That is, we might try to fit to the data set a functional relationship of the form

$$Y = \beta_0 + \beta_1 x + \beta_2 x^2 + \cdots + \beta_r x^r + e$$

where $\beta_0, \beta_1, \ldots, \beta_r$ are regression coefficients that would have to be estimated. If the data set consists of the n pairs (x_i, Y_i), $i = 1, \ldots, n$, then the least squares estimators of $\beta_0, \ldots, \beta_r$ — call them $B_0, \ldots, B_r$ — are those values that minimize

$$\sum_{i=1}^{n} (Y_i - B_0 - B_1 x_i - B_2 x_i^2 - \cdots - B_r x_i^r)^2$$

To determine these estimators, we take partial derivatives with respect to $B_0 \ldots B_r$ of the foregoing sum of squares, and then set these equal to 0 so as to determine the minimizing values. On doing so, and then rearranging the resulting equations, we obtain that the least squares estimators $B_0, B_1, \ldots, B_r$ satisfy the following set of $r + 1$ linear equations called the *normal equations*.

$$\sum_{i=1}^{n} Y_i = B_0 n + B_1 \sum_{i=1}^{n} x_i + B_2 \sum_{i=1}^{n} x_i^2 + \cdots + B_r \sum_{i=1}^{n} x_i^r$$

$$\sum_{i=1}^{n} x_i Y_i = B_0 \sum_{i=1}^{n} x_i + B_1 \sum_{i=1}^{n} x_i^2 + B_2 \sum_{i=1}^{n} x_i^3 + \cdots + B_r \sum_{i=1}^{n} x_i^{r+1}$$

$$\sum_{i=1}^{n} x_i^2 Y_i = B_0 \sum_{i=1}^{n} x_i^2 + B_1 \sum_{i=1}^{n} x_i^3 + \cdots + B_r \sum_{i=1}^{n} x_i^{r+2}$$

$$\vdots \qquad \vdots \qquad \qquad \vdots$$

$$\sum_{i=1}^{n} x_i^r Y_i = B_0 \sum_{i=1}^{n} x_i^r + B_1 \sum_{i=1}^{n} x_i^{r+1} + \cdots + B_r \sum_{i=1}^{n} x_i^{2r}$$

In fitting a polynomial to a set of data pairs, it is often possible to determine the necessary degree of the polynomial by a study of the scatter diagram. We emphasize that one should always use the lowest possible degree that appears to adequately describe the data. [Thus, for instance, whereas it is usually possible to find a polynomial of degree n that passes through all the n pairs (x_i, Y_i), $i = 1, \ldots, n$, it would be hard to ascribe much confidence to such a fit.]

Even more so than in linear regression, it is extremely risky to use a polynomial fit to predict the value of a response at an input level x_0 that is far away from the input levels $x_i, i = 1, \ldots, n$ used in finding the polynomial fit. (For one thing, the polynomial fit may be valid only in a region around the $x_i, i = 1, \ldots, n$ and not including x_0.)

EXAMPLE 9.9a Fit a polynomial to the following data.

x	Y
1	20.6
2	30.8
3	55
4	71.4
5	97.3
6	131.8
7	156.3
8	197.3
9	238.7
10	291.7

SOLUTION A plot of these data (see Figure 9.13) indicates that a quadratic relationship

$$Y = \beta_0 + \beta_1 x + \beta_2 x^2 + e$$

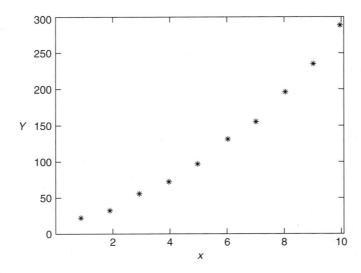

FIGURE 9.13

might hold. Since

$$\sum_i x_i = 55, \quad \sum_i x_i^2 = 385, \quad \sum_i x_i^3 = 3{,}025, \quad \sum_i x_i^4 = 25{,}333$$

$$\sum_i Y_i = 1{,}291.1, \quad \sum_i x_i Y_i = 9{,}549.3, \quad \sum_i x_i^2 Y_i = 77{,}758.9$$

the least squares estimates are the solution of the following set of equations.

$$1{,}291.1 = 10B_0 + 55B_1 + 385B_2 \qquad (9.9.1)$$
$$9{,}549.3 = 55B_0 + 385B_1 + 3{,}025B_2$$
$$77{,}758.9 = 385B_0 + 3{,}025B_1 + 25{,}333B_2$$

Solving these equations (see the remark following this example) yields that the least squares estimates are

$$B_0 = 12.59326, \quad B_1 = 6.326172, \quad B_2 = 2.122818$$

Thus, the estimated quadratic regression equation is

$$Y = 12.59 + 6.33x + 2.12x^2$$

This equation, along with the data, is plotted in Figure 9.14. ■

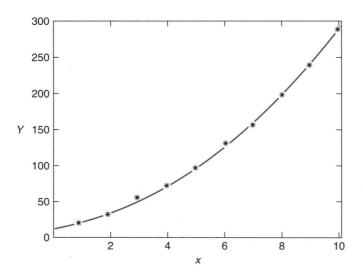

FIGURE 9.14

REMARK

In matrix notation Equation 9.9.1 can be written as

$$\begin{bmatrix} 1{,}291.1 \\ 9{,}549.3 \\ 77{,}758.9 \end{bmatrix} = \begin{bmatrix} 10 & 55 & 385 \\ 55 & 385 & 3{,}025 \\ 385 & 3{,}025 & 25{,}333 \end{bmatrix} \begin{bmatrix} B_0 \\ B_1 \\ B_2 \end{bmatrix}$$

which has the solution

$$\begin{bmatrix} B_0 \\ B_1 \\ B_2 \end{bmatrix} = \begin{bmatrix} 10 & 55 & 385 \\ 55 & 385 & 3{,}025 \\ 385 & 3{,}025 & 25{,}333 \end{bmatrix}^{-1} \begin{bmatrix} 1{,}291.1 \\ 9{,}549.3 \\ 77{,}758.9 \end{bmatrix}$$

*9.10 MULTIPLE LINEAR REGRESSION

In the majority of applications, the response of an experiment can be predicted more adequately not on the basis of a single independent input variable but on a collection of such variables. Indeed, a typical situation is one in which there are a set of, say, k input variables and the response Y is related to them by the relation

$$Y = \beta_0 + \beta_1 x_1 + \cdots + \beta_k x_k + e$$

where $x_j, j = 1, \ldots, k$ is the level of the jth input variable and e is a random error that we shall assume is normally distributed with mean 0 and (constant) variance σ^2. The parameters $\beta_0, \beta_1, \ldots, \beta_k$ and σ^2 are assumed to be unknown and must be estimated from the data, which we shall suppose will consist of the values of $Y_1, \ldots, Y_n$ where Y_i is the response level corresponding to the k input levels $x_{i1}, x_{i2}, \ldots, x_{ik}$. That is, the Y_i are related to these input levels through

$$E[Y_i] = \beta_0 + \beta_1 x_{i1} + \beta_2 x_{i2} + \cdots + \beta_k x_{ik}$$

If we let $B_0, B_1, \ldots, B_k$ denote estimators of $\beta_0, \ldots, \beta_k$, then the sum of the squared differences between the Y_i and their estimated expected values is

$$\sum_{i=1}^{n} (Y_i - B_0 - B_1 x_{i1} - B_2 x_{i2} - \cdots - B_k x_{ik})^2$$

The least squares estimators are those values of $B_0, B_1, \ldots, B_k$ that minimize the foregoing.

---

* Optional section.

To determine the least squares estimators, we repeatedly take partial derivatives of the preceding sum of squares first with respect to B_0, then to $B_1, \ldots,$ then to B_k. On equating these $k + 1$ equations to 0, we obtain the following set of equations:

$$\sum_{i=1}^{n} (Y_i - B_0 - B_1 x_{i1} - B_2 x_{i2} - \cdots - B_k x_{ik}) = 0$$

$$\sum_{i=1}^{n} x_{i1} (Y_i - B_0 - B_1 x_{i1} - \cdots - B_k x_{ik}) = 0$$

$$\sum_{i=1}^{n} x_{i2} (Y_i - B_0 - B_1 x_{i1} - \cdots - B_k x_{ik}) = 0$$

$$\vdots$$

$$\sum_{i=1}^{n} x_{ik} (Y_i - B_0 - B_1 x_{i1} - \cdots - B_i x_{ik}) = 0$$

Rewriting these equations yields that the least squares estimators $B_0, B_1, \ldots, B_k$ satisfy the following set of linear equations, called the *normal equations*:

$$\sum_{i=1}^{n} Y_i = nB_0 + B_1 \sum_{i=1}^{n} x_{i1} + B_2 \sum_{i=1}^{n} x_{i2} + \cdots + B_k \sum_{i=1}^{n} x_{ik} \qquad (9.10.1)$$

$$\sum_{i=1}^{n} x_{i1} Y_i = B_0 \sum_{i=1}^{n} x_{i1} + B_1 \sum_{i=1}^{n} x_{i1}^2 + B_2 \sum_{i=1}^{n} x_{i1} x_{i2} + \cdots + B_k \sum_{i=1}^{n} x_{i1} x_{ik}$$

$$\vdots$$

$$\sum_{i=1}^{k} x_{ik} Y_i = B_0 \sum_{i=1}^{n} x_{ik} + B_1 \sum_{i=1}^{n} x_{ik} x_{i1} + B_2 \sum_{i=1}^{n} x_{ik} x_{i2} + \cdots + B_k \sum_{i=1}^{n} x_{ik}^2$$

Before solving the normal equations, it is convenient to introduce matrix notation. If we let

$$\mathbf{Y} = \begin{bmatrix} Y_1 \\ Y_2 \\ \vdots \\ Y_n \end{bmatrix}, \qquad \mathbf{X} = \begin{bmatrix} 1 & x_{11} & x_{12} & \cdots & x_{1k} \\ 1 & x_{21} & x_{22} & \cdots & x_{2k} \\ \vdots & \vdots & \vdots & & \vdots \\ 1 & x_{n1} & x_{n2} & \cdots & x_{nk} \end{bmatrix}$$

$$\boldsymbol{\beta} = \begin{bmatrix} \beta_0 \\ \beta_1 \\ \vdots \\ \beta_k \end{bmatrix}, \qquad \mathbf{e} = \begin{bmatrix} e_1 \\ e_2 \\ \vdots \\ e_n \end{bmatrix}$$

then $\mathbf{Y}$ is an $n \times 1$, $\mathbf{X}$ an $n \times p$, $\boldsymbol{\beta}$ a $p \times 1$, and $\mathbf{e}$ an $n \times 1$ matrix where $p \equiv k + 1$.

The multiple regression model can now be written as

$$\mathbf{Y} = \mathbf{X}\boldsymbol{\beta} + \mathbf{e}$$

In addition, if we let

$$\mathbf{B} = \begin{bmatrix} B_0 \\ B_1 \\ \vdots \\ B_k \end{bmatrix}$$

be the matrix of least squares estimators, then the normal Equations 9.10.1 can be written as

$$\mathbf{X'XB} = \mathbf{X'Y} \qquad\qquad (9.10.2)$$

where $\mathbf{X'}$ is the transpose of $\mathbf{X}$.

To see that Equation 9.10.2 is equivalent to the normal Equations 9.10.1, note that

$$\mathbf{X'X} = \begin{bmatrix} 1 & 1 & \cdots & 1 \\ x_{11} & x_{21} & \cdots & x_{n1} \\ x_{12} & x_{22} & \cdots & x_{n2} \\ \vdots & \vdots & & \vdots \\ x_{1k} & x_{2k} & \cdots & x_{nk} \end{bmatrix} \begin{bmatrix} 1 & x_{11} & x_{12} & \cdots & x_{1k} \\ 1 & x_{21} & x_{22} & \cdots & x_{2k} \\ \vdots & \vdots & \vdots & & \vdots \\ 1 & x_{n1} & x_{n2} & \cdots & x_{nk} \end{bmatrix}$$

$$= \begin{bmatrix} n & \sum_i x_{i1} & \sum_i x_{i2} & \cdots & \sum_i x_{ik} \\ \sum_i x_{i1} & \sum_i x_{i1}^2 & \sum_i x_{i1}x_{i2} & \cdots & \sum_i x_{i1}x_{ik} \\ \vdots & \vdots & \vdots & & \vdots \\ \sum_i x_{ik} & \sum_i x_{ik}x_{i1} & \sum_i x_{ik}x_{i2} & \cdots & \sum_i x_{ik}^2 \end{bmatrix}$$

and

$$\mathbf{X'Y} = \begin{bmatrix} \sum_i Y_i \\ \sum_i x_{i1} Y_i \\ \vdots \\ \sum_i x_{ik} Y_i \end{bmatrix}$$

It is now easy to see that the matrix equation

$$\mathbf{X'XB} = \mathbf{X'Y}$$

is equivalent to the set of normal Equations 9.10.1. Assuming that $(\mathbf{X'X})^{-1}$ exists, which is usually the case, we obtain, upon multiplying it by both sides of the foregoing, that the least squares estimators are given by

$$\mathbf{B} = (\mathbf{X'X})^{-1}\mathbf{X'Y} \tag{9.10.3}$$

Program 9.10 computes the least squares estimates, the inverse matrix $(\mathbf{X'X})^{-1}$, and SS_R.

EXAMPLE 9.10a The data in Table 9.4 relate the suicide rate to the population size and the divorce rate at eight different locations.

TABLE 9.4

Location	Population in Thousands	Divorce Rate per 100,000	Suicide Rate per 100,000
Akron, OH	679	30.4	11.6
Anaheim, CA	1,420	34.1	16.1
Buffalo, NY	1,349	17.2	9.3
Austin, TX	296	26.8	9.1
Chicago, IL	6,975	29.1	8.4
Columbia, SC	323	18.7	7.7
Detroit, MI	4,200	32.6	11.3
Gary, IN	633	32.5	8.4

Fit a multiple linear regression model to these data. That is, fit a model of the form

$$Y = \beta_0 + \beta_1 x_1 + \beta_2 x_2 + e$$

where Y is the suicide rate, x_1 is the population, and x_2 is the divorce rate.

SOLUTION We run Program 9.10, and results are shown in Figures 9.15, 9.16, and 9.17. Thus the estimated regression line is

$$Y = 3.5073 - .0002x_1 + .2609x_2$$

The value of β_1 indicates that the population does not play a major role in predicting the suicide rate (at least when the divorce rate is also given). Perhaps the population density, rather than the actual population, would have been more useful. ■

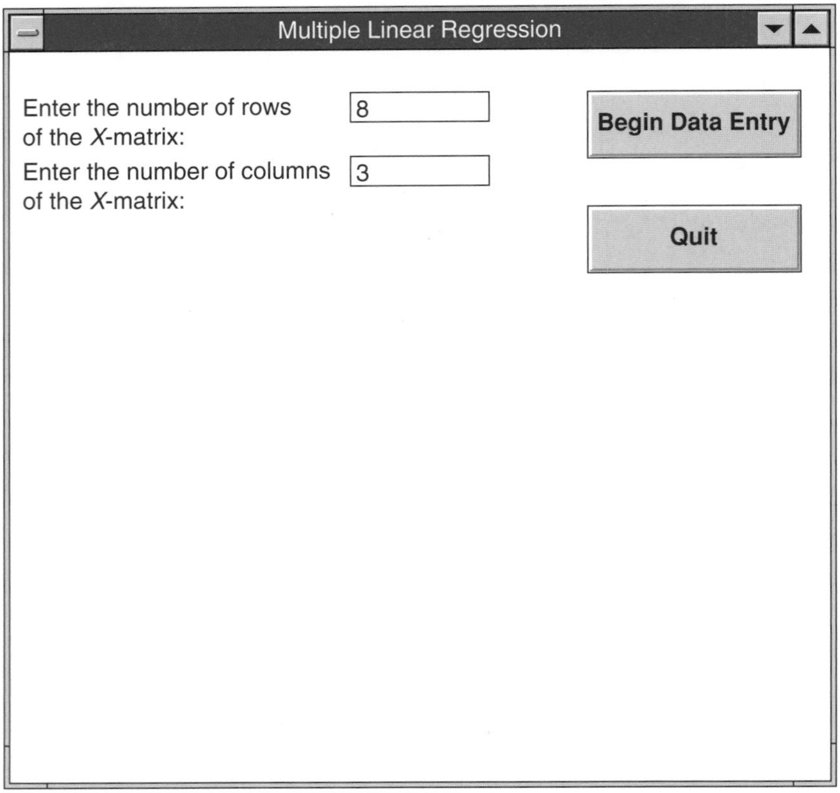

FIGURE 9.15

It follows from Equation 9.10.3 that the least squares estimators $B_0, B_1, \ldots, B_k$ — the elements of the matrix $\mathbf{B}$ — are all linear combinations of the independent normal random variables $Y_1, \ldots, Y_n$ and so will also be normally distributed. Indeed in such a situation — namely, when each member of a set of random variables can be expressed as a linear combination of independent normal random variables — we say that the set of random variables has a joint *multivariate normal distribution*.

The least squares estimators turn out to be unbiased. This can be shown as follows:

$$
\begin{aligned}
E[\mathbf{B}] &= E[(\mathbf{X}'\mathbf{X})^{-1}\mathbf{X}'\mathbf{Y}] \\
&= E[(\mathbf{X}'\mathbf{X})^{-1}\mathbf{X}'(\mathbf{X}\boldsymbol{\beta} + \mathbf{e})] \quad \text{since } \mathbf{Y} = \mathbf{X}\boldsymbol{\beta} + \mathbf{e} \\
&= E[(\mathbf{X}'\mathbf{X})^{-1}\mathbf{X}'\mathbf{X}\boldsymbol{\beta} + (\mathbf{X}'\mathbf{X})^{-1}\mathbf{X}'\mathbf{e}] \\
&= E[\boldsymbol{\beta} + (\mathbf{X}'\mathbf{X})^{-1}\mathbf{X}'\mathbf{e}] \\
&= \boldsymbol{\beta} + (\mathbf{X}'\mathbf{X})^{-1}\mathbf{X}'E[\mathbf{e}] \\
&= \boldsymbol{\beta}
\end{aligned}
$$

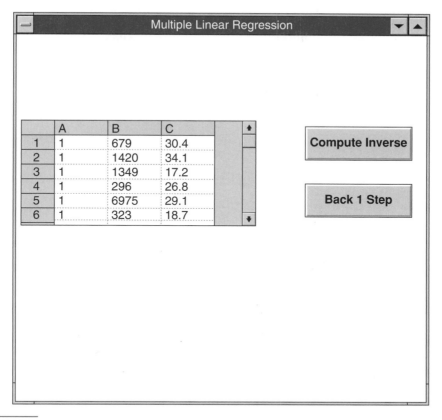

FIGURE 9.16

The variances of the least squares estimators can be obtained from the matrix $(\mathbf{X}'\mathbf{X})^{-1}$. Indeed, the values of this matrix are related to the covariances of the B_i's. Specifically, the element in the $(i + 1)$st row, $(j + 1)$st column of $(\mathbf{X}'\mathbf{X})^{-1}$ is equal to $\text{Cov}(B_i, B_j)/\sigma^2$.

To verify the preceding statement concerning $\text{Cov}(B_i, B_j)$, let

$$\mathbf{C} = (\mathbf{X}'\mathbf{X})^{-1}\mathbf{X}'$$

Since $\mathbf{X}$ is an $n \times p$ matrix and $\mathbf{X}'$ a $p \times n$ matrix, it follows that $\mathbf{X}'\mathbf{X}$ is $p \times p$, as is $(\mathbf{X}'\mathbf{X})^{-1}$, and so $\mathbf{C}$ will be a $p \times n$ matrix. Let C_{ij} denote the element in row i, column j of this matrix. Now

$$\begin{bmatrix} B_0 \\ \vdots \\ B_{i-1} \\ \vdots \\ B_k \end{bmatrix} = \mathbf{B} = \mathbf{CY} = \begin{bmatrix} C_{11} & \cdots & C_{1n} \\ \vdots & & \vdots \\ C_{i1} & \cdots & C_{in} \\ \vdots & & \vdots \\ C_{p1} & \cdots & C_{pn} \end{bmatrix} \begin{bmatrix} Y_1 \\ \vdots \\ \vdots \\ Y_n \end{bmatrix}$$

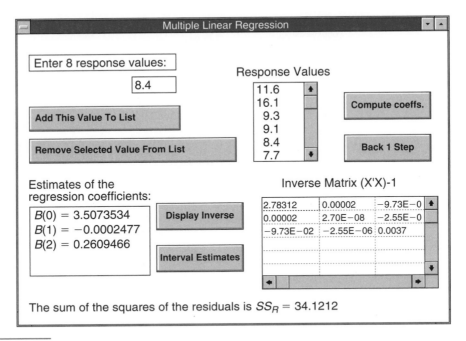

FIGURE 9.17

and so

$$B_{i-1} = \sum_{l=1}^{n} C_{il} Y_l$$

$$B_{j-1} = \sum_{r=1}^{n} C_{jr} Y_r$$

Hence

$$\mathrm{Cov}(B_{i-1}, B_{j-1}) = \mathrm{Cov}\left(\sum_{l=1}^{n} C_{il} Y_l, \sum_{r=1}^{n} C_{jr} Y_r \right)$$

$$= \sum_{r=1}^{n} \sum_{l=1}^{n} C_{il} C_{jr} \, \mathrm{Cov}(Y_l, Y_r)$$

Now Y_l and Y_r are independent when $l \neq r$, and so

$$\mathrm{Cov}(Y_l, Y_r) = \begin{cases} 0 & \text{if } l \neq r \\ \mathrm{Var}(Y_r) & \text{if } l = r \end{cases}$$

Since $\text{Var}(Y_r) = \sigma^2$, we see that

$$\text{Cov}(B_{i-1}, B_{j-1}) = \sigma^2 \sum_{r=1}^{n} C_{ir} C_{jr} \qquad (9.10.4)$$

$$= \sigma^2 (\mathbf{CC}')_{ij}$$

where $(\mathbf{CC}')_{ij}$ is the element in row i, column j of $\mathbf{CC}'$.

If we now let $\text{Cov}(\mathbf{B})$ denote the matrix of covariances — that is,

$$\text{Cov}(\mathbf{B}) = \begin{bmatrix} \text{Cov}(B_0, B_0) & \cdots & \text{Cov}(B_0, B_k) \\ \vdots & & \vdots \\ \text{Cov}(B_k, B_0) & \cdots & \text{Cov}(B_k, B_k) \end{bmatrix}$$

then it follows from Equation 9.10.4 that

$$\text{Cov}(\mathbf{B}) = \sigma^2 \mathbf{CC}' \qquad (9.10.5)$$

Now

$$\mathbf{C}' = \left((\mathbf{X}'\mathbf{X})^{-1}\mathbf{X}' \right)'$$

$$= \mathbf{X} \left((\mathbf{X}'\mathbf{X})^{-1} \right)'$$

$$= \mathbf{X}(\mathbf{X}'\mathbf{X})^{-1}$$

where the last equality follows since $(\mathbf{X}'\mathbf{X})^{-1}$ is symmetric (since $\mathbf{X}'\mathbf{X}$ is) and so is equal to its transpose. Hence

$$\mathbf{CC}' = (\mathbf{X}'\mathbf{X})^{-1}\mathbf{X}'\mathbf{X}(\mathbf{X}'\mathbf{X})^{-1}$$

$$= (\mathbf{X}'\mathbf{X})^{-1}$$

and so we can conclude from Equation 9.10.5 that

$$\text{Cov}(\mathbf{B}) = \sigma^2 (\mathbf{X}'\mathbf{X})^{-1} \qquad (9.10.6)$$

Since $\text{Cov}(B_i, B_i) = \text{Var}(B_i)$, it follows that the variances of the least squares estimators are given by σ^2 multiplied by the diagonal elements of $(\mathbf{X}'\mathbf{X})^{-1}$.

The quantity σ^2 can be estimated by using the sum of squares of the residuals. That is, if we let

$$SS_R = \sum_{i=1}^{n}(Y_i - B_0 - B_1 x_{i1} - B_2 x_{i2} - \cdots - B_k x_{ik})^2$$

then it can be shown that

$$\frac{SS_r}{\sigma^2} \sim \chi^2_{n-(k+1)}$$

and so

$$E\left[\frac{SS_R}{\sigma^2}\right] = n - k - 1$$

or

$$E[SS_R/(n - k - 1)] = \sigma^2$$

That is, $SS_R/(n - k - 1)$ is an unbiased estimator of σ^2. In addition, as in the case of simple linear regression, SS_R will be independent of the least squares estimators $B_0, B_1, \ldots, B_k$.

REMARK

If we let r_i denote the ith residual

$$r_i = Y_i - B_0 - B_1 x_{i1} - \cdots - B_k x_{ik}, \qquad i = 1, \ldots, n$$

then

$$\mathbf{r} = \mathbf{Y} - \mathbf{XB}$$

where

$$\mathbf{r} = \begin{bmatrix} r_1 \\ r_2 \\ \vdots \\ r_n \end{bmatrix}$$

Hence, we may write

$$SS_R = \sum_{i=1}^{n} r_i^2 \qquad\qquad (9.10.7)$$
$$= \mathbf{r'r}$$
$$= (\mathbf{Y} - \mathbf{XB})'(\mathbf{Y} - \mathbf{XB})$$
$$= [\mathbf{Y'} - (\mathbf{XB})'](\mathbf{Y} - \mathbf{XB})$$

$$= (\mathbf{Y'} - \mathbf{B'X'})(\mathbf{Y} - \mathbf{XB})$$
$$= \mathbf{Y'Y} - \mathbf{Y'XB} - \mathbf{B'X'Y} + \mathbf{B'X'XB}$$
$$= \mathbf{Y'Y} - \mathbf{Y'XB}$$

where the last equality follows from the normal equations

$$\mathbf{X'XB} = \mathbf{X'Y}$$

Because $\mathbf{Y'}$ is $1 \times n$, $\mathbf{X}$ is $n \times p$, and $\mathbf{B}$ is $p \times 1$, it follows that $\mathbf{Y'XB}$ is a 1×1 matrix. That is, $\mathbf{Y'XB}$ is a scalar and thus is equal to its transpose, which shows that

$$\mathbf{Y'XB} = (\mathbf{Y'XB})'$$
$$= \mathbf{B'X'Y}$$

Hence, using Equation 9.10.7 we have proven the following identity:

$$SS_R = \mathbf{Y'Y} - \mathbf{B'X'Y}$$

The foregoing is a useful computational formula for SS_R (though one must be careful of possible roundoff error when using it).

EXAMPLE 9.10b For the data of Example 9.10a, we computed that $SS_R = 34.12$. Since $n = 8, k = 2$, the estimate of σ^2 is $34.12/5 = 6.824$. ∎

EXAMPLE 9.10c The diameter of a tree at its breast height is influenced by many factors. The data in Table 9.5 relate the diameter of a particular type of eucalyptus tree to its age, average rainfall at its site, site's elevation, and the wood's mean specific gravity. (The data come from R. G. Skolmen, 1975, "Shrinkage and Specific Gravity Variation in Robusta Eucalyptus Wood Grown in Hawaii." USDA Forest Service PSW-298.)

Assuming a linear regression model of the form

$$Y = \beta_0 + \beta_1 x_1 + \beta_2 x_2 + \beta_3 x_3 + \beta_4 x_4 + e$$

where x_1 is the age, x_2 is the elevation, x_3 is the rainfall, x_4 is the specific gravity, and Y is the tree's diameter, test the hypothesis that $\beta_2 = 0$. That is, test the hypothesis that, given the other three factors, the elevation of the tree does not affect its diameter.

SOLUTION To test this hypothesis, we begin by running Program 9.10, which yields, among other things, the following:

$$(\mathbf{X'X})^{-1}_{3,3} = .379, \qquad SS_R = 19.262, \qquad B_2 = .075$$

TABLE 9.5

	Age (years)	Elevation (1,000 ft)	Rainfall (inches)	Specific Gravity	Diameter at Breast Height (inches)
1	44	1.3	250	.63	18.1
2	33	2.2	115	.59	19.6
3	33	2.2	75	.56	16.6
4	32	2.6	85	.55	16.4
5	34	2.0	100	.54	16.9
6	31	1.8	75	.59	17.0
7	33	2.2	85	.56	20.0
8	30	3.6	75	.46	16.6
9	34	1.6	225	.63	16.2
10	34	1.5	250	.60	18.5
11	33	2.2	255	.63	18.7
12	36	1.7	175	.58	19.4
13	33	2.2	75	.55	17.6
14	34	1.3	85	.57	18.3
15	37	2.6	90	.62	18.8

It now follows from Equation 9.10.6 that

$$\text{Var}(B_2) = .379\sigma^2$$

Since B_2 is normal and

$$E[B_2] = \beta_2$$

we see that

$$\frac{B_2 - \beta_2}{.616\sigma} \sim N(0, 1)$$

Replacing σ by its estimator $SS_R/10$ transforms the foregoing standard normal distribution into a t-distribution with $10(= n - k - 1)$ degrees of freedom. That is,

$$\frac{B_2 - \beta_2}{.616\sqrt{SS_R/10}} \sim t_{10}$$

Hence, if $\beta_2 = 0$ then

$$\frac{\sqrt{10/SS_R}\, B_2}{.616}$$

Since the value of the preceding statistic is $(\sqrt{10/19.262})(.075)/.616 = .088$, the p-value of the test of the hypothesis that $\beta_2 = 0$ is

$$p\text{-value} = P\{|T_{10}| > .088\}$$
$$= 2P\{T_{10} > .088\}$$
$$= .9316 \quad \text{by Program 5.8.2.A}$$

Hence, the hypothesis is accepted (and, in fact, would be accepted at any significance level less than .9316). ∎

REMARK

The quantity

$$R^2 = 1 - \frac{SS_R}{\sum_i (Y_i - \overline{Y})^2}$$

which measures the amount of reduction in the sum of squares of the residuals when using the model

$$Y = \beta_0 + \beta_1 x_1 + \cdots + \beta_n x_n + e$$

as opposed to the model

$$Y = \beta_0 + e$$

is called the *coefficient of multiple determination*.

9.10.1 PREDICTING FUTURE RESPONSES

Let us now suppose that a series of experiments is to be performed using the input levels $x_1, \ldots, x_k$. Based on our data, consisting of the prior responses $Y_1, \ldots, Y_n$, suppose we would like to estimate the mean response. Since the mean response is

$$E[Y|\mathbf{x}] = \beta_0 + \beta_1 x_1 + \cdots + \beta_k x_k$$

a point estimate of it is simply $\sum_{i=0}^{k} B_i x_i$ where $x_0 \equiv 1$.

To determine a confidence interval estimator, we need the distribution of $\sum_{i=0}^{k} B_i x_i$. Because it can be expressed as a linear combination of the independent normal random variables $Y_i, i = 1, \ldots, n$, it follows that it is also normally distributed. Its mean and variance are obtained as follows:

$$E\left[\sum_{i=0}^{k} x_i B_i\right] = \sum_{i=0}^{k} x_i E[B_i] \tag{9.10.8}$$

$$= \sum_{i=0}^{k} x_i \beta_i \quad \text{since } E[B_i] = \beta_i$$

That is, it is an unbiased estimator. Also, using the fact that the variance of a random variable is equal to the covariance between that random variable and itself, we see that

$$\text{Var}\left(\sum_{i=0}^{k} x_i B_i\right) = \text{Cov}\left(\sum_{i=0}^{k} x_i B_i, \sum_{j=0}^{k} x_j B_j\right) \tag{9.10.9}$$

$$= \sum_{i=0}^{k}\sum_{j=0}^{k} x_i x_j \text{Cov}(B_i, B_j)$$

If we let x denote the matrix

$$\mathbf{x} = \begin{bmatrix} x_0 \\ x_1 \\ \vdots \\ x_k \end{bmatrix}$$

then, recalling that $\text{Cov}(B_i, B_j)/\sigma^2$ is the element in the $(i+1)$st row and $(j+1)$st column of $(\mathbf{X'X})^{-1}$, we can express Equation 9.10.9 as

$$\text{Var}\left(\sum_{i=0}^{k} x_i B_i\right) = \mathbf{x'}(\mathbf{X'X})^{-1}\mathbf{x}\sigma^2 \tag{9.10.10}$$

Using Equations 9.10.8 and 9.10.10, we see that

$$\frac{\sum_{i=0}^{k} x_i B_i - \sum_{i=0}^{k} x_i \beta_i}{\sigma\sqrt{\mathbf{x'}(\mathbf{X'X})^{-1}\mathbf{x}}} \sim N(0, 1)$$

If we now replace σ by its estimator $\sqrt{SS_R/(n-k-1)}$ we obtain, by the usual argument, that

$$\frac{\sum_{i=0}^{k} x_i B_i - \sum_{i=0}^{k} x_i \beta_i}{\sqrt{\dfrac{SS_R}{(n-k-1)}}\sqrt{\mathbf{x'}(\mathbf{X'X})^{-1}\mathbf{x}}} \sim t_{n-k-1}$$

which gives rise to the following confidence interval estimator of $\sum_{i=0}^{k} x_i \beta_i$.

Confidence Interval Estimate of $E[Y\,|\,x] = \sum_{i=0}^{k} x_i \beta_i$, $(x_0 \equiv 1)$

A $100(1-a)$ percent confidence interval estimate of $\sum_{i=0}^{k} x_i \beta_i$ is given by

$$\sum_{i=0}^{k} x_i b_i \pm \sqrt{\frac{SS_r}{(n-k-1)}}\sqrt{\mathbf{x'}(\mathbf{X'X})^{-1}\mathbf{x}}\ \ t_{a/2, n-k-1}$$

TABLE 9.6

Hardness	Copper Content	Annealing Temperature (units of 1,000°F)
79.2	.02	1.05
64.0	.03	1.20
55.7	.03	1.25
56.3	.04	1.30
58.6	.10	1.30
84.3	.15	1.00
70.4	.15	1.10
61.3	.09	1.20
51.3	.13	1.40
49.8	.09	1.40

where $b_0, \ldots, b_k$ are the values of the least squares estimators $B_0, B_1, \ldots, B_k$, and ss_r is the value of SS_R.

EXAMPLE 9.10d A steel company is planning to produce cold reduced sheet steel consisting of .15 percent copper at an annealing temperature of 1,150 (degrees F), and is interested in estimating the average (Rockwell 30-T) hardness of a sheet. To determine this, they have collected the data shown in Table 9.6 on 10 different specimens of sheet steel having different copper contents and annealing temperatures. Estimate the average hardness and determine an interval in which it will lie with 95 percent confidence.

SOLUTION To solve this, we first run Program 9.10, which gives the results shown in Figures 9.18, 9.19, and 9.20.

Hence, a point estimate of the expected hardness of sheets containing .15 percent copper at an annealing temperature of 1,150 is 69.862. In addition, since $t_{.025,7} = 2.365$, a 95 percent confidence interval for this value is

$$69.862 \pm 4.083 \quad \blacksquare$$

When it is only a single experiment that is going to be performed at the input levels $x_1, \ldots, x_k$, we are usually more concerned with predicting the actual response than its mean value. That is, we are interested in utilizing our data set $Y_1, \ldots, Y_n$ to predict

$$Y(\mathbf{x}) = \sum_{i=0}^{k} \beta_i x_i + e, \qquad \text{where } x_0 = 1$$

A point prediction is given by $\sum_{i=0}^{k} B_i x_i$ where B_i is the least squares estimator of β_i based on the set of prior responses $Y_1, \ldots, Y_n, i = 1, \ldots, k$.

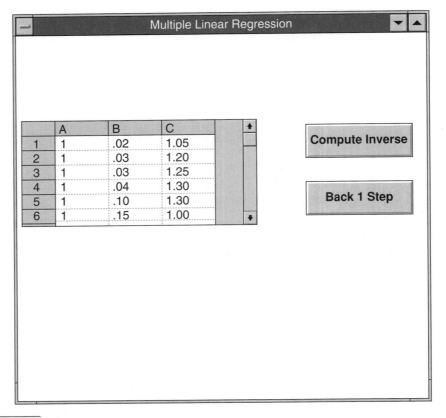

FIGURE 9.18

To determine a prediction interval for $Y(\mathbf{x})$, note first that since $B_0, \ldots, B_k$ are based on prior responses, it follows that they are independent of $Y(\mathbf{x})$. Hence, it follows that $Y(\mathbf{x}) - \sum_{i=0}^{k} B_i x_i$ is normal with mean 0 and variance given by

$$\mathrm{Var}\left[Y(\mathbf{x}) - \sum_{i=0}^{k} B_i x_i \right] = \mathrm{Var}[Y(\mathbf{x})] + \mathrm{Var}\left(\sum_{i=0}^{k} B_i x_i \right) \qquad \text{by independence}$$

$$= \sigma^2 + \sigma^2 \mathbf{x}'(\mathbf{X}'\mathbf{X})^{-1}\mathbf{x} \qquad \text{from Equation 9.10.10}$$

and so

$$\frac{Y(\mathbf{x}) - \sum_{i=0}^{k} B_i x_i}{\sigma \sqrt{1 + \mathbf{x}'(\mathbf{X}'\mathbf{X})^{-1}\mathbf{x}}} \sim N(0, 1)$$

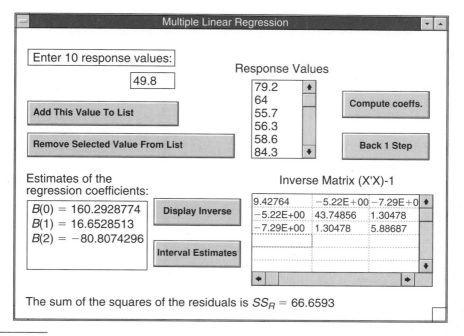

FIGURE 9.19

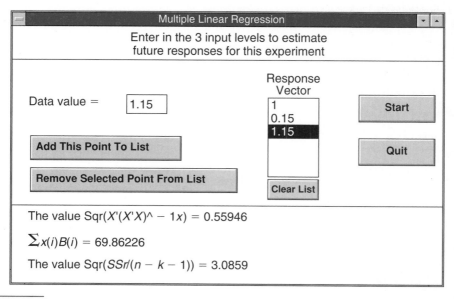

FIGURE 9.20

which yields, upon replacing σ by its estimator, that

$$\frac{Y(\mathbf{x}) - \sum\limits_{i=0}^{k} B_i x_i}{\sqrt{\dfrac{SS_R}{(n-k-1)}} \sqrt{1 + \mathbf{x}'(\mathbf{X}'\mathbf{X})^{-1}\mathbf{x}}} \sim t_{n-k-1}$$

We thus have:

Prediction Interval for $Y(\mathbf{x})$

With $100(1-a)$ percent confidence $Y(\mathbf{x})$ will lie between

$$\sum_{i=0}^{k} x_i b_i \pm \sqrt{\frac{ss_r}{(n-k-1)}} \sqrt{1 + \mathbf{x}'(\mathbf{X}'\mathbf{X})^{-1}\mathbf{x}} \;\; t_{a/2, n-k-1}$$

where $b_0, \ldots, b_k$ are the values of the least squares estimators $B_0, B_1, \ldots, B_k$, and ss_r is the value of SS_R.

EXAMPLE 9.10e   If in Example 9.10d we were interested in determining an interval in which a single steel sheet, produced with a carbon content of .15 percent and at an annealing temperature of 1,150°F, would lie, then the midpoint of the prediction interval would be as given before. However, the half-length of this prediction interval would differ from the confidence interval for the mean value by the factor $\sqrt{1.313}/\sqrt{.313} \approx 2.048$. Consequently, the 95 percent prediction interval is

$$69.862 \pm (4.083)(2.048) = 69.862 \pm 8.363 \quad \blacksquare$$

9.10.2 Dummy Variables for Categorical Data

Suppose that in determining an appropriate multiple regression model for predicting a person's blood cholesterol level a researcher has decided on the following five independent variables:

1. number of pounds overweight

2. number of pounds underweight

3. average number of hours of exercise per week

4. average number of calories due to saturated fats eaten daily

5. whether a smoker or not

Whereas each of the first four variables takes on values in some interval, the final variable is a categorical variable that indicates whether the person under consideration has or does not have a certain characteristic (which, in this case, is the characteristic of being a smoker). To determine which category the person belongs to, let

$$x_5 = \begin{cases} 1, & \text{if person is a smoker} \\ 0, & \text{if person is not a smoker} \end{cases}$$

The researcher can now try to fit the multiple regression model

$$Y = \beta_0 + \beta_1 x_1 + \beta_2 x_2 + \beta_3 x_3 + \beta_4 x_4 + \beta_5 x_5 + e$$

where x_1 is the number of pounds the individual is overweight, x_2 is the number of pounds the individual is underweight, x_3 is the average number of hours the individual exercises per week, x_4 is the average number of calories due to saturated fats that is eaten daily, x_5 is as above, and Y is the individual's cholesterol level. The variable x_5 is called a *dummy variable*, as its only purpose is to indicate whether or not the Y value is determined from data having a particular characteristic.

One might wonder at this point why a dummy variable is used rather than just running separate multiple regressions for smokers and for nonsmokers. The main reason for using a dummy variable is that we can use all the data in a single regression, thus yielding more precise estimates than if we broke the data into two parts (one for smokers and the other for nonsmokers) and then used the divided data to run separate regressions. However, it is important to understand what is being assumed when dummy variables are being used. Namely, we are assuming that if Y_s stands for the cholesterol level of a smoker, and Y_n, the cholesterol level of a nonsmoker, then for specified values of x_1, x_2, x_3, and x_4

$$E[Y_s] = \beta_0 + \beta_5 + \beta_1 x_1 + \beta_2 x_2 + \beta_3 x_3 + \beta_4 x_4$$

and

$$E[Y_n] = \beta_0 + \beta_1 x_1 + \beta_2 x_2 + \beta_3 x_3 + \beta_4 x_4$$

In other words, in using the model with a dummy variable we are assuming that if a smoker and nonsmoker had the same values for the four quantitative variables x_1, x_2, x_3, x_4 then the difference between their mean cholesterol levels would always be a constant, no matter what are the values of x_1, x_2, x_3, x_4. Thus, for instance, the dummy variable model assumes that the amount that one is overweight has the same effect on raising the expected cholesterol level on a smoker as it does on a nonsmoker. Because this might seem like a questionable assumption, it is typically preferable when the data set is large enough to use two regression models rather than combining into one model by the use of a dummy variable.

In situations where there are multiple qualitative characteristics that the researcher feels are relevant it might be necessary to utilize dummy variables, for otherwise the data set may become too fragmented to yield reliable estimates of the regression parameters. So, for instance, if the researcher felt that the sex of the person was also a relevant factor, then the researcher could utilize a multiple regression model having two dummy variables: namely x_5 and

$$x_6 = \begin{cases} 1, & \text{if person is a male} \\ 0, & \text{if person is a female} \end{cases}$$

9.11 LOGISTIC REGRESSION MODELS FOR BINARY OUTPUT DATA

In this section we consider experiments that result in either a success or a failure. We will suppose that these experiments can be performed at various levels, and that an experiment performed at level x will result in a success with probability $p(x)$, $-\infty < x < \infty$. If $p(x)$ is of the form

$$p(x) = \frac{e^{a+bx}}{1 + e^{a+bx}}$$

then the experiments are said to come from a *logistic regression* model and $p(x)$ is called the *logistics regression function*. If $b > 0$, then $p(x) = 1/[e^{-(a+bx)} + 1]$ is an increasing function that converges to 1 as $x \to \infty$; if $b < 0$, then $p(x)$ is a decreasing function that converges to 0 as $x \to \infty$. (When $b = 0$, $p(x)$ is constant.) Plots of logistics regression functions are given in Figure 9.21. Notice the s-shape of these curves.

Writing $p(x) = 1 - [1/(1 + e^{a+bx})]$ and differentiating give that

$$\frac{\partial}{\partial x} p(x) = \frac{b e^{a+bx}}{(1 + e^{a+bx})^2} = b\, p(x)[1 - p(x)]$$

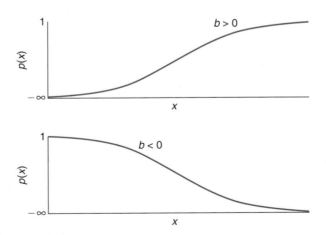

FIGURE 9.21   *Logistic regression functions.*

Thus the rate of change of $p(x)$ depends on x and is largest at those values of x for which $p(x)$ is near .5. For instance, at the value x such that $p(x) = .5$, the rate of change is $\frac{\partial}{\partial x}p(x) = .25b$, whereas at that value x for which $p(x) = .8$ the rate of change is $.16b$.

If we let $o(x)$ be the odds for success when the experiment is run at level x, then

$$o(x) = \frac{p(x)}{1 - p(x)} = e^{a+bx}$$

Thus, when $b > 0$, the odds increase exponentially in the input level x; when $b < 0$, the odds decrease exponentially in the input level x. Taking logs of the preceding shows that the log odds, called the *logit*, is a linear function:

$$\log[o(x)] = a + bx$$

The parameters a and b of the logistic regression function are assumed to be unknown and need to be estimated. This can be accomplished by using the maximum likelihood approach. That is, suppose that the experiment is to be performed at levels $x_1, \ldots, x_k$. Let Y_i be the result (either 1 if a success or 0 if a failure) of the experiment when performed at level x_i. Then, using the Bernoulli density function (that is, the binomial density for a single trial), gives

$$P\{Y_i = y_i\} = [p(x_i)]^{y_i}[1 - p(x_i)]^{1-y_i} = \left(\frac{e^{a+bx_i}}{1 + e^{a+bx_i}}\right)^{y_i}\left(\frac{1}{1 + e^{a+bx_i}}\right)^{1-y_i}, \quad y_i = 0, 1$$

Thus, the probability that the experiment at level x_i results in outcome y_i, for all $i = 1, \ldots, k$, is

$$P\{Y_i = y_i, i = 1, \ldots, k\} = \prod_i \left(\frac{e^{a+bx_i}}{1 + e^{a+bx_i}}\right)^{y_i}\left(\frac{1}{1 + e^{a+bx_i}}\right)^{1-y_i}$$

$$= \prod_i \frac{\left(e^{a+bx_i}\right)^{y_i}}{1 + e^{a+bx_i}}$$

Taking logarithms gives that

$$\log\left(P\{Y_i = y_i, i = 1, \ldots, k\}\right) = \sum_{i=1}^{k} y_i(a + bx_i) - \sum_{i=1}^{k} \log\left(1 + e^{a+bx_i}\right)$$

The maximum likelihood estimates can now be obtained by numerically finding the values of a and b that maximize the preceding likelihood. However, because the likelihood is nonlinear this requires an iterative approach; consequently, one typically resorts to specialized software to obtain the estimates.

Whereas the logistic regression model is the most frequently used model when the response data are binary, other models are often employed. For instance in situations where it is reasonable to suppose that $p(x)$, the probability of a positive response when the input level is x, is an increasing function of x, it is often supposed that $p(x)$ has the form of a specified probability distribution function. Indeed, when $b > 0$, the logistic regression model is of this form because $p(x)$ is equal to the distribution function of a logistic random variable (Section 5.9) with parameters $\mu = -a/b$, $\nu = 1/b$. Another model of this type is the *probit* model, which supposes that for some constants, $\alpha, \beta > 0$

$$p(x) = \Phi(\alpha + \beta x) = \frac{1}{\sqrt{2\pi}} \int_{-\infty}^{\alpha+\beta x} e^{-y^2/2} \, dy$$

In other words $p(x)$ is equal to the probability that a standard normal random variable is less than $\alpha + \beta x$.

EXAMPLE 9.11a   A common assumption for whether an animal becomes sick when exposed to a chemical at dosage level x is to assume a threshold model, which supposes that each animal has a random threshold and will become ill if the dosage level exceeds that threshold. The exponential distribution has sometimes been used as the threshold distribution. For instance, a model considered in Freedman and Zeisel ("From Mouse to Man: The Quantitative Assessment of Cancer Risks," *Statistical Science*, 1988, **3**, 1, 3–56) supposes that a mouse exposed to x units of DDT (measured in ppm) will contract cancer of the liver with probability

$$p(x) = 1 - e^{-ax}, \quad x > 0$$

Because of the lack of memory of the exponential distribution, this is equivalent to assuming that if the mouse who is still healthy after receiving a (partial) dosage of level x is as good as it was before receiving any dosage.

It was reported in Freedman and Zeisel that 84 of 111 mice exposed to DDT at a level of 250 ppm developed cancer. Therefore, α can be estimated from

$$1 - e^{-250\hat{\alpha}} = \frac{84}{111}$$

or

$$\hat{\alpha} = -\frac{\log(27/111)}{250} = .005655 \quad \blacksquare$$

Problems

1. The following data relate x, the moisture of a wet mix of a certain product, to Y, the density of the finished product.

x_i	Y_i
5	7.4
6	9.3
7	10.6
10	15.4
12	18.1
15	22.2
18	24.1
20	24.8

   (a) Draw a scatter diagram.
   (b) Fit a linear curve to the data.

2. The following data relate the number of units of a good that were ordered as a function of the price of the good at six different locations.

Number ordered	88	112	123	136	158	172
Price	50	40	35	30	20	15

   How many units do you think would be ordered if the price were 25?

3. The corrosion of a certain metallic substance has been studied in dry oxygen at 500 degrees centigrade. In this experiment, the gain in weight after various periods of exposure was used as a measure of the amount of oxygen that had reacted with the sample. Here are the data:

Hours	Percent Gain
1.0	.02
2.0	.03
2.5	.035
3.0	.042
3.5	.05
4.0	.054

(a) Plot a scatter diagram.
(b) Fit a linear relation.
(c) Predict the percent weight gain when the metal is exposed for 3.2 hours.

4. The following data indicate the relationship between x, the specific gravity of a wood sample, and Y, its maximum crushing strength in compression parallel to the grain.

x_i	y_i(psi)	x_i	y_i(psi)
.41	1,850	.39	1,760
.46	2,620	.41	2,500
.44	2,340	.44	2,750
.47	2,690	.43	2,730
.42	2,160	.44	3,120

(a) Plot a scatter diagram. Does a linear relationship seem reasonable?
(b) Estimate the regression coefficients.
(c) Predict the maximum crushing strength of a wood sample whose specific gravity is .43.

5. The following data indicate the gain in reading speed versus the number of weeks in the program of 10 students in a speed-reading program.

Number of Weeks	Speed Gain (wds/min)
2	21
3	42
8	102
11	130
4	52
5	57
9	105
7	85
5	62
7	90

(a) Plot a scatter diagram to see if a linear relationship is indicated.
(b) Find the least squares estimates of the regression coefficients.
(c) Estimate the expected gain of a student who plans to take the program for 7 weeks.

6. Infrared spectroscopy is often used to determine the natural rubber content of mixtures of natural and synthetic rubber. For mixtures of known percentages, the infrared spectroscopy gave the following readings:

Percentage	0	20	40	60	80	100
Reading	.734	.885	1.050	1.191	1.314	1.432

If a new mixture gives an infrared spectroscopy reading of 1.15, estimate its percentage of natural rubber.

7. The following table gives the 1996 SAT mean math and verbal scores in each state and the District of Columbia, along with the percentage of the states' graduating high school students that took the examination. Use data relating to the first 20 locations listed (Alabama to Maine) to develop a prediction of the mean student mathematics score in terms of the percentage of students that take the examination. Then compare your predicted values for the next 5 states (based on the percentage taking the exam in these states) with the actual mean math scores.

SAT Mean Scores by State, 1996 (recentered scale)

	1996		% Graduates Taking SAT
	Verbal	Math	
Alabama	565	558	8
Alaska	521	513	47
Arizona	525	521	28
Arkansas	566	550	6
California	495	511	45
Colorado	536	538	30
Connecticut	507	504	79
Delaware	508	495	66
Dist. of Columbia	489	473	50
Florida	498	496	48
Georgia	484	477	63
Hawaii	485	510	54
Idaho	543	536	15
Illinois	564	575	14
Indiana	494	494	57
Iowa	590	600	5

(continued)

	1996		% Graduates Taking
	Verbal	Math	SAT
Kansas.................	579	571	9
Kentucky	549	544	12
Louisiana	559	550	9
Maine.................	504	498	68
Maryland	507	504	64
Massachusetts	507	504	80
Michigan	557	565	11
Minnesota	582	593	9
Mississippi	569	557	4
Missouri	570	569	9
Montana...............	546	547	21
Nebraska...............	567	568	9
Nevada	508	507	31
New Hampshire	520	514	70
New Jersey	498	505	69
New Mexico............	554	548	12
New York	497	499	73
North Carolina	490	486	59
North Dakota	596	599	5
Ohio	536	535	24
Oklahoma	566	557	8
Oregon	523	521	50
Pennsylvania	498	492	71
Rhode Island	501	491	69
South Carolina	480	474	57
South Dakota	574	566	5
Tennessee	563	552	14
Texas..................	495	500	48
Utah	583	575	4
Vermont	506	500	70
Virginia................	507	496	68
Washington	519	519	47
West Virginia...........	526	506	17
Wisconsin	577	586	8
Wyoming...............	544	544	11
National Average.........	505	508	41

Source: The College Board.

8. Verify Equation 9.3.3, which states that

$$\text{Var}(A) = \frac{\sigma^2 \sum\limits_{i=1}^{n} x_i^2}{n \sum\limits_{i=1}^{n} (x_i - \bar{x})^2}$$

9. In Problem 4,

(a) Estimate the variance of an individual response.
(b) Determine a 90 percent confidence interval for the variance.

10. Verify that

$$SS_R = \frac{S_{xx}S_{YY} - S_{xY}^2}{S_{xx}}$$

11. The following table relates the number of sunspots that appeared each year from 1970 to 1983 to the number of auto accident deaths during that year. Test the hypothesis that the number of auto deaths is not affected by the number of sunspots. (The sunspot data are from Jastrow and Thompson, *Fundamentals and Frontiers of Astronomy,* and the auto death data are from *General Statistics of the U.S. 1985.*)

Year	Sunspots	Auto Accident Deaths (1,000s)
70	165	54.6
71	89	53.3
72	55	56.3
73	34	49.6
74	9	47.1
75	30	45.9
76	59	48.5
77	83	50.1
78	109	52.4
79	127	52.5
80	153	53.2
81	112	51.4
82	80	46
83	45	44.6

12. The following data set presents the heights of 12 male law school classmates whose law school examination scores were roughly equal. It also gives their first year

salaries. Each of them went into corporate law. The height is in inches and the salary in units of $1,000.

Height	Salary
64	91
65	94
66	88
67	103
69	77
70	96
72	105
72	88
74	122
74	102
75	90
76	114

(a) Do the above data establish the hypothesis that a lawyer's salary is related to his height? Use the 5 percent level of significance.

(b) What was the null hypothesis in part (a)?

13. Suppose in the simple linear regression model

$$Y = \alpha + \beta x + e$$

that $0 < \beta < 1$.

(a) Show that if $x < \alpha/(1 - \beta)$, then

$$x < E[Y] < \frac{\alpha}{1 - \beta}$$

(b) Show that if $x > \alpha/(1 - \beta)$, then

$$x > E[Y] > \frac{\alpha}{1 - \beta}$$

and conclude that $E[Y]$ is always between x and $\alpha/(1 - \beta)$.

14. A study has shown that a good model for the relationship between X and Y, the first and second year batting averages of a randomly chosen major league baseball player, is given by the equation

$$Y = .159 + .4X + e$$

where e is a normal random variable with mean 0. That is, the model is a simple linear regression with a regression toward the mean.

(a) If a player's batting average is .200 in his first year, what would you predict for the second year?

(b) If a player's batting average is .265 in his first year, what would you predict for the second year?

(c) If a player's batting average is .310 in his first year, what would you predict for the second year?

15. Experienced flight instructors have claimed that praise for an exceptionally fine landing is typically followed by a poorer landing on the next attempt, whereas criticism of a faulty landing is typically followed by an improved landing. Should we thus conclude that verbal praise tends to lower performance levels, whereas verbal criticism tends to raise them? Or is some other explanation possible?

16. Verify Equation 9.4.3.

17. The following data represent the relationship between the number of alignment errors and the number of missing rivets for 10 different aircraft.

Number of Missing Rivets $= x$	Number of Alignment Errors $= y$
13	7
15	7
10	5
22	12
30	15
7	2
25	13
16	9
20	11
15	8

(a) Plot a scatter diagram.

(b) Estimate the regression coefficients.

(c) Test the hypothesis that $\alpha = 1$.

(d) Estimate the expected number of alignment errors of a plane having 24 missing rivets.

(e) Compute a 90 percent confidence interval estimate for the quantity in (d).

18. The following are the average scores on the mathematics part of the Scholastic Aptitude Test (SAT) for some of the years from 1994 to 2009.

Year	SAT Score
1994	504
1996	508
1998	512
2000	514
2002	516
2004	518
2005	520
2007	515
2009	515

Assuming a simple linear regression model, predict the average scores in the years 1997, 2006, 2008, and 2010.

19. **(a)** Draw a scatter diagram of cigarette consumption versus death rate from bladder cancer.
    (b) Does the diagram indicate the possibility of a linear relationship?
    (c) Find the best linear fit.
    (d) If next year's average cigarette consumption is 2,500, what is your prediction of the death rate from bladder cancer?

20. **(a)** Draw a scatter diagram relating cigarette use and death rates from lung cancer.
    (b) Estimate the regression parameters α and β.
    (c) Test at the .05 level of significance the hypothesis that cigarette consumption does not affect the death rate from lung cancer.
    (d) What is the p-value of the test in part (c)?

21. **(a)** Draw a scatter diagram of cigarette use versus death rate from kidney cancer.
    (b) Estimate the regression line.
    (c) What is the p-value in the test that the slope of the regression line is 0?
    (d) Determine a 90 percent confidence interval for the mean death rate from kidney cancer in a state whose citizens smoke an average of 3,400 cigarettes per year.

22. **(a)** Draw a scatter diagram of cigarettes smoked versus death rate from leukemia.
    (b) Estimate the regression coefficients.
    (c) Test the hypothesis that there is no regression of the death rate from leukemia on the number of cigarettes used. That is, test that $\beta = 0$.
    (d) Determine a 90 percent prediction interval for the leukemia death rate in a state whose citizens smoke an average of 2,500 cigarettes.

23. **(a)** Estimate the variances in Problems 19 through 22.
    (b) Determine a 95 percent confidence interval for the variance in the data relating to lung cancer.

(c) Break up the lung cancer data into two parts — the first corresponding to states whose average cigarette consumption is less than 2,300, and the second greater. Assume a linear regression model for both sets of data. How would you test the hypothesis that the variance of a response is the same for both sets?

(d) Do the test in part (c) at the .05 level of significance.

24. Plot the standardized residuals from the data of Problem 1. What does the plot indicate about the assumptions of the linear regression model?

25. It is difficult and time consuming to measure directly the amount of protein in a liver sample. As a result, medical laboratories often make use of the fact that the amount of protein is related to the amount of light that would be absorbed by the sample. As a result, a spectrometer that emits light is shined on a solution that contains the liver sample and the amount of light absorbed is then used to estimate the amount of protein.

The above procedure was tried on five samples having known amounts of protein, with the following data resulting.

Light Absorbed	Amount of Protein (mg)
.44	2
.82	16
1.20	30
1.61	46
1.83	55

(a) Determine the coefficient of determination.

(b) Does this appear to be a reasonable way of estimating the amount of protein in a liver sample?

(c) What is the estimate of the amount of protein when the light absorbed is 1.5?

(d) Determine a prediction interval, in which we can have 90 percent confidence, for the quantity in part (c).

26. The determination of the shear strength of spot welds is relatively difficult, whereas measuring the weld diameter of spot welds is relatively simple. As a result, it would be advantageous if shear strength could be predicted from a measurement of weld diameter. The data are as follows:

(a) Draw a scatter diagram.

(b) Find the least squares estimates of the regression coefficients.

(c) Test the hypothesis that the slope of the regression line is equal to 1 at the .05 level of significance.

Shear Strength (psi)	Weld Diameter (.0001 in.)
370	400
780	800
1,210	1,250
1,560	1,600
1,980	2,000
2,450	2,500
3,070	3,100
3,550	3,600
3,940	4,000
3,950	4,000

(d) Estimate the expected value of shear strength when the weld diameter is .2500.

(e) Find a prediction interval such that, with 95 percent confidence, the value of shear strength corresponding to a weld diameter of .2250 inch will be contained in it.

(f) Plot the standardized residuals.

(g) Does the plot in part (f) support the assumptions of the model?

27. The following are the ages and weights of a random sample of 10 high school male students

Age	Weight
14	129
16	173
18	188
15	121
17	190
16	166
14	133
16	155
15	152
14	115

Assuming a simple linear regression model, give an interval that, with 95 percent confidence, will contain the average weight of all 17 year old male high school students.

28. Glass plays a key role in criminal investigations, because criminal activity often results in the breakage of windows and other glass objects. Since glass fragments often lodge in the clothing of the criminal, it is of great importance to be able

to identify such fragments as originating at the scene of the crime. Two physical properties of glass that are useful for identification purposes are its refractive index, which is relatively easy to measure, and its density, which is much more difficult to measure. The exact measurement of density is, however, greatly facilitated if one has a good estimate of this value before setting up the laboratory experiment needed to determine it exactly. Thus, it would be quite useful if one could use the refractive index of a glass fragment to estimate its density.

The following data relate the refractive index to the density for 18 pieces of glass.

Refractive Index	Density	Refractive Index	Density
1.5139	2.4801	1.5161	2.4843
1.5153	2.4819	1.5165	2.4858
1.5155	2.4791	1.5178	2.4950
1.5155	2.4796	1.5181	2.4922
1.5156	2.4773	1.5191	2.5035
1.5157	2.4811	1.5227	2.5086
1.5158	2.4765	1.5227	2.5117
1.5159	2.4781	1.5232	2.5146
1.5160	2.4909	1.5253	2.5187

(a) Predict the density of a piece of glass with a refractive index 1.52.
(b) Determine an interval that, with 95 percent confidence, will contain the density of the glass in part (a).

29. The regression model

$$Y = \beta x + e, \qquad e \sim N(0, \sigma^2)$$

is called regression through the origin since it presupposes that the expected response corresponding to the input level $x = 0$ is equal to 0. Suppose that $(x_i, Y_i), i = 1, \ldots, n$ is a data set from this model.

(a) Determine the least squares estimator B of β.
(b) What is the distribution of B?
(c) Define SS_R and give its distribution.
(d) Derive a test of $H_0 : \beta = \beta_0$ versus $H_1 : \beta \neq \beta_0$.
(e) Determine a $100(1 - \alpha)$ percent prediction interval for $Y(x_0)$, the response at input level x_0.

30. The following are the body mass index (BMI) and the systolic blood pressure of eight randomly chosen men who do not take any blood pressure medication.

BMI	Systolic Blood Pressure
20.3	116
22.0	110
26.4	131
28.2	136
31.0	144
32.6	138
17.6	122
19.4	115

Give an interval that, with 95 percent confidence, will include the systolic blood pressure of a man whose BMI is 26.0.

31. The weight and systolic blood pressure of randomly selected males in age group 25 to 30 are shown in the following table.

Subject	Weight	Systolic BP	Subject	Weight	Systolic BP
1	165	130	11	172	153
2	167	133	12	159	128
3	180	150	13	168	132
4	155	128	14	174	149
5	212	151	15	183	158
6	175	146	16	215	150
7	190	150	17	195	163
8	210	140	18	180	156
9	200	148	19	143	124
10	149	125	20	240	170

(a) Estimate the regression coefficients.
(b) Do the data support the claim that systolic blood pressure does not depend on an individual's weight?
(c) If a large number of males weighing 182 pounds have their blood pressures taken, determine an interval that, with 95 percent confidence, will contain their average blood pressure.
(d) Analyze the standardized residuals.
(e) Determine the sample correlation coefficient.

32. It has been determined that the relation between stress (S) and the number of cycles to failure (N) for a particular type of alloy is given by

$$S = \frac{A}{N^m}$$

where A and m are unknown constants. An experiment is run yielding the following data.

Stress (thousand psi)	N (million cycles to failure)
55.0	.223
50.5	.925
43.5	6.75
42.5	18.1
42.0	29.1
41.0	50.5
35.7	126
34.5	215
33.0	445
32.0	420

Estimate A and m.

33. In 1957 the Dutch industrial engineer J. R. DeJong proposed the following model for the time it takes to perform a simple manual task as a function of the number of times the task has been practiced:

$$T \approx ts^{-n}$$

where T is the time, n is the number of times the task has been practiced, and t and s are parameters depending on the task and individual. Estimate t and s for the following data set.

T	22.4	21.3	19.7	15.6	15.2	13.9	13.7
n	0	1	2	3	4	5	6

34. The chlorine residual in a swimming pool at various times after being cleaned is as given:

Time (hr)	Chlorine Residual (pt/million)
2	1.8
4	1.5
6	1.45
8	1.42
10	1.38
12	1.36

Fit a curve of the form

$$Y \approx ae^{-bx}$$

What would you predict for the chlorine residual 15 hours after a cleaning?

35. The proportion of a given heat rise that has dissipated a time t after the source is cut off is of the form

$$P = 1 - e^{-\alpha t}$$

for some unknown constant α. Given the data

| P | .07 | .21 | .32 | .38 | .40 | .45 | .51 |
| t | .1 | .2 | .3 | .4 | .5 | .6 | .7 |

estimate the value of α. Estimate the value of t at which half of the heat rise is dissipated.

36. The following data represent the bacterial count of five individuals at different times after being inoculated by a vaccine consisting of the bacteria.

Days Since Inoculation	Bacterial Count
3	121,000
6	134,000
7	147,000
8	210,000
9	330,000

(a) Fit a curve.
(b) Estimate the bacteria count of a new patient after 8 days.

37. The following data yield the amount of hydrogen present (in parts per million) in core drillings of fixed size at the following distances (in feet) from the base of a vacuum-cast ingot.

Distance	1	2	3	4	5	6	7	8	9	10
Amount	1.28	1.50	1.12	.94	.82	.75	.60	.72	.95	1.20

(a) Draw a scatter diagram.
(b) Fit a curve of the form

$$Y = \alpha + \beta x + \gamma x^2 + e$$

to the data.

38. A new drug was tested on mice to determine its effectiveness in reducing cancerous tumors. Tests were run on 10 mice, each having a tumor of size 4 grams, by varying the amount of the drug used and then determining the resulting reduction in the weight of the tumor. The data were as follows:

Coded Amount of Drug	Tumor Weight Reduction
1	.50
2	.90
3	1.20
4	1.35
5	1.50
6	1.60
7	1.53
8	1.38
9	1.21
10	.65

Estimate the maximum expected tumor reduction and the amount of the drug that attains it by fitting a quadratic regression equation of the form

$$Y = \beta_0 + \beta_1 x + \beta_2 x^2 + e$$

39. The following data represent the relation between the number of cans damaged in a boxcar shipment of cans and the speed of the boxcar at impact.

Speed	Number of Cans Damaged
3	54
3	62
3	65
5	94
5	122
5	84
6	142
7	139
7	184
8	254

(a) Analyze as a simple linear regression model.
(b) Plot the standardized residuals.
(c) Do the results of part (b) indicate any flaw in the model?
(d) If the answer to part (c) is yes, suggest a better model and estimate all resulting parameters.

40. Redo Problem 5 under the assumption that the variance of the gain in reading speed is proportional to the number of weeks in the program.

41. The following data relate the proportions of coal miners who exhibit symptoms of pneumoconiosis to the number of years of working in coal mines.

Years Working	Proportion Having Penumoconiosis
5	0
10	.0090
15	.0185
20	.0672
25	.1542
30	.1720
35	.1840
40	.2105
45	.3570
50	.4545

Estimate the probability that a coal miner who has worked for 42 years will have pneumoconiosis.

42. The following data set refers to Example 9.8c.

Number of Cars (Daily)	Number of Accidents (Monthly)
2,000	15
2,300	27
2,500	20
2,600	21
2,800	31
3,000	16
3,100	22
3,400	23
3,700	40
3,800	39
4,000	27
4,600	43
4,800	53

(a) Estimate the number of accidents in a month when the number of cars using the highway is 3,500.

(b) Use the model

$$\sqrt{Y} = \alpha + \beta x + e$$

and redo part (a).

43. The peak discharge of a river is an important parameter for many engineering design problems. Estimates of this parameter can be obtained by relating it to the watershed area (x_1) and watershed slope (x_2). Estimate the relationship based on the following data.

x_1 (m^2)	x_2 (ft/ft)	Peak Discharge (ft^3/sec)
36	.005	50
37	.040	40
45	.004	45
87	.002	110
450	.004	490
550	.001	400
1,200	.002	650
4,000	.0005	1,550

44. The sediment load in a stream is related to the size of the contributing drainage area (x_1) and the average stream discharge (x_2). Estimate this relationship using the following data.

Area ($\times 10^3$ mi^2)	Discharge (ft^3/sec)	Sediment Yield (Millions of tons/yr)
8	65	1.8
19	625	6.4
31	1,450	3.3
16	2,400	1.4
41	6,700	10.8
24	8,500	15.0
3	1,550	1.7
3	3,500	.8
3	4,300	.4
7	12,100	1.6

45. Fit a multiple linear regression equation to the following data set.

x_1	x_2	x_3	x_4	y
1	11	16	4	275
2	10	9	3	183
3	9	4	2	140
4	8	1	1	82
5	7	2	1	97
6	6	1	−1	122
7	5	4	−2	146
8	4	9	−3	246
9	3	16	−4	359
10	2	25	−5	482

46. The following data refer to Stanford heart transplants. It relates the survival time of patients that have received heart transplants to their age when the transplant occurred and to a so-called mismatch score that is supposed to be an indicator of how well the transplanted heart should fit the recipient.

Survival Time (in days)	Mismatch Score	Age
624	1.32	51.0
46	.61	42.5
64	1.89	54.6
1,350	.87	54.1
280	1.12	49.5
10	2.76	55.3
1,024	1.13	43.4
39	1.38	42.8
730	.96	58.4
136	1.62	52.0
836	1.58	45.0
60	.69	64.5

(a) Letting the dependent variable be the logarithm of the survival time, fit a regression on the independent variable's mismatch score and age.

(b) Estimate the variance of the error term.

47. (a) Fit a multiple linear regression equation to the following data set.

(b) Test the hypothesis that $\beta_0 = 0$.

(c) Test the hypothesis that $\beta_3 = 0$.

(d) Test the hypothesis that the mean response at the input levels $x_1 = x_2 = x_3 = 1$ is 8.5.

x_1	x_2	x_3	y
7.1	.68	4	41.53
9.9	.64	1	63.75
3.6	.58	1	16.38
9.3	.21	3	45.54
2.3	.89	5	15.52
4.6	.00	8	28.55
.2	.37	5	5.65
5.4	.11	3	25.02
8.2	.87	4	52.49
7.1	.00	6	38.05
4.7	.76	0	30.76
5.4	.87	8	39.69
1.7	.52	1	17.59
1.9	.31	3	13.22
9.2	.19	5	50.98

48. The tensile strength of a certain synthetic fiber is thought to be related to x_1, the percentage of cotton in the fiber, and x_2, the drying time of the fiber. A test of 10 pieces of fiber produced under different conditions yielded the following results.

Y = Tensile Strength	x_1 = Percentage of Cotton	x_2 = Drying Time
213	13	2.1
220	15	2.3
216	14	2.2
225	18	2.5
235	19	3.2
218	20	2.4
239	22	3.4
243	17	4.1
233	16	4.0
240	18	4.3

(a) Fit a multiple regression equation.

(b) Determine a 90 percent confidence interval for the mean tensile strength of a synthetic fiber having 21 percent cotton whose drying time is 3.6.

49. The time to failure of a machine component is related to the operating voltage (x_1), the motor speed in revolutions per minute (x_2), and the operating temperature (x_3).

A designed experiment is run in the research and development laboratory, and the following data, where y is the time to failure in minutes, are obtained.

y	x_1	x_2	x_3
2,145	110	750	140
2,155	110	850	180
2,220	110	1,000	140
2,225	110	1,100	180
2,260	120	750	140
2,266	120	850	180
2,334	120	1,000	140
2,340	130	1,000	180
2,212	115	840	150
2,180	115	880	150

(a) Fit a multiple regression model to these data.
(b) Estimate the error variance.
(c) Determine a 95 percent confidence interval for the mean time to failure when the operating voltage is 125, the motor speed is 900, and the operating temperature is 160.

50. Explain why, for the same data, a prediction interval for a future response always contains the corresponding confidence interval for the mean response.

51. Consider the following data set:

x_1	x_2	y
5.1	2	55.42
5.4	8	100.21
5.9	−2	27.07
6.6	12	169.95
7.5	−6	−17.93
8.6	16	197.77
9.9	−10	−25.66
11.4	20	264.18
13.1	−14	−53.88
15	24	317.84
17.1	−18	−72.53
19.4	28	385.53

(a) Fit a linear relationship between y and x_1, x_2.
(b) Determine the variance of the error term.

(c) Determine an interval that, with 95 percent confidence, will contain the response when the inputs are $x_1 = 10.2$ and $x_2 = 17$.

52. The cost of producing power per kilowatt hour is a function of the load factor and the cost of coal in cents per million Btu. The following data were obtained from 12 mills.

Load Factor (in percent)	Cost of Coal	Power Cost
84	14	4.1
81	16	4.4
73	22	5.6
74	24	5.1
67	20	5.0
87	29	5.3
77	26	5.4
76	15	4.8
69	29	6.1
82	24	5.5
90	25	4.7
88	13	3.9

(a) Estimate the relationship.
(b) Test the hypothesis that the coefficient of the load factor is equal to 0.
(c) Determine a 95 percent prediction interval for the power cost when the load factor is 85 and the coal cost is 20.

53. The following data relate the systolic blood pressure to the age (x_1) and weight (x_2) of a set of individuals of similar body type and lifestyle.

Age	Weight	Blood Pressure
25	162	112
25	184	144
42	166	138
55	150	145
30	192	152
40	155	110
66	184	118
60	202	160
38	174	108

(a) Test the hypothesis that, when an individual's weight is known, age gives no additional information in predicting blood pressure.

(b) Determine an interval that, with 95 percent confidence, will contain the average blood pressure of all individuals of the preceding type who are 45 years old and weigh 180 pounds.

(c) Determine an interval that, with 95 percent confidence, will contain the blood pressure of a given individual of the preceding type who is 45 years old and weighs 180 pounds.

54. A recently completed study attempted to relate job satisfaction to income (in 1,000s) and seniority for a random sample of 9 municipal workers. The job satisfaction value given for each worker is his or her own assessment of such, with a score of 1 being the lowest and 10 being the highest. The following data resulted.

Yearly Income	Years on the Job	Job Satisfaction
52	8	5.6
47	4	6.3
59	12	6.8
53	9	6.7
61	16	7.0
64	14	7.7
58	10	7.0
67	15	8.0
71	22	7.8

(a) Estimate the regression parameters.

(b) What qualitative conclusions can you draw about how job satisfaction changes when income remains fixed and the number of years of service increases?

(c) Predict the job satisfaction of an employee who has spent 5 years on the job and earns a yearly salary of $56,000.

55. Suppose in Problem 54 that job satisfaction was related solely to years on the job, with the following data resulting.

Years on the Job	Job Satisfaction
8	5.6
4	6.3
12	6.8
9	6.7
16	7.0
14	7.7
10	7.0
15	8.0
22	7.8

(a) Estimate the regression parameters α and β.

(b) What is the qualitative relationship between years of service and job satisfaction? That is, what appears to happen to job satisfaction as service increases?

(c) Compare your answer to part (b) with the answer you obtained in part (b) of Problem 54.

(d) What conclusion, if any, can you draw from your answer in part (c)?

56. For the logistics regression model, find the value x such that $p(x) = .5$.

57. A study of 64 prematurely born infants was interested in the relation between the gestational age (in weeks) of the infant at birth and whether the infant was breast-feeding at the time of release from the birthing hospital. The following data resulted:

Gestational Age	Frequency	Number Breast-Feeding
28	6	2
29	5	2
30	9	7
31	9	7
32	20	16
33	15	14

In the preceding, the frequency column refers to the number of babies born after the specified gestational number of weeks.

(a) Explain how the relationship between gestational age and whether the infant was breast-feeding can be analyzed via a logistics regression model.

(b) Use appropriate software to estimate the parameters for this model.

(c) Estimate the probability that a newborn with a gestational age of 29 weeks will be breast-feeding.

58. Twelve first-time heart attack victims were given a test that measures internal anger. The following data relates their scores and whether they had a second heart attack within 5 years.

Anger Score	Second Heart Attack
80	yes
77	yes
70	no
68	yes
64	no

(continued)

Anger Score	Second Heart Attack
60	yes
50	yes
46	no
40	yes
35	no
30	no
25	yes

(a) Explain how the relationship between a second heart attack and one's anger score can be analyzed via a logistics regression model.

(b) Use appropriate software to estimate the parameters for this model.

(c) Estimate the probability that a heart attack victim with an anger score of 55 will have a second attack within 5 years.

Chapter 10

ANALYSIS OF VARIANCE

10.1 INTRODUCTION

A large company is considering purchasing, in quantity, one of four different computer packages designed to teach a new programming language. Some influential people within this company have claimed that these packages are basically interchangeable in that the one chosen will have little effect on the final competence of its user. To test this hypothesis the company has decided to choose 160 of its engineers, and divide them into 4 groups of size 40. Each member in group i will then be given teaching package $i, i = 1, 2, 3, 4$, to learn the new language. When all the engineers complete their study, a comprehensive exam will be given. The company then wants to use the results of this examination to determine whether the computer teaching packages are really interchangeable or not. How can they do this?

Before answering this question, let us note that we clearly desire to be able to conclude that the teaching packages are indeed interchangeable when the average test scores in all the groups are similar and to conclude that the packages are essentially different when there is a large variation among these average test scores. However, to be able to reach such a conclusion, we should note that the method of division of the 160 engineers into 4 groups is of vital importance. For example, suppose that the members of the first group score significantly higher than those of the other groups. What can we conclude from this? Specifically, is this result due to teaching package 1 being a superior teaching package, or is it due to the fact that the engineers in group 1 are just better learners? To be able to conclude the former, it is essential that we divide the 160 engineers into the 4 groups in such a way to make it extremely unlikely that one of these groups is inherently superior. The time-tested method for doing this is to divide the engineers into 4 groups in a completely random fashion. That is, we should do it in such a way so that all possible divisions are equally likely; for in this case, it would be very unlikely that any one group would be significantly superior to any other group. So let us suppose that the division of the engineers was indeed done "at random." (Whereas it is not at all obvious how this can be accomplished, one efficient procedure is to start by arbitrarily numbering the 160 engineers. Then generate a random permutation of the integers $1, 2, \ldots, 160$ and put the engineers whose numbers are among the first 40 of the permutation into group 1, those whose numbers are among the 41st through the 80th of the permutation into group 2, and so on.)

It is now probably reasonable to suppose that the test score of a given individual should be approximately a normal random variable having parameters that depend on the package from which he was taught. Also, it is probably reasonable to suppose that whereas the average test score of an engineer will depend on the teaching package she was exposed to, the variability in the test score will result from the inherent variation of 160 different people and not from the particular package used. Thus, if we let $X_{ij}, i = 1, \ldots, 4, j = 1, \ldots, 40$, denote the test score of the jth engineer in group i, a reasonable model might be to suppose that the X_{ij} are independent random variables with X_{ij} having a normal distribution with unknown mean μ_i and unknown variance σ^2. The hypothesis that the teaching packages are interchangeable is then equivalent to the hypothesis that $\mu_1 = \mu_2 = \mu_3 = \mu_4$.

In this chapter, we present a technique that can be used to test such a hypothesis. This technique, which is rather general and can be used to make inferences about a multitude of parameters relating to population means, is known as the *analysis of variance*.

10.2   AN OVERVIEW

Whereas hypothesis tests concerning two population means were studied in Chapter 8, tests concerning multiple population means will be considered in the present chapter. In Section 10.3, we suppose that we have been provided samples of size n from m distinct populations and that we want to use these data to test the hypothesis that the m population means are equal. Since the mean of a random variable depends only on a single factor, namely, the sample the variable is from, this scenario is said to constitute a *one-way analysis of variance*. A procedure for testing the hypothesis is presented. In addition, in Section 10.3.1 we show how to obtain multiple comparisons of the $\binom{m}{2}$ differences between the pairs of population means, and in Section 10.3.2 we show how the equal means hypothesis can be tested when the m sample sizes are not all equal.

In Sections 10.4 and 10.5, we consider models that assume that there are two factors that determine the mean value of a variable. In these models, the variables can be thought of as being arranged in a rectangular array, with the mean value of a specified variable depending both on the row and on the column in which it is located. Such a model is called a *two-way analysis of variance*. In these sections we suppose that the mean value of a variable depends on its row and column in an additive fashion; specifically, that the mean of the variable in row i, column j can be written as $\mu + \alpha_i + \beta_j$. In Section 10.4, we show how to estimate these parameters, and in Section 10.5 how to test hypotheses to the effect that a given factor — either the row or the column in which a variable is located — does not affect the mean. In Section 10.6, we consider the situation where the mean of a variable is allowed to depend on its row and column in a nonlinear fashion, thus allowing for a possible *interaction* between the two factors. We show how to test the hypothesis that there is no interaction, as well as ones concerning the lack of a row effect and the lack of a column effect on the mean value of a variable.

In all of the models considered in this chapter, we assume that the data are normally distributed with the same (although unknown) variance σ^2. The analysis of variance approach for testing a null hypothesis H_0 concerning multiple parameters relating to the population means is based on deriving two estimators of the common variance σ^2. The first estimator is a valid estimator of σ^2 whether the null hypothesis is true or not, while the second one is a valid estimator only when H_0 is true. In addition, when H_0 is not true this latter estimator will tend to exceed σ^2. The test will be to compare the values of these two estimators, and to reject H_0 when the ratio of the second estimator to the first one is sufficiently large. In other words, since the two estimators should be close to each other when H_0 is true (because they both estimate σ^2 in this case) whereas the second estimator should tend to be larger than the first when H_0 is not true, it is natural to reject H_0 when the second estimator is significantly larger than the first.

We will obtain estimators of the variance σ^2 by making use of certain facts concerning chi-square random variables, which we now present. Suppose that $X_1, \ldots, X_N$ are independent normal random variables having possibly different means but a common variance σ^2, and let $\mu_i = E[X_i], i = 1, \ldots, N$. Since the variables

$$Z_i = (X_i - \mu_i)/\sigma, \qquad i = 1, \ldots, N$$

have standard normal distributions, it follows from the definition of a chi-square random variable that

$$\sum_{i=1}^{N} Z_i^2 = \sum_{i=1}^{N} (X_i - \mu_i)^2/\sigma^2 \qquad (10.2.1)$$

is a chi-square random variable with N degrees of freedom. Now, suppose that each of the values $\mu_i, i = 1, \ldots, N$, can be expressed as a linear function of a fixed set of k unknown parameters. Suppose, further, that we can determine estimators of these k parameters, which thus gives us estimators of the mean values μ_i. If we let $\hat{\mu}_i$ denote the resulting estimator of $\mu_i, i = 1, \ldots, N$, then it can be shown that the quantity

$$\sum_{i=1}^{N} (X_i - \hat{\mu}_i)^2/\sigma^2$$

will have a chi-square distribution with $N - k$ degrees of freedom.

In other words, we start with

$$\sum_{i=1}^{N} (X_i - E[X_i])^2/\sigma^2$$

which is a chi-square random variable with N degrees of freedom. If we now write each $E[X_i]$ as a linear function of k parameters and then replace each of these parameters by its estimator, then the resulting expression remains chi-square but with a degree of freedom that is reduced by 1 for each parameter that is replaced by its estimator.

For an illustration of the preceding, consider the case where all the means are known to be equal; that is,

$$E[X_i] = \mu, \qquad i = 1, \ldots, N$$

Thus $k = 1$, because there is only one parameter that needs to be estimated. Substituting $\overline{X}$, the estimator of the common mean μ, for μ_i in Equation 10.2.1, results in the quantity

$$\sum_{i=1}^{N}(X_i - \overline{X})^2/\sigma^2 \qquad\qquad (10.2.2)$$

and the conclusion is that this quantity is a chi-square random variable with $N - 1$ degrees of freedom. But in this case where all the means are equal, it follows that the data $X_1, \ldots, X_N$ constitute a sample from a normal population, and thus Equation 10.2 is equal to $(N - 1)S^2/\sigma^2$, where S^2 is the sample variance. In other words, the conclusion in this case is just the well-known result (see Section 6.5.2) that $(N - 1)S^2/\sigma^2$ is a chi-square random variable with $N - 1$ degrees of freedom.

10.3 ONE-WAY ANALYSIS OF VARIANCE

Consider m independent samples, each of size n, where the members of the ith sample — $X_{i1}, X_{i2}, \ldots, X_{in}$ — are normal random variables with unknown mean μ_i and unknown variance σ^2. That is,

$$X_{ij} \sim N(\mu_i, \sigma^2), \qquad i = 1, \ldots, m, \qquad j = 1, \ldots, n$$

We will be interested in testing

$$H_0 : \mu_1 = \mu_2 = \cdots = \mu_m$$

versus

$$H_1 : \text{not all the means are equal}$$

That is, we will be testing the null hypothesis that all the population means are equal against the alternative that at least two of them differ. One way of thinking about this is to imagine that we have m different treatments, where the result of applying treatment i on an item is a normal random variable with mean μ_i and variance σ_2. We are then interested in testing the hypothesis that all treatments have the same effect, by applying each treatment to a (different) sample of n items and then analyzing the result.

Since there are a total of nm independent normal random variables X_{ij}, it follows that the sum of the squares of their standardized versions will be a chi-square random variable with nm degrees of freedom. That is,

$$\sum_{i=1}^{m}\sum_{j=1}^{n}(X_{ij} - E[X_{ij}])^2/\sigma^2 = \sum_{i=1}^{m}\sum_{j=1}^{n}(X_{ij} - \mu_i)^2/\sigma^2 \sim \chi^2_{nm} \tag{10.3.1}$$

To obtain estimators for the m unknown parameters $\mu_1, \ldots, \mu_m$, let $X_{i.}$ denote the average of all the elements in sample i; that is,

$$X_{i.} = \sum_{j=1}^{n} X_{ij}/n$$

The variable $X_{i.}$ is the sample mean of the ith population, and as such is the estimator of the population mean μ_i, for $i = 1, \ldots, m$. Hence, if in Equation 10.3.1 we substitute the estimators $X_{i.}$ for the means μ_i, for $i = 1, \ldots, m$, then the resulting variable

$$\sum_{i=1}^{m}\sum_{j=1}^{n}(X_{ij} - X_{i.})^2/\sigma^2 \tag{10.3.2}$$

will have a chi-square distribution with $nm - m$ degrees of freedom. (Recall that 1 degree of freedom is lost for each parameter that is estimated.) Let

$$SS_W = \sum_{i=1}^{m}\sum_{j=1}^{n}(X_{ij} - X_{i.})^2$$

and so the variable in Equation 10.3.2 is SS_W/σ^2. Because the expected value of a chi-square random variable is equal to its number of degrees of freedom, it follows upon taking the expectation of the variable in 10.3.2 that

$$E[SS_W]/\sigma^2 = nm - m$$

or, equivalently,

$$E[SS_W/(nm - m)] = \sigma^2$$

We thus have our first estimator of σ^2, namely, $SS_W/(nm - m)$. Also, note that this estimator was obtained without assuming anything about the truth or falsity of the null hypothesis.

Definition

The statistic

$$SS_W = \sum_{i=1}^{m}\sum_{j=1}^{n}(X_{ij} - X_{i.})^2$$

is called the *within samples sum of squares* because it is obtained by substituting the sample population means for the population means in expression 10.3.1. The statistic

$$SS_W/(nm - m)$$

is an estimator of σ^2.

Our second estimator of σ^2 will only be a valid estimator when the null hypothesis is true. So let us assume that H_0 is true and so all the population means μ_i are equal, say, $\mu_i = \mu$ for all i. Under this condition it follows that the m sample means $X_{1.}, X_{2.}, \ldots, X_{m.}$ will all be normally distributed with the same mean μ and the same variance σ^2/n. Hence, the sum of squares of the m standardized variables

$$\frac{X_{i.} - \mu}{\sqrt{\sigma^2/n}} = \sqrt{n}(X_{i.} - \mu)/\sigma$$

will be a chi-square random variable with m degrees of freedom. That is, when H_0 is true,

$$n\sum_{i=1}^{m}(X_{i.} - \mu)^2/\sigma^2 \sim \chi_m^2 \qquad (10.3.3)$$

Now, when all the population means are equal to μ, then the estimator of μ is the average of all the nm data values. That is, the estimator of μ is $X_{..}$, given by

$$X_{..} = \frac{\displaystyle\sum_{i=1}^{m}\sum_{j=1}^{n}X_{ij}}{nm} = \frac{\displaystyle\sum_{i=1}^{m}X_{i.}}{m}$$

If we now substitute $X_{..}$ for the unknown parameter μ in expression 10.5, it follows, when H_0 is true, that the resulting quantity

$$n\sum_{i=1}^{m}(X_{i.} - X_{..})^2/\sigma^2$$

will be a chi-square random variable with $m - 1$ degrees of freedom. That is, if we define SS_b by

$$SS_b = n \sum_{i=1}^{m} (X_{i.} - X_{..})^2$$

then it follows that, when H_0 is true, SS_b/σ^2 is chi-square with $m - 1$ degrees of freedom.

From the above we obtain that when H_0 is true,

$$E[SS_b]/\sigma^2 = m - 1$$

or, equivalently,

$$E[SS_b/(m - 1)] = \sigma^2 \tag{10.3.4}$$

So, when H_0 is true, $SS_b/(m - 1)$ is also an estimator of σ^2.

Definition

The statistic

$$SS_b = n \sum_{i=1}^{m} (X_{i.} - X_{..})^2$$

is called the *between samples sum of squares*. When H_0 is true, $SS_b/(m - 1)$ is an estimator of σ^2.

Thus we have shown that

$$SS_W/(nm - m) \qquad \text{always estimates } \sigma^2$$
$$SS_b/(m - 1) \qquad \text{estimates } \sigma^2 \text{ when } H_0 \text{ is true}$$

Because it can be shown that $SS_b/(m - 1)$ will tend to exceed σ^2 when H_0 is not true,[*] it is reasonable to let the test statistic be given by

$$TS = \frac{SS_b/(m - 1)}{SS_W/(nm - m)}$$

and to reject H_0 when TS is sufficiently large.

---

[*] A proof is given at the end of this subsection.

TABLE 10.1   *Values of* $F_{r,s,.05}$

s = Degrees of Freedom for the Denominator	r = Degrees of Freedom for the Numerator			
	1	2	3	4
4	7.71	6.94	6.59	6.39
5	6.61	5.79	5.41	5.19
10	4.96	4.10	3.71	3.48

To determine how large TS needs to be to justify rejecting H_0, we use the fact that it can be shown that if H_0 is true then SS_b and SS_W are independent. It follows from this that, when H_0 is true, TS has an F-distribution with $m - 1$ numerator and $nm - m$ denominator degrees of freedom. Let $F_{m-1,nm-m,\alpha}$ denote the $100(1 - \alpha)$ percentile of this distribution — that is,

$$P\{F_{m-1,nm-m} > F_{m-1,nm-m,\alpha}\} = \alpha$$

where we are using the notation $F_{r,s}$ to represent an F-random variable with r numerator and s denominator degrees of freedom.

The significance level α test of H_0 is as follows:

$$\text{reject} \quad H_0 \quad \text{if} \quad \frac{SS_b/(m-1)}{SS_W/(nm-m)} > F_{m-1,nm-m,\alpha}$$

$$\text{do not reject} \quad H_0 \quad \text{otherwise}$$

A table of values of $F_{r,s,.05}$ for various values of r and s is presented in Table A4 of the Appendix. Part of this table is presented in Table 10.1. For instance, from Table 10.1 we see that there is a 5 percent chance that an F-random variable having 3 numerator and 10 denominator degrees of freedom will exceed 3.71.

Another way of doing the computations for the hypothesis test that all the population means are equal is by computing the p-value. If the value of the test statistic is $TS = v$, then the p-value will be given by

$$p\text{-value} = P\{F_{m-1,nm-m} \geq v\}$$

Program 10.3 will compute the value of the test statistic TS and the resulting p-value.

EXAMPLE 10.3a   An auto rental firm is using 15 identical motors that are adjusted to run at a fixed speed to test 3 different brands of gasoline. Each brand of gasoline is assigned to exactly 5 of the motors. Each motor runs on 10 gallons of gasoline until it is out of fuel.

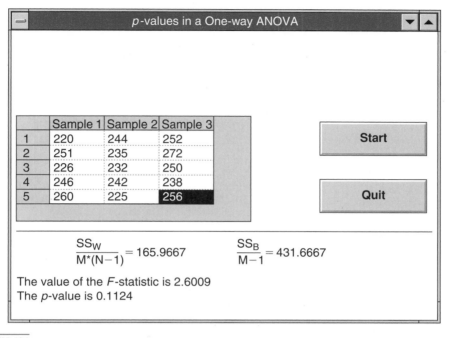

FIGURE 10.1

The following represents the total mileages obtained by the different motors:

Gas 1: 220 251 226 246 260
Gas 2: 244 235 232 242 225
Gas 3: 252 272 250 238 256

Test the hypothesis that the average mileage obtained is not affected by the type of gas used. Use the 5 percent level of significance.

SOLUTION We run Program 10.3 to obtain the results shown in Figure 10.1. Since the p-value is greater than .05, the null hypothesis that the mean mileage is the same for all 3 brands of gasoline cannot be rejected. ■

The following algebraic identity, called the *sum of squares identity*, is useful when doing the computations by hand.

The Sum of Squares Identity

$$\sum_{i=1}^{m}\sum_{j=1}^{n} X_{ij}^2 = nmX_{..}^2 + SS_b + SS_W$$

When computing by hand, the quantity SS_b defined by

$$SS_b = n \sum_{i=1}^{m} (X_{i.} - X_{..})^2$$

should be computed first. Once SS_b has been computed, SS_W can be determined from the sum of squares identity. That is, $\sum_{i=1}^{m} \sum_{j=1}^{n} X_{ij}^2$ and $X_{..}^2$ should also be computed and then SS_W determined from

$$SS_W = \sum_{i=1}^{m} \sum_{j=1}^{n} X_{ij}^2 - nmX_{..}^2 - SS_b$$

EXAMPLE 10.3b Let us do the computations of Example 10.3a by hand. The first thing to note is that subtracting a constant from each data value will not affect the value of the test statistic. So we subtract 220 from each data value to get the following information.

Gas	Mileage					$\sum_j X_{ij}$	$\sum_j X_{ij}^2$
1	0	31	6	26	40	103	3,273
2	24	15	12	22	5	78	1,454
3	32	52	30	18	36	168	6,248

Now $m = 3$ and $n = 5$ and

$$X_{1.} = 103/5 = 20.6$$
$$X_{2.} = 78/5 = 15.6$$
$$X_{3.} = 168/5 = 33.6$$

$$X_{..} = (103 + 78 + 168)/15 = 23.2667, \quad X_{..}^2 = 541.3393$$

Thus,

$$SS_b = 5[(20.6 - 23.2667)^2 + (15.6 - 23.2667)^2 + (33.6 - 23.2667)^2] = 863.3335$$

Also,

$$\sum \sum X_{ij}^2 = 3{,}273 + 1{,}454 + 6{,}248 = 10{,}975$$

and, from the sum of squares identity,

$$SS_W = 10{,}975 - 15(541.3393) - 863.3335 = 1991.5785$$

The value of the test statistic is thus

$$TS = \frac{863.3335/2}{1991.5785/12} = 2.60$$

Now, from Table A4 in the Appendix, we see that $F_{2,12,.05} = 3.89$. Hence, because the value of the test statistic does not exceed 3.89, we cannot, at the 5 percent level of significance, reject the null hypothesis that the gasolines give equal mileage. ∎

Let us now show that

$$E[SS_b/(m-1)] \geq \sigma^2$$

with equality only when H_0 is true. So, we must show that

$$E\left[\sum_{i=1}^{m}(X_{i.} - X_{..})^2/(m-1)\right] \geq \sigma^2/n$$

with equality only when H_0 is true. To verify this, let $\mu_. = \sum_{i=1}^{m} \mu_i/m$ be the average of the means. Also, for $i = 1, \ldots, m$, let

$$Y_i = X_{i.} - \mu_i + \mu_.$$

Because $X_{i.}$ is normal with mean μ_i and variance σ^2/n, it follows that Y_i is normal with mean $\mu_.$ and variance σ^2/n. Consequently, $Y_1, \ldots, Y_m$ constitutes a sample from a normal population having variance σ^2/n. Let

$$\bar{Y} = Y_. = \sum_{i=1}^{m} Y_i/m = X_{..} - \mu_. + \mu_. = X_{..}$$

be the average of these variables. Now,

$$X_{i.} - X_{..} = Y_i + \mu_i - \mu_. - Y_.$$

Consequently,

$$E\left[\sum_{i=1}^{m}(X_{i.} - X_{..})^2\right] = E\left[\sum_{i=1}^{m}(Y_i - Y_. + \mu_i - \mu_.)^2\right]$$

$$= E\left[\sum_{i=1}^{m}[(Y_i - Y_.)^2 + (\mu_i - \mu_.)^2 + 2(\mu_i - \mu_.)(Y_i - Y_.)]\right]$$

$$= E\left[\sum_{i=1}^{m}(Y_i - Y_.)^2\right] + \sum_{i=1}^{m}(\mu_i - \mu_.)^2 + 2\sum_{i=1}^{m}(\mu_i - \mu_.)E[Y_i - Y_.]$$

$$= (m-1)\sigma^2/n + \sum_{i=1}^{m}(\mu_i - \mu_.)^2 + 2\sum_{i=1}^{m}(\mu_i - \mu_.)E[\overline{Y}_i - \overline{Y}_.]$$

$$= (m-1)\sigma^2/n + \sum_{i=1}^{m}(\mu_i - \mu_.)^2$$

where the next to last equality follows because the sample variance $\sum_{i=1}^{m}(\overline{Y}_i - \overline{Y}_.)^2/(m-1)$ is an unbiased estimator of its population variance σ^2/n and the final equality because $E[\overline{Y}_i] - E[\overline{Y}_.] = \mu_. - \mu_. = 0$. Dividing by $m-1$ gives that

$$E\left[\sum_{i=1}^{m}(\overline{X}_{i.} - \overline{X}_{..})^2/(m-1)\right] = \sigma^2/n + \sum_{i=1}^{m}(\mu_i - \mu_.)^2/(m-1)$$

and the result follows because $\sum_{i=1}^{m}(\mu_i - \mu_.)^2 \geq 0$, with equality only when all the μ_i are equal.

Table 10.2 sums up the results of this section.

TABLE 10.2   *One-Way ANOVA Table*

Source of Variation	Sum of Squares	Degrees of Freedom	Value of Test Statistic
Between samples	$SS_b = n\sum_{i=1}^{m}(\overline{X}_{i.} - \overline{X}_{..})^2$	$m-1$	
Within samples	$SS_W = \sum_{i=1}^{m}\sum_{j=1}^{n}(X_{ij} - \overline{X}_{i.})^2$	$nm - m$	
			$TS = \frac{SS_b/(m-1)}{SS_W/(nm-m)}$
	Significance level α test:		
	reject H_0 if $TS \geq F_{m-1,nm-m,\alpha}$		
	do not reject otherwise		
If $TS = v$, then p-value $= P\{F_{m-1,nm-m} \geq v\}$			

10.3.1 MULTIPLE COMPARISONS OF SAMPLE MEANS

When the null hypothesis of equal means is rejected, we are often interested in a comparison of the different sample means $\mu_1, \ldots, \mu_m$. One procedure that is often used for this purpose is known as the T-method. For a specified value of α, this procedure gives joint confidence intervals for all the $\binom{m}{2}$ differences $\mu_i - \mu_j, i \neq j, i, j = 1, \ldots, m$, such that with probability $1 - \alpha$ all of the confidence intervals will contain their respective quantities $\mu_i - \mu_j$. The T-method is based on the following result:

With probability $1 - \alpha$, for every $i \neq j$
$$\overline{X}_{i.} - \overline{X}_{j.} - W < \mu_i - \mu_j < \overline{X}_{i.} - \overline{X}_{j.} + W$$

where

$$W = \frac{1}{\sqrt{n}} C(m, nm - m, \alpha) \sqrt{SS_W/(nm - m)}$$

and where the values of $C(m, nm - m, \alpha)$ are given, for $\alpha = .05$ and $\alpha = .01$, in Table A5 of the Appendix.

EXAMPLE 10.3c A college administrator claims that there is no difference in first-year grade point averages for students entering the college from any of three different city high schools. The following data give the first-year grade point averages of 12 randomly chosen students, 4 from each of the three high schools. At the 5 percent level of significance, do these data disprove the claim of the administrator? If so, determine confidence intervals for the difference in means of students from the different high schools, such that we can be 95 percent confident that all of the interval statements are valid.

School 1	School 2	School 3
3.2	3.4	2.8
3.4	3.0	2.6
3.3	3.7	3.0
3.5	3.3	2.7

SOLUTION To begin, note that there are $m = 3$ samples, each of size $n = 4$. Program 10.3 on the text disk yields the results:

$$SS_W/9 = .0431$$
$$p\text{-value} = .0046$$

so the hypothesis of equal mean scores for students from the three schools is rejected.

To determine the confidence intervals for the differences in the population means, note first that the sample means are

$$X_{1.} = 3.350, \quad X_{2.} = 3.350, \quad X_{3.} = 2.775$$

From Table A5 of the Appendix, we see that $C(3, 9, .05) = 3.95$; thus, as $W = \frac{1}{\sqrt{4}} 3.95\sqrt{.0431} = .410$, we obtain the following confidence intervals.

$$-.410 < \mu_1 - \mu_2 < .410$$
$$.165 < \mu_1 - \mu_3 < .985$$
$$.165 < \mu_2 - \mu_3 < .985$$

Hence, with 95 percent confidence, we can conclude that the mean grade point average of first-year students from high school 3 is less than the mean average of students from high school 1 or from high school 2 by an amount that is between .165 and .985, and that the difference in grade point averages of students from high schools 1 and 2 is less than .410. ■

10.3.2 One-Way Analysis of Variance with Unequal Sample Sizes

The model in the previous section supposed that there were an equal number of data points in each sample. Whereas this is certainly a desirable situation (see the Remark at the end of this section), it is not always possible to attain. So let us now suppose that we have m normal samples of respective sizes $n_1, n_2, \ldots, n_m$. That is, the data consist of the $\sum_{i=1}^{m} n_i$ independent random variables X_{ij}, $i = 1, \ldots, m$, $j = 1, \ldots, n_i$, where

$$X_{ij} \sim \mathcal{N}(\mu_i, \sigma^2)$$

Again we are interested in testing the hypothesis H_0 that all means are equal.

To derive a test of H_0, we start with the fact that

$$\sum_{i=1}^{m} \sum_{j=1}^{n_i} (X_{ij} - E[X_{ij}])^2 / \sigma^2 = \sum_{i=1}^{m} \sum_{j=1}^{n_i} (X_{ij} - \mu_i)^2 / \sigma^2$$

is a chi-square random variable with $\sum_{i=1}^{m} n_i$ degrees of freedom. Hence, upon replacing each mean μ_i by its estimator $X_{i.}$, the average of the elements in sample i, we obtain

$$\sum_{i=1}^{m} \sum_{j=1}^{n_i} (X_{ij} - X_{i.})^2 / \sigma^2$$

which is chi-square with $\sum_{i=1}^{m} n_i - m$ degrees of freedom. Therefore, letting

$$SS_W = \sum_{i=1}^{m} \sum_{j=1}^{n_i} (X_{ij} - X_{i.})^2$$

it follows that $SS_W / \left(\sum_{i=1}^{m} n_i - m \right)$ is an unbiased estimator of σ^2.

Furthermore, if H_0 is true and μ is the common mean, then the random variables $X_{i.}, i = 1, \ldots, m$ will be independent normal random variables with

$$E[X_{i.}] = \mu, \quad \text{Var}(X_{i.}) = \sigma^2 / n_i$$

As a result, when H_0 is true

$$\sum_{i=1}^{m} \frac{(X_{i.} - \mu)^2}{\sigma^2 / n_i} = \sum_{i=1}^{m} n_i (X_{i.} - \mu)^2 / \sigma^2$$

is chi-square with m degrees of freedom; therefore, replacing μ in the preceding by its estimator $X_{..}$, the average of all the X_{ij}, results in the statistic

$$\sum_{i=1}^{m} n_i (X_{i.} - X_{..})^2 / \sigma^2$$

which is chi-square with $m - 1$ degrees of freedom. Thus, letting

$$SS_b = \sum_{i=1}^{m} n_i (X_{i.} - X_{..})^2$$

it follows, when H_0 is true, that $SS_b / (m - 1)$ is also an unbiased estimator of σ^2. Because it can be shown that when H_0 is true the quantities SS_b and SS_W are independent, it follows under this condition that the statistic

$$\frac{SS_b / (m - 1)}{SS_W / \left(\sum_{i=1}^{m} n_i - m \right)}$$

is an F-random variable with $m - 1$ numerator and $\sum_{i=1}^{m} n_i - m$ denominator degrees of freedom. From this we can conclude that a significance level α test of the null hypothesis

$$H_0 : \mu_1 = \cdots = \mu_m$$

is to let $N = \sum_i n_i - m$, and then

$$\text{reject} \quad H_0 \quad \text{if} \quad \frac{SS_b / (m - 1)}{SS_W / \left(\sum_{i=1}^{m} n_i - m \right)} > F_{m-1, N, \alpha}$$

$$\text{not reject} \quad H_0 \quad \text{otherwise}$$

REMARK

When the samples are of different sizes we say that we are in the *unbalanced* case. Whenever possible it is advantageous to choose a balanced design over an unbalanced one. For one thing, the test statistic in a balanced design is relatively insensitive to slight departures from the assumption of equal population variances. (That is, the balanced design is more robust than the unbalanced one.)

10.4 TWO-FACTOR ANALYSIS OF VARIANCE: INTRODUCTION AND PARAMETER ESTIMATION

Whereas the model of Section 10.3 enabled us to study the effect of a single factor on a data set, we can also study the effects of several factors. In this section, we suppose that each data value is affected by two factors.

EXAMPLE 10.4a Four different standardized reading achievement tests were administered to each of 5 students, with the scores shown in the table resulting. Each value in this set of 20 data points is affected by two factors, namely, the exam and the student whose score on that exam is being recorded. The exam factor has 4 possible values, or *levels*, and the student factor has 5 possible levels.

			Student		
Exam	1	2	3	4	5
1	75	73	60	70	86
2	78	71	64	72	90
3	80	69	62	70	85
4	73	67	63	80	92

In general, let us suppose that there are m possible levels of the first factor and n possible levels of the second. Let X_{ij} denote the value obtained when the first factor is at level i and the second factor is at level j. We will often portray the data set in the following array of rows and columns.

$$
\begin{matrix}
X_{11} & X_{12} & \cdots & X_{1j} & \cdots & X_{1n} \\
X_{21} & X_{22} & \cdots & X_{2j} & \cdots & X_{2n} \\
X_{i1} & X_{i2} & \cdots & X_{ij} & \cdots & X_{in} \\
X_{m1} & X_{m2} & \cdots & X_{mj} & \cdots & X_{mn}
\end{matrix}
$$

Because of this we will refer to the first factor as the "row" factor, and the second factor as the "column" factor.

As in Section 10.3, we will suppose that the data $X_{ij}, i = 1, \ldots, m \; j = 1, \ldots, n$ are independent normal random variables with a common variance σ^2. However, whereas in Section 10.3 we supposed that only a single factor affected the mean value of a data point — namely, the sample to which it belongs — we will suppose in the present section that the mean value of data depends in an additive manner on both its row and its column.

If, in the model of Section 10.3, we let X_{ij} represent the value of the jth member of sample i, then that model could be symbolically represented as

$$E[X_{ij}] = \mu_i$$

However, if we let μ denote the average value of the μ_i — that is,

$$\mu = \sum_{i=1}^{m} \mu_i / m$$

then we can rewrite the model as

$$E[X_{ij}] = \mu + \alpha_i$$

where $\alpha_i = \mu_i - \mu$. With this definition of α_i as the deviation of μ_i from the average mean value, it is easy to see that

$$\sum_{i=1}^{m} \alpha_i = 0$$

A two-factor additive model can also be expressed in terms of row and column deviations. If we let $\mu_{ij} = E[X_{ij}]$, then the additive model supposes that for some constants $a_i, i = 1, \ldots, m$ and $b_j, j = 1, \ldots, n$

$$\mu_{ij} = a_i + b_j$$

Continuing our use of the "dot" (or *averaging*) notation, we let

$$\mu_{i.} = \sum_{j=1}^{n} \mu_{ij} / n, \qquad \mu_{.j} = \sum_{i=1}^{m} \mu_{ij} / m, \qquad \mu_{..} = \sum_{i=1}^{m} \sum_{j=1}^{n} \mu_{ij} / nm$$

Also, we let

$$a_. = \sum_{i=1}^{m} a_i / m, \qquad b_. = \sum_{j=1}^{n} b_j / n$$

Note that

$$\mu_{i.} = \sum_{j=1}^{n} (a_i + b_j) / n = a_i + b_.$$

Similarly,

$$\mu_{.j} = a_. + b_j, \qquad \mu_{..} = a_. + b_.$$

If we now set

$$\mu = \mu_{..} = a_. + b_.$$
$$\alpha_i = \mu_{i.} - \mu = a_i - a_.$$
$$\beta_j = \mu_{.j} - \mu = b_j - b_.$$

then the model can be written as

$$\mu_{ij} = E[X_{ij}] = \mu + \alpha_i + \beta_j$$

where

$$\sum_{i=1}^{m} \alpha_i = \sum_{j=1}^{n} \beta_j = 0$$

The value μ is called the *grand mean*, α_i is the *deviation from the grand mean due to row i*, and β_j is the *deviation from the grand mean due to column j*.

Let us now determine estimators of the parameters $\mu, \alpha_i, \beta_j, i = 1, \ldots, m, j = 1, \ldots, n$. To do so, continuing our use of "dot" notation, we let

$$X_{i.} = \sum_{j=1}^{n} X_{ij}/n = \text{average of the values in row } i$$

$$X_{.j} = \sum_{i=1}^{m} X_{ij}/m = \text{average of the values in column } j$$

$$X_{..} = \sum_{i=1}^{m}\sum_{j=1}^{n} X_{ij}/nm = \text{average of all data values}$$

Now,

$$E[X_{i.}] = \sum_{j=1}^{n} E[X_{ij}]/n$$

$$= \mu + \sum_{j=1}^{n} \alpha_i/n + \sum_{j=1}^{n} \beta_j/n$$

$$= \mu + \alpha_i \quad \text{since } \sum_{j=1}^{n} \beta_j = 0$$

Similarly, it follows that

$$E[X_{.j}] = \mu + \beta_j$$
$$E[X_{..}] = \mu$$

Because the preceding is equivalent to

$$E[X_{..}] = \mu$$
$$E[X_{i.} - X_{..}] = \alpha_i$$
$$E[X_{.j} - X_{..}] = \beta_j$$

we see that unbiased estimators of μ, α_i, β_j — call them $\hat{\mu}, \hat{\alpha}_i, \hat{\beta}_j$ — are given by

$$\hat{\mu} = X_{..}$$
$$\hat{\alpha}_i = X_{i.} - X_{..}$$
$$\hat{\beta}_j = X_{.j} - X_{..} \quad \blacksquare$$

EXAMPLE 10.4b The following data from Example 10.4a give the scores obtained when four different reading tests were given to each of five students. Use it to estimate the parameters of the model.

Examination	Student 1	2	3	4	5	Row Totals	$X_{i.}$
1	75	73	60	70	86	364	72.8
2	78	71	64	72	90	375	75
3	80	69	62	70	85	366	73.2
4	73	67	63	80	92	375	75
Column totals	306	280	249	292	353	1,480	← grand total
$X_{.j}$	76.5	70	62.25	73	88.25	$X_{..} = \dfrac{1,480}{20} = 74$	

SOLUTION The estimators are

$$\hat{\mu} = 74$$

$$
\begin{array}{ll}
\hat{\alpha}_1 = 72.8 - 74 = -1.2 & \hat{\beta}_1 = 76.5 - 74 = 2.5 \\
\hat{\alpha}_2 = 75 - 74 = 1 & \hat{\beta}_2 = 70 - 74 = -4 \\
\hat{\alpha}_3 = 73.2 - 74 = -.8 & \hat{\beta}_3 = 62.25 - 74 = -11.75 \\
\hat{\alpha}_4 = 75 - 74 = 1 & \hat{\beta}_4 = 73 - 74 = -1 \\
& \hat{\beta}_5 = 88.25 - 74 = 14.25
\end{array}
$$

Therefore, for instance, if one of the students is randomly chosen and then given a randomly chosen examination, then our estimate of the mean score that will be obtained is $\hat{\mu} = 74$. If we were told that examination i was taken, then this would increase our estimate of the mean score by the amount $\hat{\alpha}_i$; and if we were told that the student chosen was number j, then this would increase our estimate of the mean score by the amount $\hat{\beta}_j$. Thus, for instance, we would estimate that the score obtained on examination 1 by student 2 is the value of a random variable whose mean is $\hat{\mu} + \hat{\alpha}_1 + \hat{\beta}_2 = 74 - 1.2 - 4 = 68.8$. $\quad \blacksquare$

10.5 TWO-FACTOR ANALYSIS OF VARIANCE: TESTING HYPOTHESES

Consider the two-factor model in which one has data $X_{ij}, i = 1, \ldots, m$ and $j = 1, \ldots, n$. These data are assumed to be independent normal random variables with a common variance σ^2 and with mean values satisfying

$$E[X_{ij}] = \mu + \alpha_i + \beta_j$$

where

$$\sum_{i=1}^{m} \alpha_i = \sum_{j=1}^{n} \beta_j = 0$$

In this section, we will be concerned with testing the hypothesis

$$H_0 : \text{ all } \alpha_i = 0$$

against

$$H_1 : \text{ not all the } \alpha_i \text{ are equal to } 0$$

This null hypothesis states that there is no row effect, in that the value of a datum is not affected by its row factor level.

We will also be interested in testing the analogous hypothesis for columns, that is

$$H_0 : \text{ all } \beta_j \text{ are equal to } 0$$

against

$$H_1 : \text{ not all } \beta_j \text{ are equal to } 0$$

To obtain tests for the above null hypotheses, we will apply the analysis of variance approach in which two different estimators are derived for the variance σ^2. The first will always be a valid estimator, whereas the second will only be a valid estimator when the null hypothesis is true. In addition, the second estimator will tend to overestimate σ^2 when the null hypothesis is not true.

To obtain our first estimator of σ^2, we start with the fact that

$$\sum_{i=1}^{m} \sum_{j=1}^{n} (X_{ij} - E[X_{ij}])^2/\sigma^2 = \sum_{i=1}^{m} \sum_{j=1}^{n} (X_{ij} - \mu - \alpha_i - \beta_j)^2/\sigma^2$$

is chi-square with nm degrees of freedom. If in the above expression we now replace the unknown parameters $\mu, \alpha_1, \alpha_2, \ldots, \alpha_m, \beta_1, \beta_2, \ldots, \beta_n$ by their estimators $\hat{\mu}, \hat{\alpha}_1, \hat{\alpha}_2, \ldots, \hat{\alpha}_m, \hat{\beta}_1, \hat{\beta}_2, \ldots, \hat{\beta}_n$, then it turns out that the resulting expression will remain chi-square but will lose 1 degree of freedom for each parameter that is estimated. To determine how many parameters are to be estimated, we must be careful to remember that

$\sum_{i=1}^{m} \alpha_i = \sum_{j=1}^{n} \beta_j = 0$. Since the sum of all the α_i is equal to 0, it follows that once we have estimated $m-1$ of the α_i then we have also estimated the final one. Hence, only $m-1$ parameters are to be estimated in order to determine all of the estimators $\hat{\alpha}_i$. For the same reason, only $n-1$ of the β_j need be estimated to determine estimators for all n of them. Because μ also must be estimated, we see that the number of parameters that need to be estimated is $1 + m - 1 + n - 1 = n + m - 1$. As a result, it follows that

$$\sum_{i=1}^{m} \sum_{j=1}^{n} (X_{ij} - \hat{\mu} - \hat{\alpha}_i - \hat{\beta}_j)^2 / \sigma^2$$

is a chi-square random variable with $nm - (n + m - 1) = (n-1)(m-1)$ degrees of freedom.

Since $\hat{\mu} = X_{..}, \hat{\alpha}_i = X_{i.} - X_{..}, \hat{\beta}_j = X_{.j} - X_{..}$, it follows that $\hat{\mu} + \hat{\alpha}_i + \hat{\beta}_j = X_{i.} + X_{.j} - X_{..}$; thus,

$$\sum_{i=1}^{m} \sum_{j=1}^{n} (X_{ij} - X_{i.} - X_{.j} + X_{..})^2 / \sigma^2 \qquad (10.5.1)$$

is a chi-square random variable with $(n-1)(m-1)$ degrees of freedom.

Definition

The statistic SS_e defined by

$$SS_e = \sum_{i=1}^{m} \sum_{j=1}^{n} (X_{ij} - X_{i.} - X_{.j} + X_{..})^2$$

is called the *error sum of squares*.

If we think of the difference between a value and its estimated mean as being an "error," then SS_e is equal to the sum of the squares of the errors. Since SS_e/σ^2 is just the expression in 10.5.1, we see that SS_e/σ^2 is chi-square with $(n-1)(m-1)$ degrees of freedom. Because the expected value of a chi-square random variable is equal to its number of degrees of freedom, we have that

$$E[SS_e/\sigma^2] = (n-1)(m-1)$$

or

$$E[SS_e/(n-1)(m-1)] = \sigma^2$$

That is,

$$SS_e/(n-1)(m-1)$$

is an unbiased estimator of σ^2.

Suppose now that we want to test the null hypothesis that there is no row effect — that is, we want to test

$$H_0 : \text{all the } \alpha_i \text{ are equal to } 0$$

against

$$H_1 : \text{not all the } \alpha_i \text{ are equal to } 0$$

To obtain a second estimator of σ^2, consider the row averages $X_{i.}, i = 1, \ldots, m$. Note that, when H_0 is true, each α_i is equal to 0, and so

$$E[X_{i.}] = \mu + \alpha_i = \mu$$

Because each $X_{i.}$ is the average of n random variables, each having variance σ^2, it follows that

$$\mathrm{Var}(X_{i.}) = \sigma^2/n$$

Thus, we see that when H_0 is true

$$\sum_{i=1}^{m}(X_{i.} - E[X_{i.}])^2/\mathrm{Var}(X_{i.}) = n\sum_{i=1}^{m}(X_{i.} - \mu)^2/\sigma^2$$

will be chi-square with m degrees of freedom. If we now substitute $X_{..}$ (the estimator of μ) for μ in the preceding, then the resulting expression will remain chi-square but with 1 less degree of freedom. We thus have the following:

when H_0 is true

$$n\sum_{i=1}^{m}(X_{i.} - X_{..})^2/\sigma^2$$

is chi-square with $m - 1$ degrees of freedom.

Definition

The statistic SS_r is defined by

$$SS_r = n\sum_{i=1}^{m}(X_{i.} - X_{..})^2$$

and is called the *row sum of squares*.

We saw earlier that when H_0 is true, SS_r/σ^2 is chi-square with $m-1$ degrees of freedom. As a result, when H_0 is true,

$$E[SS_r/\sigma^2] = m - 1$$

or, equivalently,

$$E[SS_r/(m - 1)] = \sigma^2$$

In addition, it can be shown that $SS_r/(m - 1)$ will tend to be larger than σ^2 when H_0 is not true. Thus, once again we have obtained two estimators of σ^2. The first estimator, $SS_e/(n-1)(m-1)$, is a valid estimator whether or not the null hypothesis is true, whereas the second estimator, $SS_r/(m - 1)$, is only a valid estimator of σ^2 when H_0 is true and tends to be larger than σ^2 when H_0 is not true.

We base our test of the null hypothesis H_0 that there is no row effect, on the ratio of the two estimators of σ^2. Specifically, we use the test statistic

$$TS = \frac{SS_r/(m-1)}{SS_e/(n-1)(m-1)}$$

Because the estimators can be shown to be independent when H_0 is true, it follows that the significance level α test is to

$$\text{reject} \quad H_0 \quad \text{if} \quad TS \geq F_{m-1,(n-1)(m-1),\alpha}$$
$$\text{do not reject} \quad H_0 \quad \text{otherwise}$$

Alternatively, the test can be performed by calculating the p-value. If the value of the test statistic is v, then the p-value is given by

$$p\text{-value} = P\{F_{m-1,(n-1)(m-1)} \geq v\}$$

A similar test can be derived for testing the null hypothesis that there is no column effect — that is, that all the β_j are equal to 0. The results are summarized in Table 10.3. Program 10.5 will do the computations and give the p-value.

TABLE 10.3 *Two-Factor ANOVA*

	Sum of Squares	Degrees of Freedom
Row	$SS_r = n\sum_{i=1}^{m}(X_{i.} - X_{..})^2$	$m-1$
Column	$SS_c = m\sum_{j=1}^{n}(X_{.j} - X_{..})^2$	$n-1$
Error	$SS_e = \sum_{i=1}^{m}\sum_{j=1}^{n}(X_{ij} - X_{i.} - X_{.j} + X_{..})^2$	$(n-1)(m-1)$

Let $N = (n-1)(m-1)$

Null Hypothesis	Test Statistic	Significance Level α Test	p-value if $TS = v$
All $\alpha_i = 0$	$\dfrac{SS_r/(m-1)}{SS_e/N}$	Reject if $TS \geq F_{m-1,N,\alpha}$	$P\{F_{m-1,N} \geq v\}$
All $\beta_j = 0$	$\dfrac{SS_c/(n-1)}{SS_e/N}$	Reject if $TS \geq F_{n-1,N,\alpha}$	$P\{F_{n-1,N} \geq v\}$

EXAMPLE 10.5a The following data[*] represent the number of different macroinvertebrate species collected at 6 stations, located in the vicinity of a thermal discharge, from 1970 to 1977.

[*] Taken from Wartz and Skinner, "A 12-year macroinvertebrate study in the vicinity of 2 thermal discharges to the Susquehanna River near York Haven, PA." *Jour. of Testing and Evaluation.* Vol. 12. No. 3, May 1984, 157–163.

				Station		
Year	1	2	3	4	5	6
1970	53	35	31	37	40	43
1971	36	34	17	21	30	18
1972	47	37	17	31	45	26
1973	55	31	17	23	43	37
1974	40	32	19	26	45	37
1975	52	42	20	27	26	32
1976	39	28	21	21	36	28
1977	40	32	21	21	36	35

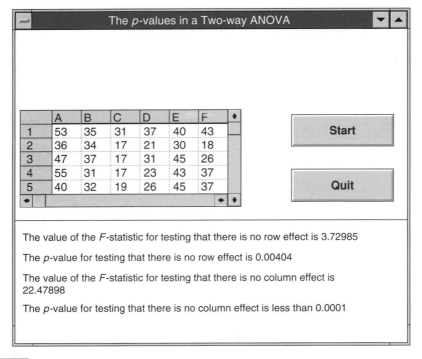

FIGURE 10.2

To test the hypotheses that the data are unchanging (a) from year to year, and (b) from station to station, run Program 10.5. Results are shown in Figure 10.2. Thus both the hypothesis that the data distribution does not depend on the year and the hypothesis that it does not depend on the station are rejected at very small significance levels. ■

10.6 TWO-WAY ANALYSIS OF VARIANCE WITH INTERACTION

In Sections 10.4 and 10.5, we considered experiments in which the distribution of the observed data depended on two factors — which we called the row and column factors. Specifically, we supposed that the mean value of X_{ij}, the data value in row i and column j, can be expressed as the sum of two terms — one depending on the row of the element and one on the column. That is, we supposed that

$$X_{ij} \sim \mathcal{N}(\mu + \alpha_i + \beta_j, \sigma^2), \quad i = 1, \ldots, m, \quad j = 1, \ldots, n$$

However, one weakness of this model is that in supposing that the row and column effects are additive, it does not allow for the possibility of a row and column interaction.

For instance, consider an experiment designed to compare the mean number of defective items produced by four different workers when using three different machines. In analyzing the resulting data, we might suppose that the incremental number of defects that resulted from using a given machine was the same for each of the workers. However, it is certainly possible that a machine could interact in a different manner with different workers. That is, there could be a worker–machine interaction that the additive model does not allow for.

To allow for the possibility of a row and column interaction, let

$$\mu_{ij} = E[X_{ij}]$$

and define the quantities $\mu, \alpha_i, \beta_j, \gamma_{ij}, i = 1, \ldots, m, j = 1, \ldots, n$ as follows:

$$\mu = \mu_{..}$$
$$\alpha_i = \mu_{i.} - \mu_{..}$$
$$\beta_j = \mu_{.j} - \mu_{..}$$
$$\gamma_{ij} = \mu_{ij} - \mu_{i.} - \mu_{.j} + \mu_{..}$$

It is immediately apparent that

$$\mu_{ij} = \mu + \alpha_i + \beta_j + \gamma_{ij}$$

and it is easy to check that

$$\sum_{i=1}^{m} \alpha_i = \sum_{j=1}^{n} \beta_j = \sum_{i=1}^{m} \gamma_{ij} = \sum_{j=1}^{n} \gamma_{ij} = 0$$

The parameter μ is the average of all nm mean values; it is called the *grand mean*. The parameter α_i is the amount by which the average of the mean values of the variables in row i exceeds the grand mean; it is called the *effect of row i*. The parameter β_j is the amount by which the average of the mean values of the variables in column j exceeds the grand mean; it is called the *effect of column j*. The parameter $\gamma_{ij} = \mu_{ij} - (\mu + \alpha_i + \beta_j)$ is the amount by which μ_{ij} exceeds the sum of the grand mean and the increments due to row i and to column j; it is thus a measure of the departure from row and column additivity of the mean value μ_{ij}, and is called the *interaction of row i and column j*.

As we shall see, in order to be able to test the hypothesis that there are no row and column interactions — that is, that all $\gamma_{ij} = 0$ — it is necessary to have more than one observation for each pair of factors. So let us suppose that we have l observations for each row and column. That is, suppose that the data are $\{X_{ijk}, i = 1, \ldots, m, j = 1, \ldots, n, k = 1, \ldots, l\}$, where X_{ijk} is the kth observation in row i and column j. Because all observations are assumed to be independent normal random variables with a common variance σ^2, the model is

$$X_{ijk} \sim \mathcal{N}(\mu + \alpha_i + \beta_j + \gamma_{ij}, \sigma^2)$$

where

$$\sum_{i=1}^{m} \alpha_i = \sum_{j=1}^{n} \beta_j = \sum_{i=1}^{m} \gamma_{ij} = \sum_{j=1}^{n} \gamma_{ij} = 0 \qquad (10.6.1)$$

We will be interested in estimating the preceding parameters and in testing the following null hypotheses:

$$H_0^r : \alpha_i = 0, \quad \text{for all } i$$
$$H_0^c : \beta_j = 0, \quad \text{for all } j$$
$$H_0^{int} : \gamma_{ij} = 0, \quad \text{for all } i, j$$

That is, H_0^r is the hypothesis of no row effect; H_0^c is the hypothesis of no column effect; and H_0^{int} is the hypothesis of no row and column interaction.

To estimate the parameters, note that it is easily verified from Equation 10.8 and the identity

$$E[X_{ijk}] = \mu_{ij} = \mu + \alpha_i + \beta_j + \gamma_{ij}$$

that

$$E[X_{ij.}] = \mu_{ij} = \mu + \alpha_i + \beta_j + \gamma_{ij}$$
$$E[X_{i..}] = \mu + \alpha_i$$
$$E[X_{j.}] = \mu + \beta_j$$
$$E[X_{...}] = \mu$$

Therefore, with a "hat" over a parameter denoting the estimator of that parameter, we obtain from the preceding that unbiased estimators are given by

$$\hat{\mu} = X_{...}$$
$$\hat{\beta}_j = X_{.j.} - X_{...}$$
$$\hat{\alpha}_i = X_{i..} - X_{...}$$
$$\hat{\gamma}_{ij} = X_{ij.} - \hat{\mu} - \hat{\beta}_j - \hat{\alpha}_i = X_{ij.} - X_{i..} - X_{.j.} + X_{...}$$

To develop tests for the null hypotheses H_0^{int}, H_0^r, and H_0^c, start with the fact that

$$\sum_{k=1}^{l}\sum_{j=1}^{n}\sum_{i=1}^{m} \frac{(X_{ijk} - \mu - \alpha_i - \beta_j - \gamma_{ij})^2}{\sigma^2}$$

is a chi-square random variable with nml degrees of freedom. Therefore,

$$\sum_{k=1}^{l}\sum_{j=1}^{n}\sum_{i=1}^{m} \frac{(X_{ijk} - \hat{\mu} - \hat{\alpha}_i - \hat{\beta}_j - \hat{\gamma}_{ij})^2}{\sigma^2}$$

will also be chi-square, but with 1 degree of freedom lost for each parameter that is estimated. Now, since $\sum_i \alpha_i = 0$, it follows that $m-1$ of the α_i need to be estimated; similarly, $n-1$ of the β_j need to be estimated. Also, since $\sum_i \gamma_{ij} = \sum_j \gamma_{ij} = 0$, it follows that if we arrange all the γ_{ij} in a rectangular array having m rows and n columns, then all the row and column sums will equal 0, and so the values of the quantities in the last row and last column will be determined by the values of all the others; hence we need only estimate $(m-1)(n-1)$ of these quantities. Because we also need to estimate μ, it follows that a total of

$$n - 1 + m - 1 + (n-1)(m-1) + 1 = nm$$

parameters need to be estimated. Since

$$\hat{\mu} + \hat{\alpha}_i + \hat{\beta}_j + \hat{\gamma}_{ij} = X_{ij.}$$

it thus follows from the preceding that if we let

$$SS_e = \sum_{k=1}^{l}\sum_{j=1}^{n}\sum_{i=1}^{m}(X_{ijk} - X_{ij.})^2$$

then

$$\frac{SS_e}{\sigma^2} \text{ is chi-square with } nm(l-1) \text{ degrees of freedom}$$

Therefore,

$$\frac{SS_e}{nm(l-1)} \text{ is an unbiased estimator of } \sigma^2$$

Suppose now that we want to test the hypothesis that there are no row and column interactions — that is, we want to test

$$H_0^{int} : \gamma_{ij} = 0, \qquad i = 1, \ldots, m, \qquad j = 1, \ldots, n$$

Now, if H_0^{int} is true, then the random variables $X_{ij.}$ will be normal with mean

$$E[X_{ij.}] = \mu + \alpha_i + \beta_j$$

Also, since each of these terms is the average of l normal random variables having variance σ^2, it follows that

$$\text{Var}(X_{ij.}) = \sigma^2/l$$

Hence, under the assumption of no interactions,

$$\sum_{j=1}^{n} \sum_{i=1}^{m} \frac{l(X_{ij.} - \mu - \alpha_i - \beta_j)^2}{\sigma^2}$$

is a chi-square random variable with nm degrees of freedom. Since a total of $1 + m - 1 + n - 1 = n + m - 1$ of the parameters $\mu, \alpha_i, i = 1, \ldots, m, \beta_j, j = 1, \ldots, n$, must be estimated, it follows that if we let

$$SS_{int} = \sum_{j=1}^{n} \sum_{i=1}^{m} l(X_{ij.} - \hat{\mu} - \hat{\alpha}_i - \hat{\beta}_j)^2 = \sum_{j=1}^{n} \sum_{i=1}^{m} l(X_{ij.} - X_{i..} - X_{.j.} + X_{...})^2$$

then, under H_0^{int},

$$\frac{SS_{int}}{\sigma^2} \text{ is chi-square with } (n-1)(m-1) \text{ degrees of freedom.}$$

Therefore, under the assumption of no interactions,

$$\frac{SS_{int}}{(n-1)(m-1)} \text{ is an unbiased estimator of } \sigma^2.$$

Because it can be shown that, under the assumption of no interactions, SS_e and SS_{int} are independent, it follows that when H_0^{int} is true

$$F_{int} = \frac{SS_{int}/(n-1)(m-1)}{SS_e/nm(l-1)}$$

is an F-random variable with $(n-1)(m-1)$ numerator and $nm(l-1)$ denominator degrees of freedom. This gives rise to the following significance level α test of

$$H_0^{int} : \text{all } \gamma_{ij} = 0$$

Namely,

$$\text{reject} \quad H_0^{int} \quad \text{if} \quad \frac{SS_{int}/(n-1)(m-1)}{SS_e/nm(l-1)} > F_{(n-1)(m-1),nm(l-1),\alpha}$$
$$\text{do not reject} \quad H_0^{int} \quad \text{otherwise}$$

Alternatively, we can compute the p-value. If $F_{int} = v$, then the p-value of the test of the null hypothesis that all interactions equal 0 is

$$p\text{-value} = P\{F_{(n-1)(m-1),nm(l-1)} > v\}$$

If we want to test the null hypothesis

$$H_0^r : \alpha_i = 0, i = 1, \ldots, m$$

then we use the fact that when H_0^r is true, $X_{i..}$ is the average of nl independent normal random variables, each with mean μ and variance σ^2. Hence, under H_0^r,

$$E[X_{i..}] = \mu, \qquad \text{Var}(X_{i..}) = \sigma^2/nl$$

and so

$$\sum_{i=1}^{m} nl\frac{(X_{i..} - \mu)^2}{\sigma^2}$$

is chi-square with m degrees of freedom. Thus, if we let

$$SS_r = \sum_{i=1}^{m} nl(X_{i..} - \hat{\mu})^2 = \sum_{i=1}^{m} nl(X_{i..} - X_{..})^2$$

then, when H_0^r is true,

$$\frac{SS_r}{\sigma^2} \text{ is chi-square with } m - 1 \text{ degrees of freedom}$$

and so

$$\frac{SS_r}{m-1} \text{ is an unbiased estimator of } \sigma^2$$

Because it can be shown that, under H_0^r, SS_e and SS_r are independent, it follows that when H_0^r is true

$$\frac{SS_r/(m-1)}{SS_e/nm(l-1)} \text{ is an } F_{m-1}, nm(l-1) \text{ random variable}$$

Thus we have the following significance level α test of

$$H_0^r : \text{all } \alpha_i = 0$$

versus

$$H_1^r : \text{at least one } \alpha_i \neq 0$$

Namely,

$$\text{reject} \quad H_0^r \quad \text{if} \quad \frac{SS_r/(m-1)}{SS_e/nm(l-1)} > F_{m-1,nm(l-1),\alpha}$$

$$\text{do not reject} \quad H_0^r \quad \text{otherwise}$$

Alternatively, if $\dfrac{SS_r/(m-1)}{SS_e/nm(l-1)} = v$, then

$$p\text{-value} = P\{F_{m-1,nm(l-1)} > v\}$$

Because an analogous result can be shown to hold when testing $H_0 : \text{all } \beta_j = 0$, we obtain the ANOVA information shown in Table 10.4.

Note that all of the preceding tests call for rejection only when their related F-statistic is large. The reason that only large (and not small) values call for rejection of the null hypothesis is that the numerator of the F-statistic will tend to be larger when H_0 is not true than when it is, whereas the distribution of the denominator will be the same whether or not H_0 is true.

Program 10.6 computes the values of the F-statistics and their associated p-values.

EXAMPLE 10.6a The life of a particular type of generator is thought to be influenced by the material used in its construction and also by the temperature at the location where it is utilized. The following table represents lifetime data on 24 generators made from three different types of materials and utilized at two different temperatures. Do the data indicate that the material and the temperature do indeed affect the lifetime of a generator? Is there evidence of an interaction effect?

Material	Temperature 10°C	Temperature 18°C
1	135, 150	50, 55
	176, 85	64, 38
2	150, 162	76, 88
	171, 120	91, 57
3	138, 111	68, 60
	140, 106	74, 51

SOLUTION Run Program 10.6 (see Figures 10.3 and 10.4). ∎

TABLE 10.4   *Two-way ANOVA with l Observations per Cell: $N = nm(l-1)$*

Source of Variation	Degrees of Freedom	Sum of Squares	F-Statistic	Level α Test	p-Value if $F = v$
Row	$m-1$	$SS_r = ln\sum_{i=1}^{m}(X_{i..} - X_{...})^2$	$F_r = \dfrac{SS_r/(m-1)}{SS_e/N}$	Reject H_0^r if $F_r > F_{m-1,N,\alpha}$	$P\{F_{m-1,N} > v\}$
Column	$n-1$	$SS_c = lm\sum_{j=1}^{n}(X_{.j.} - X_{...})^2$	$F_c = \dfrac{SS_c/(n-1)}{SS_e/N}$	Reject H_0^c if $F_c > F_{n-1,N,\alpha}$	$P\{F_{n-1,N} > v\}$
Interaction	$(n-1)(m-1)$	$SS_{int} = l\sum_{j=1}^{n} \times \sum_{i=1}^{m}(X_{ij.} - X_{i..} - X_{.j.} + X_{...})^2$	$F_{int} = \dfrac{SS_{int}/(n-1)(m-1)}{SS_e/N}$	Reject H_0^{int} if $F_{int} > F_{(n-1)(m-1),N,\alpha}$	$P\{F_{(n-1)(m-1),N} > v\}$
Error	N	$SS_e = \sum_{k=1}^{l}\sum_{j=1}^{n} \times \sum_{i=1}^{m}(X_{ijk} - X_{ij.})^2$			

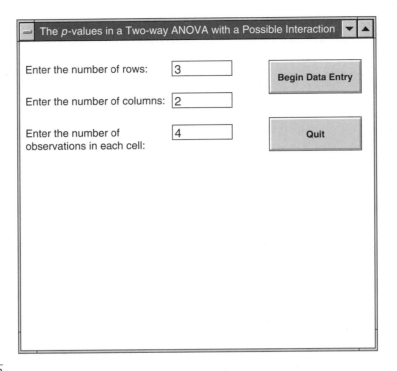

FIGURE 10.3

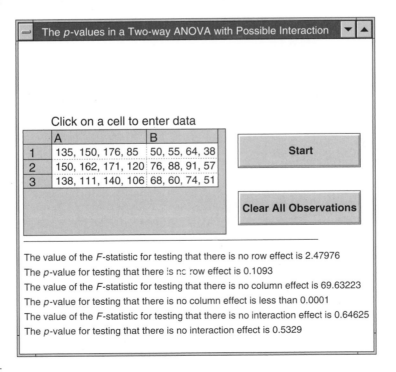

FIGURE 10.4

Problems

1. A purification process for a chemical involves passing it, in solution, through a resin on which impurities are adsorbed. A chemical engineer wishing to test the efficiency of 3 different resins took a chemical solution and broke it into 15 batches. She tested each resin 5 times and then measured the concentration of impurities after passing through the resins. Her data were as follows:

Concentration of Impurities		
Resin I	Resin II	Resin III
.046	.038	.031
.025	.035	.042
.014	.031	.020
.017	.022	.018
.043	.012	.039

Test the hypothesis that there is no difference in the efficiency of the resins.

2. We want to know what type of filter should be used over the screen of a cathode-ray oscilloscope in order to have a radar operator easily pick out targets on the presentation. A test to accomplish this has been set up. A noise is first applied to the scope to make it difficult to pick out a target. A second signal, representing the target, is put into the scope, and its intensity is increased from zero until detected by the observer. The intensity setting at which the observer first notices the target signal is then recorded. This experiment is repeated 20 times with each filter. The numerical value of each reading listed in the table of data is proportional to the target intensity at the time the operator first detects the target.

Filter No. 1	Filter No. 2	Filter No. 3
90	88	95
87	90	95
93	97	89
96	87	98
94	90	96
88	96	81
90	90	92
84	90	79
101	100	105
96	93	98
90	95	92

(continued)

Filter No. 1	Filter No. 2	Filter No. 3
82	86	85
93	89	97
90	92	90
96	98	87
87	95	90
99	102	101
101	105	100
79	85	84
98	97	102

Test, at the 5 percent level of significance, the hypothesis that the filters are the same.

3. Explain why we cannot efficiently test the hypothesis $H_0 : \mu_1 = \mu_2 = \cdots = \mu_m$ by running t-tests on all of the $\binom{m}{2}$ pairs of samples.

4. A machine shop contains 3 ovens that are used to heat metal specimens. Subject to random fluctuations, they are all supposed to heat to the same temperature. To test this hypothesis, temperatures were noted on 15 separate heatings. The following data resulted.

Oven	Temperature
1	492.4, 493.6, 498.5, 488.6, 494
2	488.5, 485.3, 482, 479.4, 478
3	502.1, 492, 497.5, 495.3, 486.7

Do the ovens appear to operate at the same temperature? Test at the 5 percent level of significance. What is the p-value?

5. Four standard chemical procedures are used to determine the magnesium content in a certain chemical compound. Each procedure is used four times on a given compound with the following data resulting.

Method			
1	2	3	4
76.42	80.41	74.20	86.20
78.62	82.26	72.68	86.04
80.40	81.15	78.84	84.36
78.20	79.20	80.32	80.68

Do the data indicate that the procedures yield equivalent results?

6. Twenty overweight individuals, each more than 40 pounds overweight, were randomly assigned to one of two diets. After 10 weeks, the total weight losses (in pounds) of the individuals on each of the diets were as follows:

Weight Loss	
Diet 1	Diet 2
22.2	24.2
23.4	16.8
24.2	14.6
16.1	13.7
9.4	19.5
12.5	17.6
18.6	11.2
32.2	9.5
8.8	30.1
7.6	21.5

Test, at the 5 percent level of significance, the hypothesis that the two diets have equal effect.

7. In a test of the ability of a certain polymer to remove toxic wastes from water, experiments were conducted at three different temperatures. The data below give the percentages of the impurities that were removed by the polymer in 21 independent attempts.

Low Temperature	Medium Temperature	High Temperature
42	36	33
41	35	44
37	32	40
29	38	36
35	39	44
40	42	37
32	34	45

Test the hypothesis that the polymer performs equally well at all three temperatures. Use the (a) 5 percent level of significance and (b) 1 percent level of significance.

8. In the one-factor analysis of variance model with n observations per sample, let S_i^2, $i = 1, \ldots, m$ denote the sample variances for the m samples. Show that

$$SS_W = (n - 1) \sum_{i=1}^{m} S_i^2$$

9. The following data relate to the ages at death of a certain species of rats that were fed 1 of 3 types of dies. Thirty rats of a type having a short life span were randomly divided into 3 groups of size 10 each. The sample means and sample variances of the ages at death (measured in months) of the 3 groups are as follows:

	Very Low Calorie	Moderate Calorie	High Calorie
Sample mean	22.4	16.8	13.7
Sample variance	24.0	23.2	17.1

Test the hypothesis, at the 5 percent level of significance, that the mean lifetime of a rat is not affected by its diet. What about at the 1 percent level?

10. Plasma bradykininogen levels are related to the body's ability to resist inflammation. In a 1968 study (Eilam, N., Johnson, P. K., Johnson, N. L., and Creger, W., "Bradykininogen levels in Hodgkin's disease," *Cancer*, **22**, pp. 631–634), levels were measured in normal patients, in patients with active Hodgkin's disease, and in patients with inactive Hodgkin's disease. The following data (in micrograms of bradykininogen per milliliter of plasma) resulted.

Normal	Active Hodgkin's Disease	Inactive Hodgkin's Disease
5.37	3.96	5.37
5.80	3.04	10.60
4.70	5.28	5.02
5.70	3.40	14.30
3.40	4.10	9.90
8.60	3.61	4.27
7.48	6.16	5.75
5.77	3.22	5.03
7.15	7.48	5.74
6.49	3.87	7.85
4.09	4.27	6.82
5.94	4.05	7.90
6.38	2.40	8.36

Test, at the 5 percent level of significance, the hypothesis that the mean bradykininogen levels are the same for all three groups.

11. A study of the trunk flexor muscle strength of 75 girls aged 3 to 7 was reported by Baldauf, K., Swenson, D., Medeiros, J., and Radtka, S., "Clinical assessment of trunk flexor muscle strength in healthy girls 3 to 7," *Physical Therapy*, **64**, pp. 1203–1208, 1984. With muscle strength graded on a scale of 0 to 5, and with 15 girls in each age group, the following sample means and sample standard deviations resulted.

Age	3	4	5	6	7
Sample mean	3.3	3.7	4.1	4.4	4.8
Sample standard deviation	.9	1.1	1.1	.9	.5

Test, at the 5 percent level of significance, the hypothesis that the mean trunk flexor strength is the same for all five age groups.

12. An emergency room physician wanted to know whether there were any differences in the amount of time it takes for three different inhaled steroids to clear a mild asthmatic attack. Over a period of weeks she randomly administered these steroids to asthma sufferers, and noted the time it took for the patients' lungs to become clear. Afterward, she discovered that 12 patients had been treated with each type of steroid, with the following sample means (in minutes) and sample variances resulting.

Steroid	$\overline{X}_i$	S_i^2
A	32	145
B	40	138
C	30	150

   (a) Test the hypothesis that the mean time to clear a mild asthmatic attack is the same for all three steroids. Use the 5 percent level of significance.
   (b) Find confidence intervals for all quantities $\mu_i - \mu_j$ that, with 95 percent confidence, are valid.

13. Five servings each of three different brands of processed meat were tested for fat content. The following data (in fat percentage per gram) resulted.

Brand	1	2	3
	32	41	36
Fat	34	32	37
content	31	33	30
	35	29	28
	33	35	33

   (a) Does the fat content differ depending on the brand?
   (b) Find confidence intervals for all quantities $\mu_i - \mu_j$ that, with 95 percent confidence, are valid.

14. A nutritionist randomly divided 15 bicyclists into 3 groups of 5 each. The first group was given a vitamin supplement to take with each of their meals during the

next 3 weeks. The second group was instructed to eat a particular type of high-fiber whole-grain cereal for the next 3 weeks. The final group was instructed to eat as they normally do. After the 3-week period elapsed, the nutritionist had each of the bicyclists ride 6 miles. The following times were recorded.

Vitamin group:	15.6	16.4	17.2	15.5	16.3
Fiber cereal group:	17.1	16.3	15.8	16.4	16.0
Control group:	15.9	17.2	16.4	15.4	16.8

(a) Are the data consistent with the hypothesis that neither the vitamin nor the fiber cereal affected the bicyclists' speeds? Use the 5 percent level of significance.

(b) Find confidence intervals for all quantities $\mu_i - \mu_j$ that, with 95 percent confidence, are valid.

15. Test the hypothesis that the following three independent samples all come from the same normal probability distribution.

Sample 1	Sample 2	Sample 3
35	29	44
37	38	52
29	34	56
27	30	
30	32	

16. For data $x_{ij}, i = 1, \ldots, m, j = 1, \ldots, n$, show that

$$x_{..} = \sum_{i=1}^{m} x_{i.}/m = \sum_{j=1}^{n} x_{.j}/n$$

17. If $x_{ij} = i + j^2$, determine

(a) $\sum_{j=1}^{3} \sum_{i=1}^{2} x_{ij}$

(b) $\sum_{i=1}^{2} \sum_{j=1}^{3} x_{ij}$

18. If $x_{ij} = a_i + b_j$, show that

$$\sum_{i=1}^{m} \sum_{j=1}^{n} x_{ij} = n \sum_{i=1}^{m} a_i + m \sum_{j=1}^{n} b_j$$

19. A study has been made on pyrethrum flowers to determine the content of pyrethrin, a chemical used in insecticides. Four methods of extracting the chemical are used,

and samples are obtained from flowers stored under three conditions: fresh flowers, flowers stored for 1 year, and flowers stored for 1 year but treated. It is assumed that there is no interaction present. The data are as follows:

Pyrethrin Content, Percent

Storage	Method			
Condition	A	B	C	D
1	1.35	1.13	1.06	.98
2	1.40	1.23	1.26	1.22
3	1.49	1.46	1.40	1.35

Suggest a model for the preceding information, and use the data to estimate its parameters.

20. The following data refer to the number of deaths per 10,000 adults in a large Eastern city in the different seasons for the years 1982 to 1986.

Year	Winter	Spring	Summer	Fall
1982	33.6	31.4	29.8	32.1
1983	32.5	30.1	28.5	29.9
1984	35.3	33.2	29.5	28.7
1985	34.4	28.6	33.9	30.1
1986	37.3	34.1	28.5	29.4

(a) Assuming a two-factor model, estimate the parameters.
(b) Test the hypothesis that death rates do not depend on the season. Use the 5 percent level of significance.
(c) Test, at the 5 percent level of significance, the hypothesis that there is no effect due to the year.

21. For the model of Problem 19:

(a) Do the methods of extraction appear to differ?
(b) Do the storage conditions affect the content? Test at the $\alpha = .05$ level of significance.

22. Three different washing machines were employed to test four different detergents. The following data give a coded score of the effectiveness of each washing.

(a) Estimate the improvement in mean value when using detergent 1 over using detergents (i) 2; (ii) 3; (iii) 4.
(b) Estimate the improvement in mean value when using machine 3 as opposed to using machine (i) 1; (ii) 2.

| | Machine | | |
	1	2	3
Detergent 1	53	50	59
Detergent 2	54	54	60
Detergent 3	56	58	62
Detergent 4	50	45	57

(c) Test the hypothesis that the detergent used does not affect the score.
(d) Test the hypothesis that the machine used does not affect the score.

Use, in both (c) and (d), the 5 percent level of significance.

23. An experiment was devised to test the effects of running 3 different types of gasoline with 3 possible types of additives. The experiment called for 9 identical motors to be run with 5 gallons for each of the pairs of gasoline and additives. The following data resulted.

Mileage Obtained

| | Additive | | |
Gasoline	1	2	3
1	124.1	131.5	127
2	126.4	130.6	128.4
3	127.2	132.7	125.6

(a) Test the hypothesis that the gasoline used does not affect the mileage.
(b) Test the hypothesis that the additives are equivalent.
(c) What assumptions are you making?

24. Suppose in Problem 6 that the 10 people placed on each diet consisted of 5 men and 5 women, with the following data.

	Diet 1	Diet 2
Women	7.6	19.5
	8.8	17.6
	12.5	16.8
	16.1	13.7
	18.6	21.5
Men	22.2	30.1
	23.4	24.2
	24.2	9.5
	32.2	14.6
	9.4	11.2

(a) Test the hypothesis that there is no interaction between gender and diet.

(b) Test the hypothesis that the diet has the same effect on men and women.

25. A researcher is interested in comparing the breaking strength of different laminated beams made from 3 different types of glue and 3 varieties of wood. To make the comparison, 5 beams of each of the 9 combinations were manufactured and then put under a stress test. The following table indicates the pressure readings at which each of the beams broke.

Wood \ Glue	G1		G2		G3	
W1	196	208	214	216	258	250
	247	216	235	240	264	248
	221		252		272	
W2	216	228	215	217	246	247
	240	224	235	219	261	250
	236		241		255	
W3	230	242	212	218	255	251
	232	244	216	224	261	258
	228		222		247	

(a) Test the hypothesis that the wood and glue effect is additive.

(b) Test the hypothesis that the wood used does not affect the breaking strength.

(c) Test the hypothesis that the glue used does not affect the breaking strength.

26. A study was made as to how the concentration of a certain drug in the blood, 24 hours after being injected, is influenced by age and gender. An analysis of the blood samples of 40 people given the drug yielded the following concentrations (in milligrams per cubic centimeter).

	Age Group			
	11–25	26–40	41–65	Over 65
Male	52	52.5	53.2	82.4
	56.6	49.6	53.6	86.2
	68.2	48.7	49.8	101.3
	82.5	44.6	50.0	92.4
	85.6	43.4	51.2	78.6
Female	68.6	60.2	58.7	82.2
	80.4	58.4	55.9	79.6
	86.2	56.2	56.0	81.4
	81.3	54.2	57.2	80.6
	77.2	61.1	60.0	82.2

(a) Test the hypothesis of no age and gender interaction.

(b) Test the hypothesis that gender does not affect the blood concentration.

(c) Test the hypothesis that age does not affect blood concentration.

27. Suppose, in Problem 23, that there has been some controversy about the assumption of no interaction between gasoline and additive used. To allow for the possibility of an interaction effect between gasoline and additive, it was decided to run 36 motors — 4 in each grouping. The following data resulted.

Gasoline	Additive		
	1	2	3
1	126.2	130.4	127
	124.8	131.6	126.6
	125.3	132.5	129.4
	127.0	128.6	130.1
2	127.2	142.1	129.5
	126.6	132.6	142.6
	125.8	128.5	140.5
	128.4	131.2	138.7
3	127.1	132.3	125.2
	128.3	134.1	123.3
	125.1	130.6	122.6
	124.9	133.0	120.9

(a) Do the data indicate an interaction effect?

(b) Do the gasolines appear to give equal results?

(c) Test whether or not there is an additive effect or whether all additives work equally well.

(d) What conclusions can you draw?

28. An experiment has been devised to test the hypothesis that an elderly person's memory retention can be improved by a set of "oxygen treatments." A group of scientists administered these treatments to men and women. The men and women were each randomly divided into 4 groups of 5 each, and the people in the ith group were given treatments over an $(i - 1)$ week interval, $i = 1, 2, 3, 4$. (The 2 groups not given any treatments served as "controls.") The treatments were set up in such a manner that all individuals thought they were receiving the oxygen treatments for the total 3 weeks. After treatment ended, a memory retention test was administered. The results (with higher scores indicating higher memory retentions) are shown in the table.

(a) Test whether or not there is an interaction effect.

(b) Test the hypothesis that the length of treatment does not affect memory retention.

(c) Is there a gender difference?

(d) A randomly chosen group of 5 elderly men, without receiving any oxygen treatment, were given the memory retention test. Their scores were 37, 35, 33, 39, 29. What conclusions can you draw?

Scores

| | Number of Weeks of Oxygen Treatment | | | |
	0	1	2	3
Men	42	39	38	42
	54	52	50	55
	46	51	47	39
	38	50	45	38
	51	47	43	51
Women	49	48	27	61
	44	51	42	55
	50	52	47	45
	45	54	53	40
	43	40	58	42

29. In a study of platelet production, 16 rats were put at an altitude of 15,000 feet, while another 16 were kept at sea level (Rand, K., Anderson, T., Lukis, G., and Creger, W., "Effect of hypoxia on platelet level in the rat," *Clinical Research,* **18,**

	Spleen Removed	Normal Spleen
Altitude	528	434
	444	331
	338	312
	342	575
	338	472
	331	444
	288	575
	319	384
Sea Level	294	272
	254	275
	352	350
	241	350
	291	466
	175	388
	241	425
	238	344

p. 178, 1970). Half of the rats in both groups had their spleens removed. The fibrinogen levels on day 21 are reported below.

(a) Test the hypothesis that there are no interactions.
(b) Test the hypothesis that there is no effect due to altitude.
(c) Test the hypothesis that there is no effect due to spleen removal. In all cases, use the 5 percent level of significance.

30. Suppose that $\mu, \alpha_1, \ldots, \alpha_m, \beta_1, \ldots, \beta_n$ and $\mu', \alpha_1', \ldots, \alpha_m', \beta_1', \ldots, \beta_n'$ are such that

$$\mu + \alpha_i + \beta_j = \mu' + \alpha_i' + \beta_j' \qquad \text{for all } i, j$$

$$\sum_i \alpha_i = \sum_i \alpha_i' = \sum_j \beta_j = \sum_j \beta_j' = 0$$

Show that

$$\mu = \mu', \; \alpha_i = \alpha_i', \; \beta_j = \beta_j'$$

for all i and j. This shows that the parameters $\mu, \alpha_1, \ldots, \alpha_m, \beta_1, \ldots, \beta_n$ in our representation of two-factor ANOVA are uniquely determined.

Chapter 11

GOODNESS OF FIT TESTS AND CATEGORICAL DATA ANALYSIS

11.1 INTRODUCTION

We are often interested in determining whether or not a particular probabilistic model is appropriate for a given random phenomenon. This determination often reduces to testing whether a given random sample comes from some specified, or partially specified, probability distribution. For example, we may *a priori* feel that the number of industrial accidents occurring daily at a particular plant should constitute a random sample from a Poisson distribution. This hypothesis can then be tested by observing the number of accidents over a sequence of days and then testing whether it is reasonable to suppose that the underlying distribution is Poisson. Statistical tests that determine whether a given probabilistic mechanism is appropriate are called *goodness of fit* tests.

The classical approach to obtaining a goodness of fit test of a null hypothesis that a sample has a specified probability distribution is to partition the possible values of the random variables into a finite number of regions. The numbers of the sample values that fall within each region are then determined and compared with the theoretical expected numbers under the specified probability distribution, and when they are significantly different the null hypothesis is rejected. The details of such a test are presented in Section 11.2, where it is assumed that the null hypothesis probability distribution is completely specified. In Section 11.3, we show how to do the analysis when some of the parameters of the null hypothesis distribution are left unspecified; that is, for instance, the null hypothesis might be that the sample distribution is a normal distribution, without specifying the mean and variance of this distribution. In Sections 11.4 and 11.5, we consider situations where each member of a population is classified according to two distinct characteristics, and we show how to use our previous analysis to test the hypothesis that the characteristics of a randomly chosen member of the population are independent. As an application, we show how to test the hypothesis that m population all have the same discrete probability distribution. Finally, in the optional section, Section 11.6, we return to the problem of testing that sample data come from a specified probability distribution, which we now assume is continuous. Rather than discretizing the data so as to be able to use the test of Section 11.2, we treat the data as given and make use of the *Kolmogorov–Smirnov test*.

11.2 GOODNESS OF FIT TESTS WHEN ALL PARAMETERS ARE SPECIFIED

Suppose that n independent random variables — $Y_1, \ldots, Y_n$, each taking on one of the values $1, 2, \ldots, k$ — are to be observed and we are interested in testing the null hypothesis that $\{p_i, i = 1, \ldots, k\}$ is the probability mass function of the Y_j. That is, if Y represents any of the Y_j, then the null hypothesis is

$$H_0 : P\{Y = i\} = p_i, \qquad i = 1, \ldots, k$$

whereas the alternative hypothesis is

$$H_1 : P\{Y = i\} \neq p_i, \qquad \text{for some } i = 1, \ldots, k$$

To test the foregoing hypothesis, let $X_i, i = 1, \ldots, k$, denote the number of the Y_j's that equal i. Then as each Y_j will independently equal i with probability $P\{Y = i\}$, it follows that, under H_0, X_i is binomial with parameters n and p_i. Hence, when H_0 is true,

$$E[X_i] = np_i$$

and so $(X_i - np_i)^2$ will be an indication as to how likely it appears that p_i indeed equals the probability that $Y = i$. When this is large, say, in relationship to np_i, then it is an indication that H_0 is not correct. Indeed such reasoning leads us to consider the following test statistic:

$$T = \sum_{i=1}^{k} \frac{(X_i - np_i)^2}{np_i} \tag{11.2.1}$$

and to reject the null hypothesis when T is large.

To determine the critical region, we need first specify a significance level α and then we must determine that critical value c such that

$$P_{H_0}\{T \geq c\} = \alpha$$

That is, we need to determine c so that the probability that the test statistic T is at least as large as c, when H_0 is true, is α. The test is then to reject the hypothesis, at the α level of significance, when $T \geq c$ and to accept when $T < c$.

It remains to determine c. The classical approach to doing so is to use the result that when n is large T will have, when H_0 is true, approximately (with the approximation becoming exact as n approaches infinity) a chi-square distribution with $k - 1$ degrees of freedom. Hence, for n large, c can be taken to equal $\chi^2_{\alpha,k-1}$; and so the approximate α level test is

$$\text{reject} \quad H_0 \quad \text{if} \quad T \geq \chi^2_{\alpha,k-1}$$

$$\text{accept} \quad H_0 \quad \text{otherwise}$$

If the observed value of T is $T = t$, then the preceding test is equivalent to rejecting H_0 if the significance level α is at least as large as the p-value given by

$$p\text{-value} = P_{H_0}\{T \geq t\}$$

$$\approx P\{\chi^2_{k-1} \geq t\}$$

where χ^2_{k-1} is a chi-square random variable with $k - 1$ degrees of freedom.

An accepted rule of thumb as to how large n need be for the foregoing to be a good approximation is that it should be large enough so that $np_i \geq 1$ for each $i, i = 1, \ldots, k$, and also at least 80 percent of the values np_i should exceed 5.

REMARKS

(a) A computationally simpler formula for T can be obtained by expanding the square in Equation 11.2.1 and using the results that $\sum_i p_i = 1$ and $\sum_i X_i = n$ (why is this true?):

$$T = \sum_{i=1}^{k} \frac{X_i^2 - 2np_i X_i + n^2 p_i^2}{np_i} \tag{11.2.2}$$

$$= \sum_i X_i^2 / np_i - 2 \sum_i X_i + n \sum_i p_i$$

$$= \sum_i X_i^2 / np_i - n$$

(b) The intuitive reason why T, which depends on the k values $X_1, \ldots, X_k$, has only $k - 1$ degrees of freedom is that 1 degree of freedom is lost because of the linear relationship $\sum_i X_i = n$.

(c) Whereas the proof that, asymptotically, T has a chi-square distribution is advanced, it can be easily shown when $k = 2$. In this case, since $X_1 + X_2 = n$, and $p_1 + p_2 = 1$, we

492 Chapter 11: Goodness of Fit Tests and Categorical Data Analysis

see that

$$
\begin{aligned}
T &= \frac{(X_1 - np_1)^2}{np_1} + \frac{(X_2 - np_2)^2}{np_2} \\
&= \frac{(X_1 - np_1)^2}{np_1} + \frac{(n - X_1 - n[1 - p_1])^2}{n(1 - p_1)} \\
&= \frac{(X_1 - np_1)^2}{np_1} + \frac{(X_1 - np_1)^2}{n(1 - p_1)} \\
&= \frac{(X_1 - np_1)^2}{np_1(1 - p_1)} \qquad \text{since} \qquad \frac{1}{p} + \frac{1}{1 - p} = \frac{1}{p(1 - p)}
\end{aligned}
$$

However, X_1 is a binomial random variable with mean np_1 and variance $np_1(1 - p_1)$ and thus, by the normal approximation to the binomial, it follows that $(X_1 - np_1)/\sqrt{np_1(1 - p_1)}$ has, for large n, approximately a standard normal distribution, and so its square has approximately a chi-square distribution with 1 degree of freedom.

EXAMPLE 11.2a In recent years, a correlation between mental and physical well-being has increasingly become accepted. An analysis of birthdays and death days of famous people could be used as further evidence in the study of this correlation. To use these data, we are supposing that being able to look forward to something betters a person's mental state, and that a famous person would probably look forward to his or her birthday because of the resulting attention, affection, and so on. If a famous person is in poor health and dying, then perhaps anticipating his birthday would "cheer him up and therefore improve his health and possibly decrease the chance that he will die shortly before his birthday." The data might therefore reveal that a famous person is less likely to die in the months before his or her birthday and more likely to die in the months afterward.

SOLUTION To test this, a sample of 1,251 (deceased) Americans was randomly chosen from *Who Was Who in America*, and their birth and death days were noted. (The data are taken from D. Phillips, "Death Day and Birthday: An Unexpected Connection," in *Statistics: A Guide to the Unknown*, Holden-Day, 1972.) The data are summarized in Table 11.1.

If the death day does not depend on the birthday, then it would seem that each of the 1,251 individuals would be equally likely to fall in any of the 12 categories. Thus, let us test the null hypothesis

$$
H_0 = p_i = \frac{1}{12}, \qquad i = 1, \ldots, 12
$$

TABLE 11.1 *Number of Deaths Before, During, and After the Birth Month*

	6 Months Before	5 Months Before	4 Months Before	3 Months Before	2 Months Before	1 Month Before	The Month	1 Month After	2 Months After	3 Months After	4 Months After	5 Months After
Number of deaths	90	100	87	96	101	86	119	118	121	114	113	106

$n = 1{,}251$
$n/12 = 104.25$

Since $np_i = 1,251/12 = 104.25$, the chi-square test statistic for this hypothesis is

$$T = \frac{(90)^2 + (100)^2 + (87)^2 + \cdots + (106)^2}{104.25} - 1,251$$

$$= 17.192$$

The p-value is

$$p\text{-value} \approx P\{\chi_{11}^2 \geq 17.192\}$$

$$= 1 - .8977 = .1023 \quad \text{by Program 5.8.1a}$$

The results of this test leave us somewhat up in the air about the hypothesis that an approaching birthday has no effect on an individual's remaining lifetime. For whereas the data are not quite strong enough (at least, at the 10 percent level of significance) to reject this hypothesis, they are certainly suggestive of its possible falsity. This raises the possibility that perhaps we should not have allowed as many as 12 data categories, and that we might have obtained a more powerful test by allowing for a fewer number of possible outcomes. For instance, let us determine what the result would have been if we had coded the data into 4 possible outcomes as follows:

$$\text{outcome } 1 = -6, -5, -4$$
$$\text{outcome } 2 = -3, -2, -1$$
$$\text{outcome } 3 = 0, 1, 2$$
$$\text{outcome } 4 = 3, 4, 5$$

That is, for instance, an individual whose death day occurred 3 months before his or her birthday would be placed in outcome 2. With this classification, the data would be as follows:

Outcome	Number of Times Occurring
1	277
2	283
3	358
4	333

$n = 1,251$
$n/4 = 312.75$

The test statistic for testing $H_0 = p_i = 1/4, i = 1, 2, 3, 4$ is

$$T = \frac{(277)^2 + (283)^2 + (358)^2 + (333)^2}{312.75} - 1.251$$

$$= 14.775$$

Hence, as $\chi^2_{.01,3} = 11.345$, the null hypothesis would be rejected even at the 1 percent level of significance. Indeed, using Program 5.8.1a yields that

$$p\text{-value} \approx P\{\chi^2_3 \geq 14.775\} = 1 - .998 = .002$$

The foregoing analysis is, however, subject to the criticism that the null hypothesis was chosen after the data were observed. Indeed, while there is nothing incorrect about using a set of data to determine the "correct way" of phrasing a null hypothesis, the additional use of those data to test that very hypothesis is certainly questionable. Therefore, to be quite certain of the conclusion to be drawn from this example, it seems prudent to choose a second random sample — coding the values as before — and again test $H_0 : p_i = 1/4, i = 1, 2, 3, 4$ (see Problem 3). ■

Program 11.2.1 can be used to quickly calculate the value of T.

EXAMPLE 11.2b A contractor who purchases a large number of fluorescent lightbulbs has been told by the manufacturer that these bulbs are not of uniform quality but rather have been produced in such a way that each bulb produced will, independently, either be of quality level A, B, C, D, or E, with respective probabilities .15, .25, .35, .20, .05. However, the contractor feels that he is receiving too many type E (the lowest quality) bulbs, and so he decides to test the producer's claim by taking the time and expense to ascertain the quality of 30 such bulbs. Suppose that he discovers that of the 30 bulbs, 3 are of quality level A, 6 are of quality level B, 9 are of quality level C, 7 are of quality level D, and 5 are of quality level E. Do these data, at the 5 percent level of significance, enable the contractor to reject the producer's claim?

SOLUTION Program 11.2.1 gives the value of the test statistic as 9.348. Therefore,

$$p\text{-value} = P_{H_0}\{T \geq 9.348\}$$
$$\approx P\{\chi^2_4 \geq 9.348\}$$
$$= 1 - .947 \quad \text{from Program 5.8.1a}$$
$$= .053$$

Thus the hypothesis would not be rejected at the 5 percent level of significance (but since it would be rejected at all significance levels above .053, the contractor should certainly remain skeptical). ■

11.2.1 DETERMINING THE CRITICAL REGION BY SIMULATION

From 1900 when Karl Pearson first showed that T has approximately (becoming exact as n approaches infinity) a chi-square distribution with $k - 1$ degrees of freedom, until relatively recently, this approximation was the only means available for determining the p-value of the goodness of fit test. However, with the advent of inexpensive, fast, and easily available computational power a second, potentially more accurate, approach has become available: namely, the use of simulation to obtain to a high level of accuracy the p-value of the test statistic.

The simulation approach is as follows. First, the value of T is determined — say, $T = t$. Now to determine whether or not to accept H_0, at a given significance level α, we need to know the probability that T would be at least as large as t when H_0 is true. To determine this probability, we simulate n independent random variables $Y_1^{(1)}, \ldots, Y_n^{(1)}$ each having the probability mass function $\{p_i, i = 1, \ldots, k\}$ — that is,

$$P\{Y_j^{(1)} = i\} = p_i, \qquad i = 1, \ldots, k, \qquad j = 1, \ldots, n$$

Now let

$$X_i^{(1)} = \text{number } j : Y_j^{(1)} = i$$

and set

$$T^{(1)} = \sum_{i=1}^{k} \frac{(X_i^{(1)} - np_i)^2}{np_i}$$

Now repeat this procedure by simulating a second set, independent of the first set, of n independent random variables $Y_1^{(2)}, \ldots, Y_n^{(2)}$ each having the probability mass function $\{p_i, i = 1, \ldots, k\}$ and then, as for the first set, determining $T^{(2)}$. Repeating this a large number, say, r, of times yields r independent random variables $T^{(1)}, T^{(2)}, \ldots, T^{(r)}$, each of which has the same distribution as does the test statistic T when H_0 is true. Hence, by the law of large numbers, the proportion of the T_i that are as large as t will be very nearly equal to the probability that T is as large as t when H_0 is true — that is,

$$\frac{\text{number } l : T^{(l)} \geq t}{r} \approx P_{H_0}\{T \geq t\}$$

In fact, by letting r be large, the foregoing can be considered to be, with high probability, almost an equality. Hence, if that proportion is less than or equal to α, then the p-value, equal to the probability of observing a T as large as t when H_0 is true, is less than α and so H_0 should be rejected.

REMARKS

(a) To utilize the foregoing simulation approach to determine whether or not to accept H_0 when T is observed, we need to specify how one can simulate, or generate, a random variable Y such that $P\{Y = i\} = p_i, i = 1, \ldots, k$. One way is as follows:

Step 1: Generate a random number U.
Step 2: If

$$p_1 + \cdots + p_{i-1} \le U < p_1 + \cdots + p_i$$

set $Y = i$ (where $p_1 + \cdots + p_{i-1} \equiv 0$ when $i = 1$). That is,

$$U < p_1 \Rightarrow Y = 1$$
$$p_1 \le U < p_1 + p_2 \Rightarrow Y = 2$$
$$\vdots$$
$$p_1 + \cdots + p_{i-1} \le U < p_1 + \cdots + p_i \Rightarrow Y = i$$
$$\vdots$$
$$p_1 + \cdots + p_{n-1} < U \Rightarrow Y = n$$

Since a random number is equivalent to a uniform $(0, 1)$ random variable, we have that

$$P\{a < U < b\} = b - a, \qquad 0 < a < b < 1$$

and so

$$P\{Y = i\} = P\{p_1 + \cdots + p_{i-1} < U < p_1 + \cdots + p_i\} = p_i$$

(b) A significant question that remains is how many simulation runs are necessary. It has been shown that the value $r = 100$ is usually sufficient at the conventional 5 percent level of significance.[*]

EXAMPLE 11.2c Let us reconsider the problem presented in Example 11.2b. A simulation study yielded the result

$$P_{H_0}\{T \le 9.52381\} = .95$$

and so the critical value should be 9.52381, which is remarkably close to $\chi^2_{.05,4} = 9.488$ given as the critical value by the chi-square approximation. This is most interesting since the rule of thumb for when the chi-square approximation can be applied — namely, that

---

[*] See Hope, A., "A Simplified Monte Carlo Significance Test Procedure," *J. of Royal Statist. Soc.*, B. 30, 582–598, 1968.

each $np_i \geq 1$ and at least 80 percent of the np_i exceed 5 — does not apply, thus raising the possibility that it is rather conservative. ■

Program 11.2.2 can be utilized to determine the p-value.

To obtain more information as to how well the chi-square approximation performs, consider the following example.

EXAMPLE 11.2d   Consider an experiment having six possible outcomes whose probabilities are hypothesized to be .1, .1, .05, .4, .2, and .15. This is to be tested by performing 40 independent replications of the experiment. If the resultant number of times that each of the six outcomes occurs is 3, 3, 5, 18, 4, 7, should the hypothesis be accepted?

SOLUTION   A direct computation, or the use of Program 11.2.1, yields that the value of the test statistic is 7.4167. Utilizing Program 5.8.1a gives the result that

$$P\{\chi_5^2 \leq 7.4167\} = .8088$$

and so

$$p\text{-value} \approx .1912$$

To check the foregoing approximation, we ran Program 11.2.2, using 10,000 simulation runs, and obtained an estimate of the p-value equal to .1843 (see Figure 11.1).

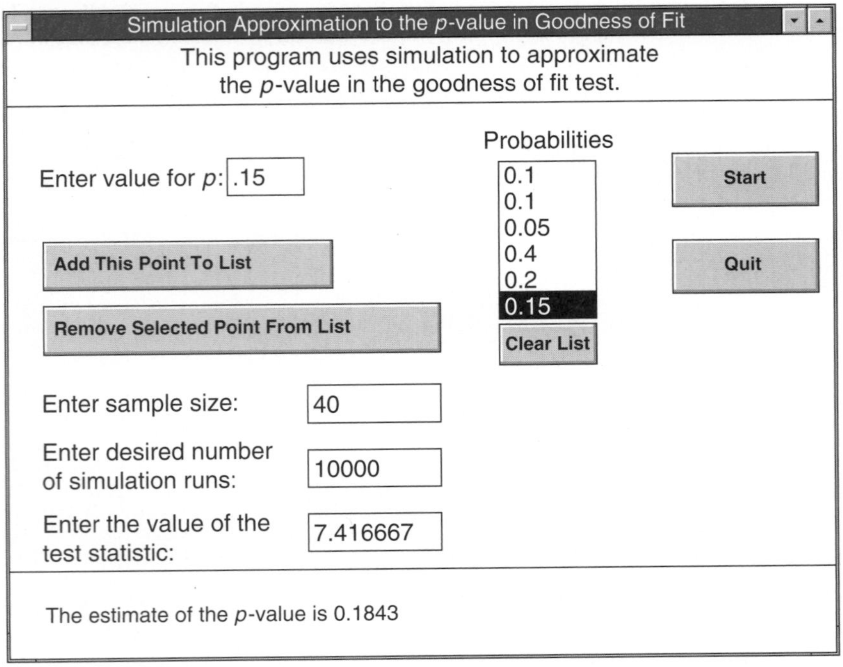

FIGURE 11.1

Since the number of the 10^4 simulated values that exceed 7.4167 is a binomial random variable with parameters $n = 10^4$ and $p = p$-value, it follows that a 90 percent confidence interval for the p-value is

$$p\text{-value} \in .1843 \pm 1.645\sqrt{.1843(.8157)/10^4}$$

That is, with 90 percent confidence

$$p\text{-value} \in (.1779, .1907) \quad \blacksquare$$

11.3 GOODNESS OF FIT TESTS WHEN SOME PARAMETERS ARE UNSPECIFIED

We can also perform goodness of fit tests of a null hypothesis that does not completely specify the probabilities $\{p_i, i = 1, \ldots, k\}$. For instance, consider the situation previously mentioned in which one is interested in testing whether the number of accidents occurring daily in a certain industrial plant is Poisson distributed with some unknown mean λ. To test this hypothesis, suppose that the daily number of accidents is recorded for n days — let $Y_1, \ldots, Y_n$ be these data. To analyze these data we must first address the difficulty that the Y_i can assume an infinite number of possible values. However, this is easily dealt with by breaking up the possible values into a finite number k of regions and then considering the region in which each Y_i falls. For instance, we might say that the outcome of the number of accidents on a given day is in region 1 if there are 0 accidents, region 2 if there is 1 accident, and region 3 if there are 2 or 3 accidents, region 4 if there are 4 or 5 accidents, and region 5 if there are more than 5 accidents. Hence, if the distribution is indeed Poisson with mean λ, then

$$
\begin{aligned}
p_1 &= P\{Y = 0\} = e^{-\lambda} \\
p_2 &= P\{Y = 1\} = \lambda e^{-\lambda} \\
p_3 &= P\{Y = 2\} + P\{Y = 3\} = \frac{e^{-\lambda}\lambda^2}{2} + \frac{e^{-\lambda}\lambda^3}{6} \\
p_4 &= P\{Y = 4\} + P\{Y = 5\} = \frac{e^{-\lambda}\lambda^4}{24} + \frac{e^{-\lambda}\lambda^5}{120} \\
p_5 &= P\{Y > 5\} = 1 - e^{-\lambda} - \lambda e^{-\lambda} - \frac{e^{-\lambda}\lambda^2}{2} - \frac{e^{-\lambda}\lambda^3}{6} - \frac{e^{-\lambda}\lambda^4}{24} - \frac{e^{-\lambda}\lambda^5}{120}
\end{aligned}
\tag{11.3.1}
$$

The second difficulty we face in obtaining a goodness of fit test results from the fact that the mean value λ is not specified. Clearly, the intuitive thing to do is to assume that H_0 is true and then estimate it from the data — say, $\hat{\lambda}$ is the estimate of λ — and then compute the test statistic

$$T = \sum_{i=1}^{k} \frac{(X_i - n\hat{p}_i)^2}{n\hat{p}_i}$$

where X_i is, as before, the number of Y_j that fall in region $i, i = 1, \ldots, k$, and $\hat{p}_i$ is the estimated probability of the event that Y_j falls in region i, which is determined by substituting $\hat{\lambda}$ for λ in expression 11.3.1 for p_i.

In general, this approach can be utilized whenever there are unspecified parameters in the null hypothesis that are needed to compute the quantities $p_i, i = 1, \ldots, k$. Suppose now that there are m such unspecified parameters and that they are to be estimated by the method of maximum likelihood. It can then be proven that when n is large, the test statistic T will have, when H_0 is true, approximately a chi-square distribution with $k - 1 - m$ degrees of freedom. (In other words, one degree of freedom is lost for each parameter that needs to be estimated.) The test is, therefore, to

$$\text{reject} \quad H_0 \quad \text{if} \quad T \geq \chi^2_{\alpha,k-1-m}$$
$$\text{accept} \quad H_0 \quad \text{otherwise}$$

An equivalent way of performing the foregoing is to first determine the value of the test statistic T, say $T = t$, and then compute

$$p\text{-value} \approx P\{\chi^2_{k-1-m} \geq t\}$$

The hypothesis would be rejected if $\alpha \geq p$-value.

EXAMPLE 11.3a Suppose the weekly number of accidents over a 30-week period is as follows:

$$
\begin{array}{cccccccccc}
8 & 0 & 0 & 1 & 3 & 4 & 0 & 2 & 12 & 5 \\
1 & 8 & 0 & 2 & 0 & 1 & 9 & 3 & 4 & 5 \\
3 & 3 & 4 & 7 & 4 & 0 & 1 & 2 & 1 & 2 \\
\end{array}
$$

Test the hypothesis that the number of accidents in a week has a Poisson distribution.

SOLUTION Since the total number of accidents in the 30 weeks is 95, the maximum likelihood estimate of the mean of the Poisson distribution is

$$\hat{\lambda} = \frac{95}{30} = 3.16667$$

Since the estimate of $P\{Y = i\}$ is then

$$P\{Y = i\} \stackrel{est}{=} \frac{e^{-\hat{\lambda}}\hat{\lambda}^i}{i!}$$

we obtain, after some computation, that with the five regions as given in the beginning of this section,

$$\hat{p}_1 = .04214$$
$$\hat{p}_2 = .13346$$
$$\hat{p}_3 = .43434$$
$$\hat{p}_4 = .28841$$
$$\hat{p}_5 = .10164$$

Using the data values $X_1 = 6, X_2 = 5, X_3 = 8, X_4 = 6, X_5 = 5$, an additional computation yields the test statistic value

$$T = \sum_{i=1}^{5} \frac{(X_i - 30\hat{p}_i)^2}{30\hat{p}_i} = 21.99156$$

To determine the p-value, we run Program 5.8.1a. This yields

$$p\text{-value} \approx P\{\chi_3^2 > 21.99\}$$
$$= 1 - .999936$$
$$= .000064$$

and so the hypothesis of an underlying Poisson distribution is rejected. (Clearly, there were too many weeks having 0 accidents for the hypothesis that the underlying distribution is Poisson with mean 3.167 to be tenable.) ∎

11.4 TESTS OF INDEPENDENCE IN CONTINGENCY TABLES

In this section, we consider problems in which each member of a population can be classified according to two distinct characteristics — which we shall denote as the X-characteristic and the Y-characteristic. We suppose that there are r possible values for the X-characteristic and s for the Y-characteristic, and let

$$P_{ij} = P\{X = i, Y = j\}$$

for $i = 1, \ldots, r, j = 1, \ldots, s$. That is, P_{ij} represents the probability that a randomly chosen member of the population will have X-characteristic i and Y-characteristic j.

The different members of the population will be assumed to be independent. Also, let

$$p_i = P\{X = i\} = \sum_{j=1}^{s} P_{ij}, \qquad i = 1, \ldots, r$$

and

$$q_j = P\{Y = j\} = \sum_{i=1}^{r} P_{ij}, \qquad j = 1, \ldots, s$$

That is, p_i is the probability that an arbitrary member of the population will have X-characteristic i, and q_j is the probability it will have Y-characteristic j.

We are interested in testing the hypothesis that a population member's X- and Y-characteristics are independent. That is, we are interested in testing

$$H_0 : P_{ij} = p_i q_j, \quad \text{for all} \quad i = 1, \ldots, r$$
$$j = 1, \ldots, s$$

against the alternative

$$H_1 : P_{ij} \neq p_i q_j, \quad \text{for some} \quad i, j \quad i = 1, \ldots, r$$
$$j = 1, \ldots, s$$

To test this hypothesis, suppose that n members of the population have been sampled, with the result that N_{ij} of them have simultaneously had X-characteristic i and Y-characteristic j, $i = 1, \ldots, r, j = 1, \ldots, s$.

Since the quantities $p_i, i = 1, \ldots, r$, and $q_j, j = 1, \ldots, s$ are not specified by the null hypothesis, they must first be estimated. Now since

$$N_i = \sum_{j=1}^{s} N_{ij}, \qquad i = 1, \ldots, r$$

represents the number of the sampled population members that have X-characteristic i, a natural (in fact, the maximum likelihood) estimator of p_i is

$$\hat{p}_i = \frac{N_i}{n}, \qquad i = 1, \ldots, r$$

Similarly, letting

$$M_j = \sum_{i=1}^{r} N_{ij}, \qquad j = 1, \ldots, s$$

denote the number of sampled members having Y-characteristic j, the estimator for q_j is

$$\hat{q}_j = \frac{M_j}{n}, \qquad j = 1, \ldots, s$$

At first glance, it may seem that we have had to use the data to estimate $r+s$ parameters. However, since the p_i's and q_j's have to sum to 1 — that is, $\sum_{i=1}^{r} p_i = \sum_{j=1}^{s} q_j = 1$ — we need estimate only $r - 1$ of the p's and $s - 1$ of the q's. (For instance, if r were equal to 2, then an estimate of p_1 would automatically provide an estimate of p_2 since $p_2 = 1 - p_1$.) Hence, we actually need estimate $r - 1 + s - 1 = r + s - 2$ parameters, and since each population member has $k = rs$ different possible values, it follows that the resulting test statistic will, for large n, have approximately a chi-square distribution with $rs - 1 - (r + s - 2) = (r - 1)(s - 1)$ degrees of freedom.

Finally, since

$$E[N_{ij}] = nP_{ij}$$
$$= np_i q_j \qquad \text{when } H_0 \text{ is true}$$

it follows that the test statistic is given by

$$T = \sum_{j=1}^{s} \sum_{i=1}^{r} \frac{(N_{ij} - n\hat{p}_i \hat{q}_j)^2}{n\hat{p}_i \hat{q}_j} = \sum_{j=1}^{s} \sum_{i=1}^{r} \frac{N_{ij}^2}{n\hat{p}_i \hat{q}_j} - n$$

and the approximate significance level α test is to

$$\begin{aligned} \text{reject} \quad & H_0 \quad \text{if} \quad T \geq \chi^2_{\alpha,(r-1)(s-1)} \\ \text{not reject} \quad & H_0 \quad \text{otherwise} \end{aligned}$$

EXAMPLE 11.4a A sample of 300 people was randomly chosen, and the sampled individuals were classified as to their gender and political affiliation, Democrat, Republican, or Independent. The following table, called a *contingency table*, displays the resulting data.

i	Democrat	Republican	Independent	Total
Women	68	56	32	156
Men	52	72	20	144
Total	120	128	52	300

Thus, for instance, the contingency table indicates that the sample of size 300 contained 68 women who classified themselves as Democrats, 56 women who classified themselves as Republicans, and 32 women who classified themselves as Independents; that is, $N_{11} = 68, N_{12} = 56$, and $N_{13} = 32$. Similarly, $N_{21} = 52, N_{22} = 72$, and $N_{23} = 20$.

Use these data to test the hypothesis that a randomly chosen individual's gender and political affiliation are independent.

SOLUTION   From the above data, we obtain that the six values of $n\hat{p}_i\hat{q}_j = N_iM_j/n$ are as follows:

$$\frac{N_1M_1}{n} = \frac{156 \times 120}{300} = 62.40$$

$$\frac{N_1M_2}{n} = \frac{156 \times 128}{300} = 66.56$$

$$\frac{N_1M_3}{n} = \frac{156 \times 52}{300} = 27.04$$

$$\frac{N_2M_1}{n} = \frac{144 \times 120}{300} = 57.60$$

$$\frac{N_2M_2}{n} = \frac{144 \times 128}{300} = 61.44$$

$$\frac{N_2M_3}{n} = \frac{144 \times 52}{300} = 24.96$$

The value of the test statistic is thus

$$TS = \frac{(68 - 62.40)^2}{62.40} + \frac{(56 - 66.56)^2}{66.56} + \frac{(32 - 27.04)^2}{27.04} + \frac{(52 - 57.60)^2}{57.60}$$
$$+ \frac{(72 - 61.44)^2}{61.44} + \frac{(20 - 24.96)^2}{24.96}$$
$$= 6.433$$

Since $(r - 1)(s - 1) = 2$, we must compare the value of TS with the critical value $\chi^2_{.05,2}$. From Table A2

$$\chi^2_{.05,2} = 5.991$$

Since $TS \geq 5.991$, the null hypothesis is rejected at the 5 percent level of significance. That is, the hypothesis that gender and political affiliation of members of the population are independent is rejected at the 5 percent level of significance.   ∎

The results of the test of independence of the characteristics of a randomly chosen member of the population can also be obtained by computing the resulting p-value. If the observed value of the test statistic is $T = t$, then the significance level α test would call for rejecting the hypothesis of independence if the p-value is less than or equal to α, where

$$p\text{-value} = P_{H_0}\{T \geq t\}$$
$$\approx P\{\chi^2_{(r-1)(s-1)} \geq t\}$$

Program 11.4 will compute the value of T.

EXAMPLE 11.4b   A company operates four machines on three separate shirts daily. The following contingency table presents the data during a 6-month time period, concerning the machine breakdowns that resulted.

Number of Breakdowns

	Machine				
	A	**B**	**C**	**D**	**Total per Shift**
Shift 1	10	12	6	7	35
Shift 2	10	24	9	10	53
Shift 3	13	20	7	10	50
Total per Machine	33	56	22	27	138

Suppose we are interested in determining whether a machine's breakdown probability during a particular shift is influenced by that shift. In other words, we are interested in testing, for an arbitrary breakdown, whether the machine causing the breakdown and the shift on which the breakdown occurred are independent.

SOLUTION   A direct computation, or the use of Program 11.4, gives that the value of the test statistic is 1.8148 (see Figure 11.2). Utilizing Program 5.8.1a then gives that

$$p\text{-value} \approx P\{\chi^2_6 \geq 1.8148\}$$
$$= 1 - .0641$$
$$= .9359$$

and so the hypothesis that the machine that causes a breakdown is independent of the shift on which the breakdown occurs is accepted.   ■

11.5 TESTS OF INDEPENDENCE IN CONTINGENCY TABLES HAVING FIXED MARGINAL TOTALS

In Example 11.4a, we were interested in determining whether gender and political affiliation were dependent in a particular population. To test this hypothesis, we first chose

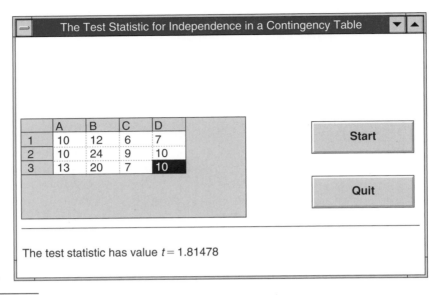

FIGURE 11.2

a random sample of people from this population and then noted their characteristics. However, another way in which we could gather data is to fix in advance the numbers of men and women in the sample and then choose random samples of those sizes from the subpopulations of men and women. That is, rather than let the numbers of women and men in the sample be determined by chance, we might decide these numbers in advance. Because doing so would result in fixed specified values for the total numbers of men and women in the sample, the resulting contingency table is often said to have *fixed margins* (since the totals are given in the margins of the table).

It turns out that even when the data are collected in the manner prescribed above, the same hypothesis test as given in Section 11.4 can still be used to test for the independence of the two characteristics. The test statistic remains

$$TS = \sum_i \sum_j \frac{(N_{ij} - \hat{e}_{ij})^2}{\hat{e}_{ij}}$$

where

$N_{ij} =$ number of members of sample who have both X-characteristic i
and Y-characteristic j

$N_i =$ number of members of sample who have X-characteristic i

$M_j =$ number of members of sample who have Y-characteristic j

and

$$\hat{e}_{ij} = n\hat{p}_i\hat{q}_j = \frac{N_i M_j}{n}$$

where n is the total size of the sample.

In addition, it is still true that when H_0 is true, TS will approximately have a chi-square distribution with $(r-1)(s-1)$ degrees of freedom. (The quantities r and s refer, of course, to the numbers of possible values of the X- and Y-characteristic, respectively.) In other words, the test of the independence hypothesis is unaffected by whether the marginal totals of one characteristic are fixed in advance or result from a random sample of the entire population.

EXAMPLE 11.5a A randomly chosen group of 20,000 nonsmokers and one of 10,000 smokers were followed over a 10-year period. The following data relate the numbers of them that developed lung cancer during that period.

	Smokers	Nonsmokers	Total
Lung cancer	62	14	76
No lung cancer	9,938	19,986	29,924
Total	10,000	20,000	30,000

Test the hypothesis that smoking and lung cancer are independent. Use the 1 percent level of significance.

SOLUTION The estimates of the expected number to fall in each ij cell when smoking and lung cancer are independent are

$$\hat{e}_{11} = \frac{(76)(10,000)}{30,000} = 25.33$$

$$\hat{e}_{12} = \frac{(76)(20,000)}{30,000} = 50.67$$

$$\hat{e}_{21} = \frac{(29,924)(10,000)}{30,000} = 9,974.67$$

$$\hat{e}_{22} = \frac{(29,924)(20,000)}{30,000} = 19,949.33$$

Therefore, the value of the test statistic is

$$TS = \frac{(62 - 25.33)^2}{25.33} + \frac{(14 - 50.67)^2}{50.67} + \frac{(9{,}938 - 9{,}974.67)^2}{9{,}974.67}$$

$$+ \frac{(19{,}986 - 19{,}949.33)^2}{19{,}949.33}$$

$$= 53.09 + 26.54 + .13 + .07 = 79.83$$

Since this is far larger than $\chi^2_{.01,1} = 6.635$, we reject the null hypothesis that whether a randomly chosen person develops lung cancer is independent of whether that person is a smoker. ∎

We now show how to use the framework of this section to test the hypothesis that m discrete population distributions are equal. Consider m separate populations, each of whose members takes on one of the values $1, \ldots, n$. Suppose that a randomly chosen member of population i will have value j with probability

$$p_{i,j}, \qquad i = 1, \ldots, m, \qquad j = 1, \ldots, n$$

and consider a test of the null hypothesis

$$H_0 : p_{1,j} = p_{2,j} = p_{3,j} = \cdots = p_{m,j}, \qquad \text{for each } j = 1, \ldots, n$$

To obtain a test of this null hypothesis, consider first the superpopulation consisting of all members of each of the m populations. Any member of this superpopulation can be classified according to two characteristics. The first characteristic specifies which of the m populations the member is from, and the second characteristic specifies its value. The hypothesis that the population distributions are equal becomes the hypothesis that, for each value, the proportion of members of each population having that value are the same. But this is exactly the same as saying that the two characteristics of a randomly chosen member of the superpopulation are independent. (That is, the value of a randomly chosen superpopulation member is independent of the population to which this member belongs.)

Therefore, we can test H_0 by randomly choosing sample members from each population. If we let M_i denote the sample size from population i and let $N_{i,j}$ denote the number of values from that sample that are equal to j, $i = 1, \ldots, m$, $j = 1, \ldots, n$, then we can test H_0 by testing for independence in the following contingency table.

	Population				
Value	1	2	i	m	Totals
1	$N_{1,1}$	$N_{2,1}\ldots$	$N_{i,1}\ldots$	$N_{m,1}$	N_1
2					
$\vdots$					
j	$N_{1,j}$	$N_{2,j}\ldots$	$N_{i,j}\ldots$	$N_{m,j}$	N_j
$\vdots$					
n	$N_{1,n}$	$N_{2,n}\ldots$	$N_{i,n}$	$N_{m,n}$	N_n
Totals	M_1	$M_2\ldots$	$M_i\ldots$	M_m	

Note that N_j denotes the number of sampled members that have value j.

EXAMPLE 11.5b A recent study reported that 500 female office workers were randomly chosen and questioned in each of four different countries. One of the questions related to whether these women often received verbal or sexual abuse on the job. The following data resulted.

Country	Number Reporting Abuse
Australia	28
Germany	30
Japan	51
United States	55

Based on these data, is it plausible that the proportions of female office workers who often feel abused at work are the same for these countries?

SOLUTION Putting the above data in the form of a contingency table gives the following.

	Country				
	1	2	3	4	Totals
Receive abuse	28	30	58	55	171
Do not receive abuse	472	470	442	445	1,829
Totals	500	500	500	500	2,000

We can now test the null hypothesis by testing for independence in the preceding contingency table. If we run Program 11.4, then the value of the test statistic and the resulting p-value are

$$TS = 19.51, \qquad p\text{-value} \approx .0002$$

Therefore, the hypothesis that the percentages of women who feel they are being abused on the job are the same for these countries is rejected at the 1 percent level of significance (and, indeed, at any significance level above .02 percent). ∎

*11.6 THE KOLMOGOROV–SMIRNOV GOODNESS OF FIT TEST FOR CONTINUOUS DATA

Suppose now that $Y_1, \ldots, Y_n$ represents sample data from a continuous distribution, and that we wish to test the null hypothesis H_0 that F is the population distribution, where F is a specified continuous distribution function. One approach to testing H_0 is to break up the set of possible values of the Y_j into k distinct intervals, say,

$$(y_0, y_1), (y_1, y_2), \ldots, (y_{k-1}, y_k), \qquad \text{where} \quad y_0 = -\infty, y_k = +\infty$$

and then consider the discretized random variables $Y_j^d, j = 1, \ldots, n$, defined by

$$Y_j^d = i \qquad \text{if } Y_j \text{ lies in the interval } (y_{i-1}, y_i)$$

The null hypothesis then implies that

$$P\{Y_j^d = i\} = F(y_i) - F(y_{i-1}), \qquad i = 1, \ldots, k$$

and this can be tested by the chi-square goodness of fit test already presented.

There is, however, another way of testing that the Y_j come from the continuous distribution function F that is generally more efficient than discretizing; it works as follows. After observing $Y_1, \ldots, Y_n$, let F_e be the empirical distribution function defined by

$$F_e(x) = \frac{\#i : Y_i \leq x}{n}$$

That is, $F_e(x)$ is the proportion of the observed values that are less than or equal to x. Because $F_e(x)$ is a natural estimator of the probability that an observation is less than or equal to x, it follows that, if the null hypothesis that F is the underlying distribution is correct, it should be close to $F(x)$. Since this is so for all x, a natural quantity on which to base a test of H_0 is the test quantity

$$D \equiv \underset{x}{\text{Maximum}} \, |F_e(x) - F(x)|$$

where the maximum is over all values of x from $-\infty$ to $+\infty$. The quantity D is called the *Kolmogorov–Smirnov test statistic*.

---

* Optional section.

To compute the value of D for a given data set $Y_j = y_j, j = 1, \ldots, n$, let $y_{(1)}, y_{(2)}, \ldots, y_{(n)}$ denote the values of the y_j in increasing order. That is,

$$y_{(j)} = j\text{th smallest of } y_1, \ldots, y_n$$

For example, if $n = 3$ and $y_1 = 3, y_2 = 5, y_3 = 1$, then $y_{(1)} = 1, y_{(2)} = 3, y_{(3)} = 5$. Since $F_e(x)$ can be written

$$
F_e(x) =
\begin{cases}
0 & \text{if } x < y_{(1)} \\
\dfrac{1}{n} & \text{if } y_{(1)} \leq x < y_{(2)} \\
\vdots & \\
\dfrac{j}{n} & \text{if } y_{(j)} \leq x < y_{(j+1)} \\
\vdots & \\
1 & \text{if } y_{(n)} \leq x
\end{cases}
$$

we see that $F_e(x)$ is constant within the intervals $(y_{(j-1)}, y_{(j)})$ and then jumps by $1/n$ at the points $y_{(1)}, \ldots, y_{(n)}$. Since $F(x)$ is an increasing function of x that is bounded by 1, it follows that the maximum value of $F_e(x) - F(x)$ is nonnegative and occurs at one of the points $y_{(j)}, j = 1, \ldots, n$ (see Figure 11.3).

That is,

$$\text{Maximum}_{x}\{F_e(x) - F(x)\} = \text{Maximum}_{j=1,\ldots,n}\left\{\frac{j}{n} - F(y_{(j)})\right\} \qquad (11.6.1)$$

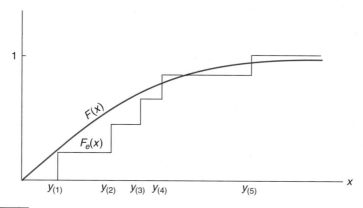

FIGURE 11.3 $n = 5$.

Similarly, the maximum value of $F(x) - F_e(x)$ is also nonnegative and occurs immediately before one of the jump points $y_{(j)}$; and so

$$\text{Maximum}_x\{F(x) - F_e(x)\} = \text{Maximum}_{j=1,\ldots,n}\left\{F(y_{(j)}) - \frac{j-1}{n}\right\} \qquad (11.6.2)$$

From Equations 11.6.1 and 11.6.2, we see that

$$D = \text{Maximum}_x |F_e(x) - F(x)|$$

$$= \text{Maximum}\{\text{Maximum}\{F_e(x) - F(x)\}, \text{Maximum}\{F(x) - F_e(x)\}\}$$

$$= \text{Maximum}\left\{\frac{j}{n} - F(y_{(j)}), F(y_{(j)}) - \frac{j-1}{n}, j = 1, \ldots, n\right\} \qquad (11.6.3)$$

Equation 11.6.3 can be used to compute the value of D.

Suppose now that the Y_j are observed and their values are such that $D = d$. Since a large value of D would appear to be inconsistent with the null hypothesis that F is the underlying distribution, it follows that the p-value for this data set is given by

$$p\text{-value} = P_F\{D \geq d\}$$

where we have written P_F to make explicit that this probability is to be computed under the assumption that H_0 is correct (and so F is the underlying distribution).

The above p-value can be approximated by a simulation that is made easier by the following proposition, which shows that $P_F\{D \geq d\}$ does not depend on the underlying distribution F. This result enables us to estimate the p-value by doing the simulation with any continuous distribution F we choose [thus allowing us to use the uniform $(0, 1)$ distribution].

PROPOSITION 11.6.1

$P_F\{D \geq d\}$ is the same for any continuous distribution F.

Proof

$$P_F\{D \geq d\} = P_F\left\{\text{Maximum}_x \left|\frac{\#i : Y_i \leq x}{n} - F(x)\right| \geq d\right\}$$

$$= P_F\left\{\text{Maximum}_x \left|\frac{\#i : F(Y_i) \leq F(x)}{n} - F(x)\right| \geq d\right\}$$

$$= P\left\{\text{Maximum}_x \left|\frac{\#i : U_i \leq F(x)}{n} - F(x)\right| \geq d\right\}$$

where $U_1, \ldots, U_n$ are independent uniform $(0, 1)$ random variables. The first equality following because F is an increasing function and so $Y \leq x$ is equivalent to $F(Y) \leq F(x)$; and the second because of the result (whose proof is left as an exercise) that if Y has the continuous distribution F then the random variable $F(Y)$ is uniform on $(0, 1)$.

Continuing the above, we see by letting $y = F(x)$ and noting that as x ranges from $-\infty$ to $+\infty$, $F(x)$ ranges from 0 to 1, that

$$P_F\{D \geq d\} = P\left\{ \underset{0 \leq y \leq 1}{\text{Maximum}} \left| \frac{\#i : U_i \leq y}{n} - y \right| \geq d \right\}$$

which shows that the distribution of D, when H_0 is true, does not depend on the actual distribution F. ∎

It follows from the above proposition that after the value of D is determined from the data, say, $D = d$, the p-value can be obtained by doing a simulation with the uniform $(0, 1)$ distribution. That is, we generate a set of n random numbers $U_1, \ldots, U_n$ and then check whether or not the inequality

$$\underset{0 \leq y \leq 1}{\text{Maximum}} \left| \frac{\#i : U_i \leq y}{n} - y \right| \geq d$$

is valid. This is then repeated many times and the proportion of times that it is valid is our estimate of the p-value of the data set. As noted earlier, the left side of the inequality can be computed by ordering the random numbers and then using the identity

$$\text{Max} \left| \frac{\#i : U_i \leq y}{n} - y \right| = \text{Max}\left\{ \frac{j}{n} - U_{(j)}, U_{(j)} - \frac{(j-1)}{n}, j = 1, \ldots, n \right\}$$

where $U_{(j)}$ is the jth smallest value of $U_1, \ldots, U_n$. For example, if $n = 3$ and $U_1 = .7$, $U_2 = .6$, $U_3 = .4$, then $U_{(1)} = .4$, $U_{(2)} = .6$, $U_{(3)} = .7$ and the value of D for this data set is

$$D = \text{Max}\left\{ \frac{1}{3} - .4, \frac{2}{3} - .6, 1 - .7, .4, .6 - \frac{1}{3}, .7 - \frac{2}{3} \right\} = .4$$

A significance level α test can be obtained by considering the quantity D^* defined by

$$D^* = (\sqrt{n} + .12 + .11/\sqrt{n})D$$

Letting d_α^* be such that

$$P_F\{D^* \geq d_\alpha^*\} = \alpha$$

then the following are accurate approximations for d_α^* for a variety of values:

$$d_{.1}^* = 1.224, \quad d_{.05}^* = 1.358, \quad d_{.025}^* = 1.480, \quad d_{.01}^* = 1.626$$

The level α test would reject the null hypothesis that F is the distribution if the observed value of D^* is at least as large as d^*_α.

EXAMPLE 11.6a Suppose we want to test the hypothesis that a given population distribution is exponential with mean 100; that is, $F(x) = 1 - e^{-x/100}$. If the (ordered) values from a sample of size 10 from this distribution are

$$66, 72, 81, 94, 112, 116, 124, 140, 145, 155$$

what conclusion can be drawn?

SOLUTION To answer the above, we first employ Equation 11.6.3 to compute the value of the Kolmogorov–Smirnov test quantity D. After some computation this gives the result $D = .4831487$, which results in

$$D^* = .48315(\sqrt{10} + 0.12 + 0.11/\sqrt{10}) = 1.603$$

Because this exceeds $d^*_{.025} = 1.480$, it follows that the null hypothesis that the data come from an exponential distribution with mean 100 would be rejected at the 2.5 percent level of significance. (On the other hand, it would not be rejected at the 1 percent level of significance.) ■

Problems

1. According to the Mendelian theory of genetics, a certain garden pea plant should produce either white, pink, or red flowers, with respective probabilities $\frac{1}{4}, \frac{1}{2}, \frac{1}{4}$. To test this theory, a sample of 564 peas was studied with the result that 141 produced white, 291 produced pink, and 132 produced red flowers. Using the chi-square approximation, what conclusion would be drawn at the 5 percent level of significance?

2. To ascertain whether a certain die was fair, 1,000 rolls of the die were recorded, with the following results.

Outcome	Number of Occurrences
1	158
2	172
3	164
4	181
5	160
6	165

Test the hypothesis that the die is fair (that is, that $p_i = \frac{1}{6}, i = 1, \ldots, 6$) at the 5 percent level of significance. Use the chi-square approximation.

3. Determine the birth and death dates of 100 famous individuals and, using the four-category approach of Example 11.2a, test the hypothesis that the death month is not affected by the birth month. Use the chi-square approximation.

4. It is believed that the daily number of electrical power failures in a certain Midwestern city is a Poisson random variable with mean 4.2. Test this hypothesis if over 150 days the number of days having i power failures is as follows:

Failures	Number of Days
0	0
1	5
2	22
3	23
4	32
5	22
6	19
7	13
8	6
9	4
10	4
11	0

5. Among 100 vacuum tubes tested, 41 had lifetimes of less than 30 hours, 31 had lifetimes between 30 and 60 hours, 13 had lifetimes between 60 and 90 hours, and 15 had lifetimes of greater than 90 hours. Are these data consistent with the hypothesis that a vacuum tube's lifetime is exponentially distributed with a mean of 50 hours?

6. The past output of a machine indicates that each unit it produces will be

top grade	with probability	.40
high grade	with probability	.30
medium grade	with probability	.20
low grade	with probability	.10

A new machine, designed to perform the same job, has produced 500 units with the following results.

top grade	234
high grade	117
medium grade	81
low grade	68

Can the difference in output be ascribed solely to chance?

7. The neutrino radiation from outer space was observed during several days. The frequencies of signals were recorded for each sidereal hour and are as given below:

Frequency of Neutrino Radiation from Outer Space

Hour Starting at	Frequency of Signals	Hour Starting at	Frequency of Signals
0	24	12	29
1	24	13	26
2	36	14	38
3	32	15	26
4	33	16	37
5	36	17	28
6	41	18	43
7	24	19	30
8	37	20	40
9	37	21	22
10	49	22	30
11	51	23	42

Test whether the signals are uniformly distributed over the 24-hour period.

8. Neutrino radiation was observed over a certain period and the number of hours in which 0, 1, 2, . . . signals were received was recorded.

Number of Signals per Hour	Number of Hours with This Frequency of Signals
0	1,924
1	541
2	103
3	17
4	1
5	1
6 or more	0

Test the hypothesis that the observations come from a population having a Poisson distribution with mean .3.

9. In a certain region, insurance data indicate that 82 percent of drivers have no accidents in a year, 15 percent have exactly 1 accident, and 3 percent have 2 or more accidents. In a random sample of 440 engineers, 366 had no accidents, 68 had exactly 1 accident, and 6 had 2 or more. Can you conclude that engineers follow an accident profile that is different from the rest of the drivers in the region?

10. A study was instigated to see if southern California earthquakes of at least moderate size (having values of at least 4.4 on the Richter scale) are more likely to occur on certain days of the week than on others. The catalogs yielded the following data on 1,100 earthquakes.

Day	Sun	Mon	Tues	Wed	Thurs	Fri	Sat
Number of Earthquakes	156	144	170	158	172	148	152

Test, at the 5 percent level, the hypothesis that an earthquake is equally likely to occur on any of the 7 days of the week.

11. Sometimes reported data fit a model so well that it makes one suspicious that the data are not being accurately reported. For instance, a friend of mine has reported that he tossed a fair coin 40,000 times and obtained 20,004 heads and 19,996 tails. Is such a result believable? Explain your reasoning.

12. Use simulation to determine the p-value and compare it with the result you obtained using the chi-square approximation in Problem 1. Let the number of simulation runs be

    (a) 1,000;
    (b) 5,000;
    (c) 10,000.

13. A sample of size 120 had a sample mean of 100 and a sample standard deviation of 15. Of these 120 data values, 3 were less than 70; 18 were between 70 and 85; 30 were between 85 and 100; 35 were between 100 and 115; 32 were between 115 and 130; and 2 were greater than 130. Test the hypothesis that the sample distribution was normal.

14. In Problem 4, test the hypothesis that the daily number of failures has a Poisson distribution.

15. A random sample of 500 migrant families was classified by region and income (in units of $1,000). The following data resulted.

Income	South	North
0–10	42	53
10–20	55	90
20–30	47	88
>30	36	89

Determine the *p*-value of the test that a family's income and region are independent.

16. The following data relate the mother's age and the birthweight (in grams) of her child.

Maternal Age	Birthweight	
	Less Than 2,500 Grams	More Than 2,500 Grams
20 years or less	10	40
Greater than 20	15	135

Test the hypothesis that the baby's birthweight is independent of the mother's age.

17. Repeat Problem 16 with all of the data values doubled — that is, with these data:

    20   80
    30   270

18. The number of infant mortalities as a function of the baby's birthweight (in grams) for 72,730 live white births in New York in 1974 is as follows:

Birthweight	Outcome at the End of 1 Year	
	Alive	Dead
Less than 2,500	4,597	618
Greater than 2,500	67,093	422

Test the hypothesis that the birthweight is independent of whether or not the baby survives its first year.

19. An experiment designed to study the relationship between hypertension and cigarette smoking yielded the following data.

	Nonsmoker	Moderate Smoker	Heavy Smoker
Hypertension	20	38	28
No hypertension	50	27	18

Test the hypothesis that whether or not an individual has hypertension is independent of how much that person smokes.

20. The following table shows the number of defective, acceptable, and superior items in samples taken both before and after the introduction of a modification in the manufacturing process.

	Defective	Acceptable	Superior
Before	25	218	22
After	9	103	14

Is this change significant at the .05 level?

21. A sample of 300 cars having cellular phones and one of 400 cars without phones were tracked for 1 year. The following table gives the number of these cars involved in accidents over that year.

	Accident	No Accident
Cellular phone	22	278
No phone	26	374

Use the above to test the hypothesis that having a cellular phone in your car and being involved in an accident are independent. Use the 5 percent level of significance.

22. To study the effect of fluoridated water supplies on tooth decay, two communities of roughly the same socioeconomic status were chosen. One of these communities had fluoridated water while the other did not. Random samples of 200 teenagers from both communities were chosen, and the numbers of cavities they had were determined. The following data resulted.

Cavities	Fluoridated Town	Nonfluoridated Town
0	154	133
1	20	18
2	14	21
3 or more	12	28

Do these data establish, at the 5 percent level of significance, that the number of dental cavities a person has is not independent of whether that person's water supply is fluoridated? What about at the 1 percent level?

23. To determine if a malpractice lawsuit is more likely to follow certain types of surgeries, random samples of three different types of surgeries were studied, and the following data resulted.

Type of Operation	Number Sampled	Number Leading to a Lawsuit
Heart surgery	400	16
Brain surgery	300	19
Appendectomy	300	7

Test the hypothesis that the percentages of the surgical operations that lead to lawsuits are the same for each of the three types.

(a) Use the 5 percent level of significance.

(b) Use the 1 percent level of significance.

24. In a famous article (S. Russell, "A red sky at night...," *Metropolitan Magazine London*, **61**, p. 15, 1926) the following data set of frequencies of sunset colors and whether each was followed by rain was presented.

Sky Color	Number of Observations	Number Followed by Rain
Red	61	26
Mainly red	194	52
Yellow	159	81
Mainly yellow	188	86
Red and yellow	194	52
Gray	302	167

Test the hypothesis that whether it rains tomorrow is independent of the color of today's sunset.

25. Data are said to be from a *lognormal* distribution with parameters μ and σ if the natural logarithms of the data are normally distributed with mean μ and standard deviation σ. Use the Kolmogorov–Smirnov test with significance level .05 to decide whether the following lifetimes (in days) of a sample of cancer-bearing mice that have been treated with a certain cancer therapy might come from a lognormal distribution with parameters $\mu = 3$ and $\sigma = 4$.

$$24, 12, 36, 40, 16, 10, 12, 30, 38, 14, 22, 18$$

NONPARAMETRIC HYPOTHESIS TESTS

12.1   INTRODUCTION

In this chapter, we shall develop some hypothesis tests in situations where the data come from a probability distribution whose underlying form is not specified. That is, it will not be assumed that the underlying distribution is normal, or exponential, or any other given type. Because no particular parametric form for the underlying distribution is assumed, such tests are called *nonparametric*.

The strength of a nonparametric test resides in the fact that it can be applied without any assumption on the form of the underlying distribution. Of course, if there is justification for assuming a particular parametric form, such as the normal, then the relevant parametric test should be employed.

In Section 12.2, we consider hypotheses concerning the median of a continuous distribution and show how the *sign test* can be used in their study. In Section 12.3, we consider the *signed rank test*, which is used to test the hypothesis that a continuous population distribution is symmetric about a specified value. In Section 12.4, we consider the two-sample problem, where one wants to use data from two separate continuous distributions to test the hypothesis that the distributions are equal, and present the *rank sum test*. Finally, in Section 12.5 we study the *runs test*, which can be used to test the hypothesis that a sequence of 0's and 1's constitutes a random sequence that does not follow any specified pattern.

12.2   THE SIGN TEST

Let $X_1, \ldots, X_n$ denote a sample from a continuous distribution F and suppose that we are interested in testing the hypothesis that the median of F, call it m, is equal to a specified value m_0. That is, consider a test of

$$H_0 : m = m_0 \qquad \text{versus} \qquad H_1 : m \neq m_0$$

where m is such that $F(m) = .5$.

This hypothesis can easily be tested by noting that each of the observations will, independently, be less than m_0 with probability $F(m_0)$. Hence, if we let

$$I_i = \begin{cases} 1 & \text{if } X_i < m_0 \\ 0 & \text{if } X_i \geq m_0 \end{cases}$$

then $I_1, \ldots, I_n$ are independent Bernoulli random variables with parameter $p = F(m_0)$; and so the null hypothesis is equivalent to stating that this Bernoulli parameter is equal to $\frac{1}{2}$. Now, if v is the observed value of $\sum_{i=1}^{n} I_i$ — that is, if v is the number of data values less than m_0 — then it follows from the results of Section 8.6 that the p-value of the test that this Bernoulli parameter is equal to $\frac{1}{2}$ is

$$p\text{-value} = 2 \min(P\{\text{Bin}(n, 1/2) \leq v\}, P\{\text{Bin}(n, 1/2) \geq v\}) \qquad (12.2.1)$$

where $\text{Bin}(n, p)$ is a binomial random variable with parameters n and p. However,

$$\begin{aligned} P\{\text{Bin}(n, p) \geq v\} &= P\{n - \text{Bin}(n, p) \leq n - v\} \\ &= P\{\text{Bin}(n, 1 - p) \leq n - v\} \quad \text{(why?)} \end{aligned}$$

and so we see from Equation 12.2.1 that the p-value is given by

$$p\text{-value} = 2 \min(P\{\text{Bin}(n, 1/2) \leq v\}, P\{\text{Bin}(n, 1/2) \leq n - v\}) \qquad (12.2.2)$$

$$= \begin{cases} 2P\{\text{Bin}(n, 1/2) \leq v\} & \text{if } v \leq \dfrac{n}{2} \\ 2P\{\text{Bin}(n, 1/2) \leq n - v\} & \text{if } v \geq \dfrac{n}{2} \end{cases}$$

Since the value of $v = \sum_{i=1}^{n} I_i$ depends on the signs of the terms $X_i - m_0$, the foregoing is called the *sign test*.

EXAMPLE 12.2a If a sample of size 200 contains 120 values that are less than m_0 and 80 values that are greater, what is the p-value of the test of the hypothesis that the median is equal to m_0?

SOLUTION From Equation 12.2.2, the p-value is equal to twice the probability that binomial random variable with parameters 200, $\frac{1}{2}$ is less than or equal to 80.

The text disk shows that

$$P\{\text{Bin}(200, .5) \leq 80\} = .00284$$

Therefore, the p-value is .00568, and so the null hypothesis would be rejected at even the 1 percent level of significance. ∎

The sign test can also be used in situations analogous to ones in which the paired t-test was previously applied. For instance, let us reconsider Example 8.4c, which is interested in testing whether or not a recently instituted industrial safety program has had an effect on the number of man-hours lost to accidents. For each of 10 plants, the data consisted of the pair X_i, Y_i, which represented, respectively, the average weekly loss at plant i before and after the program. Letting $Z_i = X_i - Y_i, i = 1, \ldots, 10$, it follows that if the program had not had any effect, then $Z_i, i = 1, \ldots, 10$, would be a sample from a distribution whose median value is 0. Since the resulting values of Z_i, — namely, $7.5, -2.3, 2.6, 3.7, 1.5, -.5, -1, 4.9, 4.8, 1.6$ — contain three whose sign is negative and seven whose sign is positive, it follows that the hypothesis that the median of Z is 0 should be rejected at significance level α if

$$\sum_{i=0}^{3} \binom{10}{i} \left(\frac{1}{2}\right)^{10} \leq \frac{\alpha}{2}$$

Since

$$\sum_{i=0}^{3} \binom{10}{i} \left(\frac{1}{2}\right)^{10} = \frac{176}{1{,}024} = .172$$

it follows that the hypothesis would be accepted at the 5 percent significance level (indeed, it would be accepted at all significance levels less than the p-value equal to .344).

Thus, the sign test does not enable us to conclude that the safety program has had any statistically significant effect, which is in contradiction to the result obtained in Example 8.4c when it was assumed that the differences were normally distributed. The reason for this disparity is that the assumption of normality allows us to take into account not only the number of values greater than 0 (which is all the sign test considers) but also the magnitude of these values. (The next test to be considered, while still being nonparametric, improves on the sign test by taking into account whether those values that most differ from the hypothesized median value m_0 tend to lie on one side of m_0 — that is, whether they tend to be primarily bigger or smaller than m_0.)

We can also use the sign test to test one-sided hypotheses about a population median. For instance, suppose that we want to test

$$H_0 : m \leq m_0 \qquad \text{versus} \qquad H_1 : m > m_0$$

where m is the population median and m_0 is some specified value. Let p denote the probability that a population value is less than m_0, and note that if the null hypothesis is true then $p \geq 1/2$, and if the alternative is true then $p < 1/2$ (see Figure 12.1).

To use the sign test to test the preceding hypothesis, choose a random sample of n members of the population. If v of them have values that are less than m_0, then the

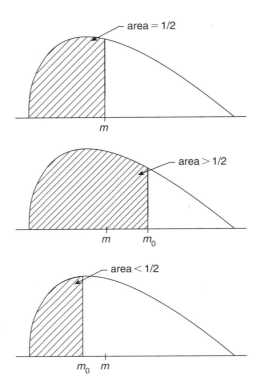

FIGURE 12.1

resulting p-value is the probability that a value of v or smaller would have occurred by chance if each element had probability $1/2$ of being less than m_0. That is,

$$p\text{-value} = P\{\text{Bin}(n, 1/2) \leq v\}$$

EXAMPLE 12.2b   A financial institution has decided to open an office in a certain community if it can be established that the median annual income of families in the community is greater than \$90,000. To obtain information, a random sample of 80 families was chosen, and the family incomes determined. If 28 of these families had annual incomes below and 52 had annual incomes above \$90,000, is this significant enough to establish, say, at the 5 percent level of significance, that the median annual income in the community is greater than \$90,000?

SOLUTION   We need to see if the data are sufficient to enable us to reject the null hypothesis when testing

$$H_0 : m \leq 90 \qquad \text{versus} \qquad H_1 : m > 90$$

The preceding is equivalent to testing

$$H_0 : p \geq 1/2 \qquad \text{versus} \qquad H_1 : p < 1/2$$

where p is the probability that a randomly chosen member of the population has an annual income of less than \$90,000. Therefore, the p-value is

$$p\text{-value} = P\{\text{Bin}(80, 1/2) \leq 28\} = .0048$$

and so the null hypothesis that the median income is less than or equal to \$90,000 is rejected. ■

A test of the one-sided null hypothesis that the median is at least m_0 is obtained similarly. If a random sample of size n is chosen, and v of the resulting values are less than m_0, then the resulting p-value is

$$p\text{-value} = P\{\text{Bin}(n, 1/2) \geq v\}$$

12.3 THE SIGNED RANK TEST

The sign test can be employed to test the hypothesis that the median of a continuous distribution F is equal to a specified value m_0. However, in many applications one is really interested in testing not only that the median is equal to m_0 but that the distribution is symmetric about m_0. That is, if X has distribution function F, then one is often interested in testing the hypothesis $H_0 : P\{X < m_0 - a\} = P\{X > m_0 + a\}$ for all $a > 0$ (see Figure 12.2). Whereas the sign test could still be employed to test the foregoing hypothesis, it suffers in that it compares only the number of data values that are less than m_0 with the number that are greater than m_0 and does not take into account whether or not one of these sets tends to be farther away from m_0 than the other. A nonparametric test that does take this into account is the so-called *signed rank* test. It is described as follows.

Let $Y_i = X_i - m_0, i = 1, \ldots, n$ and rank (that is, order) the absolute values $|Y_1|, |Y_2|, \ldots, |Y_n|$. Set, for $j = 1, \ldots, n$,

$$I_j = \begin{cases} 1 & \text{if the } j\text{th smallest value comes from a data value that is smaller} \\ & \text{than } m_0 \\ 0 & \text{otherwise} \end{cases}$$

Now, whereas $\sum_{j=1}^{n} I_j$ represents the test statistic for the sign test, the signed rank test uses the statistic $T = \sum_{j=1}^{n} jI_j$. That is, like the sign test it considers those data values that are less than m_0, but rather than giving equal weight to each such value it gives larger weights to those data values that are farthest away from m_0.

EXAMPLE 12.3a If $n = 4$, $m_0 = 2$, and the data values are $X_1 = 4.2$, $X_2 = 1.8$, $X_3 = 5.3$, $X_4 = 1.7$, then the rankings of $|X_i - 2|$ are .2, .3, 2.2, 3.3. Since the first of these values — namely, .2 — comes from the data point X_2, which is less than 2, it follows that

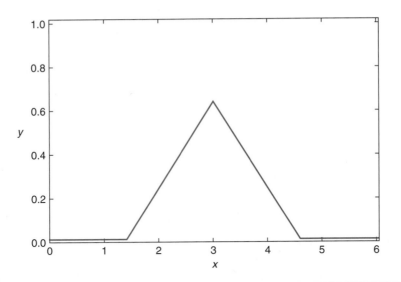

$$f(x) = \begin{cases} \max\{0, .4(x-3) + \sqrt{.4}\} & x \le 3 \\ \max\{0, -.4(x-3) + \sqrt{.4}\} & x > 3 \end{cases}$$

$I_1 = 1$. Similarly, $I_2 = 1$, and I_3 and I_4 equal 0. Hence, the value of the test statistic is $T = 1 + 2 = 3$. ∎

When H_0 is true, the mean and variance of the test statistic T are easily computed. This is accomplished by noting that, since the distribution of $Y_j = X_j - m_0$ is symmetric about 0, for any given value of $|Y_j|$ — say, $|Y_j| = y$ — it is equally likely that either $Y_j = y$ or $Y_j = -y$. From this fact it can be seen that under H_0, $I_1, \ldots, I_n$ will be independent random variables such that

$$P\{I_j = 1\} = \tfrac{1}{2} = P\{I_j = 0\}, \quad j = 1, \ldots, n$$

Hence, we can conclude that under H_0,

$$E[T] = E\left[\sum_{j=1}^{n} jI_j \right]$$

$$= \sum_{j=1}^{n} \frac{j}{2} = \frac{n(n+1)}{4} \qquad (12.3.1)$$

$$\text{Var}(T) = \text{Var}\left(\sum_{j=1}^{n} jI_j\right)$$

$$= \sum_{j=1}^{n} j^2 \, \text{Var}(I_j)$$

$$= \sum_{j=1}^{n} \frac{j^2}{4} = \frac{n(n+1)(2n+1)}{24} \tag{12.3.2}$$

where the fact that the variance of the Bernoulli random variable I_j is $\frac{1}{2}(1 - \frac{1}{2}) = \frac{1}{4}$ is used.

It can be shown that for moderately large values of n ($n > 25$ is often quoted as being sufficient) T will, when H_0 is true, have approximately a normal distribution with mean and variance as given by Equations 12.3.1 and 12.3.2. Although this approximation can be used to derive an approximate level α test of H_0 (which has been the usual approach until the recent advent of fast and cheap computational power), we shall not pursue this approach but rather will determine the p-value for given test data by an explicit computation of the relevant probabilities. This is accomplished as follows.

Suppose we desire a significance level α test of H_0. Since the alternative hypothesis is that the median is not equal to m_0, a two-sided test is called for. That is, if the observed value of T is equal to t, then H_0 should be rejected if either

$$P_{H_0}\{T \le t\} \le \frac{\alpha}{2} \qquad \text{or} \qquad P_{H_0}\{T \ge t\} \le \frac{\alpha}{2} \tag{12.3.3}$$

The p-value of the test data when $T = t$ is given by

$$p\text{-value} = 2 \min(P_{H_0}\{T \le t\}, P_{H_0}\{T \ge t\}) \tag{12.3.4}$$

That is, if $T = t$, the signed rank test calls for rejection of the null hypothesis if the significance level α is at least as large as this p-value. The amount of computation necessary to compute the p-value can be reduced by utilizing the following equality (whose proof will be given at the end of the section).

$$P_{H_0}\{T \ge t\} = P_{H_0}\left\{T \le \frac{n(n+1)}{2} - t\right\}$$

Using Equation 12.3.4, the p-value is given by

$$p\text{-value} = 2 \min\left(P_{H_0}\{T \le t\}, P_{H_0}\left\{T \le \frac{n(n+1)}{2} - t\right\}\right)$$

$$= 2P_{H_0}\{T \le t^*\}$$

where

$$t^* = \min\left(t, \frac{n(n+1)}{2} - t\right)$$

It remains to compute $P_{H_0}\{T \le t^*\}$. To do so, let $P_k(i)$ denote the probability, under H_0, that the signed rank statistic T will be less than or equal to i when the sample size is k. We will determine a recursive formula for $P_k(i)$ starting with $k = 1$. When $k = 1$, since there is only a single data value, which, when H_0 is true, is equally likely to be either less than or greater than m_0, it follows that T is equally likely to be either 0 or 1. Thus

$$P_1(i) = \begin{cases} 0 & i < 0 \\ \frac{1}{2} & i = 0 \\ 1 & i \ge 1 \end{cases} \tag{12.3.5}$$

Now suppose the sample size is k. To compute $P_k(i)$, we condition on the value of I_k as follows:

$$P_k(i) = P_{H_0}\left\{\sum_{j=1}^{k} jI_j \le i\right\}$$

$$= P_{H_0}\left\{\sum_{j=1}^{k} jI_j \le i | I_k = 1\right\} P_{H_0}\{I_k = 1\}$$

$$+ P_{H_0}\left\{\sum_{j=1}^{k} jI_j \le i | I_k = 0\right\} P_{H_0}\{I_k = 0\}$$

$$= P_{H_0}\left\{\sum_{j=1}^{k-1} jI_j \le i - k | I_k = 1\right\} P_{H_0}\{I_k = 1\}$$

$$+ P_{H_0}\left\{\sum_{j=1}^{k-1} jI_j \le i | I_k = 0\right\} P_{H_0}\{I_k = 0\}$$

$$= P_{H_0}\left\{\sum_{j=1}^{k-1} jI_j \le i - k\right\} P_{H_0}\{I_k = 1\} + P_{H_0}\left\{\sum_{j=1}^{k-1} jI_j \le i\right\} P_{H_0}\{I_k = 0\}$$

where the last equality utilized the independence of $I_1, \ldots, I_{k-1}$ and I_k (when H_0 is true). Now $\sum_{j=1}^{k-1} jI_j$ has the same distribution as the signed rank statistic of a sample of size $k - 1$, and since

$$P_{H_0}\{I_k = 1\} = P_{H_0}\{I_k = 0\} = \frac{1}{2}$$

we see that

$$P_k(i) = \tfrac{1}{2}P_{k-1}(i-k) + \tfrac{1}{2}P_{k-1}(i) \qquad (12.3.6)$$

Starting with Equation 12.3.5, the recursion given by Equation 12.3.6 can be successfully employed to compute $P_2(\cdot)$, then $P_3(\cdot)$, and so on, stopping when the desired value $P_n(t^*)$ has been obtained.

EXAMPLE 12.3b For the data of Example 12.3a,

$$t^* = \min\left(3, \frac{4\cdot 5}{2} - 3\right) = 3$$

Hence the p-value is $2P_4(3)$, which is computed as follows:

$$P_2(0) = \tfrac{1}{2}[P_1(-2) + P_1(0)] = \tfrac{1}{4}$$
$$P_2(1) = \tfrac{1}{2}[P_1(-1) + P_1(1)] = \tfrac{1}{2}$$
$$P_2(2) = \tfrac{1}{2}[P_1(0) + P_1(2)] = \tfrac{3}{4}$$
$$P_2(3) = \tfrac{1}{2}[P_1(1) + P_1(3)] = 1$$
$$P_3(0) = \tfrac{1}{2}[P_2(-3) + P_2(0)] = \tfrac{1}{8} \qquad \text{since } P_2(-3) = 0$$
$$P_3(1) = \tfrac{1}{2}[P_2(-2) + P_2(1)] = \tfrac{1}{4}$$
$$P_3(2) = \tfrac{1}{2}[P_2(-1) + P_2(2)] = \tfrac{3}{8}$$
$$P_3(3) = \tfrac{1}{2}[P_2(0) + P_2(3)] = \tfrac{5}{8}$$
$$P_4(0) = \tfrac{1}{2}[P_3(-4) + P_3(0)] = \tfrac{1}{16}$$
$$P_4(1) = \tfrac{1}{2}[P_3(-3) + P_3(1)] = \tfrac{1}{8}$$
$$P_4(2) = \tfrac{1}{2}[P_3(-2) + P_3(2)] = \tfrac{3}{16}$$
$$P_4(3) = \tfrac{1}{2}[P_3(-1) + P_3(3)] = \tfrac{5}{16} \quad \blacksquare$$

Program 12.3 will use the recursion in Equations 12.3.5 and 12.3.6 to compute the p-value of the signed rank test data. The input needed is the sample size n and the value of test statistic T.

EXAMPLE 12.3c Suppose we are interested in determining whether a certain population has an underlying probability distribution that is symmetric about 0. If a sample of size 20 from this population results in a signed rank test statistic of value 142, what conclusion can we draw at the 10 percent level of significance?

SOLUTION Running Program 12.3 yields that

$$p\text{-value} = .177$$

Thus the hypothesis that the population distribution is symmetric about 0 is accepted at the $\alpha = .10$ level of significance. ∎

We end this section with a proof of the equality

$$P_{H_0}\{T \geq t\} = P_{H_0}\left\{T \leq \frac{n(n+1)}{2} - t\right\}$$

To verify the foregoing, note first that $1 - I_j$ will equal 1 if the jth smallest value of $|Y_1|, \ldots, |Y_n|$ comes from a data value larger than m_0, and it will equal 0 otherwise. Hence, if we let

$$T^1 = \sum_{j=1}^{n} j(1 - I_j)$$

then T^1 will represent the sum of the ranks of the $|Y_j|$ that correspond to data values larger than m_0. By symmetry, T^1 will have, under H_0, the same distribution as T. Now

$$T^1 = \sum_{j=1}^{n} j - \sum_{j=1}^{n} jI_j = \frac{n(n+1)}{2} - T$$

and so

$$P\{T \geq t\} = P\{T^1 \geq t\} \qquad \text{since } T \text{ and } T^1 \text{ have the same distribution}$$

$$= P\left\{\frac{n(n+1)}{2} - T \geq t\right\}$$

$$= P\left\{T \leq \frac{n(n+1)}{2} - t\right\}$$

REMARK ON TIES

Since we have assumed that the population distribution is continuous, there is no possibility of ties — that is, with probability 1, all observations will have different values. However, since in practice all measurements are quantized, ties are always a distinct possibility. If ties do occur, then the weights given to the values less than m_0 should be the average of the different weights they could have had if the values had differed slightly. For instance, if $m_0 = 0$ and the data values are 2, 4, 7, −5, −7, then the ordered absolute values are 2, 4, 5, 7, 7. Since 7 has rank both 4 and 5, the value of the test statistic

T is $T = 3 + 4.5 = 7.5$. The p-value should be computed as when we assumed that all values were distinct. (Although technically this is not correct, the discrepancy is usually minor.)

12.4 THE TWO-SAMPLE PROBLEM

Suppose that one is considering two different methods for producing items having measurable characteristics with an interest in determining whether the two methods result in statistically identical items.

To attack this problem let $X_1, \ldots, X_n$ denote a sample of the measurable values of n items produced by method 1, and, similarly, let $Y_1, \ldots, Y_m$ be the corresponding value of m items produced by method 2. If we let F and G, both assumed to be continuous, denote the distribution functions of the two samples, respectively, then the hypothesis we wish to test is $H_0 : F = G$.

One procedure for testing H_0 — which is known by such names as the rank sum test, the Mann-Whitney test, or the Wilcoxon test — calls initially for ranking, or ordering, the $n + m$ data values $X_1, \ldots, X_n, Y_1, \ldots, Y_m$. Since we are assuming that F and G are continuous, this ranking will be unique — that is, there will be no ties. Give the smallest data value rank 1, the second smallest rank 2, ..., and the $(n + m)$th smallest rank $n + m$. Now, for $i = 1, \ldots, n$, let

$$R_i = \text{rank of the data value } X_i$$

The rank sum test utilizes the test statistic T equal to the sum of the ranks from the first sample — that is,

$$T = \sum_{i=1}^{n} R_i$$

EXAMPLE 12.4a An experiment designed to compare two treatments against corrosion yielded the following data in pieces of wire subjected to the two treatments.

Treatment 1	65.2, 67.1, 69.4, 78.2, 74, 80.3
Treatment 2	59.4, 72.1, 68, 66.2, 58.5

(The data represent the maximum depth of pits in units of one thousandth of an inch.) The ordered values are 58.5, 59.4, 65.2*, 66.2, 67.1*, 68, 69.4*, 72.1, 74*, 78.2*, 80.3* with an asterisk noting that the data value was from sample 1. Hence, the value of the test statistic is $T = 3 + 5 + 7 + 9 + 10 + 11 = 45$. ■

Suppose that we desire a significance level α test of H_0. If the observed value of T is $T = t$, then H_0 should be rejected if either

$$P_{H_0}\{T \leq t\} \leq \frac{\alpha}{2} \qquad \text{or} \qquad P_{H_0}\{T \geq t\} \leq \frac{\alpha}{2} \qquad (12.4.1)$$

That is, the hypothesis that the two samples are equivalent should be rejected if the sum of the ranks from the first sample is either too small or too large to be explained by chance.

Since for integral t,

$$P\{T \geq t\} = 1 - P\{T < t\}$$
$$= 1 - P\{T \leq t - 1\}$$

it follows from Equation 12.4.1 that H_0 should be rejected if either

$$P_{H_0}\{T \leq t\} \leq \frac{\alpha}{2} \qquad \text{or} \qquad P_{H_0}\{T \leq t - 1\} \geq 1 - \frac{\alpha}{2} \qquad (12.4.2)$$

To compute the probabilities in Equation 12.4.2, let $P(N, M, K)$ denote the probability that the sum of the ranks of the first sample will be less than or equal to K when the sample sizes are N and M and H_0 is true. We will now determine a recursive formula for $P(N, M, K)$, which will then allow us to obtain the desired quantities $P(n, m, t) = P_{H_0}\{T \leq t\}$ and $P(n, m, t - 1)$.

To compute the probability that the sum of the ranks of the first sample is less than or equal to K when N and M are the sample sizes and H_0 is true, let us condition on whether the largest of the $N + M$ data values belongs to the first or second sample. If it belongs to the first sample, then the sum of the ranks of this sample is equal to $N + M$ plus the sum of the ranks of the other $N - 1$ values from the first sample. Hence this sum will be less than or equal to K if the sum of the ranks of the other $N - 1$ values is less than or equal to $K - (N + M)$. But since the remaining $N - 1 + M$ — that is, all but the largest — values all come from the same distribution (when H_0 is true), it follows that the sum of the ranks of $N - 1$ of them will be less than or equal to $K - (N + M)$ with probability $P(N - 1, M, K - N - M)$. By a similar argument we can show that, given that the largest value is from the second sample, the sum of the ranks of the first sample will be less than or equal to K with probability $P(N, M - 1, K)$. Also, since the largest value is equally likely to be any of the $N + M$ values $X_1, \ldots, X_N, Y_1, \ldots, Y_M$, it follows that it will come from the first sample with probability $N/(N + M)$. Putting these together, we thus obtain that

$$P(N, M, K) = \frac{N}{N + M} P(N - 1, M, K - N - M)$$
$$+ \frac{M}{N + M} P(N, M - 1, K) \qquad (12.4.3)$$

Starting with the boundary condition

$$P(1, 0, K) = \begin{cases} 0 & K \le 0 \\ 1 & K > 0 \end{cases}, \qquad P(0, 1, K) = \begin{cases} 0 & K < 0 \\ 1 & K \ge 0 \end{cases}$$

Equation 12.4.3 can be solved recursively to obtain $P(n, m, t - 1)$ and $P(n, m, t)$.

EXAMPLE 12.4b Suppose we wanted to determine $P(2, 1, 3)$. We use Equation 12.4.3 as follows:

$$P(2, 1, 3) = \tfrac{2}{3}P(1, 1, 0) + \tfrac{1}{3}P(2, 0, 3)$$

and

$$P(1, 1, 0) = \tfrac{1}{2}P(0, 1, -2) + \tfrac{1}{2}P(1, 0, 0) = 0$$
$$P(2, 0, 3) = P(1, 0, 1)$$
$$= P(0, 0, 0) = 1$$

Hence,

$$P(2, 1, 3) = \tfrac{1}{3}$$

which checks since in order for the sum of the ranks of the two X values to be less than or equal to 3, the largest of the values X_1, X_2, Y_1, must be Y_1, which, when H_0 is true, has probability $\tfrac{1}{3}$. ∎

Since the rank sum test calls for rejection when either

$$2P(n, m, t) \le \alpha \qquad \text{or} \qquad \alpha \ge 2[1 - P(n, m, t - 1)]$$

it follows that the p-value of the test statistic when $T = t$ is

$$p\text{-value} = 2\min\{P(n, m, t), 1 - P(n, m, t - 1)\}$$

Program 12.4 uses the recursion in Equation 12.4.3 to compute the p-value for the rank sum test. The input needed is the sizes of the first and second samples and the sum of the ranks of the elements of the first sample. Whereas either sample can be designated as the first sample, the program will run fastest if the first sample is the one whose sum of ranks is smallest.

EXAMPLE 12.4c In Example 12.4a, the sizes of the two samples are 5 and 6, respectively, and the sum of the ranks of the first sample is 21. Running Program 12.4 yields the result:

$$p\text{-value} = .1255 \quad ∎$$

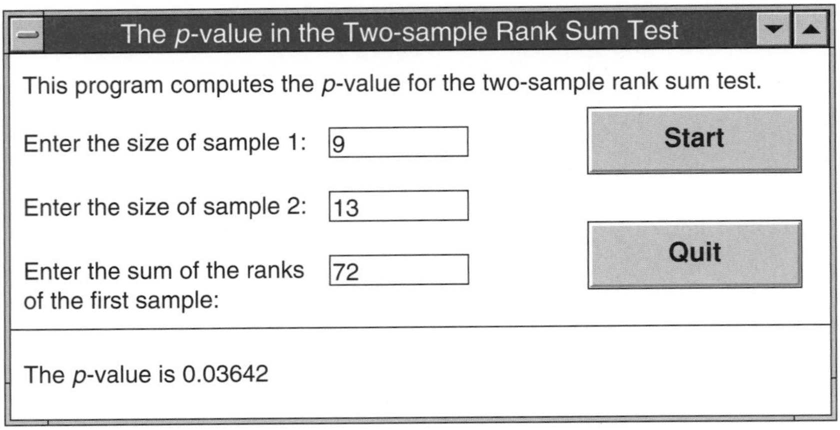

FIGURE 12.3

EXAMPLE 12.4d Suppose that in testing whether 2 production methods yield identical results, 9 items are produced using the first method and 13 using the second. If, among all 22 items, the sum of the ranks of the 9 items produced by method 1 is 72, what conclusions would you draw?

SOLUTION Run Program 12.4 to obtain the result shown in Figure 12.3. Thus, the hypothesis of identical results would be rejected at the 5 percent level of significance. ■

It remains to compute the value of the test statistic T. It is quite efficient to compute T directly by first using a standard computer science algorithm (such as quicksort) to sort, or order, the $n + m$ values. Another approach, easily programmed, although efficient for only small values of n and m, uses the following identity.

PROPOSITION 12.4.1 For $i = 1, \ldots, n, j = 1, \ldots, m$, let

$$W_{ij} = \begin{cases} 1 & \text{if } X_i > Y_j \\ 0 & \text{otherwise} \end{cases}$$

Then

$$T = \frac{n(n+1)}{2} + \sum_{i=1}^{n} \sum_{j=1}^{m} W_{ij}$$

Proof

Consider the values $X_1, \ldots, X_n$ of the first sample and order them. Let $X_{(i)}$ denote the ith smallest, $i = 1, \ldots, n$. Now consider the rank of $X_{(i)}$ among all $n + m$ data values.

This is given by

$$\text{rank of } X_{(i)} = i + \text{ number } j\colon Y_j < X_{(i)}$$

Summing over i gives

$$\sum_{i=1}^{n} \text{rank of } X_{(i)} = \sum_{i=1}^{n} i + \sum_{i=1}^{n} (\text{number } j\colon Y_j < X_{(i)}) \qquad (12.4.4)$$

But since the order in which we add terms does not change the sum obtained, we see that

$$\sum_{i=1}^{n} \text{rank of } X_{(i)} = \sum_{i=1}^{n} \text{rank of } X_i = T \qquad (12.4.5)$$

$$\sum_{i=1}^{n} (\text{number } j : Y_j < X_{(i)}) = \sum_{i=1}^{n} (\text{number } j : Y_j < X_i)$$

Hence, from Equations 12.4.4 and 12.4.5, we obtain that

$$T = \sum_{i=1}^{n} i + \sum_{i=1}^{n} (\text{number } j : Y_j < X_i)$$

$$= \frac{n(n+1)}{2} + \sum_{i=1}^{n} \sum_{j=1}^{m} W_{ij} \quad \blacksquare$$

*12.4.1 THE CLASSICAL APPROXIMATION AND SIMULATION

The difficulty with employing the recursion in Equation 12.4.3 to compute the p-value of the two-sample sum of rank test statistic is that the amount of computation grows enormously as the sample sizes increase. For instance, if $n = m = 200$, then even if we choose the test statistic to be the smaller sum of ranks, since the sum of all the ranks is $1+2+\cdots+400 = 80{,}200$, it is possible that the test statistic could have a value as large as $40{,}100$. Hence, there can be as many as 1.604×10^9 values of $P(N, M, K)$ that would have to be computed to determine the p-value. Thus, for large sample sizes the approach based on the recursion in Equation 12.4.3 is not viable. Two approximate methods that can be utilized in such cases are (a) a classical method based on approximating the distribution of the test statistic and (b) simulation.

(a) *The Classical Approximation*   When the null hypothesis is true and so $F = G$, it follows that all $n + m$ data values come from the same distribution and thus all $(n + m)!$ possible rankings of the values $X_1, \ldots, X_n, Y_1, \ldots, Y_m$ are equally likely.

 Simulation will be covered in Chapter 15.

From this it follows that choosing the n rankings of the first sample is probabilistically equivalent to randomly choosing n of the (possible rank) values $1, 2, \ldots, n + m$. Using this, it can be shown that T has a mean and variance given by

$$E_{H_0}[T] = \frac{n(n + m + 1)}{2}$$

$$\text{Var}_{H_0}(T) = \frac{nm(n + m + 1)}{12}$$

In addition, it can be shown that when both n and m are of moderate size (both being greater than 7 should suffice) T has, under H_0, approximately a normal distribution. Hence, when H_0 is true

$$\frac{T - \dfrac{n(n + m + 1)}{2}}{\sqrt{\dfrac{nm(n + m + 1)}{12}}} \stackrel{.}{\sim} \mathcal{N}(0, 1) \tag{12.4.6}$$

If we let d denote the absolute value of the difference between the observed value of T and its mean value given above, then based on Equation 12.4.6 the approximate p-value is

$$p\text{-value} = P_{H_0}\{|T - E_{H_0}[T]| > d\}$$

$$\approx P\left\{|Z| > d/\sqrt{\frac{nm(n + m + 1)}{12}}\right\} \qquad \text{where } Z \sim \mathcal{N}(0, 1)$$

$$= 2P\left\{Z > d/\sqrt{\frac{nm(n + m + 1)}{12}}\right\}$$

EXAMPLE 12.4e In Example 12.4a, $n = 5, m = 6$, and the test statistic's value is 21. Since

$$\frac{n(n + m + 1)}{2} = 30$$

$$\frac{nm(n + m + 1)}{12} = 30$$

we have that $d = 9$ and so

$$p\text{-value} \approx 2P\left\{Z > \frac{9}{\sqrt{30}}\right\}$$

$$= 2P\{Z > 1.643108\}$$

$$= 2(1 - .9498)$$
$$= .1004$$

which can be compared with the exact value, as given in Example 12.4c, of .1225.

In Example 12.4d, $n = 9$, $m = 13$, and so

$$\frac{n(n + m + 1)}{2} = 103.5$$
$$\frac{nm(n + m + 1)}{12} = 224.25$$

Since $T = 72$, we have that

$$d = |72 - 103.5| = 31.5$$

Thus, the approximate p-value is

$$p\text{-value} \approx 2P\left\{Z > \frac{31.5}{\sqrt{224.25}}\right\}$$
$$= 2P\{Z > 2.103509\}$$
$$= 2(1 - .9823) = .0354$$

which is quite close to the exact p-value (as given in Example 12.4d) of .0364.

Thus, in the two examples considered, the normal approximation worked quite well in the second example — where the guideline that both sample sizes should exceed 7 held — and not so well in the first example — where the guideline did not hold. ∎

(b) *Simulation* If the observed value of the test statistic is $T = t$, then the p-value is given by

$$p\text{-value} = 2 \min \left\{P_{H_0}\{T \geq t\}, P_{H_0}\{T \leq t\}\right\}$$

We can approximate this value by continually simulating a random selection of n of the values $1, 2, \ldots, n + m$ — noting on each occasion the sum of the n values. The value of $P_{H_0}\{T \geq t\}$ can be approximated by the proportion of time that the sum obtained is greater than or equal to t, and $P_{H_0}\{T \leq t\}$ by the proportion of time that it is less than or equal to t.

A Chapter 12 text disk program approximates the p-value by performing the preceding simulation. The program will run most efficiently when the sample of smallest size is designated as the first sample.

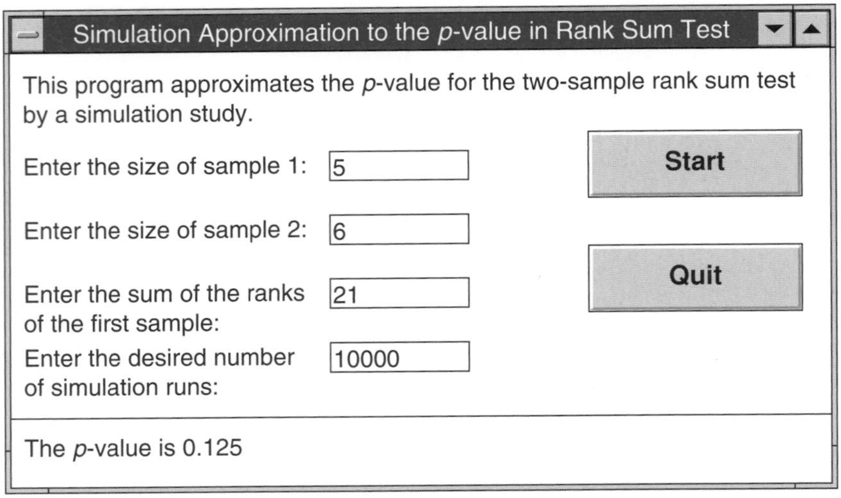

FIGURE 12.4

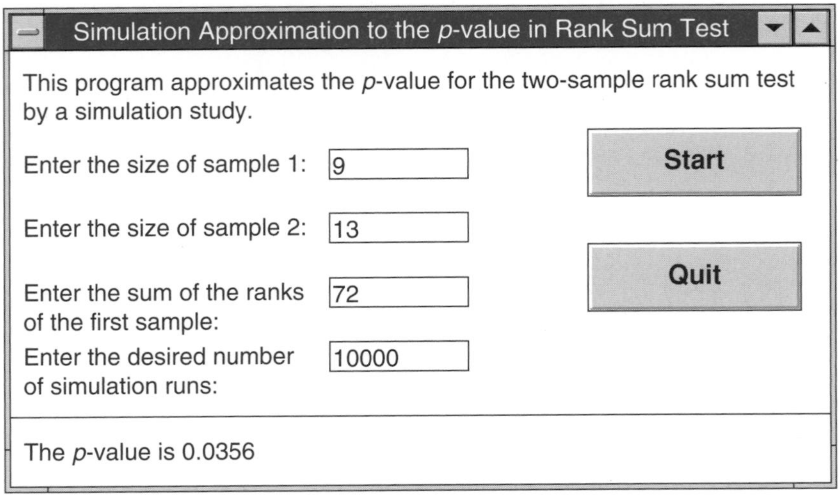

FIGURE 12.5

EXAMPLE 12.4f Running the text disk program on the data of Example 12.4c yields Figure 12.4, which is quite close to the exact value of .1225. Running the program using the data of Example 12.4d yields Figure 12.5, which is again quite close to the exact value of .0364. ∎

Both of the approximation methods work quite well. The normal approximation, when n and m both exceed 7, is usually quite accurate and requires almost no computational time. The simulation approach, on the other hand, can require a great deal of computational time. However, if an immediate answer is not required and great accuracy is desired, then simulation, by running a large number of cases, can be made accurate to an arbitrarily prescribed precision.

12.4.2 TESTING THE EQUALITY OF MULTIPLE PROBABILITY DISTRIBUTIONS

Whereas the preceding sections showed how to test the hypothesis that two population distributions are identical, we are sometimes faced with the situation where there are more than two populations. So suppose there are k populations, that F_i is the distribution function of some measurable value of the elements of population i, and that we are interested in testing the null hypothesis

$$H_0: \quad F_1 = F_2 = \cdots = F_k$$

against the alternative

$$H_1: \quad \text{not all of the } F_i \text{ are equal}$$

To test the preceding null hypothesis, suppose that independent samples are drawn from each of the k populations. Let n_i denote the size of the sample chosen from population $i, i = 1, \ldots, k$. and let $N = \sum_{i=1}^{k} n_i$ denote the total number of data values obtained. Now, rank these N data values from the smallest to largest, and let R_i denote the sum of the ranks of the n_i data values from population i, $i = 1, \ldots, k$.

Now, when H_0 is true, the rank of any individual data value is equally likely to be any of the ranks $1, \ldots, N$, and thus the expected value of its rank is $\frac{1+2+\cdots+N}{N} = \frac{N+1}{2}$. Consequently, with $\bar{r} = \frac{N+1}{2}$, it follows when H_0 is true that the expected sum of the ranks of the n_i data values from population i is $n_i \bar{r}$. That is, when H_0 is true

$$E[R_i] = n_i \bar{r}.$$

Drawing inspiration from the goodness of fit test, let us consider the test statistic

$$T = \sum_{i=1}^{k} \frac{(R_i - n_i \bar{r})^2}{n_i \bar{r}}$$

and use a test that rejects the null hypothesis when T is large. Now,

$$
\begin{aligned}
T &= \frac{1}{\bar{r}} \sum_{i=1}^{k} \frac{R_i^2 - 2R_i n_i \bar{r} + n_i^2 \bar{r}^2}{n_i} \\
&= \frac{1}{\bar{r}} \sum_{i=1}^{k} \frac{R_i^2}{n_i} - 2 \sum_{i=1}^{k} R_i + \bar{r} \sum_{i=1}^{k} n_i \\
&= \frac{1}{\bar{r}} \sum_{i=1}^{k} \frac{R_i^2}{n_i} - N\bar{r}
\end{aligned}
$$

where the final equality used that $\sum_{i=1}^{k} R_i$ is the sum of the ranks of all $N = \sum_i n_i$ data values and so

$$\sum_{i=1}^{k} R_i = 1 + 2 + \cdots + N = \frac{N(N+1)}{2} = N\bar{r}$$

Hence, rejecting H_0 when T is large is equivalent to rejecting H_0 when $\sum_{i=1}^{k} R_i^2/n_i$ is large. So we might as well let the test statistic be

$$TS = \sum_{i=1}^{k} \frac{R_i^2}{n_i}$$

To determine the appropriate α level significance test, we need the distribution of TS when H_0 is true. While its exact distribution is rather complicated, we can use the result that, when H_0 is true and all n_i are at least 5, the distribution of

$$\frac{12}{N(N+1)} TS - 3(N+1)$$

is approximately that of a chi-squared random variable with $k-1$ degrees of freedom. Using this, we see that an approximate significance level α test of the null hypothesis that all distributions are identical is to

$$\text{reject } H_0 \quad \text{if} \quad \frac{12}{N(N+1)} TS - 3(N+1) \geq \chi_{k-1,\alpha}^2$$

For even more accuracy, simulation can be used. To implement it, one should first compute the value of TS, say that it is equal to t. The resulting p-value is

$$p\text{-value} = P_{H_0}\{TS \geq t\}$$

To determine the preceding by a simulation, one should continually simulate $TS = \sum_i R_i^2/n_i$ under the assumption that H_0 is true and then use the fraction of the simulated values that are at least t as the estimate of the p-value. Because, under H_0, all possible orderings of the N data values are equal likely, one could simulate TS by generating a permutation of $1, \ldots, N$ that is equally likely to be any of the $N!$ permutations. (See Example 15.2b for an efficient way to simulate such a *random permutation*.) One can then let the first n_1 values of the permutation be the ranks of the first sample, the next n_2 values be the ranks of the second sample, and so on. That is, if $p_1, \ldots, p_N$ is the generated value of the permutation, then with $s_0 = 0$, $s_j = n_1 + \cdots + n_j, j \geq 1$, the simulated values of $R_1, \ldots, R_k$ would be

$$R_i = \sum_{j=s_{i-1}+1}^{s_i} p_j, \quad i = 1, \ldots, k$$

The preceding is known as the *Kruskal–Wallis* test.

EXAMPLE 12.4g The following data give the number of visitors to a medium size Los Angeles library on Tuesdays, Wednesdays, and Thursdays of 10 successive weeks.

Tuesday visitors	721, 660, 622, 738, 820, 707, 672, 589, 902, 688
Wednesday visitors	604, 626, 744, 802, 691, 665, 711, 715, 661, 729
Thursday visitors	642, 480, 705, 584, 713, 654, 704, 522, 683, 708

Are these data consistent with the hypothesis that the distributions of the number of visitors for the three days are identical?

SOLUTION Ordering the $N = 30$ data values, gives that the sum of the ranks of the three samples are

$$R_1 = 176, \quad R_2 = 175, \quad R_3 = 114$$

Therefore,

$$\frac{12}{N(N+1)} TS - 3(N+1) = \frac{12}{30 \cdot 31} \frac{176^2 + 175^2 + 114^2}{10} - 93 = 3.254$$

Because $\chi^2_{2,.05} = 5.99$, the null hypothesis that the distributions of the number of visitors for each of the three weekdays are identical cannot be rejected at the 5 percent level of significance. Indeed, the resulting p-value is

$$p\text{-value} = P\{\chi^2_2 \geq 3.254\} = .1965 \quad \blacksquare$$

12.5 THE RUNS TEST FOR RANDOMNESS

A basic assumption in much of statistics is that a set of data constitutes a random sample from some population. However, it is sometimes the case that the data are not generated by a truly random process but by one that may follow a trend or a type of cyclical pattern. In this section, we will consider a test — called the runs test — of the hypothesis H_0 that a given data set constitutes a random sample.

To begin, let us suppose that each of the data values is either a 0 or a 1. That is, we shall assume that each data value can be dichotomized as being either a success or a failure. Let $X_1, \ldots, X_N$ denote the set of data. Any consecutive sequence of either 1's or 0's is called a *run*. For instance, the data set

$$1\ 0\ 0\ 1\ 1\ 1\ 0\ 0\ 1\ 0\ 1\ 1\ 1\ 1\ 0\ 1\ 0\ 0\ 0\ 0\ 1\ 1$$

contains 11 runs — 6 runs of 1 and 5 runs of 0. Suppose that the data set $X_1, \ldots, X_N$ contains n 1's and m 0's, where $n + m = N$, and let R denote the number of runs. Now, if H_0 were true, then $X_1, \ldots, X_N$ would be equally likely to be any of the $N!/(n!m!)$ permutations of n 1's and m 0's, and therefore, given a total of n 1's and m 0's, it follows that, under H_0, the probability mass function of R, the number of runs is given by

$$P_{H_0}\{R = k\} = \frac{\text{number of permutations of } n \text{ 1's and } m \text{ 0's resulting in } k \text{ runs}}{\dbinom{n+m}{n}}$$

This number of permutations can be explicitly determined and it can be shown that

$$P_{H_0}\{R = 2k\} = 2\frac{\dbinom{m-1}{k-1}\dbinom{n-1}{k-1}}{\dbinom{m+n}{n}}$$

$$\text{(12.5.1)}$$

$$P_{H_0}\{R = 2k+1\} = \frac{\dbinom{m-1}{k-1}\dbinom{n-1}{k} + \dbinom{m-1}{k}\dbinom{n-1}{k-1}}{\dbinom{n+m}{n}}$$

If the data contain n 1's and m 0's, then the runs test calls for rejection of the hypothesis that the data constitutes a random sample if the observed number of runs is either too large or too small to be explained by chance. Specifically, if the observed number of runs is r, then the p-value of the runs test is

$$p\text{-value} = 2\min(P_{H_0}\{R \geq r\}, P_{H_0}\{R \leq r\})$$

Program 12.5 uses Equation 12.5.1 to compute the p-value.

EXAMPLE 12.5a The following is the result of the last 30 games played by an athletic team, with W signifying a win and L a loss.

$$W\ W\ W\ L\ W\ W\ L\ W\ W\ L\ W\ L\ W\ W\ L\ W\ W\ W\ W\ L\ W\ L\ W\ W\ W\ L\ W\ L\ W\ L$$

Are these data consistent with pure randomness?

SOLUTION To test the hypothesis of randomness, note that the data, which consist of 20 W's and 10 L's, contain 20 runs. To see whether this justifies rejection at, say, the 5 percent level of significance, we run Program 12.5 and observe the results in Figure 12.6. Therefore, the hypothesis of randomness would be rejected at the 5 percent level of significance. (The striking thing about these data is that the team always came back to win after losing a game, which would be quite unlikely if all outcomes containing 20 wins and 10 losses were equally likely.) ∎

The above can also be used to test for randomness when the data values are not just 0's and 1's. To test whether the data $X_1, \ldots, X_N$ constitute a random sample, let s-med denote the sample median. Also let n denote the number of data values that are less than

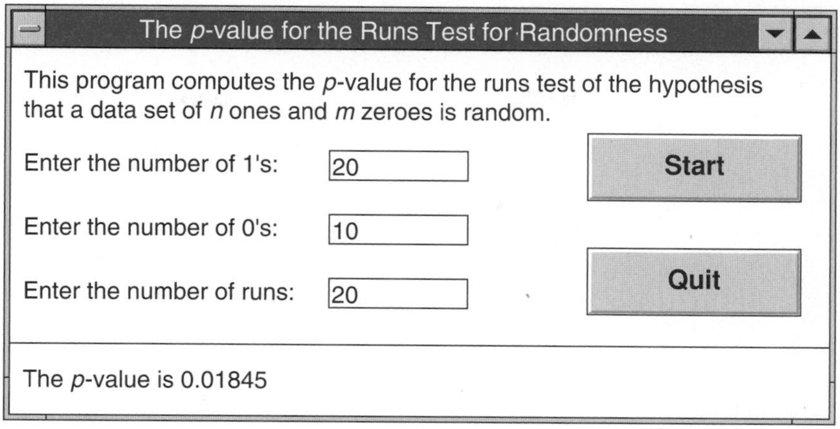

The p-value for the Runs Test for Randomness

This program computes the p-value for the runs test of the hypothesis that a data set of n ones and m zeroes is random.

Enter the number of 1's: 20

Enter the number of 0's: 10

Enter the number of runs: 20

Start

Quit

The p-value is 0.01845

FIGURE 12.6

or equal to s-med and m the number that are greater. (Thus, if N is even and all data values are distinct, then $n = m = N/2$.) Define $I_1, \ldots, I_N$ by

$$I_j = \begin{cases} 1 & \text{if } X_j \leq \text{s-med} \\ 0 & \text{otherwise} \end{cases}$$

Now, if the original data constituted a random sample, then the number of runs in $I_1, \ldots, I_N$ would have a probability mass function given by Equation 12.5.1. Thus, it follows that we can use the preceding runs test on the data values $I_1, \ldots, I_N$ to test that the original data are random.

EXAMPLE 12.5b   The lifetime of 19 successively produced storage batteries is as follows:

145 152 148 155 176 134 184 132 145 162 165

185 174 198 179 194 201 169 182

The sample median is the 10th smallest value — namely, 169. The data indicating whether the successive values are less than or equal to or greater than 169 are as follows:

1 1 1 1 0 1 0 1 1 1 1 0 0 0 0 0 0 1 0

Hence, the number of runs is 8. To determine if this value is statistically significant, we run Program 12.5 (with $n = 10$, $m = 9$) to obtain the result:

$$p\text{-value} = .357$$

Thus the hypothesis of randomness is accepted.   ∎

It can be shown that, when n and m are both large and H_0 is true, R will have approximately a normal distribution with mean and standard deviation given by

$$\mu = \frac{2nm}{n+m} + 1 \quad \text{and} \quad \sigma = \sqrt{\frac{2nm(2nm-n-m)}{(n+m)^2(n+m-1)}} \qquad (12.5.2)$$

Therefore, when n and m are both large

$$P_{H_0}\{R \leq r\} = P_{H_0}\left\{\frac{R-\mu}{\sigma} \leq \frac{r-\mu}{\sigma}\right\}$$

$$\approx P\left\{Z \leq \frac{r-\mu}{\sigma}\right\}, \qquad Z \sim \mathcal{N}(0,1)$$

$$= \Phi\left(\frac{r-\mu}{\sigma}\right)$$

and, similarly,

$$P_{H_0}\{R \geq r\} \approx 1 - \Phi\left(\frac{r-\mu}{\sigma}\right)$$

Hence, for large n and m, the p-value of the runs test for randomness is approximately given by

$$p\text{-value} \approx 2\min\left\{\Phi\left(\frac{r-\mu}{\sigma}\right), 1 - \Phi\left(\frac{r-\mu}{\sigma}\right)\right\}$$

where μ and σ are given by Equation 12.5.2 and r is the observed number of runs.

EXAMPLE 12.5c Suppose that a sequence of sixty 1's and sixty 0's resulted in 75 runs. Since

$$\mu = 61 \quad \text{and} \quad \sigma = \sqrt{\frac{3{,}540}{119}} = 5.454$$

we see that the approximate p-value is

$$p\text{-value} \approx 2\min\{\Phi(2.567), 1 - \Phi(2.567)\}$$
$$= 2 \times (1 - .9949)$$
$$= .0102$$

On the other hand, by running Program 12.5 we obtain that the exact p-value is

$$p\text{-value} = .0130$$

If the number of runs was equal to 70 rather than 75, then the approximate p-value would be

$$p\text{-value} \approx 2[1 - \Phi(1.650)] = .0990$$

as opposed to the exact value of

$$p\text{-value} = .1189 \quad \blacksquare$$

Problems

1. A new medicine against hypertension was tested on 18 patients. After 40 days of treatment, the following changes of the diastolic blood pressure were observed.

$$-5, \quad -1, \quad +2, \quad +8, \quad -25, \quad +1, \quad +5, \quad -12, \quad -16$$
$$-9, \quad -8, \quad -18, \quad -5, \quad -22, \quad +4, \quad -21, \quad -15, \quad -11$$

Use the sign test to determine if the medicine has an effect on blood pressure. What is the p-value?

2. An engineering firm is involved in selecting a computer system, and the choice has been narrowed to two manufacturers. The firm submits eight problems to the two computer manufacturers and has each manufacturer measure the number of seconds required to solve the design problem with the manufacturer's software. The times for the eight design problems are given below.

Design problem	1	2	3	4	5	6	7	8
Time with computer A	15	32	17	26	42	29	12	38
Time with computer B	22	29	1	23	46	25	19	47

Determine the p-value of the sign test when testing the hypothesis that there is no difference in the distribution of the time it takes the two types of software to solve problems.

3. The published figure for the median systolic blood pressure of middle-aged men is 128. To determine if there has been any change in this value, a random sample of 100 men has been selected. Test the hypothesis that the median is equal to 128 if

(a) 60 men have readings above 128;
(b) 70 men have readings above 128;
(c) 80 men have readings above 128.

In each case, determine the p-value.

4. To test the hypothesis that the median weight of 16-year-old females from Los Angeles is at least 110 pounds, a random sample of 200 such females was chosen. If 120 females weighed less than 110 pounds, does this discredit the hypothesis? Use the 5 percent level of significance. What is the p-value?

5. In 2004, the national median salary of all U.S. financial accountants was $124,400. A recent random sample of 14 financial accountants showed 2007 incomes of (in units of $1,000)

$$125.5, \ 130.3, \ 133.0, \ 102.6, \ 198.0, \ 232.5, \ 106.8,$$

$$114.5, \ 122.0, \ 100.0, \ 118.8, \ 108.6, \ 312.7, \ 125.5$$

Use these data to test the hypothesis that the median salary of financial accountants in 2007 was not greater than in 2004. What is the p-value?

6. An experiment was initiated to study the effect of a newly developed gasoline detergent on automobile mileage. The following data, representing mileage per gallon before and after the detergent was added for each of eight cars, resulted.

Car	Mileage without Additive	Mileage with Additive
1	24.2	23.5
2	30.4	29.6
3	32.7	32.3
4	19.8	17.6
5	25.0	25.3
6	24.9	25.4
7	22.2	20.6
8	21.5	20.7

Find the p-value of the test of the hypothesis that mileage is not affected by the additive when

(a) the sign test is used;
(b) the signed rank test is used.

7. Determine the p-value when using the signed rank statistic in Problems 1 and 2.

8. Twelve patients having high albumin content in their blood were treated with a medicine. Their blood content of albumin was measured before and after treatment. The measured values are shown in the table.

Blood Content of Albumin[a]

Patient	Before Treatment	After Treatment
1	5.02	4.66
2	5.08	5.15
3	4.75	4.30
4	5.25	5.07
5	4.80	5.38
6	5.77	5.10
7	4.85	4.80
8	5.09	4.91
9	6.05	5.22
10	4.77	4.50
11	4.85	4.85
12	5.24	4.56

[a] Values given in grams per 100 ml.

Is the effect of the medicine significant at the 5 percent level?

(a) Use the sign test.

(b) Use the signed rank test.

9. An engineer claims that painting the exterior of a particular aircraft affects its cruising speed. To check this, the next 10 aircraft off the assembly line were flown to determine cruising speed prior to painting, and were then painted and reflown. The following data resulted.

	Cruising Speed (knots)	
Aircraft	Not Painted	Painted
1	426.1	416.7
2	418.4	403.2
3	424.4	420.1
4	438.5	431.0
5	440.6	432.6
6	421.8	404.2
7	412.2	398.3
8	409.8	405.4
9	427.5	422.8
10	441.2	444.8

Do the data uphold the engineer's claim?

10. Ten pairs of duplicate spectrochemical determinations for nickel are presented below. The readings in column 2 were taken with one type of measuring instrument and those in column 3 were taken with another type.

Sample	Duplicates	
1	1.94	2.00
2	1.99	2.09
3	1.98	1.95
4	2.07	2.03
5	2.03	2.08
6	1.96	1.98
7	1.95	2.03
8	1.96	2.03
9	1.92	2.01
10	2.00	2.12

Test the hypothesis, at the 5 percent level of significance, that the two measuring instruments give equivalent results.

11. Let $X_1, \ldots, X_n$ be a sample from the continuous distribution F having median m; and suppose we are interested in testing the hypothesis $H_0 : m = m_0$ against the one-sided alternative $H_1 : m > m_0$. Present the one-sided analog of the signed rank test. Explain how the p-value would be computed.

12. In a study of bilingual coding, 12 bilingual (French and English) college students are divided into two groups. Each group reads an article written in French, and each answers a series of 25 multiple-choice questions covering the content of the article. For one group the questions are written in French; the other takes the examination in English. The score (total correct) for the two groups is:

Examination in French	11	12	16	22	25	25
Examination in English	10	13	17	19	21	24

Is this evidence at the 5 percent significance level that there is difficulty in transferring information from one language to another?

13. Fifteen cities, of roughly equal size, are chosen for a traffic safety study. Eight of them are randomly chosen, and in these cities a series of newspaper articles dealing with traffic safety is run over a 1-month period. The number of traffic accidents reported in the month following this campaign is as follows:

Treatment group	19	31	39	45	47	66	74	81
Control group	28	36	44	49	52	52	60	

Determine the exact p-value when testing the hypothesis that the articles have not had any effect.

14. Determine the p-value in Problem 13 by

   (a) using the normal approximation;
   (b) using a simulation study.

15. The following are the weights of random samples of adult males from different political affiliations.

   Republicans: 204, 178, 195, 187, 240, 182, 152, 166
   Democrats: 175, 200, 168, 192, 156, 164, 180, 138

   We want to use these data to test the null hypothesis that the two distributions are identical.

   (a) Find the exact p-value.
   (b) Determine the p-value obtained when using the normal approximation.

16. In a 1943 experiment (Whitlock, H. V., and Bliss, D. H., "A bioassay technique for antihelminthics," *Journal of Parasitology*, **29**, pp. 48–58, 10), albino rats were used to study the effectiveness of carbon tetrachloride as a treatment for worms. Each rat received an injection of worm larvae. After 8 days, the rats were randomly divided into 2 groups of 5 each; each rat in the first group received a dose of .032 cc of carbon tetrachloride, whereas the dosage for each rat in the second group was .063 cc. Two days later the rats were killed, and the number of adult worms in each rat was determined. The numbers detected in the group receiving the .032 dosage were

$$421, 462, 400, 378, 413$$

   whereas they were

$$207, 17, 412, 74, 116$$

   for those receiving the .063 dosage. Do the data prove that the larger dosage is more effective than the smaller?

17. In a 10-year study of the dispersal patterns of beavers (Sun, L. and Muller-Schwarze, D., "Statistical resampling methods in biology: A case study of beaver dispersal patterns," *American Journal of Mathematical and Management Sciences*, **16**, pp. 463–502, 1996) a total of 332 beavers were trapped in Allegheny State Park in southwestern New York. The beavers were tagged (so as to be identifiable when later caught) and released. Over time a total of 32 of them, 9 female and

23 male, were discovered to have resettled in other sites. The following data give the dispersal distances (in kilometers) between these beavers' original and resettled sites for the females and for the males.

Females: .660, .984, .984, 1.992, 4.368, 6.960, 10.656, 21.600, 31.680
Males: .288, .312, .456,   .528,   .576, .720,   .792,   .984, 1.224,
           1.584, 2.304, 2.328, 2.496, 2.688, 3.096, 3.408, 4.296, 4.884,
           5.928, 6.192, 6.384, 13.224, 27.600

Do the data prove that the dispersal distances are gender related?

18. The following data give the numbers of people who visit a local health clinic in the day following

   (1) a Saturday win by the local university football team;
   (2) a Saturday loss by the team;
   (3) a Saturday when the team does not play.

   Number following a win   71, 66, 62, 79, 80, 70, 66, 59, 89, 68
   Number following a loss   64, 62, 75, 81, 69, 67, 73, 71, 69, 74
   Number when no game   49, 48, 70, 58, 73, 65, 55, 52, 68, 74

   Do these data prove that the resulting number of clinic visits depends on what happens with the football team? Test at the 5 percent level.

19. A production run of 50 items resulted in 11 defectives, with the defectives occurring on the following items (where the items are numbered by their order of production): 8, 12, 13, 14, 31, 32, 37, 38, 40, 41, 42. Can we conclude that the successive items did not constitute a random sample?

20. The following data represent the successive quality levels of 25 articles: 100, 110, 122, 132, 99, 96, 88, 75, 45, 211, 154, 143, 161, 142, 99, 111, 105, 133, 142, 150, 153, 121, 126, 117, 155. Does it appear that these data are a random sample from some population?

21. Can we use the runs test if we consider whether each data value is less than or greater than some predetermined value rather than the value s-med?

22. The following table (taken from Quinn, W. H., Neal, T. V., and Antuñez de Mayolo, S. E., 1987, "El Niño occurrences over the past four-and-a-half centuries," *Journal of Geophysical Research*, **92** (C13), pp. 14,449–14,461) gives the years and magnitude (either moderate or strong) of major El Niño years between 1800 and 1987. Use it to test the hypothesis that the successive El Niño magnitudes constitute a random sample.

Year and Magnitude (0 = moderate, 1 = strong) of Major El Niño Events, 1800–1987

Year	Magnitude	Year	Magnitude	Year	Magnitude
1803	1	1866	0	1918	0
1806	0	1867	0	1923	0
1812	0	1871	1	1925	1
1814	1	1874	0	1930	0
1817	0	1877	1	1932	1
1819	0	1880	0	1939	0
1821	0	1884	1	1940	1
1824	0	1887	0	1943	0
1828	1	1891	1	1951	0
1832	0	1896	0	1953	0
1837	0	1899	1	1957	1
1844	1	1902	0	1965	0
1850	0	1905	0	1972	1
1854	0	1907	0	1976	0
1857	0	1911	1	1982	1
1860	0	1914	0	1987	0
1864	1	1917	1		

Chapter 13

QUALITY CONTROL

13.1 INTRODUCTION

Almost every manufacturing process results in some random variation in the items it produces. That is, no matter how stringently the process is being controlled, there is always going to be some variation between the items produced. This variation is called *chance variation* and is considered to be inherent to the process. However, there is another type of variation that sometimes appears. This variation, far from being inherent to the process, is due to some *assignable cause* and usually results in an adverse effect on the quality of the items produced. For instance, this latter variation may be caused by a faulty machine setting, or by poor quality of the raw materials presently being used, or by incorrect software, or human error, or any other of a large number of possibilities. When the only variation present is due to chance, and not to assignable cause, we say that the process is in control, and a key problem is to determine whether a process is in or is *out of control*.

The determination of whether a process is in or out of control is greatly facilitated by the use of *control charts*, which are determined by two numbers — the upper and lower control limits. To employ such a chart, the data generated by the manufacturing process are divided into subgroups and subgroup statistics — such as the subgroup average and subgroup standard deviation — are computed. When the subgroup statistic does not fall within the upper and lower control limit, we conclude that the process is out of control.

In Sections 13.2 and 13.3, we suppose that the successive items produced have measurable characteristics, whose mean and variance are fixed when the process is in control. We show how to construct control charts based on subgroup averages (in Section 13.2) and on subgroup standard deviations (in Section 13.3). In Section 13.4, we suppose that rather than having a measurable characteristic, each item is judged by an *attribute* — that is, it is classified as either acceptable or unacceptable. Then we show how to construct control charts that can be used to indicate a change in the quality of the items produced. In Section 13.5, we consider control charts in situations where each item produced has a random number of defects. Finally, in Section 13.6 we consider more sophisticated types of control charts — ones that don't consider each subgroup value in isolation but rather take into account the values of other subgroups. Three different control charts of this type — known as moving average, exponential weighted moving average, and cumulative sum control charts — are presented in Section 13.6.

13.2 CONTROL CHARTS FOR AVERAGE VALUES: THE $\overline{X}$ CONTROL CHART

Suppose that when the process is in control the successive items produced have measurable characteristics that are independent, normal random variables with mean μ and variance σ^2. However, due to special circumstances, suppose that the process may go out of control and start producing items having a different distribution. We would like to be able to recognize when this occurs so as to stop the process, find out what is wrong, and fix it.

Let $X_1, X_2, \ldots$ denote the measurable characteristics of the successive items produced. To determine when the process goes out of control, we start by breaking the data up into subgroups of some fixed size — call it n. The value of n is chosen so as to yield uniformity within subgroups. That is, n may be chosen so that all data items within a subgroup were produced on the same day, or on the same shift, or using the same settings, and so on. In other words, the value of n is chosen so that it is reasonable that a shift in distribution would occur between and not within subgroups. Typical values of n are 4, 5, or 6.

Let $\overline{X}_i$, $i = 1, 2, \ldots$ denote the average of the ith subgroup. That is,

$$\overline{X}_1 = \frac{X_1 + \cdots + X_n}{n}$$

$$\overline{X}_2 = \frac{X_{n+1} + \cdots + X_{2n}}{n}$$

$$\overline{X}_3 = \frac{X_{2n+1} + \cdots + X_{3n}}{n}$$

and so on. Since, when in control, each of the X_i have mean μ and variance σ^2, it follows that

$$E(\overline{X}_i) = \mu$$

$$\mathrm{Var}(\overline{X}_i) = \frac{\sigma^2}{n}$$

and so

$$\frac{\overline{X}_i - \mu}{\sqrt{\dfrac{\sigma^2}{n}}} \sim \mathcal{N}(0, 1)$$

That is, if the process is in control throughout the production of subgroup i, then $\sqrt{n}(\overline{X}_i - \mu)/\sigma$ has a standard normal distribution. Now it follows that a standard normal random variable Z will almost always be between -3 and $+3$. (Indeed, $P\{-3 < Z < 3\} = .9973$.) Hence, if the process is in control throughout the production of the items in subgroup i, then we would certainly expect that

$$-3 < \sqrt{n}\,\frac{\overline{X}_i - \mu}{\sigma} < 3$$

or, equivalently, that

$$\mu - \frac{3\sigma}{\sqrt{n}} < \overline{X}_i < \mu + \frac{3\sigma}{\sqrt{n}}$$

The values

$$\text{UCL} \equiv \mu + \frac{3\sigma}{\sqrt{n}}$$

and

$$\text{LCL} \equiv \mu - \frac{3\sigma}{\sqrt{n}}$$

are called, respectively, the *upper* and *lower control limits*.

The $\overline{X}$ control chart, which is designed to detect a change in the average value of an item produced, is obtained by plotting the successive subgroup averages $\overline{X}_i$ and declaring that the process is out of control the first time $\overline{X}_i$ does not fall between LCL and UCL (see Figure 13.1).

EXAMPLE 13.2a A manufacturer produces steel shafts having diameters that should be normally distributed with mean 3 mm and standard deviation .1 mm. Successive samples of four shafts have yielded the following sample averages in millimeters.

Sample	$\overline{X}$	Sample	$\overline{X}$
1	3.01	6	3.02
2	2.97	7	3.10
3	3.12	8	3.14
4	2.99	9	3.09
5	3.03	10	3.20

What conclusion should be drawn?

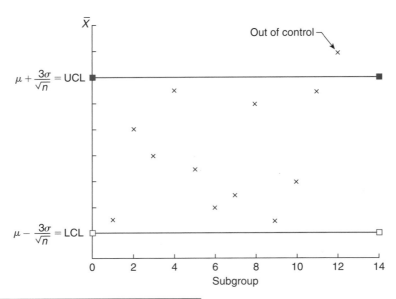

FIGURE 13.1   *Control chart for $\overline{X}$, n = size of subgroup.*

SOLUTION When in control the successive diameters have mean $\mu = 3$ and standard deviation $\sigma = .1$, and so with $n = 4$ the control limits are

$$\text{LCL} = 3 - \frac{3(.1)}{\sqrt{4}} = 2.85, \qquad \text{UCL} = 3 + \frac{3(.1)}{\sqrt{4}} = 3.15$$

Because sample number 10 falls above the upper control limit, it appears that there is reason to suspect that the mean diameter of shafts now differs from 3. (Clearly, judging from the results of Samples 6 through 10 it appears to have increased beyond 3.) ∎

REMARKS

(a) The foregoing supposes that when the process is in control the underlying distribution is normal. However, even if this is not the case, by the central limit theorem it follows that the subgroup averages should have a distribution that is roughly normal and so would be unlikely to differ from its mean by more than 3 standard deviations.

(b) It is frequently the case that we do not determine the measurable qualities of all the items produced but only those of a randomly chosen subset of items. If this is so then it is natural to select, as a subgroup, items that are produced at roughly the same time.

It is important to note that even when the process is in control there is a chance — namely, .0027 — that a subgroup average will fall outside the control limit and so one would incorrectly stop the process and hunt for the nonexistent source of trouble.

Let us now suppose that the process has just gone out of control by a change in the mean value of an item from μ to $\mu + a$ where $a > 0$. How long will it take (assuming things do not change again) until the chart will indicate that the process is now out of control? To answer this, note that a subgroup average will be within the control limits if

$$-3 < \sqrt{n}\,\frac{\bar{X} - \mu}{\sigma} < 3$$

or, equivalently, if

$$-3 - \frac{a\sqrt{n}}{\sigma} < \sqrt{n}\,\frac{\bar{X} - \mu}{\sigma} - \frac{a\sqrt{n}}{\sigma} < 3 - \frac{a\sqrt{n}}{\sigma}$$

or

$$-3 - \frac{a\sqrt{n}}{\sigma} < \sqrt{n}\,\frac{\bar{X} - \mu - a}{\sigma} < 3 - \frac{a\sqrt{n}}{\sigma}$$

Hence, since $\bar{X}$ is normal with mean $\mu + a$ and variance σ^2/n — and so $\sqrt{n}(\bar{X} - \mu - a)/\sigma$ has a standard normal distribution — the probability that it will fall within the control limits is

$$P\left\{-3 - \frac{a\sqrt{n}}{\sigma} < Z < 3 - \frac{a\sqrt{n}}{\sigma}\right\} = \Phi\left(3 - \frac{a\sqrt{n}}{\sigma}\right) - \Phi\left(-3 - \frac{a\sqrt{n}}{\sigma}\right)$$

$$\approx \Phi\left(3 - \frac{a\sqrt{n}}{\sigma}\right)$$

and so the probability that it falls outside is approximately $1 - \Phi(3 - a\sqrt{n}/\sigma)$. For instance, if the subgroup size is $n = 4$, then an increase in the mean value of 1 standard deviation — that is, $a = \sigma$ — will result in the subgroup average falling outside of the control limits with probability $1 - \Phi(1) = .1587$. Because each subgroup average will independently fall outside the control limits with probability $1 - \Phi(3 - a\sqrt{n}/\sigma)$, it follows that the number of subgroups that will be needed to detect this shift has a geometric distribution with mean $\{1 - \Phi(3 - a\sqrt{n}/\sigma)\}^{-1}$. (In the case mentioned before with $n = 4$, the number of subgroups one would have to chart to detect a change in the mean of 1 standard deviation has a geometric distribution with mean $1/.158 \approx 6.3$.)

13.2.1 CASE OF UNKNOWN μ AND σ

If one is just starting up a control chart and does not have reliable historical data, then μ and σ would not be known and would have to be estimated. To do so, we employ k of the subgroups where k should be chosen so that $k \geq 20$ and $nk \geq 100$. If $\bar{X}_i, i = 1, \ldots, k$ is the average of the ith subgroup, then it is natural to estimate μ by $\bar{\bar{X}}$ the average of these subgroup averages. That is,

$$\bar{\bar{X}} = \frac{\bar{X}_1 + \cdots + \bar{X}_k}{k}$$

To estimate σ, let S_i denote the sample standard deviation of the ith subgroup, $i = 1, \ldots, k$. That is,

$$S_1 = \sqrt{\sum_{i=1}^{n} \frac{(X_i - \overline{X}_1)^2}{n-1}}$$

$$S_2 = \sqrt{\sum_{i=1}^{n} \frac{(X_{n+i} - \overline{X}_2)^2}{n-1}}$$

$$\vdots$$

$$S_k = \sqrt{\sum_{i=1}^{n} \frac{(X_{(k-1)n+i} - \overline{X}_k)^2}{n-1}}$$

Let

$$\overline{S} = (S_1 + \cdots + S_k)/k$$

The statistic $\overline{S}$ will not be an unbiased estimator of σ — that is, $E[\overline{S}] \neq \sigma$. To transform it into an unbiased estimator, we must first compute $E[\overline{S}]$, which is accomplished as follows:

$$E[\overline{S}] = \frac{E[S_1] + \cdots + E[S_k]}{k} \tag{13.2.1}$$

$$= E[S_1]$$

where the last equality follows since $S_1, \ldots, S_k$ are independent and identically distributed (and thus have the same mean). To compute $E[S_1]$, we make use of the following fundamental result about normal samples — namely, that

$$\frac{(n-1)S_1^2}{\sigma^2} = \sum_{i=1}^{n} \frac{(X_i - \overline{X}_1)^2}{\sigma^2} \sim \chi_{n-1}^2 \tag{13.2.2}$$

Now it is not difficult to show (see Problem 3) that

$$E[\sqrt{Y}] = \frac{\sqrt{2}\Gamma(n/2)}{\Gamma(\frac{n-1}{2})} \quad \text{when } Y \sim \chi_{n-1}^2 \tag{13.2.3}$$

Since

$$E[\sqrt{(n-1)S^2/\sigma^2}] = \sqrt{n-1}\,E[S_1]/\sigma$$

we see from Equations 13.2.2 and 13.2.3 that

$$E[S_1] = \frac{\sqrt{2}\Gamma(n/2)\sigma}{\sqrt{n-1}\Gamma(\frac{n-1}{2})}$$

Hence, if we set

$$c(n) = \frac{\sqrt{2}\Gamma(n/2)}{\sqrt{n-1}\Gamma(\frac{n-1}{2})}$$

then it follows from Equation 13.2.1 that $\bar{S}/c(n)$ is an unbiased estimator of σ.

Table 13.1 presents the values of $c(n)$ for $n = 2$ through $n = 10$.

TECHNICAL REMARK

In determining the values in Table 13.1, the computation of $\Gamma(n/2)$ and $\Gamma(n - \frac{1}{2})$ was based on the recursive formula

$$\Gamma(a) = (a - 1)\Gamma(a - 1)$$

TABLE 13.1   Values of c(n)

$c(2)$	=	.7978849
$c(3)$	=	.8862266
$c(4)$	=	.9213181
$c(5)$	=	.9399851
$c(6)$	=	.9515332
$c(7)$	=	.9593684
$c(8)$	=	.9650309
$c(9)$	=	.9693103
$c(10)$	=	.9726596

which was established in Section 5.7. This recursion yields that, for integer n,

$$\Gamma(n) = (n-1)(n-2)\cdots 3 \cdot 2 \cdot 1 \cdot \Gamma(1)$$

$$= (n-1)! \quad \text{since } \Gamma(1) = \int_0^\infty e^{-x}\, dx = 1$$

The recursion also yields that

$$\Gamma\left(\frac{n+1}{2}\right) = \left(\frac{n-1}{2}\right)\left(n-\frac{3}{2}\right)\cdots\frac{3}{2}\cdot\frac{1}{2}\cdot\Gamma\left(\frac{1}{2}\right)$$

with

$$\Gamma\left(\frac{1}{2}\right) = \int_0^\infty e^{-x} x^{-1/2}\, dx$$

$$= \int_0^\infty e^{-y^2/2} \frac{\sqrt{2}}{y} y\, dy \quad \text{by } x = \frac{y^2}{2} \quad dx = y\, dy$$

$$= \sqrt{2} \int_0^\infty e^{-y^2/2}\, dy$$

$$= 2\sqrt{\pi}\, \frac{1}{\sqrt{2\pi}} \int_0^\infty e^{-y^2/2}\, dy$$

$$= 2\sqrt{\pi}\, P[N(0, 1) > 0]$$

$$= \sqrt{\pi}$$

The preceding estimates for μ and σ make use of all k subgroups and thus are reasonable only if the process has remained in control throughout. To check this, we compute the control limits based on these estimates of μ and σ, namely,

$$\text{LCL} = \overline{\overline{X}} - \frac{3\overline{S}}{c(n)\sqrt{n}} \tag{13.2.4}$$

$$\text{UCL} = \overline{\overline{X}} + \frac{3\overline{S}}{c(n)\sqrt{n}}$$

We now check that each of the subgroup averages $\overline{X}_i$ falls within these lower and upper limits. Any subgroup whose average value does not fall within the limits is removed (we suppose that the process was temporarily out of control) and the estimates are recomputed. We then again check that all the remaining subgroup averages fall within the control limits. If not, then they are removed, and so on. Of course, if too many of the subgroup averages fall outside the control limits, then it is clear that no control has yet been established.

EXAMPLE 13.2b   Let us reconsider Example 13.2a under the new supposition that the process is just beginning and so μ and σ are unknown. Also suppose that the sample standard deviations were as follows:

	$\overline{X}$	S		$\overline{X}$	S
1	3.01	.12	6	3.02	.08
2	2.97	.14	7	3.10	.15
3	3.12	.08	8	3.14	.16
4	2.99	.11	9	3.09	.13
5	3.03	.09	10	3.20	.16

Since $\overline{\overline{X}} = 3.067, \bar{S} = .122, c(4) = .9213$, the control limits are

$$LCL = 3.067 - \frac{3(.122)}{2 \times .9213} = 2.868$$

$$UCL = 3.067 + \frac{3(.122)}{2 \times .9213} = 3.266$$

Since all the $\bar{X}_i$ fall within these limits, we suppose that the process is in control with $\mu = 3.067$ and $\sigma = \bar{S}/c(4) = .1324$.

Suppose now that the values of the items produced are supposed to fall within the specifications $3 \pm .1$. Assuming that the process remains in control and that the foregoing are accurate estimates of the true mean and standard deviation, what proportion of the items will meet the desired specifications?

SOLUTION To answer the foregoing, we note that when $\mu = 3.067$ and $\sigma = .1324$,

$$P\{2.9 \leq X \leq 3.1\} = P\left\{\frac{2.9 - 3.067}{.1324} \leq \frac{X - 3.067}{.1324} \leq \frac{3.1 - 3.067}{.1324}\right\}$$

$$= \Phi(.2492) - \Phi(-1.2613)$$

$$= .5984 - (1 - .8964)$$

$$= .4948$$

Hence, 49 percent of the items produced will meet the specifications. ■

REMARKS

(a) The estimator $\overline{\overline{X}}$ is equal to the average of all nk measurements and is thus the obvious estimator of μ. However, it may not immediately be clear why the sample standard deviation of all the nk measurements, namely,

$$S \equiv \sqrt{\sum_{i=1}^{nk} \frac{(X_i - \overline{\overline{X}})^2}{nk - 1}}$$

is not used as the initial estimator of σ. The reason it is not is that the process may not have been in control throughout the first k subgroups, and thus this latter estimator could be far away from the true value. Also, it often happens that a process goes out of control by an occurrence that results in a change of its mean value μ while leaving its standard deviation unchanged. In such a case, the subgroup sample deviations would still be estimators of σ, whereas the entire sample standard deviation would not. Indeed, even in the case where the process appears to be in control throughout, the estimator of σ presented is preferred over the sample standard deviation S. The reason for this is that we cannot be certain that the mean has not changed throughout this time. That is, even though all the subgroup averages fall within the control limits, and so we have concluded that the process is in control, there is no assurance that there are no assignable causes of variation present (which might have resulted in a change in the mean that has not yet been picked up by the chart). It merely means that for practical purposes it pays to act as if the process was in control and let it continue to produce items. However, since we realize that some assignable cause of variation might be present, it has been argued that $\overline{S}/c(n)$ is a "safer" estimator than the sample standard deviation. That is, although it is not quite as good when the process has really been in control throughout, it could be a lot better if there had been some small shifts in the mean.

(b) In the past, an estimator of σ based on subgroup ranges — defined as the difference between the largest and smallest value in the subgroup — has been employed. This was done to keep the necessary computations simple (it is clearly much easier to compute the range than it is to compute the subgroup's sample standard deviation). However, with modern-day computational power this should no longer be a consideration, and since the standard deviation estimator both has smaller variance than the range estimator and is more robust (in the sense that it would still yield a reasonable estimate of the population standard deviation even when the underlying distribution is not normal), we will not consider the latter estimator in this text.

13.3 S-CONTROL CHARTS

The $\overline{X}$ control charts presented in the previous section are designed to pick up changes in the population mean. In cases where one is also concerned about possible changes in the population variance, we can utilize an S-control chart.

As before, suppose that, when in control, the items produced have a measurable characteristic that is normally distributed with mean μ and variance σ^2. If S_i is the sample standard deviation for the ith subgroup, that is,

$$S_i = \sqrt{\sum_{j=1}^{n} \frac{(X_{(i-1)n+j} - \overline{X}_i)^2}{(n-1)}}$$

then, as was shown in Section 13.2.1,

$$E[S_i] = c(n)\sigma \tag{13.3.1}$$

In addition,

$$\begin{aligned} \text{Var}(S_i) &= E[S_i^2] - (E[S_i])^2 \\ &= \sigma^2 - c^2(n)\sigma^2 \\ &= \sigma^2[1 - c^2(n)] \end{aligned} \tag{13.3.2}$$

where the next to last equality follows from Equation 13.2.2 and the fact that the expected value of a chi-square random variable is equal to its degrees of freedom parameter.

On using the fact that, when in control, S_i has the distribution of a constant (equal to $\sigma/\sqrt{n-1}$) times the square root of a chi-square random variable with $n-1$ degrees of freedom, it can be shown that S_i will, with probability near to 1, be within 3 standard deviations of its mean. That is,

$$P\{E[S_i] - 3\sqrt{\text{Var}(S_i)} < S_i < E[S_i] + 3\sqrt{\text{Var}(S_i)}\} \approx .99$$

Thus, using the formulas 13.3.1 and 13.3.2 for $E[S_i]$ and $\text{Var}(S_i)$, it is natural to set the upper and lower control limits for the S chart by

$$\text{UCL} = \sigma[c(n) + 3\sqrt{1 - c^2(n)}] \qquad (13.3.3)$$

$$\text{LCL} = \sigma[c(n) - 3\sqrt{1 - c^2(n)}]$$

The successive values of S_i should be plotted to make certain they fall within the upper and lower control limits. When a value falls outside, the process should be stopped and declared to be out of control.

When one is just starting up a control chart and σ is unknown, it can be estimated from $\overline{S}/c(n)$. Using the foregoing, the estimated control limits would then be

$$\text{UCL} = \overline{S}[1 + 3\sqrt{1/c^2(n) - 1}] \qquad (13.3.4)$$

$$\text{LCL} = \overline{S}[1 - 3\sqrt{1/c^2(n) - 1}]$$

As in the case of starting up an $\overline{X}$ control chart, it should then be checked that the k subgroup standard deviations $S_1, S_2, \ldots, S_k$ all fall within these control limits. If any of them falls outside, then those subgroups should be discarded and $\overline{S}$ recomputed.

EXAMPLE 13.3a The following are the $\overline{X}$ and S values for 20 subgroups of size 5 for a recently started process.

Subgroup	$\overline{X}$	S	Subgroup	$\overline{X}$	S	Subgroup	$\overline{X}$	S	Subgroup	$\overline{X}$	S
1	35.1	4.2	6	36.4	4.5	11	38.1	4.2	16	41.3	8.2
2	33.2	4.4	7	35.9	3.4	12	37.6	3.9	17	35.7	8.1
3	31.7	2.5	8	38.4	5.1	13	38.8	3.2	18	36.3	4.2
4	35.4	3.2	9	35.7	3.8	14	34.3	4.0	19	35.4	4.1
5	34.5	2.6	10	27.2	6.2	15	43.2	3.5	20	34.6	3.7

Since $\overline{\overline{X}} = 35.94$, $\overline{S} = 4.35$, $c(5) = .9400$, we see from Equations 13.2.4 and 13.3.4 that the preliminary upper and lower control limits for $\overline{X}$ and S are

$$\text{UCL}(\overline{X}) = 42.149$$
$$\text{LCL}(\overline{X}) = 29.731$$
$$\text{UCL}(S) = 9.087$$
$$\text{LCL}(S) = -.386$$

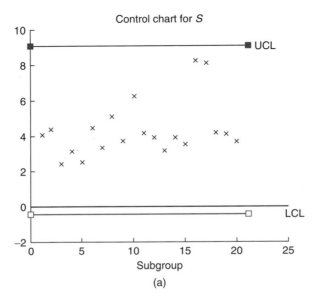

(a)

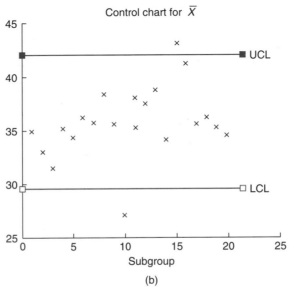

(b)

FIGURE 13.2

The control charts for $\overline{X}$ and S with the preceding control limits are shown in Figures 13.2a and 13.2b. Since $\overline{X}_{10}$ and $\overline{X}_{15}$ fall outside the $\overline{X}$ control limits, these subgroups must be eliminated and the control limits recomputed. We leave the necessary computations as an exercise. ∎

13.4 CONTROL CHARTS FOR THE FRACTION DEFECTIVE

The $\overline{X}$ and S-control charts can be used when the data are measurements whose values can vary continuously over a region. There are also situations in which the items produced have quality characteristics that are classified as either being defective or nondefective. Control charts can also be constructed in this latter situation.

Let us suppose that when the process is in control each item produced will independently be defective with probability p. If we let X denote the number of defective items in a subgroup of n items, then assuming control, X will be a binomial random variable with parameters (n, p). If $F = X/n$ is the fraction of the subgroup that is defective, then assuming the process is in control, its mean and standard deviation are given by

$$E[F] = \frac{E[X]}{n} = \frac{np}{n} = p$$

$$\sqrt{\text{Var}(F)} = \sqrt{\frac{\text{Var}(X)}{n^2}} = \sqrt{\frac{np(1-p)}{n^2}} = \sqrt{\frac{p(1-p)}{n}}$$

Hence, when the process is in control the fraction defective in a subgroup of size n should, with high probability, be between the limits

$$\text{LCL} = p - 3\sqrt{\frac{p(1-p)}{n}}, \quad \text{UCL} = p + 3\sqrt{\frac{p(1-p)}{n}}$$

The subgroup size n is usually much larger than the typical values of between 4 and 10 used in $\overline{X}$ and S charts. The main reason for this is that if p is small and n is not of reasonable size, then most of the subgroups will have zero defects even when the process goes out of control. Thus, it would take longer than it would if n were chosen so that np were not close to zero to detect a shift in quality.

To start such a control chart it is, of course, necessary first to estimate p. To do so, choose k of the subgroups, where again one should try to take $k \geq 20$, and let F_i denote the fraction of the ith subgroup that are defective. The estimate of p is given by $\overline{F}$ defined by

$$\overline{F} = \frac{F_1 + \cdots + F_k}{k}$$

Since nF_i is equal to the number of defectives in subgroup i, we see that $\overline{F}_k$ can also be expressed as

$$\overline{F} = \frac{nF_1 + \cdots + nF_k}{nk}$$

$$= \frac{\text{total number of defectives in all the subgroups}}{\text{number of items in the subgroups}}$$

In other words, the estimate of p is just the proportion of items inspected that are defective. The upper and lower control limits are now given by

$$\text{LCL} = \overline{F} - 3\sqrt{\frac{\overline{F}(1-\overline{F})}{n}}, \qquad \text{UCL} = \overline{F} + 3\sqrt{\frac{\overline{F}(1-\overline{F})}{n}}$$

We should now check whether the subgroup fractions $F_1, F_2, \ldots, F_k$ fall within these control limits. If some of them fall outside, then the corresponding subgroups should be eliminated and $\overline{F}$ recomputed.

EXAMPLE 13.4a Successive samples of 50 screws are drawn from the hourly production of an automatic screw machine, with each screw being rated as either acceptable or defective. This is done for 20 such samples with the following data resulting.

Subgroup	Defectives	F	Subgroup	Defectives	F
1	6	.12	11	1	.02
2	5	.10	12	3	.06
3	3	.06	13	2	.04
4	0	.00	14	0	.00
5	1	.02	15	1	.02
6	2	.04	16	1	.02
7	1	.02	17	0	.00
8	0	.00	18	2	.04
9	2	.04	19	1	.02
10	1	.02	20	2	.04

We can compute the trial control limits as follows:

$$\overline{F} = \frac{\text{total number defectives}}{\text{total number items}} = \frac{34}{1{,}000} = .034$$

and so

$$UCL = .034 + 3\sqrt{\frac{(.034)(.968)}{50}} = .1109$$

$$LCL = .034 - 3\sqrt{\frac{(.034)(.966)}{50}} = -.0429$$

Since the proportion of defectives in the first subgroup falls outside the upper control limit, we eliminate that subgroup and recompute $\overline{F}$ as

$$\overline{F} = \frac{34 - 6}{950} = .0295$$

The new upper and lower control limits are $.0295 \pm \sqrt{(.0295)(1 - .0295)/50}$, or

$$LCL = -.0423, \quad UCL = .1013$$

Since the remaining subgroups all have fraction defectives that fall within the control limits, we can accept that, when in control, the fraction of defective items in a subgroup should be below .1013. ∎

REMARK

Note that we are attempting to detect any change in quality even when this change results in improved quality. That is, we regard the process as being "out of control" even when the probability of a defective item decreases. The reason for this is that it is important to notice any change in quality, for either better or worse, to be able to evaluate the reason for the change. In other words, if an improvement in product quality occurs, then it is important to analyze the production process to determine the reason for the improvement. (That is, what are we doing right?)

13.5 CONTROL CHARTS FOR NUMBER OF DEFECTS

In this section, we consider situations in which the data are the numbers of defects in units that consist of an item or group of items. For instance, it could be the number of defective rivets in an airplane wing, or the number of defective computer chips that are produced daily by a given company. Because it is often the case that there are a large number of possible things that can be defective, with each of these having a small probability of actually being defective, it is probably reasonable to assume that the resulting number of defects has a Poisson distribution.[*] So let us suppose that, when the process is in control, the number of defects per unit has a Poisson distribution with mean λ.

---

[*] See Section 5.2 for a theoretical explanation.

If we let X_i denote the number of defects in the ith unit, then, since the variance of a Poisson random variable is equal to its mean, when the process is in control

$$E[X_i] = \lambda, \qquad \text{Var}(X_i) = \lambda$$

Hence, when in control each X_i should with high probability be within $\lambda \pm 3\sqrt{\lambda}$, and so the upper and lower control limits are given by

$$\text{UCL} = \lambda + 3\sqrt{\lambda}, \qquad \text{LCL} = \lambda - 3\sqrt{\lambda}$$

As before, when the control chart is started and λ is unknown, a sample of k units should be used to estimate λ by

$$\overline{X} = (X_1 + \cdots + X_k)/k$$

This results in trial control limits

$$\overline{X} + 3\sqrt{\overline{X}} \quad \text{and} \quad \overline{X} - 3\sqrt{\overline{X}}$$

If all the $X_i, i = 1, \ldots, k$ fall within these limits, then we suppose that the process is in control with $\lambda = \overline{X}$. If some fall outside, then these points are eliminated and we recompute $\overline{X}$, and so on.

In situations where the mean number of defects per item (or per day) is small, one should combine items (days) and use as data the number of defects in a given number — say, n — of items (or days). Since the sum of independent Poisson random variables remains a Poisson random variable, the data values will be Poisson distributed with a larger mean value λ. Such combining of items is useful when the mean number of defects per item is less than 25.

To obtain a feel for the advantage in combining items, suppose that the mean number of defects per item is 4 when the process is under control, and suppose that something occurs that results in this value changing from 4 to 6, that is, an increase of 1 standard deviation occurs. Let us see how many items will be produced, on average, until the process is declared out of control when the successive data consist of the number of defects in n items.

Since the number of defects in a sample of n items is, when under control, Poisson distributed with mean and variance equal to $4n$, the control limits are $4n \pm 3\sqrt{4n}$ or $4n \pm 6\sqrt{n}$. Now if the mean number of defects per item changes to 6, then a data value will be Poisson with mean $6n$ and so the probability that it will fall outside the control limits — call it $p(n)$ — is given by

$$p(n) = P\{Y > 4n + 6\sqrt{n}\} + P\{Y < 4n - 6\sqrt{n}\}$$

when Y is Poisson with mean $6n$. Now

$$p(n) \approx P\{Y > 4n + 6\sqrt{n}\}$$

$$= P\left\{\frac{Y - 6n}{\sqrt{6n}} > \frac{6\sqrt{n} - 2n}{\sqrt{6n}}\right\}$$

$$\approx P\left\{Z > \frac{6\sqrt{n} - 2n}{\sqrt{6n}}\right\} \quad \text{where } Z \sim N(0, 1)$$

$$= 1 - \Phi\left(\sqrt{6} - 2\sqrt{\frac{n}{6}}\right)$$

Because each data value will be outside the control limits with probability $p(n)$, it follows that the number of data values needed to obtain one outside the limits is a geometric random variable with parameter $p(n)$, and thus has mean $1/p(n)$. Finally, since there are n items for each data value, it follows that the number of items produced before the process is seen to be out of control has mean value $n/p(n)$:

Average number of items produced while out of control $= n/(1 - \Phi(\sqrt{6} - \sqrt{\frac{2n}{3}}))$

We plot this for various n in Table 13.2. Since larger values of n are better when the process is in control (because the average number of items produced before the process is incorrectly said to be out of control is approximately $n/.0027$), it is clear from Table 13.2 that one should combine at least 9 of the items. This would mean that each data value (equal to the number of defects in the combined set) would have mean at least $9 \times 4 = 36$.

TABLE 13.2

n	Average Number of Items
1	19.6
2	20.66
3	19.80
4	19.32
5	18.80
6	18.18
7	18.13
8	18.02
9	18
10	18.18
11	18.33
12	18.51

EXAMPLE 13.5a The following data represent the number of defects discovered at a factory on successive units of 10 cars each.

Cars	Defects	Cars	Defects	Cars	Defects	Cars	Defects
1	141	6	74	11	63	16	68
2	162	7	85	12	74	17	95
3	150	8	95	13	103	18	81
4	111	9	76	14	81	19	102
5	92	10	68	15	94	20	73

Does it appear that the production process was in control throughout?

SOLUTION Since $\overline{X} = 94.4$, it follows that the trial control limits are

$$\text{LCL} = 94.4 - 3\sqrt{94.4} = 65.25$$

$$\text{UCL} = 94.4 + 3\sqrt{94.4} = 123.55$$

Since the first three data values are larger than UCL, they are removed and the sample mean recomputed. This yields

$$\overline{X} = \frac{(94.4)20 - (141 + 162 + 150)}{17} = 84.41$$

and so the new trial control limits are

$$\text{LCL} = 84.41 - 3\sqrt{84.41} = 56.85$$

$$\text{UCL} = 84.41 + 3\sqrt{84.41} = 111.97$$

At this point since all remaining 17 data values fall within the limits, we could declare that the process is now in control with a mean value of 84.41. However, because it seems that the mean number of defects was initially high before settling into control, it seems quite plausible that the data value X_4 also originated before the process was in control. Thus, it would seem prudent in this situation to also eliminate X_4 and recompute. Based on the remaining 16 data values, we obtain that

$$\overline{X} = 82.56$$

$$\text{LCL} = 82.56 - 3\sqrt{82.56} = 55.30$$

$$\text{UCL} = 82.56 + 3\sqrt{82.56} = 109.82$$

and so it appears that the process is now in control with a mean value of 82.56. ■

13.6 OTHER CONTROL CHARTS FOR DETECTING CHANGES IN THE POPULATION MEAN

The major weakness of the $\overline{X}$ control chart presented in Section 13.2 is that it is relatively insensitive to small changes in the population mean. That is, when such a change occurs, since each plotted value is based on only a single subgroup and so tends to have a relatively large variance, it takes, on average, a large number of plotted values to detect the change. One way to remedy this weakness is to allow each plotted value to depend not only on the most recent subgroup average but on some of the other subgroup averages as well. Three approaches for doing this that have been found to be quite effective are based on (1) moving averages, (2) exponentially weighted moving averages, and (3) cumulative sum control charts.

13.6.1 Moving-Average Control Charts

The moving-average control chart of span size k is obtained by continually plotting the average of the k most recent subgroups. That is, the moving average at time t, call it M_t, is defined by

$$M_t = \frac{\overline{X}_t + \overline{X}_{t-1} + \cdots + \overline{X}_{t-k+1}}{k}$$

where $\overline{X}_i$ is the average of the values of subgroup i. The successive computations can be easily performed by noting that

$$kM_t = \overline{X}_t + \overline{X}_{t-1} + \cdots + \overline{X}_{t-k+1}$$

and, substituting $t + 1$ for t,

$$kM_{t+1} = \overline{X}_{t+1} + \overline{X}_t + \cdots + \overline{X}_{t-k+2}$$

Subtraction now yields that

$$kM_{t+1} - kM_t = \overline{X}_{t+1} - \overline{X}_{t-k+1}$$

or

$$M_{t+1} = M_t + \frac{\overline{X}_{t+1} - \overline{X}_{t-k+1}}{k}$$

In words, the moving average at time $t + 1$ is equal to the moving average at time t plus $1/k$ times the difference between the newly added and the deleted value in the moving average. For values of t less than k, M_t is defined as the average of the first t subgroups. That is,

$$M_t = \frac{\overline{X}_1 + \cdots + \overline{X}_t}{t} \qquad \text{if } t < k$$

Suppose now that when the process is in control the successive values come from a normal population with mean μ and variance σ^2. Therefore, if n is the subgroup size, it follows that $\overline{X}_i$ is normal with mean μ and variance σ^2/n. From this we see that the average of m of the $\overline{X}_i$ will be normal with mean μ and variance given by $\text{Var}(\overline{X}_i)/m = \sigma^2/nm$ and, therefore, when the process is in control

$$E[M_t] = \mu$$

$$\text{Var}(M_t) = \begin{cases} \sigma^2/nt & \text{if } t < k \\ \sigma^2/nk & \text{otherwise} \end{cases}$$

Because a normal random variable is almost always within 3 standard deviations of its mean, we have the following upper and lower control limits for M_t:

$$\text{UCL} = \begin{cases} \mu + 3\sigma/\sqrt{nt} & \text{if } t < k \\ \mu + 3\sigma/\sqrt{nk} & \text{otherwise} \end{cases}$$

$$\text{LCL} = \begin{cases} \mu - 3\sigma/\sqrt{nt} & \text{if } t < k \\ \mu - 3\sigma/\sqrt{nk} & \text{otherwise} \end{cases}$$

In other words, aside from the first $k - 1$ moving averages, the process will be declared out of control whenever a moving average differs from μ by more than $3\sigma/\sqrt{nk}$.

EXAMPLE 13.6a When a certain manufacturing process is in control, it produces items whose values are normally distributed with mean 10 and standard deviation 2. The following simulated data represent the values of 25 subgroup averages of size 5 from a normal population with mean 11 and standard deviation 2. That is, these data represent the subgroup averages after the process has gone out of control with its mean value increasing from 10 to 11. Table 13.3 presents these 25 values along with the moving averages based on span size $k = 8$ as well as the upper and lower control limits. The lower and upper control limits for $t > 8$ are 9.051318 and 10.94868.

As the reader can see, the first moving average to fall outside its control limits occurred at time 11, with other such occurrences at times 12, 13, 14, 16, and 25. (It is interesting to note that the usual control chart — that is, the moving average with $k = 1$ — would have declared the process out of control at time 7 since $\overline{X}_7$ was so large. However, this is the only point where this chart would have indicated a lack of control (see Figure 13.3).

TABLE 13.3

t	$\overline{X}_t$	M_t	LCL	UCL
1	9.617728	9.617728	7.316719	12.68328
2	10.25437	9.936049	8.102634	11.89737
3	9.876195	9.913098	8.450807	11.54919
4	10.79338	10.13317	8.658359	11.34164
5	10.60699	10.22793	8.8	11.2
6	10.48396	10.2706	8.904554	11.09545
7	13.33961	10.70903	8.95815	11.01419
8	9.462969	10.55328	9.051318	10.94868
			$\vdots$	$\vdots$
9	10.14556	10.61926		
10	11.66342	10.79539		
*11	11.55484	11.00634		
*12	11.26203	11.06492		
*13	12.31473	11.27839		
*14	9.220009	11.1204		
15	11.25206	10.85945		
16	10.48662	10.98741		
17	9.025091	10.84735		
18	9.693386	10.6011		
19	11.45989	10.58923		
20	12.44213	10.73674		
21	11.18981	10.59613		
22	11.56674	10.88947		
23	9.869849	10.71669		
24	12.11311	10.92		
*25	11.48656	11.22768		

* = Out of control.

There is an inverse relationship between the size of the change in the mean value that one wants to guard against and the appropriate moving-average span size k. That is, the smaller this change is, the larger k ought to be. ■

13.6.2 EXPONENTIALLY WEIGHTED MOVING-AVERAGE CONTROL CHARTS

The moving-average control chart of Section 13.6.1 considered at each time t a weighted average of all subgroup averages up to that time, with the k most recent values being given weight $1/k$ and the others given weight 0. Since this appears to be a most effective procedure for detecting small changes in the population mean, it raises the possibility that other sets of weights might also be successfully employed. One set of weights that is often utilized is obtained by decreasing the weight of each earlier subgroup average by a constant factor.

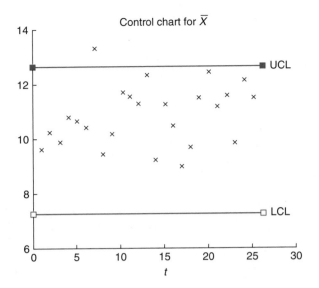

Control chart for $\overline{X}$

FIGURE 13.3

Let

$$W_t = \alpha \overline{X}_t + (1 - \alpha) W_{t-1} \qquad (13.6.1)$$

where α is a constant between 0 and 1, and where

$$W_0 = \mu$$

The sequence of values $W_t, t = 0, 1, 2, \ldots$ is called an *exponentially weighted moving average*. To understand why it has been given that name, note that if we continually substitute for the W term on the right side of Equation 13.6.1, we obtain that

$$W_t = \alpha \overline{X}_t + (1 - \alpha)[\alpha \overline{X}_{t-1} + (1 - \alpha) W_{t-2}] \qquad (13.6.2)$$

$$= \alpha \overline{X}_t + \alpha(1 - \alpha)\overline{X}_{t-1} + (1 - \alpha)^2 W_{t-2}$$

$$= \alpha \overline{X}_t + \alpha(1 - \alpha)\overline{X}_{t-1} + (1 - \alpha)^2[\alpha \overline{X}_{t-2} + (1 - \alpha) W_{t-3}]$$

$$= \alpha \overline{X}_t + \alpha(1 - \alpha)\overline{X}_{t-1} + \alpha(1 - \alpha)^2 \overline{X}_{t-2} + (1 - \alpha)^3 W_{t-3}$$

$$\vdots$$

$$= \alpha \overline{X}_t + \alpha(1 - \alpha)\overline{X}_{t-1} + \alpha(1 - \alpha)^2 \overline{X}_{t-2} + \cdots$$

$$+ \alpha(1 - \alpha)^{t-1}\overline{X}_1 + (1 - \alpha)^t \mu$$

where the foregoing used the fact that $W_0 = \mu$. Thus we see from Equation 13.6.2 that W_t is a weighted average of all the subgroup averages up to time t, giving weight α to the most recent subgroup and then successively decreasing the weight of earlier subgroup averages by the constant factor $1 - \alpha$, and then giving weight $(1 - \alpha)^t$ to the in-control population mean.

The smaller the value of α, the more even the successive weights. For instance, if $\alpha = .1$ then the initial weight is .1 and the successive weights decrease by the factor .9; that is, the weights are .1, .09, .081, .073, .066, .059, and so on. On the other hand, if one chooses, say, $\alpha = .4$, then the successive weights are .4, .24, .144, .087, .052,.... Since the successive weights $\alpha(1 - \alpha)^{i-1}$, $i = 1, 2, \ldots$, can be written as

$$\alpha(1 - \alpha)^{i-1} = \overline{\alpha}e^{-\beta i}$$

where

$$\overline{\alpha} = \frac{\alpha}{1 - \alpha}, \qquad \beta = -\log(1 - \alpha)$$

we say that the successively older data values are "exponentially weighted" (see Figure 13.4).

To compute the mean and variance of the W_t, recall that, when in control, the subgroup averages $\overline{X}_i$ are independent normal random variables each having mean μ and variance σ^2/n. Therefore, using Equation 13.6.2, we see that

$$E[W_t] = \mu[\alpha + \alpha(1 - \alpha) + \alpha(1 - \alpha)^2 + \cdots + \alpha(1 - \alpha)^{t-1} + (1 - \alpha)^t]$$
$$= \frac{\mu\alpha[1 - (1 - \alpha)^t]}{1 - (1 - \alpha)} + \mu(1 - \alpha)^t$$
$$= \mu$$

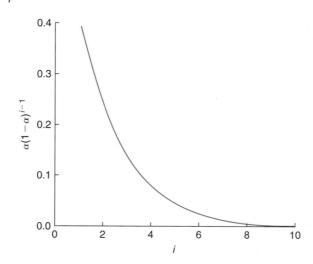

FIGURE 13.4   *Plot of* $\alpha(1 - \alpha)^{i-1}$ *when* $\alpha = .4$.

To determine the variance, we again use Equation 13.6.2:

$$\text{Var}(W_t) = \frac{\sigma^2}{n}\left\{\alpha^2 + [\alpha(1-\alpha)]^2 + [\alpha(1-\alpha)^2]^2 + \cdots + [\alpha(1-\alpha)^{t-1}]^2\right\}$$

$$= \frac{\sigma^2}{n}\alpha^2[1 + \beta + \beta^2 + \cdots + \beta^{t-1}] \qquad \text{where } \beta = (1-\alpha)^2$$

$$= \frac{\sigma^2\alpha^2[1 - (1-\alpha)^{2t}]}{n[1 - (1-\alpha)^2]}$$

$$= \frac{\sigma^2\alpha[1 - (1-\alpha)^{2t}]}{n(2-\alpha)}$$

Hence, when t is large we see that, provided that the process has remained in control throughout,

$$E[W_t] = \mu$$

$$\text{Var}(W_t) \approx \frac{\sigma^2\alpha}{n(2-\alpha)} \qquad \text{since } (1-\alpha)^{2t} \approx 0$$

Thus, the upper and lower control limits for W_t are given by

$$\text{UCL} = \mu + 3\sigma\sqrt{\frac{\alpha}{n(2-\alpha)}}$$

$$\text{LCL} = \mu - 3\sigma\sqrt{\frac{\alpha}{n(2-\alpha)}}$$

Note that the preceding control limits are the same as those in a moving-average control chart with span k (after the initial k values) when

$$\frac{3\sigma}{\sqrt{nk}} = 3\sigma\sqrt{\frac{\alpha}{n(2-\alpha)}}$$

or, equivalently, when

$$k = \frac{2-\alpha}{\alpha} \qquad \text{or} \qquad \alpha = \frac{2}{k+1}$$

EXAMPLE 13.6b A repair shop will send a worker to a caller's home to repair electronic equipment. Upon receiving a request, it dispatches a worker who is instructed to call in when the job is completed. Historical data indicate that the time from when the server is dispatched until he or she calls is a normal random variable with mean 62 minutes and standard deviation 24 minutes. To keep aware of any changes in this distribution, the repair shop plots a standard exponentially weighted moving-average (EWMA) control chart with each data value being the average of 4 successive times, and with a weighting factor of $\alpha = .25$. If the present value of the chart is 60 and the following are the next 16 subgroup averages, what can we conclude?

$$48, 52, 70, 62, 57, 81, 56, 59, 77, 82, 78, 80, 74, 82, 68, 84$$

SOLUTION Starting with $W_0 = 60$, the successive values of $W_1, \ldots, W_{16}$ can be obtained from the formula

$$W_t = .25\overline{X}_t + .75W_{t-1}$$

This gives

$$W_1 = (.25)(48) + (.75)(60) = 57$$
$$W_2 = (.25)(52) + (.75)(57) = 55.75$$
$$W_3 = (.25)(70) + (.75)(55.75) = 59.31$$
$$W_4 = (.25)(62) + (.75)(59.31) = 59.98$$
$$W_5 = (.25)(57) + (.75)(59.98) = 59.24$$
$$W_6 = (.25)(81) + (.75)(59.24) = 64.68$$

and so on, with the following being the values of W_7 through W_{16}:

$$62.50, 61.61, 65.48, 69.60, 71.70, 73.78, 73.83, 75.87, 73.90, 76.43$$

Since

$$3\sqrt{\frac{.25}{1.75}\frac{24}{\sqrt{4}}} = 13.61$$

the control limits of the standard EWMA control chart with weighting factor $\alpha = .25$ are

$$\text{LCL} = 62 - 13.61 = 48.39$$
$$\text{UCL} = 62 + 13.61 = 75.61$$

Thus, the EWMA control chart would have declared the system out of control after determining W_{14} (and also after W_{16}). On the other hand, since a subgroup standard deviation is $\sigma/\sqrt{n} = 12$, it is interesting that no data value differed from $\mu = 62$ by even as much as 2 subgroup standard deviations, and so the standard $\overline{X}$ control chart would not have declared the system out of control. ∎

EXAMPLE 13.6c　Consider the data of Example 13.6a but now use an exponentially weighted moving-average control chart with $\alpha = 2/9$. This gives rise to the following data set.

t	$\bar{X}_t$	W_t	t	$\bar{X}_t$	W_t
1	9.617728	9.915051	14	9.220009	10.84522
2	10.25437	9.990456	15	11.25206	10.93563
3	9.867195	9.963064	16	10.48662	10.83585
4	10.79338	10.14758	17	9.025091	10.43346
5	10.60699	10.24967	18	9.693386	10.269
6	10.48396	10.30174	19	11.45989	10.53364
*7	13.33961	10.97682	*20	12.44213	10.95775
8	9.462969	10.64041	*21	11.18981	11.00932
9	10.14556	10.53044	*22	11.56674	11.13319
10	11.66342	10.78221	23	9.869849	10.85245
*11	11.55484	10.95391	*24	12.11311	11.13259
*12	11.26203	11.02238	*25	11.48656	11.21125
*13	12.31473	11.30957			

* = *Out of control.*

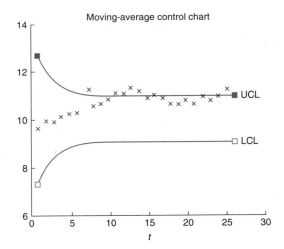

FIGURE 13.5

Since

$$UCL = 10.94868$$
$$LCL = 9.051318$$

we see that the process could be declared out of control as early as $t = 7$ (see Figure 13.5). ■

13.6.3 CUMULATIVE SUM CONTROL CHARTS

The major competitor to the moving-average type of control chart for detecting a small-to moderate-sized change in the mean is the cumulative sum (often reduced to cu-sum) control chart.

Suppose, as before, that $\overline{X}_1, \overline{X}_2, \ldots$ represent successive averages of subgroups of size n and that when the process is in control these random variables have mean μ and standard deviation $\sigma/\sqrt{n}$. Initially, suppose that we are only interested in determining when an increase in the mean value occurs. The (one-sided) cumulative sum control chart for detecting an increase in the mean operates as follows: Choose positive constants d and B, and let

$$Y_j = \overline{X}_j - \mu - d\sigma/\sqrt{n}, \quad j \geq 1$$

Note that when the process is in control, and so $E[\overline{X}_j] = \mu$,

$$E[Y_j] = -d\sigma/\sqrt{n} < 0$$

Now, let

$$S_0 = 0$$
$$S_{j+1} = \max\{S_j + Y_{j+1}, 0\}, \quad j \geq 0$$

The cumulative sum control chart having parameters d and B continually plots S_j, and declares that the mean value has increased at the first j such that

$$S_j > B\sigma/\sqrt{n}$$

To understand the rationale behind this control chart, suppose that we had decided to continually plot the sum of all the random variables Y_i that have been observed so far. That is, suppose we had decided to plot the successive values of P_j, where

$$P_j = \sum_{i=1}^{j} Y_i$$

which can also be written as

$$P_0 = 0$$
$$P_{j+1} = P_j + Y_{j+1}, \quad j \geq 0$$

Now, when the system has always been in control, all of the Y_i have a negative expected value, and thus we would expect their sum to be negative. Hence, if the value of P_j ever became large — say, greater than $B\sigma/\sqrt{n}$ — then this would be strong evidence that the process has gone out of control (by having an increase in the mean value of a produced item). The difficulty, however, is that if the system goes out of control only after some large time, then the value of P_j at that time will most likely be strongly negative (since up to then we would have been summing random variables having a negative mean), and thus it would take a long time for its value to exceed $B\sigma/\sqrt{n}$. Therefore, to keep the sum from becoming very negative while the process is in control, the cumulative sum control chart employs the simple trick of resetting its value to 0 whenever it becomes negative. That is, the quantity S_j is the cumulative sum of all of the Y_i up to time j, with the exception that any time this sum becomes negative its value is reset to 0.

EXAMPLE 13.6d   Suppose that the mean and standard deviation of a subgroup average are $\mu = 30$ and $\sigma/\sqrt{n} = 8$, respectively, and consider the cumulative sum control chart with $d = .5, B = 5$. If the first eight subgroup averages are

$$29, 33, 35, 42, 36, 44, 43, 45$$

then the successive values of $Y_j = \overline{X}_j - 30 - 4 = \overline{X}_j - 34$ are

$$Y_1 = -5, Y_2 = -1, Y_3 = 1, Y_4 = 8, Y_5 = 2, Y_6 = 10, Y_7 = 9, Y_8 = 11$$

Therefore,

$$S_1 = \max\{-5, 0\} = 0$$
$$S_2 = \max\{-1, 0\} = 0$$
$$S_3 = \max\{1, 0\} = 1$$
$$S_4 = \max\{9, 0\} = 9$$
$$S_5 = \max\{11, 0\} = 11$$
$$S_6 = \max\{21, 0\} = 21$$
$$S_7 = \max\{30, 0\} = 30$$
$$S_8 = \max\{41, 0\} = 41$$

Since the control limit is

$$B\sigma/\sqrt{n} = 5(8) = 40$$

the cumulative sum chart would declare that the mean has increased after observing the eighth subgroup average.   ∎

To detect either a positive or a negative change in the mean, we employ two one-sided cumulative sum charts simultaneously. We begin by noting that a decrease in $E[X_i]$ is equivalent to an increase in $E[-X_i]$. Hence, we can detect a decrease in the mean value of an item by running a one-sided cumulative sum chart on the negatives of the subgroup averages. That is, for specified values d and B, not only do we plot the quantities S_j as before, but, in addition, we let

$$W_j = -\overline{X}_j - (-\mu) - d\sigma/\sqrt{n} = \mu - \overline{X}_j - d\sigma/\sqrt{n}$$

and then also plot the values T_j, where

$$T_0 = 0$$
$$T_{j+1} = \max\{T_j + W_{j+1}, 0\}, \qquad j \geq 0$$

The first time that either S_j or T_j exceeds $B\sigma/\sqrt{n}$, the process is said to be out of control.

Summing up, the following steps result in a cumulative sum control chart for detecting a change in the mean value of a produced item: Choose positive constants d and B; use the successive subgroup averages to determine the values of S_j and T_j; declare the process out of control the first time that either exceeds $B\sigma/\sqrt{n}$. Three common choices of the pair of values d and B are $d = .25, B = 8.00$, or $d = .50, B = 4.77$, or $d = 1, B = 2.49$. Any of these choices results in a control rule that has approximately the same false alarm rate as does the $\overline{X}$ control chart that declares the process out of control the first time a subgroup average differs from μ by more than $3\sigma/\sqrt{n}$. As a general rule of thumb, the smaller the change in mean that one wants to guard against, the smaller should be the chosen value of d.

Problems

1. Assume that items produced are supposed to be normally distributed with mean 35 and standard deviation 3. To monitor this process, subgroups of size 5 are sampled. If the following represents the averages of the first 20 subgroups, does it appear that the process was in control?

Subgroup No.	$\overline{X}$	Subgroup No.	$\overline{X}$
1	34.0	6	32.2
2	31.6	7	33.0
3	30.8	8	32.6
4	33.0	9	33.8
5	35.0	10	35.8

(continued)

Subgroup No.	$\overline{X}$	Subgroup No.	$\overline{X}$
11	35.8	16	31.6
12	35.8	17	33.0
13	34.0	18	33.2
14	35.0	19	31.8
15	33.8	20	35.6

2. Suppose that a process is in control with $\mu = 14$ and $\sigma = 2$. An $\overline{X}$ control chart based on subgroups of size 5 is employed. If a shift in the mean of 2.2 units occurs, what is the probability that the next subgroup average will fall outside the control limits? On average, how many subgroups will have to be looked at in order to detect this shift?

3. If Y has a chi-square distribution with $n - 1$ degrees of freedom, show that

$$E[\sqrt{Y}] = \sqrt{2}\frac{\Gamma(n/2)}{\Gamma[(n-1)/2]}$$

(*Hint*: Write

$$E[\sqrt{Y}] = \int_0^\infty \sqrt{y} f_{\chi_{n-1}^2}(y)\, dy$$

$$= \int_0^\infty \sqrt{y}\, \frac{e^{-y/2}y^{(n-1)/2-1}\, dy}{2^{(n-1)/2}\Gamma\left[\dfrac{(n-1)}{2}\right]}$$

$$= \int_0^\infty \frac{e^{-y/2}y^{n/2-1}\, dy}{2^{(n-1)/2}\Gamma\left[\dfrac{(n-1)}{2}\right]}$$

Now make the transformation $x = y/2$.)

4. Samples of size 5 are taken at regular intervals from a production process, and the values of the sample averages and sample standard deviations are calculated. Suppose that the sum of the $\overline{X}$ and S values for the first 25 samples are given by

$$\sum \overline{X}_i = 357.2, \qquad \sum S_i = 4.88$$

(a) Assuming control, determine the control limits for an $\overline{X}$ control chart.
(b) Suppose that the measurable values of the items produced are supposed to be within the limits $14.3 \pm .45$. Assuming that the process remains in control with a mean and variance that is approximately equal to the estimates derived,

approximately what percentage of the items produced will fall within the specification limits?

5. Determine the revised $\overline{X}$ and S-control limits for the data in Example 13.3a.

6. In Problem 4, determine the control limits for an S-control chart.

7. The following are $\overline{X}$ and S values for 20 subgroups of size 5.

Subgroup	$\overline{X}$	S	Subgroup	$\overline{X}$	S	Subgroup	$\overline{X}$	S
1	33.8	5.1	8	36.1	4.1	15	35.6	4.8
2	37.2	5.4	9	38.2	7.3	16	36.4	4.6
3	40.4	6.1	10	32.4	6.6	17	37.2	6.1
4	39.3	5.5	11	29.7	5.1	18	31.3	5.7
5	41.1	5.2	12	31.6	5.3	19	33.6	5.5
6	40.4	4.8	13	38.4	5.8	20	36.7	4.2
7	35.0	5.0	14	40.2	6.4			

(a) Determine trial control limits for an $\overline{X}$ control chart.
(b) Determine trial control limits for an S-control chart.
(c) Does it appear that the process was in control throughout?
(d) If your answer in part (c) is no, suggest values for upper and lower control limits to be used with succeeding subgroups.
(e) If each item is supposed to have a value within 35 ± 10, what is your estimate of the percentage of items that will fall within this specification?

8. Control charts for $\overline{X}$ and S are maintained on the shear strength of spot welds. After 30 subgroups of size 4, $\sum \overline{X}_i = 12{,}660$ and $\sum S_i = 500$. Assume that the process is in control.

(a) What are the $\overline{X}$ control limits?
(b) What are the S-control limits?
(c) Estimate the standard deviation for the process.
(d) If the minimum specification for this weld is 400 pounds, what percentage of the welds will not meet the minimum specification?

9. Control charts for $\overline{X}$ and S are maintained on resistors (in ohms). The subgroup size is 4. The values of $\overline{X}$ and S are computed for each subgroup. After 20 subgroups, $\sum \overline{X}_i = 8{,}620$ and $\sum S_i = 450$.

(a) Compute the values of the limits for the $\overline{X}$ and S charts.
(b) Estimate the value of σ on the assumption that the process is in statistical control.
(c) If the specification limits are 430 ± 30, what conclusions can you draw regarding the ability of the process to produce items within these specifications?

(d) If μ is increased by 60, what is the probability of a subgroup average falling outside the control limits?

10. The following data refer to the amounts by which the diameters of $\frac{1}{4}$ inch ball bearings differ from $\frac{1}{4}$ inch in units of .001 inches. The subgroup size is $n = 5$.

Subgroup	Data Values				
1	2.5	.5	2.0	−1.2	1.4
2	.2	.3	.5	1.1	1.5
3	1.5	1.3	1.2	−1.0	.7
4	.2	.5	−2.0	.0	−1.3
5	−.2	.1	.3	−.6	.5
6	1.1	−.5	.6	.5	.2
7	1.1	−1.0	−1.2	1.3	.1
8	.2	−1.5	−.5	1.5	.3
9	−2.0	−1.5	1.6	1.4	.1
10	−.5	3.2	−.1	−1.0	−1.5
11	.1	1.5	−.2	.3	2.1
12	.0	−2.0	−.5	.6	−.5
13	−1.0	−.5	−.5	−1.0	.2
14	.5	1.3	−1.2	−.5	−2.7
15	1.1	.8	1.5	−1.5	1.2

(a) Set up trial control limits for $\overline{X}$ and S-control charts.
(b) Does the process appear to have been in control throughout the sampling?
(c) If the answer to part (b) is no, construct revised control limits.

11. Samples of $n = 6$ items are taken from a manufacturing process at regular intervals. A normally distributed quality characteristic is measured, and $\overline{X}$ and S values are calculated for each sample. After 50 subgroups have been analyzed, we have

$$\sum_{i=1}^{50} \overline{X}_i = 970 \quad \text{and} \quad \sum_{i=1}^{50} S_i = 85$$

(a) Compute the control limit for the $\overline{X}$ and S-control charts. Assume that all points on both charts plot within the control limits.
(b) If the specification limits are 19 ± 4.0, what are your conclusions regarding the ability of the process to produce items conforming to specifications?

12. The following data present the number of defective bearing and seal assemblies in samples of size 100.

Sample Number	Number of Defectives	Sample Number	Number of Defectives
1	5	11	4
2	2	12	10
3	1	13	0
4	5	14	8
5	9	15	3
6	4	16	6
7	3	17	2
8	3	18	1
9	2	19	6
10	5	20	10

Does it appear that the process was in control throughout? If not, determine revised control limits if possible.

13. The following data represent the results of inspecting all personal computers produced at a given plant during the past 12 days.

Day	Number of Units	Number Defective
1	80	5
2	110	7
3	90	4
4	80	9
5	100	12
6	90	10
7	80	4
8	70	3
9	80	5
10	90	6
11	90	5
12	110	7

Does the process appear to have been in control? Determine control limits for future production.

14. Suppose that when a process is in control each item will be defective with probability .04. Suppose that your control chart calls for taking daily samples of size 500. What is the probability that, if the probability of a defective item should suddenly shift to .08, your control chart would detect this shift on the next sample?

15. The following data represent the number of defective chips produced on the last 15 days: 121, 133, 98, 85, 101, 78, 66, 82, 90, 78, 85, 81, 100, 75, 89. Would

you conclude that the process has been in control throughout these 15 days? What control limits would you advise using for future production?

16. Surface defects have been counted on 25 rectangular steel plates, and the data are shown below. Set up a control chart. Does the process producing the plates appear to be in statistical control?

Plate Number	Number of Defects	Plate Number	Number of Defects
1	2	14	10
2	3	15	2
3	4	16	2
4	3	17	6
5	1	18	5
6	2	19	4
7	5	20	6
8	0	21	3
9	2	22	7
10	5	23	0
11	1	24	2
12	7	25	4
13	8		

17. The following data represent 25 successive subgroup averages and moving averages of span size 5 of these subgroup averages. The data are generated by a process that, when in control, produces normally distributed items having mean 30 and variance 40. The subgroups are of size 4. Would you judge that the process has been in control throughout?

$\bar{X}_t$	M_t	$\bar{X}_t$	M_t
35.62938	35.62938	35.80945	32.34106
39.13018	37.37978	30.9136	33.1748
29.45974	34.73976	30.54829	32.47771
32.5872	34.20162	36.39414	33.17019
30.06041	33.37338	27.62703	32.2585
26.54353	31.55621	34.02624	31.90186
37.75199	31.28057	27.81629	31.2824
26.88128	30.76488	26.99926	30.57259
32.4807	30.74358	32.44703	29.78317
26.7449	30.08048	38.53433	31.96463
34.03377	31.57853	28.53698	30.86678
32.93174	30.61448	28.65725	31.03497
32.18547	31.67531		

18. The data shown below give subgroup averages and moving averages of the values from Problem 17. The span of the moving averages is $k = 8$. When in control the subgroup averages are normally distributed with mean 50 and variance 5. What can you conclude?

$\overline{X}_t$	M_t
50.79806	50.79806
46.21413	48.50609
51.85793	49.62337
50.27771	49.78696
53.81512	50.59259
50.67635	50.60655
51.39083	50.71859
51.65246	50.83533
52.15607	51.00508
54.57523	52.05022
53.08497	52.2036
55.02968	52.79759
54.25338	52.85237
50.48405	52.82834
50.34928	52.69814
50.86896	52.6002
52.03695	52.58531
53.23255	52.41748
48.12588	51.79759
52.23154	51.44783

19. Redo Problem 17 by employing an exponential weighted moving average control chart with $\alpha = \frac{1}{3}$.

20. Analyze the data of Problem 18 with an exponential weighted moving-average control chart having $\alpha = \frac{2}{9}$.

21. Explain why a moving-average control chart with span size k must use different control limits for the first $k - 1$ moving averages, whereas an exponentially weighted moving-average control chart can use the same control limits throughout. [*Hint:* Argue that $\text{Var}(M_t)$ decreases in t, whereas $\text{Var}(W_t)$ increases, and explain why this is relevant.]

22. Repeat Problem 17, this time using a cumulative sum control chart with
    (a) $d = .25, B = 8$;
    (b) $d = .5, B = 4.77$.

23. Repeat Problem 18, this time using a cumulative sum control chart with $d = 1$ and $B = 2.49$.

LIFE TESTING

14.1 INTRODUCTION

In this chapter, we consider a population of items having lifetimes that are assumed to be independent random variables with a common distribution that is specified up to an unknown parameter. The problem of interest will be to use whatever data are available to estimate this parameter.

In Section 14.2, we introduce the concept of the hazard (or failure) rate function — a useful engineering concept that can be utilized to specify lifetime distributions. In Section 14.3, we suppose that the underlying life distribution is exponential and show how to obtain estimates (point, interval, and Bayesian) of its mean under a variety of sampling plans. In Section 14.4, we develop a test of the hypothesis that two exponentially distributed populations have a common mean. In Section 14.5, we consider two approaches to estimating the parameters of a Weibull distribution.

14.2 HAZARD RATE FUNCTIONS

Consider a positive continuous random variable X, that we interpret as being the lifetime of some item, having distribution function F and density f. The *hazard rate* (sometimes called the *failure rate*) function $\lambda(t)$ of F is defined by

$$\lambda(t) = \frac{f(t)}{1 - F(t)}$$

To interpret $\lambda(t)$, suppose that the item has survived for t hours and we desire the probability that it will not survive for an additional time dt. That is, consider

---

* Optional chapter.

$P\{X \in (t, t + dt) \mid X > t\}$. Now

$$P\{X \in (t, t + dt) | X > t\} = \frac{P\{X \in (t, t + dt), X > t\}}{P\{X > t\}}$$

$$= \frac{P\{X \in (t, t + dt)\}}{P\{X > t\}}$$

$$\approx \frac{f(t)}{1 - F(t)} dt$$

That is, $\lambda(t)$ represents the conditional probability intensity that an item of age t will fail in the next moment.

Suppose now that the lifetime distribution is exponential. Then, by the memoryless property of the exponential distribution it follows that the distribution of remaining life for a t-year-old item is the same as for a new item. Hence $\lambda(t)$ should be constant, which is verified as follows:

$$\lambda(t) = \frac{f(t)}{1 - F(t)}$$

$$= \frac{\lambda e^{-\lambda t}}{e^{-\lambda t}}$$

$$= \lambda$$

Thus, the failure rate function for the exponential distribution is constant. The parameter λ is often referred to as the *rate* of the distribution.

We now show that the failure rate function $\lambda(t), t \geq 0$, uniquely determines the distribution F. To show this, note that by definition

$$\lambda(s) = \frac{f(s)}{1 - F(s)}$$

$$= \frac{\frac{d}{ds} F(s)}{1 - F(s)}$$

$$= \frac{d}{ds}\{-\log[1 - F(s)]\}$$

Integrating both sides of this equation from 0 to t yields

$$\int_0^t \lambda(s)\, ds = -\log[1 - F(t)] + \log[1 - F(0)]$$

$$= -\log[1 - F(t)] \quad \text{since } F(0) = 0$$

which implies that

$$1 - F(t) = \exp\left\{-\int_0^t \lambda(s)\,ds\right\} \tag{14.2.1}$$

Hence a distribution function of a positive continuous random variable can be specified by giving its hazard rate function. For instance, if a random variable has a linear hazard rate function — that is, if

$$\lambda(t) = a + bt$$

then its distribution function is given by

$$F(t) = 1 - e^{-at - bt^2/2}$$

and differentiation yields that its density is

$$f(t) = (a + bt)e^{-(at + bt^2/2)}, \qquad t \geq 0$$

When $a = 0$, the foregoing is known as the *Rayleigh density function*.

EXAMPLE 14.2a One often hears that the death rate of a person who smokes is, at each age, twice that of a nonsmoker. What does this mean? Does it mean that a nonsmoker has twice the probability of surviving a given number of years as does a smoker of the same age?

SOLUTION If $\lambda_s(t)$ denotes the hazard rate of a smoker of age t and $\lambda_n(t)$ that of a nonsmoker of age t, then the foregoing is equivalent to the statement that

$$\lambda_s(t) = 2\lambda_n(t)$$

The probability that an A-year-old nonsmoker will survive until age $B, A < B$, is

$$
\begin{aligned}
P\{&A\text{-year-old nonsmoker reaches age } B\} \\
&= P\{\text{nonsmoker's lifetime} > B \mid \text{nonsmoker's lifetime} > A\} \\
&= \frac{1 - F_{\text{non}}(B)}{1 - F_{\text{non}}(A)} \\
&= \frac{\exp\left\{-\int_0^B \lambda_n(t)\,dt\right\}}{\exp\left\{-\int_0^A \lambda_n(t)\,dt\right\}} \qquad \text{from Equation 14.2.1} \\
&= \exp\left\{-\int_A^B \lambda_n(t)\,dt\right\}
\end{aligned}
$$

whereas the corresponding probability for a smoker is, by the same reasoning,

$$P\{A\text{-year-old smoker reaches age }B\} = \exp\left\{-\int_A^B \lambda_s(t)\,dt\right\}$$

$$= \exp\left\{-2\int_A^B \lambda_n(t)\,dt\right\}$$

$$= \left[\exp\left\{-\int_A^B \lambda_n(t)\,dt\right\}\right]^2$$

In other words, of two individuals of the same age, one of whom is a smoker and the other a nonsmoker, the probability that the smoker survives to any given age is the *square* (not one-half) of the corresponding probability for a nonsmoker. For instance, if $\lambda_n(t) = 1/20, 50 \le t \le 60$, then the probability that a 50-year-old nonsmoker reaches age 60 is $e^{-1/2} = .607$, whereas the corresponding probability for a smoker is $e^{-1} = .368$.   ■

REMARK ON TERMINOLOGY

We will say that X has failure rate function $\lambda(t)$ when more precisely we mean that the distribution function of X has failure rate function $\lambda(t)$.

14.3   THE EXPONENTIAL DISTRIBUTION IN LIFE TESTING

14.3.1 SIMULTANEOUS TESTING — STOPPING AT THE rTH FAILURE

Suppose that we are testing items whose life distribution is exponential with unknown mean θ. We put n independent items simultaneously on test and stop the experiment when there have been a total of r, $r \le n$, failures. The problem is to then use the observed data to estimate the mean θ.

The observed data will be the following:

$$Data: \quad x_1 \le x_2 \le \cdots \le x_r, \quad i_1, i_2, \ldots, i_r \qquad (14.3.1)$$

with the interpretation that the jth item to fail was item i_j and it failed at time x_j. Thus, if we let $X_i, i = 1, \ldots, n$ denote the lifetime of component i, then the data will be as given in Equation 14.3.1 if

$$X_{i_1} = x_1, X_{i_2} = x_2, \ldots, X_{i_r} = x_r$$

other $n - r$ of the X_j are all greater than x_r

Now the probability density of X_{i_j} is

$$f_{X_{i_j}}(x_j) = \frac{1}{\theta}e^{-x_j/\theta}, \quad j = 1, \ldots, r$$

and so, by independence, the joint probability density of $X_{i_j}, j = 1, \ldots, r$ is

$$f_{X_{i_1}, \ldots, X_{i_r}}(x_1, \ldots, x_r) = \prod_{j=1}^{r} \frac{1}{\theta} e^{-x_j/\theta}$$

Also, the probability that the other $n - r$ of the Xs are all greater than x_r is, again using independence,

$$P\{X_j > x_r \text{ for } j \neq i_1 \text{ or } i_2 \ldots \text{ or } i_r\} = (e^{-x_r/\theta})^{n-r}$$

Hence, we see that the *likelihood* of the observed data — call it $L(x_1, \ldots, x_r, i_1, \ldots, i_r)$ — is, for $x_1 \leq x_2 \leq \cdots \leq x_r$,

$$
\begin{aligned}
L(x_1, \ldots, x_r, i_1, \ldots, i_r) & \qquad\qquad\qquad\qquad (14.3.2)\\
&= f_{X_{i_1}, X_{i_2}, \ldots, X_{i_r}}(x_1, \ldots, x_r) P\{X_j > x_r, j \neq i_1, \ldots, i_r\}\\
&= \frac{1}{\theta} e^{-x_1/\theta} \cdots \frac{1}{\theta} e^{-x_r/\theta} (e^{-x_r/\theta})^{n-r}\\
&= \frac{1}{\theta^r} \exp\left\{ -\frac{\sum_{i=1}^{r} x_i}{\theta} - \frac{(n-r)x_r}{\theta} \right\}
\end{aligned}
$$

REMARK

The likelihood in Equation 14.3.2 not only specifies that the first r failures occur at times $x_1 \leq x_2 \leq \cdots \leq x_r$ but also that the r items to fail were, in order, $i_1, i_2, \ldots, i_r$. If we only desired the density function of the first r failure times, then since there are $n(n-1) \cdots (n-(r-1)) = n!/(n-r)!$ possible (ordered) choices of the first r items to fail, it follows that the joint density is, for $x_1 \leq x_2 \leq \cdots \leq x_r$,

$$f(x_1, x_2, \ldots, x_r) = \frac{n!}{(n-r)! \, \theta^r} \exp\left\{ -\frac{\sum_{i=1}^{r} x_i}{\theta} - \frac{(n-r)}{\theta} x_r \right\}$$

To obtain the maximum likelihood estimator of θ, we take the logarithm of both sides of Equation 14.3.2. This yields

$$\log L(x_1, \ldots, x_r, i_1, \ldots, i_r) = -r \log \theta - \frac{\sum_{i=1}^{r} x_i}{\theta} - \frac{(n-r)x_r}{\theta}$$

and so

$$\frac{\partial}{\partial \theta} \log L(x_1, \ldots, x_r, i_1, \ldots, i_r) = -\frac{r}{\theta} + \frac{\sum\limits_{i=1}^{r} x_i}{\theta^2} + \frac{(n-r)x_r}{\theta^2}$$

Equating to 0 and solving yields that $\hat{\theta}$, the maximum likelihood estimate, is given by

$$\hat{\theta} = \frac{\sum\limits_{i=1}^{r} x_i + (n-r)x_r}{r}$$

Hence, if we let $X_{(i)}$ denote the time at which the ith failure occurs ($X_{(i)}$ is called the ith *order statistic*), then the maximum likelihood estimator of θ is

$$\hat{\theta} = \frac{\sum\limits_{i=1}^{r} X_{(i)} + (n-r)X_{(r)}}{r} \qquad (14.3.3)$$

$$= \frac{\tau}{r}$$

where τ, defined to equal the numerator in Equation 14.3.3, is called the *total-time-on-test statistic*. We call it this since the ith item to fail functions for a time $X_{(i)}$ (and then fails), $i = 1, \ldots, r$, whereas the other $n - r$ items function throughout the test (which lasts for a time $X_{(r)}$). Hence the sum of the times that all the items are on test is equal to τ.

To obtain a confidence interval for θ, we will determine the distribution of τ, the total time on test. Recalling that $X_{(i)}$ is the time of the ith failure, $i = 1, \ldots, r$, we will start by rewriting the expression for τ. To write an expression for τ, rather than summing the total time on test of each of the items, let us ask how much additional time on test was generated between each successive failure. That is, let us denote by $Y_i, i = 1, \ldots, r$, the additional time on test generated between the $(i-1)$st and ith failure. Now up to the first $X_{(1)}$ time units (as all n items are functioning throughout this interval), the total time on test is

$$Y_1 = nX_{(1)}$$

Between the first and second failures, there are a total of $n - 1$ functioning items, and so

$$Y_2 = (n-1)(X_{(2)} - X_{(1)})$$

In general, we have

$$Y_1 = nX_{(1)}$$
$$Y_2 = (n-1)(X_{(2)} - X_{(1)})$$
$$\vdots$$
$$Y_j = (n-j+1)(X_{(j)} - X_{(j-1)})$$
$$\vdots$$
$$Y_r = (n-r+1)(X_{(r)} - X_{(r-1)})$$

and

$$\tau = \sum_{j=1}^{r} Y_j$$

The importance of the foregoing representation for τ follows from the fact that the distributions of the Y_j's are easily obtained as follows. Since $X_{(1)}$, the time of the first failure, is the minimum of n independent exponential lifetimes, each having rate $1/\theta$, it follows from Proposition 5.6.1 that it is itself exponentially distributed with rate n/θ. That is, $X_{(1)}$ is exponential with mean θ/n, and so $nX_{(1)}$ is exponential with mean θ. Also, at the moment when the first failure occurs, the remaining $n-1$ functioning items are, by the memoryless property of the exponential, as good as new and so each will have an additional life that is exponential with mean θ; hence, the additional time until one of them fails is exponential with rate $(n-1)/\theta$. That is, independent of $X_{(1)}$, $X_{(2)} - X_{(1)}$ is exponential with mean $\theta/(n-1)$ and so $Y_2 = (n-1)(X_{(2)} - X_{(1)})$ is exponential with mean θ. Indeed, continuing this argument leads us to the following conclusion:

$$Y_1, \ldots, Y_r \text{ are independent exponential}$$
random variables each having mean θ \hfill (14.3.4)

Hence, since the sum of independent and identically distributed exponential random variables has a gamma distribution (Corollary 5.7.2), we see that

$$\tau \sim \text{gamma}(r, 1/\theta)$$

That is, τ has a gamma distribution with parameters r and $1/\theta$. Equivalently, by recalling that a gamma random variable with parameters $(r, 1/\theta)$ is equivalent to $\theta/2$ times a chi-square random variable with $2r$ degrees of freedom (see Section 5.8.1), we obtain that

$$\frac{2\tau}{\theta} \sim \chi^2_{2r} \qquad (14.3.5)$$

That is, $2\tau/\theta$ has a chi-square distribution with $2r$ degrees of freedom. Hence,

$$P\{\chi^2_{1-\alpha/2,2r} < 2\tau/\theta < \chi^2_{\alpha/2,2r}\} = 1 - \alpha$$

and so a $100(1-\alpha)$ percent confidence interval for θ is

$$\theta \in \left(\frac{2\tau}{\chi^2_{\alpha/2,2r}}, \frac{2\tau}{\chi^2_{1-\alpha/2,2r}}\right) \tag{14.3.6}$$

One-sided confidence intervals can be similarly obtained.

EXAMPLE 14.3a A sample of 50 transistors is simultaneously put on a test that is to be ended when the 15th failure occurs. If the total time on test of all transistors is equal to 525 hours, determine a 95 percent confidence interval for the mean lifetime of a transistor. Assume that the underlying distribution is exponential.

SOLUTION From Program 5.8.1b,

$$\chi^2_{.025,30} = 46.98, \quad \chi^2_{.975,30} = 16.89$$

and so, using Equation 14.3.6, we can assert with 95 percent confidence that

$$\theta \in (22.35, 62.17) \quad \blacksquare$$

In testing a hypothesis about θ, we can use Equation 14.3.6 to determine the p-value of the test data. For instance, suppose we are interested in the one-sided test of

$$H_0 : \theta \geq \theta_0$$

versus the alternative

$$H_1 : \theta < \theta_0$$

This can be tested by first computing the value of the test statistic $2\tau/\theta_0$ — call this value v — and then computing the probability that a chi-square random variable with $2r$ degrees of freedom would be as small as v. This probability is the p-value in the sense that it represents the (maximal) probability that such a small value of $2\tau/\theta_0$ would have been observed if H_0 were true. The hypothesis should then be rejected at all significance levels at least as large as this p-value.

EXAMPLE 14.3b A producer of batteries claims that the lifetimes of the items it manufactures are exponentially distributed with a mean life of at least 150 hours. To test this claim, 100 batteries are simultaneously put on a test that is slated to end when the 20th failure occurs. If, at the end of the experiment, the total test time of all the 100 batteries is equal to 1,800, should the manufacturer's claim be accepted?

SOLUTION Since $2\tau/\theta_0 = 3{,}600/150 = 24$, the p-value is

$$p\text{-value} = P\{\chi^2_{40} \leq 24\}$$
$$= .021 \quad \text{from Program 5.8.1a}$$

Hence, the manufacturer's claim should be rejected at the 5 percent level of significance (indeed at any significance level at least as large as .021). ■

It follows from Equation 14.3.5 that the accuracy of the estimator τ/r depends only on r and not on n, the number of items put on test. The importance of n resides in the fact that by choosing it large enough we can ensure that the test is, with high probability, of short duration. In fact, the moments of $X_{(r)}$, the time at which the test ends, are easily obtained. Since, with $X_{(0)} \equiv 0$,

$$X_{(j)} - X_{(j-1)} = \frac{Y_j}{n - j + 1}, \quad j = 1, \ldots, r$$

it follows upon summing that

$$X_{(r)} = \sum_{j=1}^{r} \frac{Y_j}{n - j + 1}$$

Hence, from Equation 14.3.4, $X_{(r)}$ is the sum of r independent exponentials having respective means $\theta/n, \theta/(n-1), \ldots, \theta/(n-r+1)$. Using this, we see that

$$E[X_{(r)}] = \sum_{j=1}^{r} \frac{\theta}{n - j + 1} = \theta \sum_{j=n-r+1}^{n} \frac{1}{j} \qquad (14.3.7)$$

$$\text{Var}(X_{(r)}) = \sum_{j=1}^{r} \left(\frac{\theta}{n - j + 1}\right)^2 = \theta^2 \sum_{j=n-r+1}^{n} \frac{1}{j^2}$$

where the second equality uses the fact that the variance of an exponential is equal to the square of its mean. For large n, we can approximate the preceding sums as follows:

$$\sum_{j=n-r+1}^{n} \frac{1}{j} \approx \int_{n-r+1}^{n} \frac{dx}{x} = \log\left(\frac{n}{n - r + 1}\right)$$

$$\sum_{j=n-r+1}^{n} \frac{1}{j^2} \approx \int_{n-r+1}^{n} \frac{dx}{x^2} = \frac{1}{n - r + 1} - \frac{1}{n} = \frac{r - 1}{n(n - r + 1)}$$

Thus, for instance, if in Example 14.3b the true mean life was 120 hours, then the expectation and variance of the length of the test are approximately given by

$$E[X_{(20)}] \approx 120 \log\left(\frac{100}{81}\right) = 25.29$$

$$\mathrm{Var}(X_{(20)}) \approx (120)^2 \frac{19}{100(81)} = 33.78$$

14.3.2 Sequential Testing

Suppose now that we have an infinite supply of items, each of whose lifetime is exponential with an unknown mean θ, which are to be tested sequentially, in that the first item is put on test and on its failure the second is put on test, and so on. That is, as soon as an item fails, it is immediately replaced on life test by the next item. We suppose that at some fixed time T the text ends.

The observed data will consist of the following:

$$Data: \quad r, x_1, x_2, \ldots, x_r$$

with the interpretation that there has been a total of r failures with the ith item on test having functioned for a time x_i. Now the foregoing will be the observed data if

$$X_i = x_i, \quad i = 1, \ldots, r, \quad \sum_{i=1}^{r} x_i < T \qquad (14.3.8)$$

$$X_{r+1} > T - \sum_{i=1}^{r} x_i$$

where X_i is the functional lifetime of the ith item to be put in use. This follows since in order for there to be r failures, the rth failure must occur before time T — and so $\sum_{i=1}^{r} X_i < T$ — and the functional life of the $(r+1)$st item must exceed $T - \sum_{i=1}^{r} X_i$ (see Figure 14.1).

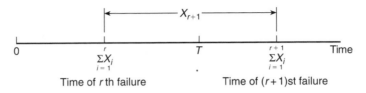

FIGURE 14.1   *r failures by time T.*

From Equation 14.3.8, we obtain that the likelihood of the data $r, x_1, \ldots, x_r$ is as follows:

$$f(r, x_1, \ldots, x_r | \theta)$$

$$= f_{X_1, \ldots, X_r}(x_1, \ldots, x_r) P \left\{ X_{r+1} > T - \sum_{i=1}^{r} x_i \right\}, \quad \sum_{i=1}^{r} x_i < T$$

$$= \frac{1}{\theta^r} e^{-\sum_{i=1}^{r} x_i/\theta} e^{-(T - \sum_{i=1}^{r} x_i)/\theta}$$

$$= \frac{1}{\theta^r} e^{-T/\theta}$$

Therefore,

$$\log f(r, x_1, \ldots, x_r | \theta) = -r \log \theta - \frac{T}{\theta}$$

and so

$$\frac{\partial}{\partial \theta} \log f(r, x_1, \ldots, x_r | \theta) = -\frac{r}{\theta} + \frac{T}{\theta^2}$$

On equating to 0 and solving, we obtain that the maximum likelihood estimate for θ is

$$\hat{\theta} = \frac{T}{r}$$

Since T is the total time on test of all items, it follows once again that the maximum likelihood estimate of the unknown exponential mean is equal to the total time on test divided by the number of observed failures in this time.

If we let $N(T)$ denote the number of failures by time T, then the maximum likelihood estimator of θ is $T/N(T)$. Suppose now that the observed value of $N(T)$ is $N(T) = r$. To determine a $100(1 - \alpha)$ percent confidence interval estimate for θ, we will first determine the values θ_L and θ_U, which are such that

$$P_{\theta_U}\{N(T) \geq r\} = \frac{\alpha}{2}, \quad P_{\theta_L}\{N(T) \leq r\} = \frac{\alpha}{2}$$

where by $P_\theta(A)$ we mean that we are computing the probability of the event A under the supposition that θ is the true mean. The $100(1 - \alpha)$ percent confidence interval estimate for θ is

$$\theta \in (\theta_L, \theta_U)$$

To understand why those values of θ for which either $\theta < \theta_L$ or $\theta > \theta_U$ are not included in the confidence interval, note that $P_\theta\{N(T) \geq r\}$ decreases and $P_\theta\{N(T) \leq r\}$

increases in θ (why?). Hence,

$$\text{if } \theta < \theta_L, \quad \text{then } P_\theta\{N(T) \le r\} < P_{\theta_L}\{N(T) \le r\} = \frac{\alpha}{2}$$

$$\text{if } \theta > \theta_U, \quad \text{then } P_\theta\{N(T) \ge r\} < P_{\theta_U}\{N(T) \ge r\} = \frac{\alpha}{2}$$

It remains to determine θ_L and θ_U. To do so, note first that the event that $N(T) \ge r$ is equivalent to the statement that the rth failure occurs before or at time T. That is,

$$N(T) \ge r \Leftrightarrow X_1 + \cdots + X_r \le T$$

and so

$$
\begin{aligned}
P_\theta\{N(T) \ge r\} &= P_\theta\{X_1 + \cdots + X_r \le T\} \\
&= P\{\Gamma(r, 1/\theta) \le T\} \\
&= P\left\{\frac{\theta}{2}\chi^2_{2r} \le T\right\} \\
&= P\{\chi^2_{2r} \le 2T/\theta\}
\end{aligned}
$$

Hence, upon evaluating the foregoing at $\theta = \theta_U$, and using the fact that $P\{\chi^2_{2r} \le \chi^2_{1-\alpha/2,2r}\} = \alpha/2$, we obtain that

$$\frac{\alpha}{2} = P\left\{\chi^2_{2r} \le \frac{2T}{\theta_U}\right\}$$

and that

$$\frac{2T}{\theta_U} = \chi^2_{1-\alpha/2,2r}$$

or

$$\theta_U = 2T/\chi^2_{1-\alpha/2,2r}$$

Similarly, we can show that

$$\theta_L = 2T/\chi^2_{\alpha/2,2r}$$

and thus the $100(1-\alpha)$ percent confidence interval estimate for θ is

$$\theta \in (2T/\chi^2_{\alpha/2,2r}, \quad 2T/\chi^2_{1-\alpha/2,2r})$$

EXAMPLE 14.3c   If a one-at-a-time sequential test yields 10 failures in the fixed time of $T = 500$ hours, then the maximum likelihood estimate of θ is $500/10 = 50$ hours. A 95 percent confidence interval estimate of θ is

$$\theta \in (1{,}000/\chi^2_{.025,20}, \; 1{,}000/\chi^2_{.975,20})$$

Running Program 5.8.1b yields that

$$\chi^2_{.025,20} = 34.17, \quad \chi^2_{.975,20} = 9.66$$

and so, with 95 percent confidence,

$$\theta \in (29.27, 103.52) \quad \blacksquare$$

If we wanted to test the hypothesis

$$H_0 : \theta = \theta_0$$

versus the alternative

$$H_1 : \theta \neq \theta_0$$

then we would first determine the value of $N(T)$. If $N(T) = r$, then the hypothesis would be rejected provided either

$$P_{\theta_0}\{N(T) \leq r\} \leq \frac{\alpha}{2} \quad \text{or} \quad P_{\theta_0}\{N(T) \geq r\} \leq \frac{\alpha}{2}$$

In other words, H_0 would be rejected at all significance levels greater than or equal to the p-value given by

$$p\text{-value} = 2 \min(P_{\theta_0}\{N(T) \geq r\}, P_{\theta_0}\{N(T) \leq r\})$$
$$p\text{-value} = 2 \min(P_{\theta_0}\{N(T) \geq r\}, 1 - P_{\theta_0}\{N(T) \geq r + 1\})$$
$$= 2 \min \left(P\left\{ \chi^2_{2r} \leq \frac{2T}{\theta_0} \right\}, 1 - P\left\{ \chi^2_{2(r+1)} \leq \frac{2T}{\theta_0} \right\} \right)$$

The p-value for a one-sided test is similarly obtained.

The chi-square probabilities in the foregoing can be computed by making use of Program 5.8.1a.

EXAMPLE 14.3d   A company claims that the mean lifetimes of the semiconductors it produces is at least 25 hours. To substantiate this claim, an independent testing service has decided to sequentially test, one at a time, the company's semiconductors for 600 hours. If 30 semiconductors failed during this period, what can we say about the validity of the company's claim? Test at the 10 percent level.

SOLUTION This is a one-sided test of

$$H_0 : \theta \geq 25 \quad \text{versus} \quad H_1 : \theta < 25$$

The relevant probability for determining the p-value is the probability that there would have been as many as 30 failures if the mean life were 25. That is,

$$
\begin{aligned}
p\text{-value} &= P_{25}\{N(600) \geq 30\} \\
&= P\{\chi^2_{60} \leq 1{,}200/25\} \\
&= .132 \quad \text{from Program 5.8.1a}
\end{aligned}
$$

Thus, H_0 would be accepted when the significance level is .10. ■

14.3.3 SIMULTANEOUS TESTING — STOPPING BY A FIXED TIME

Suppose again that we are testing items whose life distributions are independent exponential random variables with a common unknown mean θ. As in Section 14.3.1, the n items are simultaneously put on test, but now we suppose that the test is to stop either at some fixed time T or whenever all n items have failed — whichever occurs first. The problem is to use the observed data to estimate θ.

The observed data will be as follows:

$$Data: \quad i_1, i_2, \ldots, i_r, \quad x_1, x_2, \ldots, x_r$$

with the interpretation that the preceding results when the r items numbered $i_1, \ldots, i_r$ are observed to fail at respective times $x_1, \ldots, x_r$ and the other $n - r$ items have not failed by time T.

Since an item will not have failed by time T if and only if its lifetime is greater than T, we see that the likelihood of the foregoing data is

$$
\begin{aligned}
f(i_1, \ldots, i_r, x_1, \ldots, x_r) &= f_{X_{i_1}, \ldots, X_{i_r}}(x_1, \ldots, x_r) P\{X_j > T, j \neq i_1, \ldots, i_r\} \\
&= \frac{1}{\theta} e^{-x_1/\theta} \cdots \frac{1}{\theta} e^{-x_r/\theta} (e^{-T/\theta})^{n-r} \\
&= \frac{1}{\theta^r} \exp\left\{ -\frac{\sum\limits_{i=1}^{r} x_i}{\theta} - \frac{(n-r)T}{\theta} \right\}
\end{aligned}
$$

To obtain the maximum likelihood estimates, take logs to obtain

$$\log f(i_1, \ldots, i_r, x_1, \ldots, x_r) = -r \log \theta - \frac{\sum\limits_{1}^{r} x_i}{\theta} - \frac{(n-r)T}{\theta}$$

Hence,

$$\frac{\partial}{\partial \theta} \log f(i_1, \ldots, i_r, x_1, \ldots, x_r) = -\frac{r}{\theta} + \frac{\sum\limits_{1}^{r} x_i + (n-r)T}{\theta^2}$$

Equating to 0 and solving yields that $\hat{\theta}$, the maximum likelihood estimate, is given by

$$\hat{\theta} = \frac{\sum\limits_{i=1}^{r} x_i + (n-r)T}{r}$$

Hence, if we let R denote the number of items that fail by time T and let $X_{(i)}$ be the ith smallest of the failure times, $i = 1, \ldots, R$, then the maximum likelihood estimator of θ is

$$\hat{\theta} = \frac{\sum\limits_{i=1}^{R} X_{(i)} + (n-R)T}{R}$$

Let τ denote the sum of the times that all items are on life test. Then, because the R items that fail are on test for times $X_{(1)}, \ldots, X_{(R)}$ whereas the $n - R$ nonfailed items are all on test for time T, it follows that

$$\tau = \sum_{i=1}^{R} X_{(i)} + (n-R)T$$

and thus we can write the maximum likelihood estimator as

$$\hat{\theta} = \frac{\tau}{R}$$

In words, the maximum likelihood estimator of the mean life is (as in the life testing procedures of Sections 14.3.1 and 14.3.2) equal to the total time on test divided by the number of items observed to fail.

REMARK

As the reader may possibly have surmised, it turns out that for all possible life testing schemes for the exponential distribution, the maximum likelihood estimator of the unknown mean θ will always be equal to the total time on test divided by the number of observed failures. To see why this is true, consider *any* testing situation and suppose that the outcome of the data is that r items are observed to fail after having been on test for times $x_1, \ldots, x_r$, respectively, and that s items have not yet failed when the test ends — at which time they had been on test for respective times $y_1, \ldots, y_s$. The likelihood of this outcome will be

$$\text{likelihood} = K\frac{1}{\theta}e^{-x_1/\theta} \cdots \frac{1}{\theta}e^{-x_r/\theta}e^{-y_1/\theta} \cdots e^{-y_s/\theta}$$

$$= \frac{K}{\theta^r}\exp\left\{\frac{-\left(\sum\limits_{i=1}^{r}x_i + \sum\limits_{i=1}^{s}y_i\right)}{\theta}\right\} \qquad (14.3.9)$$

where K, which is a function of the testing scheme and the data, does not depend on θ. (For instance, K may relate to a testing procedure in which the decision as to when to stop depends not only on the observed data but is allowed to be random.) It follows from the foregoing that the maximum likelihood estimate of θ will be

$$\hat{\theta} = \frac{\sum\limits_{i=1}^{r}x_i + \sum\limits_{i=1}^{s}y_i}{r} \qquad (14.3.10)$$

But $\sum_{i=1}^{r}x_i + \sum_{i=1}^{s}y_i$ is just the total-time-on-test statistic and so the maximum likelihood estimator of θ is indeed the total time on test divided by the number of observed failures in that time.

The distribution of τ/R is rather complicated for the life testing scheme described in this section[*] and thus we will not be able to easily derive a confidence interval estimator for θ. Indeed, we will not further pursue this problem but rather will consider the Bayesian approach to estimating θ.

14.3.4 THE BAYESIAN APPROACH

Suppose that items having independent and identically distributed exponential lifetimes with an unknown mean θ are put on life test. Then, as noted in the remark given in Section 14.3.3, the likelihood of the data can be expressed as

$$f(\text{data}|\theta) = \frac{K}{\theta^r}e^{-t/\theta}$$

where t is the total time on test — that is, the sum of the time on test of all items used — and r is the number of observed failures for the given data.

Let $\lambda = 1/\theta$ denote the rate of the exponential distribution. In the Bayesian approach, it is more convenient to work with the rate λ rather than its reciprocal. From the

---

[*] For instance, for the scheme considered, τ and R are not only both random but are also dependent.

foregoing we see that

$$f(\text{data}|\lambda) = K\lambda^r e^{-\lambda t}$$

If we suppose prior to testing, that λ is distributed according to the prior density $g(\lambda)$, then the posterior density of λ given the observed data is as follows:

$$f(\lambda|\text{data}) = \frac{f(\text{data}|\lambda)g(\lambda)}{\int f(\text{data}|\lambda)g(\lambda)\,d\lambda}$$

$$= \frac{\lambda^r e^{-\lambda t} g(\lambda)}{\int \lambda^r e^{-\lambda t} g(\lambda)\,d\lambda} \qquad (14.3.11)$$

The preceding posterior density becomes particularly convenient to work with when g is a gamma density function with parameters, say, (b, a) — that is, when

$$g(\lambda) = \frac{ae^{-a\lambda}(a\lambda)^{b-1}}{\Gamma(b)}, \qquad \lambda > 0$$

for some nonnegative constants a and b. Indeed for this choice of g we have from Equation 14.3.11 that

$$f(\lambda|\text{data}) = Ce^{-(a+t)\lambda}\lambda^{r+b-1}$$

$$= Ke^{-(a+t)\lambda}[(a+t)\lambda]^{b+r-1}$$

where C and K do not depend on λ. Because we recognize the preceding as the gamma density with parameters $(b+r, a+t)$, we can rewrite it as

$$f(\lambda|\text{data}) = \frac{(a+t)e^{-(a+t)\lambda}[(a+t)\lambda]^{b+r-1}}{\Gamma(b+r)}, \qquad \lambda > 0$$

In other words, if the prior distribution of λ is gamma with parameters (b, a), then no matter what the testing scheme, the (posterior) conditional distribution of λ given the data is gamma with parameters $(b + R, a + \tau)$, where τ and R represent respectively the total-time-on-test statistic and the number of observed failures. Because the mean of a gamma random variable with parameters (b, a) is equal to b/a (see Section 5.7), we can conclude that $E[\lambda|\text{data}]$, the Bayes estimator of λ, is

$$E[\lambda|\text{data}] = \frac{b+R}{a+\tau}$$

EXAMPLE 14.3e Suppose that 20 items having an exponential life distribution with an unknown rate λ are put on life test at various times. When the test is ended, there have been 10 observed failures — their lifetimes being (in hours) 5, 7, 6.2, 8.1, 7.9, 15, 18,

3.9, 4.6, 5.8. The 10 items that did not fail had, at the time the test was terminated, been on test for times (in hours) 3, 3.2, 4.1, 1.8, 1.6, 2.7, 1.2, 5.4, 10.3, 1.5. If prior to the testing it was felt that λ could be viewed as being a gamma random variable with parameters (2, 20), what is the Bayes estimator of λ?

SOLUTION Since

$$\tau = 116.1, \quad R = 10$$

it follows that the Bayes estimate of λ is

$$E[\lambda | \text{data}] = \frac{12}{136.1} = .088 \quad \blacksquare$$

REMARK

As we have seen, the choice of a gamma prior distribution for the rate of an exponential distribution makes the resulting computations quite simple. Whereas, from an applied viewpoint, this is not a sufficient rationale, such a choice is often made with one justification being that the flexibility in fixing the two parameters of the gamma prior usually enables one to reasonably approximate their true prior feelings.

14.4 A TWO-SAMPLE PROBLEM

A company has set up two separate plants to produce vacuum tubes. The company supposes that tubes produced at Plant I function for an exponentially distributed time with an unknown mean θ_1 whereas those produced at Plant II function for an exponentially distributed time with unknown mean θ_2. To test the hypothesis that there is no difference between the two plants (at least in regard to the lifetimes of the tubes they produce), the company samples n tubes from Plant I and m from Plant II and then utilizes these tubes to determine their lifetimes. How can they thus determine whether the two plants are indeed identical?

If we let $X_1, \ldots, X_n$ denote the lifetimes of the n tubes produced at Plant I and $Y_1, \ldots, Y_m$ denote the lifetimes of the m tubes produced at Plant II, then the problem is to test the hypothesis that $\theta_1 = \theta_2$ when the $X_i, i = 1, \ldots, n$ are a random sample from an exponential distribution with mean θ_1 and the $Y_i, i = 1, \ldots, m$ are a random sample from an exponential distribution with mean θ_2. Moreover, the two samples are supposed to be independent.

To develop a test of the hypothesis that $\theta_1 = \theta_2$, let us begin by noting that $\sum_{i=1}^{n} X_i$ and $\sum_{i=1}^{m} Y_i$ (being the sum of independent and identically distributed exponentials) are independent gamma random variables with respective parameters $(n, 1/\theta_1)$ and $(m, 1/\theta_2)$.

Hence, by the equivalence of the gamma and chi-square distribution it follows that

$$\frac{2}{\theta_1} \sum_{i=1}^{n} X_i \sim \chi_{2n}^2$$

$$\frac{2}{\theta_2} \sum_{i=1}^{m} Y_i \sim \chi_{2m}^2$$

Hence, it follows from the definition of the F-distribution that

$$\frac{\frac{2}{2n\theta_1} \sum_{i=1}^{n} X_i}{\frac{2}{2m\theta_2} \sum_{i=1}^{m} Y_i} \sim F_{n,m}$$

That is, if $\overline{X}$ and $\overline{Y}$ are the two sample means, respectively, then

$$\frac{\theta_2 \overline{X}}{\theta_1 \overline{Y}} \quad \text{has an } F\text{-distribution with } n \text{ and } m \text{ degrees of freedom}$$

Hence, when the hypothesis $\theta_1 = \theta_2$ is true, we see that $\overline{X}/\overline{Y}$ has an F-distribution with n and m degrees of freedom. This suggests the following test of the hypothesis that $\theta_1 = \theta_2$.

Test: $H_0 : \theta_1 = \theta_2$ vs. alternative $H_1 : \theta_1 \neq \theta_2$
Step 1:   Choose a significance level α.
Step 2:   Determine the value of the test statistic $\overline{X}/\overline{Y}$ — say its value is v.
Step 3:   Compute $P\{F \leq v\}$ where $F \sim F_{n,m}$. If this probability is either less than $\alpha/2$ (which occurs when $\overline{X}$ is significantly less than $\overline{Y}$) or greater than $1 - \alpha/2$ (which occurs when $\overline{X}$ is significantly greater than $\overline{Y}$), then the hypothesis is rejected.

In other words, the p-value of the test data is given by

$$p\text{-value} = 2 \min(P\{F \leq v\}, 1 - P\{F \leq v\})$$

EXAMPLE 14.4a   Test the hypothesis, at the 5 percent level of significance, that the lifetimes of items produced at two given plants have the same exponential life distribution if a sample of size 10 from the first plant has a total lifetime of 420 hours whereas a sample of 15 from the second plant has a total lifetime of 510 hours.

SOLUTION   The value of the test statistic $\overline{X}/\overline{Y}$ is $42/34 = 1.2353$. To compute the probability that an F-random variable with parameters 10, 15 is less than this value, we run Program 5.8.3a to obtain that

$$P\{F_{10,15} < 1.2353\} = .6554$$

Because the p-value is equal to $2(1 - .6554) = .6892$, we cannot reject H_0.   ∎

14.5 THE WEIBULL DISTRIBUTION IN LIFE TESTING

Whereas the exponential distribution arises as the life distribution when the hazard rate function $\lambda(t)$ is assumed to be constant over time, there are many situations in which it is more realistic to suppose that $\lambda(t)$ either increases or decreases over time. One example of such a hazard rate function is given by

$$\lambda(t) = \alpha\beta t^{\beta-1}, \quad t > 0 \tag{14.5.1}$$

where α and β are positive constants. The distribution whose hazard rate function is given by Equation 14.5.1 is called the *Weibull* distribution with parameters (α, β). Note that $\lambda(t)$ increases when $\beta > 1$, decreases when $\beta < 1$, and is constant (reducing to the exponential) when $\beta = 1$.

The Weibull distribution function is obtained from Equation 14.5.1 as follows:

$$F(t) = 1 - \exp\left\{-\int_0^t \lambda(s)\,ds\right\}, \quad t > 0$$
$$= 1 - \exp\{-\alpha t^\beta\}$$

Differentiating yields its density function:

$$f(t) = \alpha\beta t^{\beta-1}\exp\{-\alpha t^\beta\}, \quad t > 0 \tag{14.5.2}$$

This density is plotted for a variety of values of α and β in Figure 14.2.

Suppose now that $X_1, \ldots, X_n$ are independent Weibull random variables each having parameters (α, β), which are assumed unknown. To estimate α and β, we can employ the maximum likelihood approach. Equation 14.5.2 yields the likelihood, given by

$$f(x_1, \ldots, x_n) = \alpha^n\beta^n x_1^{\beta-1}\cdots x_n^{\beta-1}\exp\left\{-\alpha\sum_{i=1}^n x_i^\beta\right\}$$

Hence,

$$\log f(x_1, \ldots, x_n) = n\log\alpha + n\log\beta + (\beta - 1)\sum_{i=1}^n \log x_i - \alpha\sum_{i=1}^n x_i^\beta$$

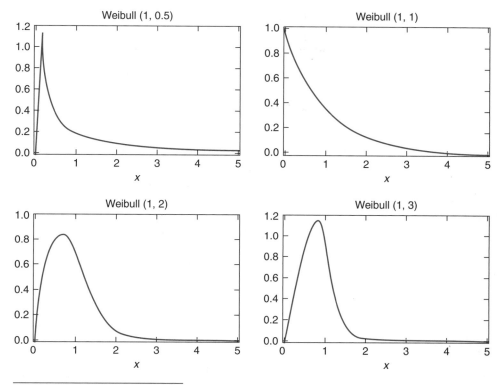

FIGURE 14.2 *Weibull density functions.*

and

$$\frac{\partial}{\partial \alpha} \log f(x_1, \ldots, x_n) = \frac{n}{\alpha} - \sum_{i=1}^{n} x_i^{\beta}$$

$$\frac{\partial}{\partial \beta} \log f(x_1, \ldots, x_n) = \frac{n}{\beta} + \sum_{i=1}^{n} \log x_i - \alpha \sum_{i=1}^{n} x_i^{\beta} \log x_i$$

Equating to zero shows that the maximum likelihood estimates $\hat{\alpha}$ and $\hat{\beta}$ are the solutions of

$$\frac{n}{\hat{\alpha}} = \sum_{i=1}^{n} x_i^{\hat{\beta}}$$

$$\frac{n}{\hat{\beta}} + \sum_{i=1}^{n} \log x_i = \hat{\alpha} \sum_{i=1}^{n} x_i^{\hat{\beta}} \log x_i$$

or, equivalently,

$$\hat{\alpha} = \frac{n}{\sum\limits_{i=1}^{n} x_i^{\hat{\beta}}}$$

$$n + \hat{\beta} \log\left(\prod_{i=1}^{n} x_i\right) = \frac{n\hat{\beta} \sum\limits_{i=1}^{n} x_i^{\hat{\beta}} \log x_i}{\sum\limits_{i=1}^{n} x_i^{\hat{\beta}}}$$

This latter equation can then be solved numerically for $\hat{\beta}$, which will then also determine $\hat{\alpha}$. However, rather than pursuing this approach any further, let us consider a second approach, which is not only computationally easier but appears, as indicated by a simulation study, to yield more accurate estimates.

14.5.1 PARAMETER ESTIMATION BY LEAST SQUARES

Let $X_1, \ldots, X_n$ be a sample from the distribution

$$F(x) = 1 - e^{-\alpha x^\beta}, \quad x \geq 0$$

Note that

$$\log(1 - F(x)) = -\alpha x^\beta$$

or

$$\log\left(\frac{1}{1 - F(x)}\right) = \alpha x^\beta$$

and so

$$\log\log\left(\frac{1}{1 - F(x)}\right) = \beta \log x + \log \alpha \qquad (14.5.3)$$

Now let $X_{(1)} < X_{(2)} < \cdots < X_{(n)}$ denote the ordered sample values — that is, for $i = 1, \ldots, n$,

$$X_{(i)} = i\text{th smallest of } X_1, \ldots, X_n$$

and suppose that the data results in $X_{(i)} = x_{(i)}$. If we were able to approximate the quantities $\log\log(1/[1 - F(x_{(i)})])$ — say, by the values $y_1, \ldots, y_n$ — then from Equation 14.5.3, we could conclude that

$$y_i \approx \beta \log x_{(i)} + \log \alpha, \quad i = 1, \ldots, n \tag{14.5.4}$$

We could then choose α and β to minimize the sum of the squared errors — that is, α and β are chosen to

$$\underset{\alpha, \beta}{\text{minimize}} \sum_{i=1}^{n} (y_i - \beta \log x_{(i)} - \log \alpha)^2$$

Indeed, using Proposition 9.2.1 we obtain that the preceding minimum is attained when $\alpha = \hat{\alpha}, \beta = \hat{\beta}$ where

$$\hat{\beta} = \frac{\displaystyle\sum_{i=1}^{n} y_i \log x_{(i)} - n \overline{\log x} \bar{y}}{\displaystyle\sum_{i=1}^{n} (\log x_{(i)})^2 - n (\overline{\log x})^2}$$

$$\log \hat{\alpha} = \bar{y} - \beta \overline{\log x}$$

where

$$\overline{\log x} = \sum_{i=1}^{n} (\log x_{(i)}) \Big/ n, \quad \bar{y} = \sum_{i=1}^{n} y_i \Big/ n$$

To utilize the foregoing, we need to be able to determine values y_i that approximate $\log \log(1/[1 - F(x_{(i)})]) = \log[-\log(1 - F(x_{(i)}))], i = 1, \ldots, n$. We now present two different methods for doing this.

Method 1: This method uses the fact that

$$E[F(X_{(i)})] = \frac{i}{(n+1)} \tag{14.5.5}$$

and then approximates $F(x_{(i)})$ by $E[F(X_{(i)})]$. Thus, this method calls for using

$$y_i = \log\{-\log(1 - E[F(X_{(i)})])\} \tag{14.5.6}$$
$$= \log\left\{-\log\left(1 - \frac{i}{(n+1)}\right)\right\}$$
$$= \log\left\{-\log\left(\frac{n+1-i}{n+1}\right)\right\}$$

Method 2: This method uses the fact that

$$E[-\log(1 - F(X_{(i)}))] = \frac{1}{n} + \frac{1}{n-1} + \frac{1}{n-2} + \cdots + \frac{1}{n-i+1} \qquad (14.5.7)$$

and then approximates $-\log(1 - F(x_{(i)}))$ by the foregoing. Thus, this second method calls for setting

$$y_i = \log\left[\frac{1}{n} + \frac{1}{(n-1)} + \cdots + \frac{1}{(n-i+1)}\right] \qquad (14.5.8)$$

REMARKS

(a) It is not, at present, clear which method provides superior estimates of the parameters of the Weibull distribution, and extensive simulation studies will be necessary to determine this.

(b) Proofs of equalities 14.5.5 and 14.5.7 [which hold whenever $X_{(i)}$ is the ith smallest of a sample of size n from any continuous distribution F] are outlined in Problems 28–30.

Problems

1. A random variable whose distribution function is given by

$$F(t) = 1 - \exp\{-\alpha t^\beta\}, \quad t \geq 0$$

is said to have a Weibull distribution with parameters α, β. Compute its failure rate function.

2. If X and Y are independent random variables having failure rate functions $\lambda_x(t)$ and $\lambda_y(t)$, show that the failure rate function of $Z = \min(X, Y)$ is

$$\lambda_z(t) = \lambda_x(t) + \lambda_y(t)$$

3. The lung cancer rate of a t-year-old male smoker, $\lambda(t)$, is such that

$$\lambda(t) = .027 + .025 \left(\frac{t - 40}{10}\right)^4, \quad t \geq 40$$

Assuming that a 40-year-old male smoker survives all other hazards, what is the probability that he survives to (a) age 50, (b) age 60, without contracting lung cancer? In the foregoing we are assuming that he remains a smoker throughout his life.

4. Suppose the life distribution of an item has failure rate function $\lambda(t) = t^3, 0 < t < \infty$.

   (a) What is the probability that the item survives to age 2?
   (b) What is the probability that the item's life is between .4 and 1.4?
   (c) What is the mean life of the item?
   (d) What is the probability a 1-year-old item will survive to age 2?

5. A continuous life distribution is said to be an IFR (increasing failure rate) distribution if its failure rate function $\lambda(t)$ is nondecreasing in t.

   (a) Show that the gamma distribution with density

   $$f(t) = \lambda^2 t e^{-\lambda t}, \quad t > 0$$

   is IFR.
   (b) Show, more generally, that the gamma distribution with parameters α, λ is IFR whenever $\alpha \geq 1$.
   Hint: Write

   $$\lambda(t) = \left[\frac{\int_t^\infty \lambda e^{-\lambda s}(\lambda s)^{\alpha-1}\, ds}{\lambda e^{-\lambda t}(\lambda t)^{\alpha-1}} \right]^{-1}$$

6. Show that the uniform distribution on (a, b) is an IFR distribution.

7. For the model of Section 14.3.1, explain how the following figure can be used to show that

   $$\tau = \sum_{j=1}^{r} Y_j$$

   where

   $$Y_j = (n - j + 1)(X_{(j)} - X_{(j-1)})$$

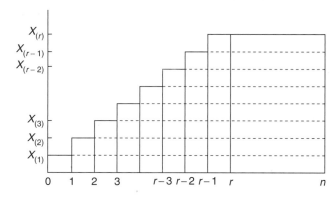

   (*Hint*: Argue that both τ and $\sum_{j=1}^{r} Y_j$ equal the total area of the figure shown.)

8. When 30 transistors were simultaneously put on a life test that was to be terminated when the 10th failure occurred, the observed failure times were (in hours)
   4.1, 7.3, 13.2, 18.8, 24.5, 30.8, 38.1, 45.5, 53, 62.2. Assume an exponential life
   distribution.

   (a) What is the maximum likelihood estimate of the mean life of a transistor?
   (b) Compute a 95 percent two-sided confidence interval for the mean life of a
       transistor.
   (c) Determine a value c that we can assert, with 95 percent confidence, is less
       than the mean transistor life.
   (d) Test at the $\alpha = .10$ level of significance the hypothesis that the mean lifetime
       is 7.5 hours versus the alternative that it is not 7.5 hours.

9. Consider a test of $H_0 : \theta = \theta_0$ versus $H_1 : \theta \neq \theta_0$ for the model of Section 14.3.1. Suppose that the observed value of $2\tau/\theta_0$ is v. Show that the
   hypothesis should be rejected at significance level α whenever α is less than the
   p-value given by

   $$p\text{-value} = 2\min(P\{\chi^2_{2r} < v\}, 1 - P\{\chi^2_{2r} < v\})$$

   where χ^2_{2r} is a chi-square random variable with $2r$ degrees of freedom.

10. Suppose 30 items are put on test that is scheduled to stop when the 8th failure
    occurs. If the failure times are, in hours, .35, .73, .99, 1.40, 1.45, 1.83, 2.20,
    2.72, test, at the 5 percent level of significance, the hypothesis that the mean life
    is equal to 10 hours. Assume that the underlying distribution is exponential.

11. Suppose that 20 items are to be put on test that is to be terminated when the
    10th failure occurs. If the lifetime distribution is exponential with mean 10 hours,
    compute the following quantities.

    (a) The mean length of the testing period.
    (b) The variance of the testing period.

12. Vacuum tubes produced at a certain plant are assumed to have an underlying
    exponential life distribution having an unknown mean θ. To estimate θ it has
    been decided to put a certain number n of tubes on test and to stop the test at
    the 10th failure. If the plant officials want the mean length of the testing period
    to be 3 hours when the value of θ is $\theta = 20$, approximately how large should
    n be?

13. A one-at-a-time sequential life testing scheme is scheduled to run for 300 hours.
    A total of 16 items fail within that time. Assuming an exponential life distribution
    with unknown mean θ (measured in hours):

    (a) Determine the maximum likelihood estimate of θ.

(b) Test at the 5 percent level of significance the hypothesis that $\theta = 20$ versus the alternative that $\theta \neq 20$.

(c) Determine a 95 percent confidence interval for θ.

14. Using the fact that a Poisson process results when the times between successive events are independent and identically distributed exponential random variables, show that

$$P\{X \geq n\} = F_{\chi^2_{2n}}(x)$$

when X is a Poisson random variable with mean $x/2$ and $F_{\chi^2_{2n}}$ is the chi-square distribution function with $2n$ degrees of freedom. (*Hint*: Use the results of Section 14.3.2.)

15. From a sample of items having an exponential life distribution with unknown mean θ, items are tested in sequence. The testing continues until either the rth failure occurs or after a time T elapses.

(a) Determine the likelihood function.

(b) Verify that the maximum likelihood estimator of θ is equal to the total time on test of all items divided by the number of observed failures.

16. Verify that the maximum likelihood estimate corresponding to Equation 14.3.9 is given by Equation 14.3.10.

17. A testing laboratory has facilities to simultaneously life test 5 components. The lab tested a sample of 10 components from a common exponential distribution by initially putting 5 on test and then replacing any failed component by one still waiting to be tested. The test was designed to end either at 200 hours or when all 10 components had failed. If there were a total of 9 failures occurring at times 15, 28.2, 46, 62.2, 76, 86, 128, 153, 197, what is the maximum likelihood estimate of the mean life of a component?

18. Suppose that the remission time, in weeks, of leukemia patients that have undergone a certain type of chemotherapy treatment is an exponential random variable having an unknown mean θ. A group of 20 such patients is being monitored and, at present, their remission times are (in weeks) 1.2, 1.8*, 2.2, 4.1, 5.6, 8.4, 11.8*, 13.4*, 16.2, 21.7, 29*, 41, 42*, 42.4*, 49.3, 60.5, 61*, 94, 98, 99.2* where an asterisk next to the data means that the patient's remission is continuing, whereas a data point without an asterisk means that the remission ended at that time. What is the maximum likelihood estimate of θ?

19. In Problem 17, suppose that prior to the testing phase and based on past experience one felt that the value of $\lambda = 1/\theta$ could be thought of as the outcome of a gamma random variable with parameters 1, 100. What is the Bayes estimate of λ?

20. What is the Bayes estimate of $\lambda = 1/\theta$ in Problem 18 if the prior distribution on λ is exponential with mean $1/30$?

21. The following data represent failure times, in minutes, for two types of electrical insulation subject to a certain voltage stress.

Type I	212, 88.5, 122.3, 116.4, 125, 132, 66
Type II	34.6, 54, 162, 49, 78, 121, 128

Test the hypothesis that the two sets of data come from the same exponential distribution.

22. Suppose that the life distributions of two types of transistors are both exponential. To test the equality of means of these two distributions, n_1 type 1 transistors are simultaneously put on a life test that is scheduled to end when there have been a total of r_1 failures. Similarly, n_2 type 2 transistors are simultaneously put on a life test that is to end when there have been r_2 failures.

   (a) Using results from Section 14.3.1, show how the hypothesis that the means are equal can be tested by using a test statistic that, when the means are equal, has an F-distribution with $2r_1$ and $2r_2$ degrees of freedom.

   (b) Suppose $n_1 = 20$, $r_1 = 10$ and $n_2 = 10$, $r_2 = 7$ with the following data resulting.

   Type 1 failures at times:

   $$10.4, 23.2, 31.4, 45, 61.1, 69.6, 81.3, 95.2, 112, 129.4$$

   Type 2 failures at times:

   $$6.1, 13.8, 21.2, 31.6, 46.4, 66.7, 92.4$$

   What is the smallest significance level α for which the hypothesis of equal means would be rejected? (That is, what is the p-value of the test data?)

23. If X is a Weibull random variable with parameters (α, β), show that

$$E[X] = \alpha^{-1/\beta}\Gamma(1 + 1/\beta)$$

where $\Gamma(y)$ is the gamma function defined by

$$\Gamma(y) = \int_0^\infty e^{-x}x^{y-1}\, dx$$

Hint: Write

$$E[X] = \int_0^\infty t\alpha\beta t^{\beta-1} \exp\{-\alpha t^\beta\}\, dt$$

and make the change of variables

$$x = \alpha t^{\beta}, \quad dx = \alpha \beta t^{\beta-1} \, dt$$

24. Show that if X is a Weibull random variable with parameters (α, β), then

$$\text{Var}(X) = \alpha^{-2/\beta} \left[\Gamma\left(1 + \frac{2}{\beta}\right) - \left(\Gamma\left(1 + \frac{1}{\beta}\right)\right)^2 \right]$$

25. If the following are the sample data from a Weibull population having unknown parameters α and β, determine the least square estimates of these quantities, using either of the methods presented.

    Data: 15.4, 16.8, 6.2, 10.6, 21.4, 18.2, 1.6, 12.5, 19.4, 17

26. Show that if X is a Weibull random variable with parameters (α, β), then αX^{β} is an exponential random variable with mean 1.

27. If U is uniformly distributed on $(0, 1)$ — that is, U is a random number — show that $[-(1/\alpha) \log U]^{1/\beta}$ is a Weibull random variable with parameters (α, β).

The next three problems are concerned with verifying Equations 14.5.5 and 14.5.7.

28. If X is a continuous random variable having distribution function F, show that

    (a) $F(X)$ is uniformly distributed on $(0, 1)$;
    (b) $1 - F(X)$ is uniformly distributed on $(0, 1)$.

29. Let $X_{(i)}$ denote ith smallest of a sample of size n from a continuous distribution function F. Also, let $U_{(i)}$ denote the ith smallest from a sample of size n from a uniform $(0, 1)$ distribution.

    (a) Argue that the density function of $U_{(i)}$ is given by

$$f_{U_{(i)}}(t) = \frac{n!}{(n-i)!(i-1)!} t^{i-1}(1-t)^{n-i}, \quad 0 < t < 1$$

    [*Hint*: In order for the ith smallest of n uniform $(0, 1)$ random variables to equal t, how many must be less than t and how many must be greater? Also, in how many ways can a set of n elements be broken into three subsets of respective sizes $i - 1, 1$, and $n - i$?]

(b) Use part (a) to show that $E[U(i)] = i/(n+1)$. [*Hint*: To evaluate the resulting integral, use the fact that the density in part (a) must integrate to 1.]

(c) Use part (b) and Problem 28a to conclude that $E[F(X_{(i)})] = i/(n+1)$.

30. If U is uniformly distributed on $(0, 1)$, show that $-\log U$ has an exponential distribution with mean 1. Now use Equation 14.3.7 and the results of the previous problems to establish Equation 14.5.7.

Chapter 15

SIMULATION, BOOTSTRAP STATISTICAL METHODS, AND PERMUTATION TESTS

15.1 INTRODUCTION

In this chapter we introduce two powerful modern statistical techniques: bootstrap statistical methods and permutation tests. Both are nonparametric procedures in the sense that they make no specific assumptions about the form of any underlying probability distributions. Bootstrap methods enable us to measure the efficacy of an estimator of a parameter, while permutation tests yield new ways to test certain statistical hypotheses. Both, however, require a large amount of computation in their implementation. The most efficient and effective way of doing the needed computation uses simulation, the third topic of this chapter.

In Section 15.2 we introduce random numbers, which are the keys to a simulation. We show how random numbers can be used to generate random permutations and random subsets. In Section 15.2.1 we present the Monte Carlo simulation method for approximating expectations. In Section 15.3 we introduce the method of bootstrap statistics and show how the needed analysis can be done by applying the Monte Carlos simulation method. In Section 15.4 we discuss permutation tests, which are nonparametric tests for determining whether a sequence of data comes from a single population distribution. In the remaining sections we return to the study of simulation. In Sections 15.5 and 15.6 we show how random numbers can be used to generate the values of arbitrarily distributed discrete and continuous random variables, and in Section 15.7 we consider the question of when to end a Monte Carlo simulation study.

15.2 RANDOM NUMBERS

The value of a uniform $(0, 1)$ random variable is called a *random number*. Whereas in the past, mechanical devices have often been used to generate random numbers, today we commonly use random number generators to generate a sequence of pseudo random numbers. Such random number generators start with an initial value x_0, called the *seed*, and then recursively determine values by first specifying positive integers a, c, and m and then letting

$$x_{n+1} = (ax_n + c) \ \text{modulo} \ m, \ n \geq 0$$

where the preceding means that x_{n+1} is the remainder obtained when $ax_n + c$ is divided by m. Thus each x_n is one of the values $0, 1, \ldots, m - 1$, and the quantity x_n/m is taken as the random number. It can be shown that for suitable choices of a, c, and m, the preceding gives rise to a sequence of numbers that looks as if it was generated by observing the values of independent uniform $(0, 1)$ random variables. For this reason we call the numbers x_n/m, $n \geq 1$, *pseudo random numbers*.

EXAMPLE 15.2a If $a = 3$, $c = 7$, $m = 23$, then with $x_0 = 2$

$$x_1 = 3(2) + 7 \quad \text{modulo } 23 = 13$$
$$x_2 = 3(13) + 7 \quad \text{modulo } 23 = 0$$
$$x_3 = 3(0) + 7 \quad \text{modulo } 23 = 7$$
$$x_4 = 3(7) + 7 \quad \text{modulo } 23 = 5$$
$$x_5 = 3(5) + 7 \quad \text{modulo } 23 = 22$$

and so on. Consequently, using the seed $x_0 = 2$, the pseudo random numbers obtained are $13/23$, 0, $7/23$, $5/23$, $22/23, \ldots$ ■

Most computers have built-in random number generators, and we shall take as our starting point in simulation that we can generate the values of pseudo random numbers; moreover, we will act as if these pseudo random numbers were actually true random numbers. That is, we will act as if the sequence of random numbers were actually a sequence of values of a sample from the uniform $(0, 1)$ distribution.

Random numbers are the key to any simulation study. This is illustrated in our next example, which is concerned with generating a random permutation.

EXAMPLE 15.2b Suppose we want to generate a permutation of the numbers $1, 2, \ldots, n$ in such a manner that all $n!$ possible permutations are equally likely. To accomplish this we can first randomly choose one of the numbers $1, 2, \ldots, n$ and put that number in position n. We can then randomly choose one of the remaining $n-1$ numbers and put that number in position $n-1$, and then randomly choose one of the remaining $n-2$ numbers and put that number in position $n-2$, and so on (where "randomly choose" means that each of the possible choices is equally likely to be made). However, so that we do not have to directly consider exactly which elements remain to be placed, it is convenient and effective to keep the numbers in an ordered list and then randomly choose the position of the number rather than the number itself. That is, starting with any permutation $r_1, r_2, \ldots, r_n$ of the numbers $1, 2, \ldots, n$, randomly choose one of the positions $1, \ldots, n$ and then interchange the number in that position with the number in position n. Then randomly choose one of the positions $1, \ldots, n-1$, and interchange the number in that position with the number in position $n-1$. Then randomly choose one of the positions $1, \ldots, n-2$, and interchange the number in that position with the number in position $n-2$, and so on.

To implement the preceding we need to be able to generate a random variable that is equally likely to take on any of the values $1, \ldots, k$. To accomplish this, let U denote a random number — that is, U is uniformly distributed over $(0, 1)$ — and let $\text{Int}(kU)$ be the integer part of kU — that is, it is the largest integer less than or equal to kU. Then, for $i = 1, \ldots, k$

$$
\begin{aligned}
P[\text{Int}(kU) + 1 = i] &= P[\text{Int}(kU) = i - 1] \\
&= P(i - 1 \le kU < i) \\
&= P(\tfrac{i-1}{k} \le U < \tfrac{i}{k}) \\
&= 1/k
\end{aligned}
$$

Thus, $\text{Int}(kU) + 1$ is equally likely to take on any of the values $1, \ldots, k$.

The algorithm for generating a random permutation of the numbers $1, 2, \ldots, n$ can now be written as follows:

1. Let $r_1, r_2, \ldots, r_n$ be any permutation of the numbers $1, 2, \ldots, n$. (For instance, we could have $r_j = j, j = 1, \ldots, n$.)
2. Set $k = n$. (The number to be put in position k is to be determined.)
3. Generate a random number U and let $I = \text{Int}(kU) + 1$.
4. Interchange the values of r_I and r_k.
5. Let $k = k - 1$.
6. If $k > 1$, go to Step 3; if $k = 1$, go to step 7.
7. $r_1, \ldots, r_n$ is the desired permutation.

For instance, suppose $n = 4$ and the initial permutation is $1, 2, 3, 4$. If the first value of I — which is equally likely to be any of the numbers $1, 2, 3, 4$ — is 2, then the number in position 2 is interchanged with the number in position 4 to give the new permutation $1, 4, 3, 2$. If the next value of I — which is equally likely to be any of the numbers $1, 2, 3$ — is 3, then the number in position 3 is interchanged with the one in position 3, so the permutation remains $1, 4, 3, 2$. If the final value of I — which is equally likely to be any of the numbers $1, 2$ — is 1, then the number in position 1 is interchanged with the one in position 2 to give the final permutation $4, 1, 3, 2$.

An important property of the preceding algorithm is that it can be used to generate a random subset of size r from the set $\{1, 2, \ldots, n\}$. For $r \le n/2$, just follow the preceding algorithm until the elements in the final r positions (that is, in positions $n, n-1, \ldots, n - r + 1$) are specified, and then take the numbers in these positions as the random subset of size r. For $r > n/2$, rather than directly choosing the r numbers to be in the subset, it is quicker to choose the $n - r$ numbers that are not in the subset. So in this case, follow the preceding algorithm until the final $n - r$ positions are filled, and then take the numbers that remain as the random subset of size r. ■

15.2.1 THE MONTE CARLO SIMULATION APPROACH

Suppose we want to compute the expected value of a statistic $h(X_1, X_2, \ldots, X_n)$ when $X_1, X_2, \ldots, X_n$ are independent and identically distributed random variables having density function $f(x)$. Using that the joint density function of $X_1, X_2, \ldots, X_n$ is

$$f(x_1, \ldots, x_n) = f(x_1)f(x_2) \cdots f(x_n)$$

we can write that

$$E[h(X_1, X_2, \ldots, X_n)] = \int \int \cdots \int h(x_1, \ldots, x_n)f(x_1)f(x_2) \cdots f(x_n)\, dx_1\, dx_2 \cdots dx_n$$

The difficulty, however, with the preceding formula is that it is often impossible to analytically compute the preceding multiple integral and also difficult to numerically evaluate it to within a specified accuracy. One approach that remains is to approximate $E[h(X_1, X_2, \ldots, X_n)]$ by a simulation.

To accomplish this approximation, start by generating the values of n independent random variables $X_1^1, X_2^1, \ldots, X_n^1$, each having density function f, and then compute

$$Y_1 = h(X_1^1, X_2^1, \ldots, X_n^1)$$

Now generate the values of a second set of n independent random variables having density function f that are also independent of the first set. Calling this second set of random variables $X_1^2, X_2^2, \ldots, X_n^2$, compute

$$Y_2 = h(X_1^2, X_2^2, \ldots, X_n^2)$$

Continue doing this until you have generated r sets of n independent random variables having density function f, and have computed the corresponding values of Y. In this way, we would have generated values of r independent and identically random variables $Y_i = h(X_1^i, X_2^i, \ldots, X_n^i)$, $i = 1, \ldots, r$. Now, by the strong law of large numbers

$$\lim_{r \to \infty} \frac{Y_1 + \cdots + Y_r}{r} = E[Y_i] = E[h(X_1, X_2, \ldots, X_n)]$$

and so we can use the average of the generated values of the Y_i's as an estimate of $E[h(X_1, X_2, \ldots, X_n)]$. This approximation method is called the *Monte Carlo* simulation approach. Each time we generate a new value of Y we say that a new *simulation run* has been completed.

Of course in order to make use of the preceding approach we need to be able to generate random variables having a specified density function. Although at present we only know how to do this for a uniform random variable — by using a random number generator — this will suffice for the needed computations both in the bootstrap method and in running permutation tests. As a result, the next two sections will present these topics. We will then return to the simulation question of how to generate random variables having arbitrary distributions, as well as how to determine when to end a simulation study, in the final sections of this chapter.

15.3 THE BOOTSTRAP METHOD

Let $X_1, \ldots, X_n$ be a sample from a population having distribution F, and suppose we want to use this sample to estimate a parameter θ of F. For instance, θ could be the common mean or variance of the X_i. Suppose we have an estimator $d = d(X_1, \ldots, X_n)$ of θ and we would like to evaluate how good an estimator of θ it is. One measure of the worth of $d(X_1, \ldots, X_n)$ as an estimator of θ is its mean square error, defined as

$$MSE_F(d) = E_F[(d(X_1, \ldots, X_n) - \theta)^2]$$

That is, $MSE_F(d)$ is the expected square of the distance between the estimator $d(X_1, \ldots, X_n)$ and the parameter θ, where we use the notation MSE_F and E_F to indicate that the expected value is to be computed under the assumption that $X_1, \ldots, X_n$ are independent random variables having distribution function F. How can this quantity be estimated?

EXAMPLE 15.3a If $\theta = E[X_i]$ is the mean of the distribution F, and $d(X_1, \ldots, X_n) = \bar{X}_n = \sum_{i=1}^{n} X_i/n$ is the sample mean of the data values $X_1, \ldots, X_n$, then because

$$E_F[d(X_1, \ldots, X_n)] = E_F[\bar{X}_n] = E_F[X_i] = \theta$$

it follows that

$$MSE_F(\bar{X}_n) = E_F[(\bar{X}_n - \theta)^2] = \text{Var}_F(\bar{X}_n) = \sigma^2/n$$

where $\sigma^2 = \text{Var}_F(X_i)$. Thus, in this case, $MSE_F(\bar{X}_n)$ can be estimated by the quantity S_n^2/n, where

$$S_n^2 = \frac{1}{n-1} \sum_{i=1}^{n} (X_i - \bar{X}_n)^2$$

is the sample variance of the data values $X_1, \ldots, X_n$, and can be used to estimate the population variance σ^2. ∎

Whereas in the preceding example it was easy to estimate the mean square error of the sample mean as an estimator of a population mean, what if we initially wanted to estimate the population variance? That is, what if $\theta = \text{Var}_F(X_i)$. In this case we can use the sample variance as the estimator. However, while it was easy to come up with this estimator $d(X_1, \ldots, X_n) = S_n^2$, it is not easy to see how to estimate its mean square error. One way is to use the approach of bootstrap statistics, which we now present.

To estimate the mean square error of the estimator $d(X_1, \ldots, X_n)$ of the parameter θ, suppose that the data values are $X_i = x_i$, $i = 1, \ldots, n$. For any x, let $F_e(x)$ denote the proportion of the data values that are less than or equal to x. That is,

$$F_e(x) = \frac{\text{number of } i \leq n : x_i \leq x}{n}$$

For instance, if $n = 5$ and $X_1 = 5$, $X_2 = 3$, $X_3 = 9$, $X_4 = 2$, and $X_5 = 6$, then

$$F_e(x) = \begin{cases} 0 & \text{if} & x < 2 \\ 1/5 & \text{if} & 2 \leq x < 3 \\ 2/5 & \text{if} & 3 \leq x < 5 \\ 3/5 & \text{if} & 5 \leq x < 6 \\ 4/5 & \text{if} & 6 \leq x < 9 \\ 1 & \text{if} & x \geq 9 \end{cases}$$

The function $F_e(x)$ is called the *empirical distribution function*. When the values $x_1, \ldots, x_n$ are all distinct, F_e is the distribution function of a random variable X_e that is equally likely to be any of the values $x_1, \ldots, x_n$. That is, if the data values are all distinct, then F_e is the distribution function of the random variable X_e such that

$$P(X_e = x_i) = 1/n, \quad i = 1, \ldots, n$$

When the data values are not all distinct, then F_e is the distribution function of the random variable X_e whose probability of being equal to any specified data value is the number of times that value appears in the data set divided by n. For instance, if $n = 3$ and $x_1 = x_2 = 1, x_3 = 2$ then X_e is a random variable that takes on the value 1 with probability 2/3 and 2 with probability 1/3. With this understanding about the weight put on a distinct value, we will still say that F_e is the distribution function of a random variable that is equally likely to be any of the values $x_1, x_2, \ldots, x_n$.

Now, for any value of x, each of the data values X_i, $i = 1, \ldots, n$, will be less than or equal to x with probability $F(x)$. Hence, by the strong law of large numbers it follows that the proportion of them that are less than or equal to x will, with probability 1, converge to $F(x)$ as n goes to infinity. Thus, for n large, $F_e(x)$ should be close to $F(x)$, indicating that the empirical distribution function F_e can be used as an estimator of the population distribution function F.

Now let θ_e have the same relationship to the distribution F_e as θ has to the distribution F. For instance, if θ is the variance of a random variable X having distribution F, then θ_e is the variance of a random variable X_e having distribution F_e. Now, if F_e is close to F, then it almost always follows that θ_e will be close to θ. (Technically speaking, this will be true provided that θ is a continuous function of the distribution F.) For these reasons we can approximate the mean square error of the estimator $d(X_1, \ldots, X_n)$ of θ as follows:

$$MSE_F(d) = E_F[(d(X_1, \ldots, X_n) - \theta)^2] \approx E_{F_e}[(d(X_1, \ldots, X_n) - \theta_e)^2]$$

where by E_{F_e} we mean that the expectation is to be taken under the assumption that $X_1, \ldots, X_n$ are independent random variables, each having distribution function F_e. That is, each of $X_1, \ldots, X_n$ is equally likely to be any of the values $x_1, \ldots, x_n$.

The quantity

$$MSE_{F_e}(d) = E_{F_e}[(d(X_1, \ldots, X_n) - \theta_e)^2]$$

is called the *bootstrap estimate of the mean square error* of $d(X_1, \ldots, X_n)$ as an estimator of θ.

Let us now see how well $MSE(F_e)$ estimates $MSE(F)$ in the one case where its use as an estimator is not needed—namely, when estimating the mean of a distribution by the sample mean.

EXAMPLE 15.3b Consider Example 15.3a, where $\bar{X}_n = \sum_{i=1}^n X_i/n$ is used as an estimator of the mean of the distribution F. Because X_e puts equal weight on each of the data values $x_1, \ldots, x_n$, it follows, when θ_e is the mean of this distribution, that

$$\theta_e = E[X_e] = \sum_{i=1}^n x_i P(X_e = x_i) = \frac{1}{n} \sum_{i=1}^n x_i = \bar{x}_n$$

Because

$$E_{F_e}[\sum_{i=1}^n X_i/n] = E_{F_e}[X] = \theta_e = \bar{x}_n$$

it follows that

$$MSE_{F_e}(\bar{X}_n) = E_{F_e}[(\sum_{i=1}^n X_i/n - \bar{x}_n)^2]$$

$$= \text{Var}_{F_e}(\sum_{i=1}^n X_i/n)$$

$$= \frac{1}{n} \text{Var}_{F_e}(X)$$

Now,

$$\text{Var}_{F_e}(X) = E_{F_e}[(X - \bar{x}_n)^2]$$

$$= \sum_{i=1}^{n} (x_i - \bar{x}_n)^2 P_{F_e}(X = x_i)$$

$$= \frac{1}{n} \sum_{i=1}^{n} (x_i - \bar{x}_n)^2$$

Thus, we have shown that

$$MSE_{F_e}(\bar{X}_n) = \frac{1}{n^2} \sum_{i=1}^{n} (x_i - \bar{x}_n)^2$$

As the usual estimator of $MSE_F(\bar{X}_n) = \frac{1}{n} \text{Var}_F(X)$ is S_n^2/n, whose observed value is $\frac{1}{n(n-1)} \sum_{i=1}^{n} (x_i - \bar{x}_n)^2$, we see that the bootstrap estimate is almost identical to the usual estimate in this case. ∎

As previously noted, if the data values are $X_i = x_i, i = 1, \ldots, n$, then the empirical distribution function F_e puts equal weight $1/n$ on each of the points x_i; consequently, it is usually easy to compute the value of θ_e. To compute the bootstrap estimate of the mean square error of the estimator $d(X_1, \ldots, X_n)$ of θ, we then have to compute

$$MSE_{F_e}(d) = E_{F_e}[(d(X_1, \ldots, X_n) - \theta_e)^2]$$

However, since the preceding expectation is to be computed under the assumption that $X_1, \ldots, X_n$ are all distributed according to F_e, it follows that the vector $(X_1, \ldots, X_n)$ is equally likely to be any of the n^n possible values $(x_{i_1}, x_{i_2}, \ldots, x_{i_n})$, where each i_j is one of the values $1, \ldots, n$. Consequently an exact computation of $MSE_{F_e}(d)$ is prohibitive unless n is small.

It is, however, easy to approximate $MSE_{F_e}(d)$ by a simulation. To do so, we generate n independent random variables $X_1^1, \ldots, X_n^1$ having distribution F_e and use them to compute the value of

$$Y_1 = (d(X_1^1, \ldots, X_n^1) - \theta_e)^2$$

We then repeat this process and generate a second set of n independent random variables $X_1^2, \ldots, X_n^2$ having distribution F_e and use them to compute the value of

$$Y_2 = (d(X_1^2, \ldots, X_n^2) - \theta_e)^2$$

This is then repeated a large number of times, say r, to obtain the values $Y_1, \ldots, Y_r$. The average of these values, $\sum_{i=1}^{r} Y_i/r$, would be the approximation of $MSE_{F_e}(d)$, which would then be used as the estimate of $MSE_F(d)$.

REMARK

It is easy to generate a random variable X having distribution F_e. Just generate a random number U; let $I = \text{Int}(nU) + 1$, so that I is equally likely to be any of the values $1, \ldots, n$; and then set

$$X = x_I$$

EXAMPLE 15.3c Suppose we use the sample variance $S_n^2 = \sum_{i=1}^n (X_i - \bar{X}_n)^2/(n-1)$ of a sample of size n from the distribution F as an estimator of σ^2, the variance of the distribution F. To estimate the mean square error of the sample variance, let the observed data be $X_i = x_i, i = 1, \ldots, n$.

Because the distribution F_e puts equal weight on all of the values $x_i, i = 1, \ldots, n$, it follows that

$$E_{F_e}[X] = \sum_{i=1}^n x_i P_{F_e}(X = x_i) = \sum_{i=1}^n x_i/n = \bar{x}_n$$

showing that θ_e, the variance of the distribution F_e, is given by

$$\theta_e = \text{Var}_{F_e}(X) = E_{F_e}[(X - \bar{x}_n)^2] = \sum_{i=1}^n (x_i - \bar{x}_n)^2/n$$

Consequently,

$$MSE_{F_e}(S_n^2) = E_{F_e}[(S_n^2 - \theta_e)^2] = E_{F_e}\left[\left(\frac{\sum_{i=1}^n (X_i - \bar{X}_n)^2}{n-1} - \theta_e\right)^2\right]$$

To approximate $MSE_{F_e}(S_n^2)$, we use simulation.

For instance, suppose $n = 8$ and the data values were $x_1 = 5, x_2 = 9, x_3 = 12, x_4 = 8, x_5 = 7, x_6 = 15, x_7 = 3, x_8 = 6$. Then

$$\bar{x}_8 = 8.125, \qquad \theta_e = \sum_{i=1}^8 (x_i - \bar{x}_8)^2/8 \approx 13.11$$

In the following approach for the simulation-based approximation of $MSE_{F_e}(S_n^2)$, the $x_i, i = 1, \ldots, 8$ are as given in the preceding. There are to be a total of r simulation runs, with the variable N representing the number of the current simulation run. In each run we generate the values of 8 random variables $X_M, M = 1, \ldots, 8$, distributed according to F_e. The quantities S and SS represent running totals of, respectively, the sum of the X_M and the sum of the squares of the X_M so far generated in the run. When the run is completed, the sample variance SV is computed by using the identity

$$\frac{\sum_{i=1}^8 (X_i - \bar{X}_8)^2}{7} = \frac{\sum_{i=1}^8 X_i^2 - 8\bar{X}_8^2}{7} = \frac{\sum_{i=1}^8 X_i^2 - (\sum_{i=1}^8 X_i)^2/8}{7} = \frac{SS - S^2/8}{7}$$

The squared difference between SV and $\theta_e = 13.11$ is computed and then added to T, the sum of the $N - 1$ previous squared differences. When r runs have been completed the simulation is ended; the average of the squared differences between the sample variances and θ_e is the simulation-based approximation to $MSE_{F_e}(S_n^2)$.

1. Let $T = 0, N = 1$
2. Let $M = 1$
3. $S = 0$, $SS = 0$
4. Generate a random number U
5. Set $I = \text{Int}(8U) + 1$
6. $S = S + x_I$
7. $SS = SS + x_I^2$
8. If $M < 8$, set $M = M + 1$ and go to 4
9. $SV = (SS - S^2/8)/7$
10. Let $T = T + (SV - 13.11)^2$
11. If $N < r$, set $N = N + 1$ and go to 2
12. If $N = r$, return T/r as the approximation to $MSE_{F_e}(S_n^2)$ ∎

Now suppose we wanted to estimate not the mean square error of the estimator but rather the probability that the estimator of θ will be within h of the actual value of θ. That is, suppose we want to estimate

$$p_h \equiv P_F(|d(X_1, \ldots, X_n) - \theta| \leq h)$$

To obtain an estimator of the preceding, we use that

$$P_F(|d(X_1, \ldots, X_n) - \theta| \leq h) \approx P_{F_e}(|d(X_1, \ldots, X_n) - \theta_e| \leq h)$$

and then employ simulation to estimate the right side of the preceding. That is, after the data $X_1, \ldots, X_n$ are observed to take on the values $X_i = x_i, i = 1, \ldots, n$, we let F_e be the empirical distribution. That is, F_e is the distribution function of a random variable that is equally likely to take on any of the values $x_1, \ldots, x_n$. We next compute the value of θ_e. We then continually generate sets of n independent random variables from the distribution F_e. For each set of values obtained, we compute d evaluated at these values, and check whether this quantity is within h of θ_e. The fraction of times that it is within h is our simulation-based estimate of

$$P_{F_e}(|d(X_1, \ldots, X_n) - \theta_e| \leq h)$$

and is also what we use to estimate p_h.

More specifically, we use the original data to obtain F_e and the resulting value of θ_e. We then decide on the number of simulation runs (typically between 10^4 and 10^5 will suffice, but see Section 15.7 for specifics on how to determine the number of runs that should be performed). With r runs, we need to generate r sets of n independent random variables from the distribution F_e. With $x_{i,1}, \ldots, x_{i,n}$ being the ith set of values generated, we compute the value of $d_i = d(x_{i,1}, \ldots, x_{i,n})$. The proportion of the values of $i, i = 1, \ldots, r$, for which $|d_i - \theta_e| \leq h$ is our estimate of $p_h \equiv P_F(|d(X_1, \ldots, X_n) - \theta| \leq h)$.

EXAMPLE 15.3d   The following are the PSAT math scores of a random sample of 16 students from a certain school district.

$$522, 474, 644, 708, 466, 534, 422, 480, 502, 655, 418, 464, 600, 412, 530, 564$$

Use them to estimate

(a) the average score of all students in the district;

(b) the probability that the estimator of the district average will be within 5 of the actual district average;

(c) the probability that the estimator of the district average will be within 10 of the actual district average.

SOLUTION   We suppose that the data constitute a random sample from a distribution F with mean $\theta(F) = \mu$. The natural estimator of μ is the sample average $\bar{X}$, yielding the estimate

$$\theta_e = \bar{x} = 524.7$$

The probability p_h, that the sample mean of a sample of size 16 will be within h of the population mean, is estimated by

$$P_{F_e}(|\bar{X}_{16} - \theta_e| \leq h) = P_{F_e}(|\bar{X}_{16} - 524.7| \leq h)$$

where $\bar{X}_{16}$ is the average of a sample of size 16 from the distribution that puts probability $1/16$ on each of the original 16 data values. A simulation based on 10^5 simulation runs— with each run generating a sample of size 16 from F_e—yielded the estimates .1801 and .3542 for $h = 5$ and $h = 10$, respectively.

Because we are estimating the mean of the distribution by the sample average, we could also approximate the probability p_h by making use of the central limit theorem. With μ and σ being the mean and standard deviation of F, the probability that the sample mean of a sample of size 16 is within h of μ can be approximated by using the fact that $\bar{X}_{16}$ approximately has a normal distribution with mean μ and variance $\sigma^2/16$. Consequently, with Z being a standard normal random variable

$$P(-h \leq \bar{X}_{16} - \mu \leq h) = P(\frac{-h}{\sigma/4} \leq \frac{\bar{X}_{16} - \mu}{\sigma/4} \leq \frac{d}{\sigma/4})$$
$$\approx P(-4h/\sigma \leq Z \leq 4h/\sigma)$$
$$= 2\,\Phi(4h/\sigma) - 1$$

An easy calculation gives that the sample standard deviation of the 16 data values is $s = 89.1$. Taking this value as an approximation of σ yields that

$$2\,\Phi(4h/\sigma) - 1 \approx 2\,\Phi(4h/89.1) - 1$$

Thus, the estimate of the probability that the sample mean is within 5 of the population mean is $2\Phi(.2245) - 1 = .1776$, whereas the estimate that it is within 10 of the population mean is $2\Phi(.4490) - 1 = .3466$, which are quite close to the ones obtained by the nonparametric bootstrap approach. However, it should be noted that the central limit theorem approximation would not be available to us if we were estimating some other parameter of the distribution aside from its mean.

For instance, suppose we wanted to use the 16 data values to estimate σ, the standard deviation of the scores of all the student in the district. Using the sample standard deviation as the estimator yields the estimate $s = 89.1$. Now suppose we wanted to estimate the probability that our estimator will be within 10 of σ. That is, suppose we wanted to estimate the probability that the sample standard deviation of a sample of size 16 from the distribution F will be within 10 of the actual standard deviation of F. To do so, we estimate this by the probability that the sample standard deviation of a sample of size 16 from the empirical distribution F_e is within 10 of the standard deviation of F_e. Now, because the distribution F_e puts equal weight on each of the 16 values $x_1, \ldots, x_{16}$, its mean is $\bar{x}$ and its standard deviation is

$$\sigma_e = \sqrt{E_{F_e}[(X - \bar{x})^2]} = \sqrt{\frac{1}{16}\sum_{i=1}^{16}(x_i - \bar{x})^2} = 89.1\sqrt{15/16} = 86.27$$

Consequently the estimate of the probability that the sample standard deviation differs from σ by at most 10 is

$$P_{F_e}(|S_{16} - \sigma_e| \le 10) = P_{F_e}(|S_{16} - 86.27| \le 10)$$

where S_{16} is the sample standard deviation of a sample of size 16 from the distribution F_e. This probability can be approximated by a simulation. Indeed, a simulation performed with 10^5 runs yielded the result

$$P_{F_e}(|S_{16} - 86.27| \le 10) \approx .5424$$

so there is roughly a 54 percent chance that the actual standard deviation of all student scores is within 79.1 and 99.1.   ∎

15.4 PERMUTATION TESTS

Suppose we want to test the null hypothesis H_0 that the data $X_1, \ldots, X_N$ is a sample from some unspecified distribution. *Permutation tests* are tests of this hypothesis in which the p-value is computed conditional on knowing the set $\mathcal{S}$ of data values observed but without knowing which data value corresponds to X_1, which corresponds to X_2 and so on. For instance, if $N = 3$ and $X_1 = 5, X_2 = 7, X_3 = 2$, then the p-value is computed conditional on the information that the set of data values is $\mathcal{S} = \{2, 5, 7\}$. The computation of the

p-value makes use of the fact that, conditional on the set of data values $\mathcal{S}$, each of the $N!$ possible ways of assigning these N values to the original data is equally likely when the null hypothesis is true. That is, suppose that $N = 3$ and the set of data values is, as in the preceding, $\mathcal{S} = \{2, 5, 7\}$. Now the null hypothesis H_0 states that X_1, X_2, X_3 are independent and identically distributed. Consequently, if H_0 is true then, given the data set $\mathcal{S}$, it follows that the vector (X_1, X_2, X_3) is equally likely to equal any of the $3!$ permutations of the values $2, 5, 7$.

The implementation of a permutation test is as follows. Depending on the alternative hypothesis, a test statistic $T(X_1, \ldots, X_N)$ is chosen. Suppose, for the moment, that large values of the test statistic are evidence for the alternative hypothesis. The data values are then observed, say that $X_i = x_i, i = 1, \ldots, N$, and the value of $T(x_1, \ldots, x_N)$ is calculated. Now let $\mathcal{S} = \{x_1, \ldots, x_N\}$ be the unordered set consisting of the N observed values. Then, if the value of the test statistic is $T(x_1, \ldots, x_N) = t$, the resulting p-value of the null hypothesis that results from these data is

$$p\text{-value} = P_{H_0}(T(X_1, \ldots, X_N) \geq t | \mathcal{S} = \{x_1, \ldots, x_N\})$$

Now, under H_0, $X_1, \ldots, X_N$ is equally likely to equal any of the $N!$ permutations of $x_1, \ldots, x_N$. Consequently, letting $I_1, \ldots, I_N$ be a random vector that is equally likely to be any of the $N!$ permutations of $1, \ldots, N$, we can write the preceding p-value as

$$p\text{-value} = P\{T(x_{I_1}, x_{I_2}, \ldots, x_{I_N}) \geq t\}$$
$$= \frac{\text{number of permutations } (i_1, \ldots, i_N) : T(x_{i_1}, x_{i_2}, \ldots, x_{i_N}) \geq t}{N!}$$

For an illustration, suppose we are to observe data over N weeks, with X_i being the data value observed in week $i, i = 1, \ldots, N$, and that we want to use these data to test the null hypothesis

$$H_0 : X_1, \ldots, X_N \text{ are independent and identically distributed}$$

against

$$H_1 : X_i \text{ tends to increase as } i \text{ increases}$$

Now if the null hypothesis is true and the data are independent and identically distributed, then, conditional on knowing the set of values $X_1, \ldots, X_N$, but not knowing which value corresponds to X_1 or which corresponds to X_2 and so on, the statistic $\sum_{j=1}^{N} jX_j$ would be distributed as if we randomly paired up the two data sets $\{1, \ldots, N\}$ and $\{X_1, \ldots, X_N\}$ and then summed the products of the N paired values. On the other hand, if the alternative hypothesis were true, then $\sum_{j=1}^{N} jX_j$ would tend to be larger than if we just randomly paired the values $1, \ldots, N$ with the values $X_1, \ldots, X_N$ and then summed the products of the N pairs. This is because the sum of the paired values of two sets of equal size is largest when the largest values are paired with each other, the second largest are paired with each other, and so on. (In statistical terms the correlation coefficient of data pairs

$(j, X_j), j = 1, \ldots, N$ is large when the X_j tend to increase as j increases.) Consequently, one possible permutation test of H_0 versus H_1 is to

1. Observe the data values—say that $X_j = x_j, j = 1, \ldots, N$
2. Let $t = \sum_{j=1}^{N} j x_j$
3. Determine the p-value given by

$$p\text{-value} = P(\sum_{j=1}^{N} I_j x_j \geq t)$$

where $I_1, \ldots, I_N$ is equally likely to be any of the $N!$ permutations of $1, \ldots, N$.

The p-value in the preceding can be approximated by a simulation that uses the method of Example 15.2b to generate random permutations.

EXAMPLE 15.4a To determine if the weekly sales of DVD players is on a downward trend, the manager of a large electronics store has been tracking such sales for the past 12 weeks, with the following sales figures from week 1 to week 12 (the current week) resulting:

$$22, 24, 20, 18, 16, 14, 15, 15, 13, 17, 12, 14$$

Are the data strong enough to reject the null hypothesis that the distribution of sales is unchanging in time, and so enable the manager to conclude that there is a downward trend in sales?

SOLUTION Let the null hypothesis be that the distribution of sales is unchanged over time, and let the alternative hypothesis be that there is a downward trend in sales. Thus, if the alternative hypothesis is true then there would be a negative correlation between X_j, the sales during week j, and j. So a relatively small value of $\sum_{j=1}^{12} j X_j$ would be evidence in favor of the alternative hypothesis. Now, with x_j equal to the observed value of X_j, the sales data give that

$$\sum_{j=1}^{12} j x_j = 1{,}178$$

Hence, the p-value of the permutation test of the null hypothesis that the data come from the same distribution versus the alternative that the data tend to be decreasing in time is given by

$$p\text{-value} = P(\sum_{j=1}^{12} I_j x_j \leq 1{,}178)$$

where $I_1, \ldots, I_{12}$ is equally likely to be any of the 12! permutations of $1, \ldots, 12$. A simulation, using 10^5 runs, yielded that

$$p\text{-value} \approx .00039$$

leading us to reject the null hypothesis that the distribution is unchanging over time. ∎

Although $\sum_{j=1}^{N} jX_j$ is the test statistic most commonly used to test the null hypothesis that $X_1, \ldots, X_n$ are independent and identically distributed against the alternative that X_j tends to increase as j increases, it is not the only possibility. Indeed, we could have chosen any test statistic of the form $\sum_{j=1}^{N} a_j X_j$, where $a_1 < a_2 < \ldots < a_n$. (For instance, we could have chosen $a_j = j^2$.) Analogous to the preceding, the value of the statistic would first be determined, say it is t. Because the alternative hypothesis will tend to make $\sum_{j=1}^{N} a_j X_j$ larger than it would be under the null hypothesis—since large values of the a_j would tend to be paired with large data values when the alternative hypothesis is true—we would again want to reject the null hypothesis when t is large. Consequently, the resulting p-value would be

$$p\text{-value} = P\left(\sum_{j=1}^{N} a_{I_j} x_j \geq t \right)$$

where $I_1, \ldots, I_N$ is equally likely to be any of the $N!$ permutations of $1, \ldots, N$.

Depending on the alternative hypothesis, we could choose other constants $a_j, j = 1, \ldots, N$ to test the null hypothesis that the data values are independent and identically distributed. For instance, if the alternative was that the data tended to be higher in the middle values and lower in the extremes, then we could let the test statistic be of the form $T = \sum_{j=1}^{N} a_j X_j$, where $a_1, \ldots, a_N$ is such that its middle values tend to be larger than its earlier or later values. For instance, we could use $a_j = j(N - j), j = 1, \ldots, N$. As this would again make it more likely that larger data values are paired with larger constants when the alternative hypothesis is true, we would again want to reject the null hypothesis when T is large.

15.4.1 NORMAL APPROXIMATIONS IN PERMUTATION TESTS

Although not as accurate as doing a simulation, the p-value of a permutation test can be approximated by assuming that the test statistic is approximately normally distributed. Now, under the null hypothesis that the data values are independent and identically distributed, it follows that, given the data set $\mathcal{S} = \{x_1, \ldots, x_N\}$, the random variable X_i is equally likely to be any of these N values and the random vector $(X_i, X_j), i \neq j$ is equally likely to take on any of the $N(N - 1)$ values $x_k x_r, r \neq k$. Consequently, given $\mathcal{S} = \{x_1, \ldots, x_N\}$,

$$E[X_i] = \frac{1}{N} \sum_{i=1}^{N} x_i = \bar{x}$$

$$E[X_i^2] = \frac{1}{N} \sum_{i=1}^{N} x_i^2$$

$$E[X_i X_j] = \frac{1}{N(N - 1)} \sum_{k} \sum_{r \neq k} x_k x_r$$

$$= \frac{1}{N(N-1)} \left(\sum_k \sum_r x_k x_r - \sum_k \sum_{r=k} x_k x_r \right)$$

$$= \frac{1}{N(N-1)} \left(\sum_k x_k \sum_r x_r - \sum_k x_k^2 \right)$$

$$= \frac{1}{N(N-1)} \left(N^2 \bar{x}^2 - \sum_{k=1}^N x_k^2 \right)$$

So, with $v = \mathrm{Var}(X_i)$ and $c = \mathrm{Cov}(X_i, X_j)$, $i \neq j$, the preceding yields

$$E[X_i] = \bar{x}$$

$$v = \mathrm{Var}(X_i) = \frac{1}{N} \sum_{i=1}^N x_i^2 - \bar{x}^2$$

$$c = \mathrm{Cov}(X_i, X_j) = \frac{1}{N(N-1)} \left(N^2 \bar{x}^2 - \sum_{k=1}^N x_k^2 \right) - \bar{x}^2$$

$$= \frac{\bar{x}^2}{N-1} - \frac{1}{N(N-1)} \sum_{k=1}^N x_k^2$$

$$= \frac{1}{N-1} \left(\bar{x}^2 - \sum_{k=1}^N x_k^2/N \right)$$

which also shows that

$$v - c = \frac{\sum_{i=1}^N x_i^2 - N\bar{x}^2}{N-1}$$

Therefore, when H_0 is true, the test statistic $T = \sum_{j=1}^N j X_j$ has mean

$$E[T] = \frac{N(N+1)}{2} \bar{x}$$

and variance

$$\mathrm{Var}(T) = \mathrm{Var}\left(\sum_{j=1}^N j X_j \right)$$

$$= \sum_{j=1}^N \mathrm{Var}(j X_j) + \sum_i \sum_{j \neq i} \mathrm{Cov}(i X_i, j X_j)$$

$$= v \sum_{j=1}^N j^2 + c \sum_i \sum_{j \neq i} ij$$

$$= v \sum_{j=1}^{N} j^2 + c \left(\sum_i \sum_j ij - \sum_i \sum_{j=i} ij \right)$$

$$= v \sum_{j=1}^{N} j^2 + c \left(\sum_{i=1}^{N} i \sum_{j=1}^{N} j - \sum_{i=1}^{N} i^2 \right)$$

$$= (v - c) \sum_{j=1}^{N} j^2 + \frac{cN^2(N+1)^2}{4}$$

$$= (v - c) \frac{N(N+1)(2N+1)}{6} + \frac{cN^2(N+1)^2}{4}$$

Using the preceding we can approximate the p-value of a permutation test by assuming that the distribution of T, when H_0 is true, is approximately normal.

EXAMPLE 15.4b Again consider Example 15.4a. A calculation yields that, under H_0,

$$E[T] = 1,300 \quad \text{Var}(T) = 1,958.81$$

Thus, the normal approximation yields that

$$p\text{-value} = P_{H_0}(T \leq 1,178)$$
$$= P_{H_0}\left(\frac{T - 1,300}{\sqrt{1,958.81}} \leq \frac{1,178 - 1,300}{\sqrt{1,958.81}} \right)$$
$$\approx \Phi(-2.757)$$
$$= .0029$$

which is quite close to the value given by the simulation.

Let us now suppose that whereas the set of 12 data values was as before, they now appeared in the order

$$22, 14, 14, 16, 24, 20, 18, 15, 17, 15, 12, 13$$

With these data, the value of the test statistic is $\sum_{j=1}^{12} j X_j = 1,233$, and the normal approximation yields that

$$p\text{-value} = P_{H_0}(T \leq 1,233)$$
$$= P_{H_0}\left(\frac{T - 1,300}{\sqrt{1,958.81}} \leq \frac{1,233 - 1,300}{\sqrt{1,958.81}} \right)$$
$$\approx \Phi(-1.514)$$
$$= .065$$

Finally, suppose again that the set of 12 data values was as before, but suppose that they now appeared in the order

$$22, 14, 14, 16, 24, 13, 18, 15, 17, 15, 12, 20$$

In this case, the value of the test statistic is $\sum_{j=1}^{12} jX_j = 1275$. Thus, the normal approximation yields that

$$p\text{-value} = P_{H_0}(T \leq 1{,}275)$$
$$= P_{H_0}\left(\frac{T - 1{,}300}{\sqrt{1{,}958.81}} \leq \frac{1{,}275 - 1{,}300}{\sqrt{1{,}958.81}}\right)$$
$$\approx \Phi(-.565)$$
$$\approx .286$$

A simulation of 10^5 runs yielded values quite similar to the preceding. The simulation gave

$$P_{H_0}(T \leq 1{,}233) \approx .068$$

and

$$P_{H_0}(T \leq 1{,}275) \approx .299$$

which are quite close to the values given by the normal approximation. ∎

EXAMPLE 15.4c For another indication as to the validity of a normal approximation, suppose that $N = 4$, with the data appearing in the following order:

$$13, 7, 5, 3$$

Suppose that we want to use these data to test the null hypothesis that the data are a sample from some distribution against the alternative hypothesis that the data tend to be decreasing. The value of the test statistic is $T = \sum_{j=1}^{r} jX_j = 54$. An easy computation yields

$$c = -4.667, \quad v = 14$$

showing that, under H_0,

$$E[T] = 70, \quad \text{Var}(T) = 93.33$$

Consequently, with Z being a standard normal random, the normal approximation yields that

$$p\text{-value} = P_{H_0}(T \leq 54)$$
$$= P_{H_0}\left(\frac{T - 70}{\sqrt{93.33}} \leq \frac{54 - 70}{\sqrt{93.33}}\right)$$
$$\approx P(Z \leq -1.656)$$
$$= .049$$

whereas the exact value is

$$p\text{-value} = P_{H_0}(T \le 54) = 1/4! \approx .042 \quad \blacksquare$$

15.4.2 TWO-SAMPLE PERMUTATION TESTS

Permutation tests are also useful in the two-sample problems where we test whether samples from two populations have the same underlying distribution. Specifically, let $X_1, \ldots, X_n$ be a sample from an unknown population distribution F, and let $X_{n+1}, \ldots, X_{n+m}$ be an independent sample from an unknown population distribution G, and suppose we want to use these data to test the hypothesis that the two population distributions are identical against the alternative hypothesis that data from the second distribution tend to be larger than those from the first. That is, we want to use these data to test the null hypothesis

$$H_0 : F = G$$

against the alternative

$$H_1 : \text{ data from } G \text{ tend to be larger than data from } F$$

If the data values are $X_i = x_i, i = 1, \ldots, n+m$, then a permutation test of the preceding null hypothesis is done conditional on knowing $\mathcal{S} = \{x_1, \ldots, x_{n+m}\}$, the set of these $n + m$ numbers in no particular order. Then if H_0 is true, and so all $n + m$ random variables $X_1, \ldots, X_{n+m}$ are independent and identically distributed, then given the set of values $\mathcal{S}$, each subset of size n of this set is equally likely to be the set of the data values of $X_1, \ldots, X_n$. Because the alternative hypothesis is that data from the population distribution F tend to be smaller than data from the population distribution G, a reasonable test would be to reject the null hypothesis if the sum of the data values from the population distribution F is smaller than might be expected by chance when n values are randomly chosen from the data set $\mathcal{S}$. More specifically, we can test H_0 by computing $\sum_{i=1}^{n} x_i$; say its value is t. Then the p-value of this permutation test of H_0 versus H_1 would equal the probability that a random selection of n of the values $x_1, \ldots, x_{n+m}$ would be less than or equal to t. That is,

$$p\text{-value} = P\left(\sum_{i \in R} x_i \le t\right)$$

where R is equally likely to be any of the $\binom{n+m}{n}$ subsets of size n from the set $\{1, 2, \ldots, n+m\}$. Whereas an exact computation of the preceding is possible only when $\binom{n+m}{n}$ is small, a precise approximation is easily obtained by simulation. In each simulation run we use the method of Example 15.2a to randomly generate a subset of n of the values $1, \ldots, n + m$. If R is the subset obtained, then we check whether $\sum_{i \in R} x_i$ is less than or equal to t. The fraction of simulation runs for which this is the case is our estimate of the preceding p-value.

REMARK

We can again use a normal approximation, rather than a simulation, to estimate the p-value. Starting with the random subset R, equally likely to be any of the $\binom{n+m}{n}$ subsets of size n from the set $\{1, 2, \ldots, n+m\}$, let, for $i = 1, \ldots, n+m$,

$$I_i = \begin{cases} 1, & \text{if } i \in R \\ 0, & \text{if } i \notin R \end{cases}$$

Then,

$$\sum_{i \in R} x_i = \sum_{i=1}^{n+m} x_i I_i$$

By a similar analysis as in Section 15.4.1, we can now show that

$$E\left[\sum_{i \in R} x_i\right] = E\left[\sum_{i=1}^{n+m} x_i I_i\right] = n\bar{x}$$

and

$$\text{Var}\left(\sum_{i \in R} x_i\right) = \text{Var}\left(\sum_{i=1}^{n+m} x_i I_i\right) = \frac{nm}{n+m-1}\left(\frac{\sum_{i=1}^{n+m} x_i^2}{n+m} - \bar{x}^2\right)$$

where $\bar{x} = \sum_{i=1}^{n+m} x_i/(n+m)$.

15.5 GENERATING DISCRETE RANDOM VARIABLES

Suppose we want to generate the value of a random variable X having probability mass function

$$P(X = x_i) = p_i, i = 1, \ldots, \sum_i p_i = 1$$

To generate the value of X, generate a random number U and set

$$X = x_i \quad \text{if} \quad p_1 + \ldots p_{i-1} < U \leq p_1 + \ldots p_{i-1} + p_i$$

That is,

$$X = \begin{cases} x_1, & \text{if } U \leq p_1 \\ x_2, & \text{if } p_1 < U \leq p_1 + p_2 \\ x_3, & \text{if } p_1 + p_2 < U \leq p_1 + p_2 + p_3 \\ \cdot \\ \cdot \\ \cdot \\ x_i, & \text{if } p_1 + \ldots + p_{i-1} < U \leq p_1 + \ldots + p_{i-1} + p_i \\ \cdot \\ \cdot \\ \cdot \end{cases}$$

Because U is uniformly distributed on $(0, 1)$, it follows that for $0 < a < b < 1$

$$P(a < U \leq b) = b - a$$

Consequently,

$$P\left(\sum_{j=1}^{i-1} p_j < U \leq \sum_{j=1}^{i} p_j \right) = p_i$$

which shows that X has the desired probability mass function. This method of generating X is called the *discrete inverse transform method*.

EXAMPLE 15.5a To generate a Bernoulli random variable X such that

$$P(X = 1) = p = 1 - P(X = 0)$$

generate a random number U, and set

$$X = \begin{cases} 1, & \text{if } U \leq p \\ 0, & \text{if } U > p. \end{cases} \quad \blacksquare$$

EXAMPLE 15.5b Suppose now that we wanted to generate a binomial random variable X with parameters n and p. Recalling that X represents the number of successes in n independent trials when each trial is a success with probability p, we can generate X by generating the results of the n trials. That is, we can generate n random numbers $U_1, \ldots, U_n$, say that trial i is a success if $U_i \leq p$, and then set

$$X = \text{number of } i : U_i \leq p$$

Another possibility is to use the inverse transform method.

To efficiently use the inverse transform method we need an efficient method to recursively compute the values

$$p_i = P(X = i) = \binom{n}{i} p^i (1 - p)^{n-i}, \ i = 0, \ldots, n$$

This is accomplished by first noting that

$$\frac{\binom{n}{i+1}}{\binom{n}{i}} = \frac{n!}{(n-i-1)!\,(i+1)!} \frac{(n-i)!\,i!}{n!}$$

$$= \frac{n-i}{i+1}$$

which yields that

$$\frac{p_{i+1}}{p_i} = \frac{n-i}{i+1} \frac{p^{i+1}(1-p)^{n-i-1}}{p^i(1-p)^{n-i}}$$

$$= \frac{n-i}{i+1} \frac{p}{1-p}$$

Thus,

$$p_{i+1} = \frac{n-i}{i+1} \frac{p}{1-p} p_i$$

Using the preceding, we are now ready to give the inverse transform method for generating a binomial (n, p) random variable X. In the following, i represents the possible value of X, the variable P is the probability that $X = i$, and the variable F is the probability that $X \le i$. (That is, for given i, $P = p_i$ and $F = \sum_{j=0}^{i} p_j$.) Also, let $\alpha = p_0 = (1-p)^n$, and let $b = \frac{p}{1-p}$.

1. Set $i = 0$, $P = \alpha$, $F = \alpha$
2. Generate a random number U
3. If $U \le F$ set $X = i$ and stop
4. $P = \frac{n-i}{i+1} b P$
5. $F = F + P$
6. $i = i + 1$
7. Go to 3

(In the preceding, when we say that $P = \frac{n-i}{i+1} b P$, we don't mean this literally as an algebraic identity; rather we mean that the value of P is to be changed. Its new value is its old value multiplied by $\frac{n-i}{i+1} b$. Similarly, when we write $F = F + P$ we mean that the value of F is to be changed by adding P to its old value.)

Because the algorithm first checks whether $X = 0$, then whether $X = 1$, and so on, it follows that the number of iterations needed (that is, the number of times that it goes to step 3) is one more than the final value of X. So, on average, this algorithm requires $E[X + 1] = np + 1$ iterations to generate the value of X. ■

15.6 GENERATING CONTINUOUS RANDOM VARIABLES

Let F be the distribution function of a continuous random variable. For any u between 0 and 1, the quantity $F^{-1}(u)$ is defined to be that value x such that $F(x) = u$. That is, $F(F^{-1}(u)) = u$. Because the distribution function of a continuous random variable is strictly increasing, it follows that there is a unique value of $F^{-1}(u)$. We call F^{-1} the inverse function of F.

A general method for generating a continuous random variable having distribution function F, known as the *inverse transformation method*, is based on the following proposition.

PROPOSITION 15.6.1 Let U be a uniform $(0, 1)$ random variable. For any continuous distribution function F, if we define

$$X = F^{-1}(U)$$

then X has distribution function F.

Proof

Because a distribution function F is nondecreasing, it follows that for any numbers a and b the inequality $a \leq b$ is equivalent to the inequality $F(a) \leq F(b)$. Consequently,

$$
\begin{aligned}
P(F^{-1}(U) \leq x) &= P(F(F^{-1}(U)) \leq F(x)) \\
&= P(U \leq F(x)) \\
&= F(x)
\end{aligned}
$$

thus showing that $F^{-1}(U)$ has distribution F. ∎

EXAMPLE 15.6a (Generating an Exponential Random Variable)   Let

$$F(x) = 1 - e^{-\lambda x}, \quad x \geq 0$$

be the distribution function of an exponential random variable with parameter λ. Then $F^{-1}(u)$ is that value x such that

$$u = F(x) = 1 - e^{-\lambda x}$$

or, equivalently,

$$e^{-\lambda x} = 1 - u$$

or

$$-\lambda x = \log(1 - u)$$

or

$$x = -\frac{1}{\lambda} \log(1 - u)$$

So, by Proposition 15.6.1, we can generate an exponential random variable X with parameter λ by generating a uniform $(0, 1)$ random variable U and setting

$$X = -\frac{1}{\lambda} \log(1 - U)$$

Because $1 - U$ is also a uniform $(0, 1)$ random variable, it follows that $-\frac{1}{\lambda} \log(1 - U)$ and $-\frac{1}{\lambda} \log(U)$ have the same distribution, thus showing that

$$X = -\frac{1}{\lambda} \log(U)$$

is also exponential with parameter λ. ∎

15.6.1 Generating a Normal Random Variable

Because inverting the distribution function of a normal random variable is computationally involved, special methods are used for generating normal random variables. The following one is known as the *Box-Muller method*.

To begin, suppose that X and Y are independent standard normal random variables, so their joint density function is

$$f(x,y) = \frac{1}{\sqrt{2\pi}} e^{-x^2/2} \frac{1}{\sqrt{2\pi}} e^{-y^2/2} = \frac{1}{2\pi} e^{-(x^2+y^2)/2} , \quad -\infty < x, y < \infty$$

Let R, Θ be the polar coordinates of the point (X, Y). Now $R^2 = X^2 + Y^2$ is, by definition, a chi-square random variable with 2 degrees of freedom, and as shown in Section 5.8.1.1 this distribution is the same as an exponential distribution with parameter $1/2$ (that is, with mean 2). Consequently, the density function of R^2 is

$$f_{R^2}(r) = \frac{1}{2} e^{-r/2} , \quad 0 < r < \infty$$

Consider now the conditional joint density function of X, Y given that $R^2 = r$. Because

$$f(x,y) = \frac{1}{2\pi} e^{-r/2} \quad \text{when} \quad x^2 + y^2 = r$$

is a constant when $x^2 + y^2 = r$, it is intuitive (and can be proven) that conditional on $R^2 = r$, the vector X, Y is uniformly distributed on the circumference of the circle of radius $\sqrt{r}$. But this implies that, conditional on $R^2 = r$, the polar coordinate Θ of the point (X, Y) is uniformly distributed over $(0, 2\pi)$. Because this is true for all r, it follows that the polar coordinates R and Θ are independent, with R distributed as the square root of an exponential random variable with mean 2, and Θ being a uniform random variable on $(0, 2\pi)$.

Using the preceding, we can generate independent standard normal random variables X and Y by first generating their polar coordinates R and Θ. Because $-\log(U)$ is exponential with mean 1, we can generate the polar coordinates of (X, Y) by generating independent uniform $(0, 1)$ random variables U_1 and U_2 and then setting

$$R^2 = -2 \log(U_1)$$

and

$$\Theta = 2\pi U_2$$

Using the formula for going from the polar coordinates R, Θ back to the rectangular coordinates

$$X = R\cos(\Theta), \quad Y = R\sin(\Theta)$$

shows that

$$X = \sqrt{-2\log(U_1)}\cos(2\pi U_2)$$
$$Y = \sqrt{-2\log(U_1)}\sin(2\pi U_2)$$

are independent standard normal random variables.

To generate normal random variables with mean μ and variance σ^2, just generate the independent standard normals X and Y and then take the variables $\mu + \sigma X$ and $\mu + \sigma Y$.

15.7 DETERMINING THE NUMBER OF SIMULATION RUNS IN A MONTE CARLO STUDY

Suppose we are going to generate r independent and identically distributed random variables $Y_1, \ldots, Y_r$ having mean μ, so as to use

$$\bar{Y}_r = \sum_{i=1}^{r} Y_i/r$$

as an estimator of μ. Now, with σ^2 being the variance of the Y_i, it follows by the central limit theorem that $\bar{Y}_r$ will approximately have a normal distribution with mean μ and variance σ^2/r. Consequently, we can be 95 percent certain that μ will lie in the interval

$$(\bar{Y}_r - 1.96\,\sigma/\sqrt{r}, \quad \bar{Y}_r + 1.96\,\sigma/\sqrt{r}).$$

(More generally, we can be $100(1 - \alpha)$ percent confident that μ will be between $\bar{Y}_r \pm z_{\alpha/2}\,\sigma/\sqrt{r}$.)

Thus, if σ^2 were known we could choose r to give ourselves the desired level of accuracy. However, it is almost always the case that σ^2, like μ, will be unknown. To get around this difficulty, we can do a two-stage simulation experiment. In the first stage, we generate k runs where k is typically much smaller than the number we expect to use in the study. Doing these runs generates the values of the random variables $Y_1, \ldots, Y_k$. We then use the sample variance of these values,

$$S_k^2 = \frac{1}{k-1}\sum_{i=1}^{k}(Y_i - \bar{Y}_k)^2$$

to estimate σ^2. Then, acting as if that were the actual value of σ^2, we determine an appropriate value for r. Then, in the second stage of the simulation, we generate an additional $r - k$ runs.

Problems

1. If $x_0 = 5$, and

$$x_n = 3\,x_{n-1} \quad \text{mod } 5$$

find $x_1, x_2, \ldots, x_{10}$.

2. Another method of generating a random permutation, different from the one given in Example 15.2b, is to successively generate a random permutation of the numbers $1, 2, \ldots, n$ starting with $n = 1$, then $n = 2$, and so on. (Of course, the random permutation when $n = 1$ is 1.) Once we have a random permutation of the numbers $1, \ldots, n - 1$ — call it $P_1, P_2, \ldots, P_{n-1}$ — the random permutation of the numbers $1, \ldots, n$ is obtained by starting with the permutation $P_1, P_2, \ldots, P_{n-1}, n$, then interchanging the element in position n (namely, n) with the element in a randomly chosen position that is equally likely to be any of the positions $1, 2, \ldots, n$.

   (a) Write an algorithm that accomplishes the preceding.
   (b) Verify when $n = 2$ and when $n = 3$ that all $n!$ possible permutations are equally likely.

3. Suppose that we are to observe the independent and identically distributed vectors $(X_1, Y_1), (X_2, Y_2), \ldots, (X_n, Y_n)$, and that we want to use these data to estimate $\theta \equiv E[X_1]/E[Y_1]$.

   (a) Give an estimator of θ.
   (b) Explain how you could estimate the mean square error of this estimator.

4. Suppose that $X_1, \ldots, X_n$ is a sample from a distribution whose variance σ^2 is unknown. Suppose we are planning to estimate σ^2 by the sample variance $S^2 = \sum_{i=1}^{n} (X_i - \bar{X})^2/(n - 1)$, and we want to use the bootstrap technique to estimate $\text{Var}(S^2)$.

   (a) If $n = 2$ and $X_1 = 1$ and $X_2 = 3$, what is the bootstrap estimate of $\text{Var}(S^2)$?
   (b) If $n = 15$, and the data values are

$$5, 4, 9, 6, 21, 17, 11, 20, 7, 10, 21, 15, 13, 8, 6$$

use simulation to obtain the bootstrap estimate of $\text{Var}(S^2)$.

5. Let $X_1, \ldots, X_8$ be independent and identically distributed random variables with mean μ. Let

$$p = P\left(\sum_{i=1}^{8} X_i/8 < \mu\right)$$

Estimate p if the values of the X_i are $5, 2, 8, 6, 24, 6, 9, 4$.

6. The following are a student's weekly exam scores. Do they prove that the student improved (as far as exam score) as the semester progressed?

$$68, 64, 72, 80, 72, 84, 76, 86, 94, 92$$

7. A baseball player has the reputation of starting slowly at the beginning of a season but then continually improving as the season progresses. Do the following data, which indicate the number of hits he has in consecutive five-game strings of the season, strongly validate the player's reputation?

$$8, 3, 7, 12, 4, 7, 13, 6, 0, 9, 12, 4, 4, 6, 10$$

8. A group of 16 mice were exposed to 300 rads of radiation at the age of 5 weeks. The group was then randomly divided into two subgroups. Mice in the first subgroup lived in a normal laboratory environment, whereas those from the second subgroup were raised in a special germ-free environment. The following data give the lifetimes, in days, of the mice in each group:

   Group 1 lifetimes: $133, 145, 156, 159, 164, 202, 208, 222$
   Group 2 lifetimes: $145, 148, 157, 171, 178, 191, 200, 204$

   Use a permutation test to test the hypothesis that the lifetime distributions are identical. Use the normal approximation to approximate the p-value.

9. Do Problem 13 in Chapter 12 by using a permutation test. Use the normal approximation to approximate the p-value.

10. Do Problem 16 in Chapter 12 by using a permutation test. Use the normal approximation to approximate the p-value.

11. Write an algorithm, similar to what was done in the text to generate a binomial random variable, that uses the discrete inverse transform algorithm to generate a Poisson random variable with mean λ.

12. Show that the discrete inverse transform algorithm for generating a geometric random variable with parameter p reduces to the following:

   1. Generate a random number U
   2. Set $X = \text{Int}(\frac{\log(1-U)}{\log(1-p)}) + 1$

Give a second algorithm for generating a geometric random variable with parameter p that takes into account the probabilistic interpretation of such a random variable.

13. Give a method for generating a random variable having density function

$$f(x) = e^x/(e - 1), \quad 0 < x < 1$$

14. Give a method for generating a random variable having distribution function

$$F(x) = x^n, \quad 0 < x < 1$$

15. Give a method for generating a random variable having distribution function

$$F(x) = \frac{1}{2}(x + x^2), \quad 0 < x < 1$$

16. Suppose that the following are the generated values of 20 random variables from the distribution F, whose mean μ is unknown:

$$5, 4, 9, 6, 21, 12, 7, 14, 17, 11, 20, 7, 10, 21, 15, 26, 9, 13, 8, 6$$

How many additional random variables from F will we need to generate if we want to be 99 percent certain that our estimate of μ is correct to within ± 0.1?

APPENDIX OF TABLES

TABLE AI   *Standard Normal Distribution Function:* $\Phi(x) = \dfrac{1}{\sqrt{2\pi}} \displaystyle\int_{-\infty}^{x} e^{-y^2/2}\, dy$

x	.00	.01	.02	.03	.04	.05	.06	.07	.08	.09
.0	.5000	.5040	.5080	.5120	.5160	.5199	.5239	.5279	.5319	.5359
.1	.5398	.5438	.5478	.5517	.5557	.5596	.5636	.5675	.5714	.5753
.2	.5793	.5832	.5871	.5910	.5948	.5987	.6026	.6064	.6103	.6141
.3	.6179	.6217	.6255	.6293	.6331	.6368	.6406	.6443	.6480	.6517
.4	.6554	.6591	.6628	.6664	.6700	.6736	.6772	.6808	.6844	.6879
.5	.6915	.6950	.6985	.7019	.7054	.7088	.7123	.7157	.7190	.7224
.6	.7257	.7291	.7324	.7357	.7389	.7422	.7454	.7486	.7517	.7549
.7	.7580	.7611	.7642	.7673	.7704	.7734	.7764	.7794	.7823	.7852
.8	.7881	.7910	.7939	.7967	.7995	.8023	.8051	.8078	.8106	.8133
.9	.8159	.8186	.8212	.8238	.8264	.8289	.8315	.8340	.8365	.8389
1.0	.8413	.8438	.8461	.8485	.8508	.8531	.8554	.8577	.8599	.8621
1.1	.8643	.8665	.8686	.8708	.8729	.8749	.8770	.8790	.8810	.8830
1.2	.8849	.8869	.8888	.8907	.8925	.8944	.8962	.8980	.8997	.9015
1.3	.9032	.9049	.9066	.9082	.9099	.9115	.9131	.9147	.9162	.9177
1.4	.9192	.9207	.9222	.9236	.9251	.9265	.9279	.9292	.9306	.9319
1.5	.9332	.9345	.9357	.9370	.9382	.9394	.9406	.9418	.9429	.9441
1.6	.9452	.9463	.9474	.9484	.9495	.9505	.9515	.9525	.9535	.9545
1.7	.9554	.9564	.9573	.9582	.9591	.9599	.9608	.9616	.9625	.9633
1.8	.9641	.9649	.9656	.9664	.9671	.9678	.9686	.9693	.9699	.9706
1.9	.9713	.9719	.9726	.9732	.9738	.9744	.9750	.9756	.9761	.9767
2.0	.9772	.9778	.9783	.9788	.9793	.9798	.9803	.9808	.9812	.9817
2.1	.9821	.9826	.9830	.9834	.9838	.9842	.9846	.9850	.9854	.9857
2.2	.9861	.9864	.9868	.9871	.9875	.9878	.9881	.9884	.9887	.9890
2.3	.9893	.9896	.9898	.9901	.9904	.9906	.9909	.9911	.9913	.9916
2.4	.9918	.9920	.9922	.9925	.9927	.9929	.9931	.9932	.9934	.9936
2.5	.9938	.9940	.9941	.9943	.9945	.9946	.9948	.9949	.9951	.9952
2.6	.9953	.9955	.9956	.9957	.9959	.9960	.9961	.9962	.9963	.9964
2.7	.9965	.9966	.9967	.9968	.9969	.9970	.9971	.9972	.9973	.9974
2.8	.9974	.9975	.9976	.9977	.9977	.9978	.9979	.9979	.9980	.9981
2.9	.9981	.9982	.9982	.9983	.9984	.9984	.9985	.9985	.9986	.9986
3.0	.9987	.9987	.9987	.9988	.9988	.9989	.9989	.9989	.9990	.9990
3.1	.9990	.9991	.9991	.9991	.9992	.9992	.9992	.9992	.9993	.9993
3.2	.9993	.9993	.9994	.9994	.9994	.9994	.9994	.9995	.9995	.9995
3.3	.9995	.9995	.9995	.9996	.9996	.9996	.9996	.9996	.9996	.9997
3.4	.9997	.9997	.9997	.9997	.9997	.9997	.9997	.9997	.9997	.9998

TABLE A2 *Values of $x_{\alpha,n}^2$*

n	$\alpha = .995$	$\alpha = .99$	$\alpha = .975$	$\alpha = .95$	$\alpha = .05$	$\alpha = .025$	$\alpha = .01$	$\alpha = .005$
1	.0000393	.000157	.000982	.00393	3.841	5.024	6.635	7.879
2	.0100	.0201	.0506	.103	5.991	7.378	9.210	10.597
3	.0717	.115	.216	.352	7.815	9.348	11.345	12.838
4	.207	.297	.484	.711	9.488	11.143	13.277	14.860
5	.412	.554	.831	1.145	11.070	12.832	13.086	16.750
6	.676	.872	1.237	1.635	12.592	14.449	16.812	18.548
7	.989	1.239	1.690	2.167	14.067	16.013	18.475	20.278
8	1.344	1.646	2.180	2.733	15.507	17.535	20.090	21.955
9	1.735	2.088	2.700	3.325	16.919	19.023	21.666	23.589
10	2.156	2.558	3.247	3.940	18.307	20.483	23.209	25.188
11	2.603	3.053	3.816	4.575	19.675	21.920	24.725	26.757
12	3.074	3.571	4.404	5.226	21.026	23.337	26.217	28.300
13	3.565	4.107	5.009	5.892	22.362	24.736	27.688	29.819
14	4.075	4.660	5.629	6.571	23.685	26.119	29.141	31.319
15	4.601	5.229	6.262	7.261	24.996	27.488	30.578	32.801
16	5.142	5.812	6.908	7.962	26.296	28.845	32.000	34.267
17	5.697	6.408	7.564	8.672	27.587	30.191	33.409	35.718
18	6.265	7.015	8.231	9.390	28.869	31.526	34.805	37.156
19	6.844	7.633	8.907	10.117	30.144	32.852	36.191	38.582
20	7.434	8.260	9.591	10.851	31.410	34.170	37.566	39.997
21	8.034	8.897	10.283	11.591	32.671	35.479	38.932	41.401
22	8.643	9.542	10.982	12.338	33.924	36.781	40.289	42.796
23	9.260	10.196	11.689	13.091	35.172	38.076	41.638	44.181
24	9.886	10.856	12.401	13.484	36.415	39.364	42.980	45.558
25	10.520	11.524	13.120	14.611	37.652	40.646	44.314	46.928
26	11.160	12.198	13.844	15.379	38.885	41.923	45.642	48.290
27	11.808	12.879	14.573	16.151	40.113	43.194	46.963	49.645
28	12.461	13.565	15.308	16.928	41.337	44.461	48.278	50.993
29	13.121	14.256	16.047	17.708	42.557	45.772	49.588	52.336
30	13.787	14.953	16.791	18.493	43.773	46.979	50.892	53.672

Other chi-square probabilities:

$x_{.9,9}^2 = 4.2$ $P\{x_{16}^2 < 14.3\} = .425$ $P\{x_{11}^2 < 17.1875\} = .8976.$

TABLE A3   *Values of $t_{\alpha,n}$*

n	$\alpha = .10$	$\alpha = .05$	$\alpha = .025$	$\alpha = .01$	$\alpha = .005$
1	3.078	6.314	12.706	31.821	63.657
2	1.886	2.920	4.303	6.965	9.925
3	1.638	2.353	3.182	4.541	5.841
4	1.533	2.132	2.776	3.474	4.604
5	1.476	2.015	2.571	3.365	4.032
6	1.440	1.943	2.447	3.143	3.707
7	1.415	1.895	2.365	2.998	3.499
8	1.397	1.860	2.306	2.896	3.355
9	1.383	1.833	2.262	2.821	3.250
10	1.372	1.812	2.228	2.764	3.169
11	1.363	1.796	2.201	2.718	3.106
12	1.356	1.782	2.179	2.681	3.055
13	1.350	1.771	2.160	2.650	3.012
14	1.345	1.761	2.145	2.624	2.977
15	1.341	1.753	2.131	2.602	2.947
16	1.337	1.746	2.120	2.583	2.921
17	1.333	1.740	2.110	2.567	2.898
18	1.330	1.734	2.101	2.552	2.878
19	1.328	1.729	2.093	2.539	2.861
20	1.325	1.725	2.086	2.528	2.845
21	1.323	1.721	2.080	2.518	2.831
22	1.321	1.717	2.074	2.508	2.819
23	1.319	1.714	2.069	2.500	2.807
24	1.318	1.711	2.064	2.492	2.797
25	1.316	1.708	2.060	2.485	2.787
26	1.315	1.706	2.056	2.479	2.779
27	1.314	1.703	2.052	2.473	2.771
28	1.313	1.701	2.048	2.467	2.763
29	1.311	1.699	2.045	2.462	2.756
∞	1.282	1.645	1.960	2.326	2.576

Other t probabilities:
$P\{T_8 < 2.541\} = .9825$ $P\{T_8 < 2.7\} = .9864$ $P\{T_{11} < .7635\} = .77$ $P\{T_{11} < .934\} = .81$ $P\{T_{11} < 1.66\} = .94$ $P\{T_{12} < 2.8\} = .984$.

TABLE A4   *Values of* $F_{.05,n,m}$

m = Degrees of Freedom for Denominator	n = Degrees of Freedom for Numerator				
	1	2	3	4	5
1	161	200	216	225	230
2	18.50	19.00	19.20	19.20	19.30
3	10.10	9.55	9.28	9.12	9.01
4	7.71	6.94	6.59	6.39	6.26
5	6.61	5.79	5.41	5.19	5.05
6	5.99	5.14	4.76	4.53	4.39
7	5.59	4.74	4.35	4.12	3.97
8	5.32	4.46	4.07	3.84	3.69
9	5.12	4.26	3.86	3.63	3.48
10	4.96	4.10	3.71	3.48	3.33
11	4.84	3.98	3.59	3.36	3.20
12	4.75	3.89	3.49	3.26	3.11
13	4.67	3.81	3.41	3.18	3.03
14	4.60	3.74	3.34	3.11	2.96
15	4.54	3.68	3.29	3.06	2.90
16	4.49	3.63	3.24	3.01	2.85
17	3.45	3.59	3.20	2.96	2.81
18	4.41	3.55	3.16	2.93	2.77
19	4.38	3.52	3.13	2.90	2.74
20	4.35	3.49	3.10	2.87	2.71
21	4.32	3.47	3.07	2.84	2.68
22	4.30	3.44	3.05	2.82	2.66
23	4.28	3.42	3.03	2.80	2.64
24	4.26	3.40	3.01	2.78	2.62
25	4.24	3.39	2.99	2.76	2.60
30	4.17	3.32	2.92	2.69	2.53
40	4.08	3.23	2.84	2.61	2.45
60	4.00	3.15	2.76	2.53	2.37
120	3.92	3.07	2.68	2.45	2.29
∞	3.84	3.00	2.60	2.37	2.21

Other F probabilities:

$F_{.1,7,5} = .337$ $P\{F_{7,7} < 1.376\} = .316$ $P\{F_{20,14} < 2.461\} = .911$ $P\{F_{9,4} < .5\} = .1782.$

TABLE A5 · *Values of C(m, d, α)*

							m				
d	α	2	3	4	5	6	7	8	9	10	11
5	.05	3.64	4.60	5.22	5.67	6.03	6.33	6.58	6.80	6.99	7.17
	.01	5.70	6.98	7.80	8.42	8.91	9.32	9.67	9.97	10.24	10.48
6	.05	3.46	4.34	4.90	5.30	5.63	5.90	6.12	6.32	6.49	6.65
	.01	5.24	6.33	7.03	7.56	7.97	8.32	8.61	8.87	9.10	9.30
7	.05	3.34	4.16	4.68	5.06	5.36	5.61	5.82	6.00	6.16	6.30
	.01	4.95	5.92	6.54	7.01	7.37	7.68	7.94	8.17	8.37	8.55
8	.05	3.26	4.04	4.53	4.89	5.17	5.40	5.60	5.77	5.92	6.05
	.01	4.75	5.64	6.20	6.62	6.96	7.24	7.47	7.68	7.86	8.03
9	.05	3.20	3.95	4.41	4.76	5.02	5.24	5.43	5.59	5.74	5.87
	.01	4.60	5.43	5.96	6.35	6.66	6.91	7.13	7.33	7.49	7.65
10	.05	3.15	3.88	4.33	4.65	4.91	5.12	5.30	5.46	5.60	5.72
	.01	4.48	5.27	5.77	6.14	6.43	6.67	6.87	7.05	7.21	7.36
11	.05	3.11	3.82	4.26	4.57	4.82	5.03	5.20	5.35	5.49	5.61
	.01	4.39	5.15	5.62	5.97	6.25	6.48	6.67	6.84	6.99	7.13
12	.05	3.08	3.77	4.20	4.51	4.75	4.95	5.12	5.27	5.39	5.51
	.01	4.32	5.05	5.50	5.84	6.10	6.32	6.51	6.67	6.81	6.94
13	.05	3.06	3.73	4.15	4.45	4.69	4.88	5.05	5.19	5.32	5.43
	.01	4.26	4.96	5.40	5.73	5.98	6.19	6.37	6.53	6.67	6.79
14	.05	3.03	3.70	4.11	4.41	4.64	4.83	4.99	5.13	5.25	5.36
	.01	4.21	4.89	5.32	5.63	5.88	6.08	6.26	6.41	6.54	6.66
15	.05	3.01	3.67	4.08	4.37	4.59	4.78	4.94	5.08	5.20	5.31
	.01	4.17	4.84	5.25	5.56	5.80	5.99	6.16	6.31	6.44	6.55
16	.05	3.00	3.65	4.05	4.33	4.56	4.74	4.90	5.03	5.15	5.26
	.01	4.13	4.79	5.19	5.49	5.72	5.92	6.08	6.22	6.35	6.46
17	.05	2.98	3.63	4.02	4.30	4.52	4.70	4.86	4.99	5.11	5.21
	.01	4.10	4.74	5.14	5.43	5.66	5.85	6.01	6.15	6.27	6.38
18	.05	2.97	3.61	4.00	4.28	4.49	4.67	4.82	4.96	5.07	5.17
	.01	4.07	4.70	5.09	5.38	5.60	5.79	5.94	6.08	6.20	6.31
19	.05	2.96	3.59	3.98	4.25	4.47	4.65	4.79	4.92	5.04	5.14
	.01	4.05	4.67	5.05	5.33	5.55	5.73	5.89	6.02	6.14	6.25
20	.05	2.95	3.58	3.96	4.23	4.45	4.62	4.77	4.90	5.01	5.11
	.01	4.02	4.64	5.02	5.29	5.51	5.69	5.84	5.97	6.09	6.19
24	.05	2.92	3.53	3.90	4.17	4.37	4.54	4.68	4.81	4.92	5.01
	.01	3.96	4.55	4.91	5.17	5.37	5.54	5.69	5.81	5.92	6.02
30	.05	2.89	3.49	3.85	4.10	4.30	4.46	4.60	4.72	4.82	4.92
	.01	3.89	4.45	4.80	5.05	5.24	5.40	5.54	5.65	5.76	5.85
40	.05	2.86	3.44	3.79	4.04	4.23	4.39	4.52	4.63	4.73	4.82
	.01	3.82	4.37	4.70	4.93	5.11	5.26	5.39	5.50	5.60	5.69
60	.05	2.83	3.40	3.74	3.98	4.16	4.31	4.44	4.55	4.65	4.73
	.01	3.76	4.28	4.59	4.82	4.99	5.13	5.25	5.36	5.45	5.53
120	.05	2.80	3.36	3.68	3.92	4.10	4.24	4.36	4.47	4.56	4.64
	.01	3.70	4.20	4.50	4.71	4.87	5.01	5.12	5.21	5.30	5.37
∞	.05	2.77	3.31	3.63	3.86	4.03	4.17	4.29	4.39	4.47	4.55
	.01	3.64	4.12	4.40	4.60	4.76	4.88	4.99	5.08	5.16	5.23

Index